HANDBOOK OF SERVICE BUSIN

Handbook of Service Business

Management, Marketing, Innovation and Internationalisation

Edited by

John R. Bryson

Professor of Enterprise and Competitiveness and Director of the City-Region Economic Development Institute, Birmingham Business School, University of Birmingham, UK

Peter W. Daniels

Emeritus Professor of Geography, School of Geography, Earth and Environmental Sciences, University of Birmingham, UK

Edward Elgar PUBLISHING

Cheltenham, UK • Northampton, MA, USA

© John R. Bryson and Peter W. Daniels 2015

All rights reserved. No part of this publication may be reproduced, stored in a retrieval system or transmitted in any form or by any means, electronic, mechanical or photocopying, recording, or otherwise without the prior permission of the publisher.

Published by
Edward Elgar Publishing Limited
The Lypiatts
15 Lansdown Road
Cheltenham
Glos GL50 2JA
UK

Edward Elgar Publishing, Inc.
William Pratt House
9 Dewey Court
Northampton
Massachusetts 01060
USA

Paperback edition 2016

A catalogue record for this book
is available from the British Library

Library of Congress Control Number: 2014954952

This book is available electronically in the **Elgar**online
Business subject collection
DOI 10.4337/9781781000410

ISBN 978 1 78100 040 3 (cased)
ISBN 978 1 78100 041 0 (eBook)
ISBN 978 1 78643 494 4 (paperback)

Typeset by Servis Filmsetting Ltd, Stockport, Cheshire
Printed and bound in Great Britain by TJ International Ltd, Padstow

Contents

List of contributors	vii
Preface: the structure of the Handbook	xiii
Acknowledgements	xv

1	Service business: growth, innovation, competitiveness *John R. Bryson and Peter W. Daniels*	1

PART I UNDERSTANDING SERVICE BUSINESS: DISCIPLINARY PERSPECTIVES

2	Growth and productivity in EU service sectors *Mary O'Mahony*	21
3	Service research and economic geography *Patrik Ström*	35
4	The new scientific study of service *Paul P. Maglio and Cheryl A. Kieliszewski*	49
5	The role of the Big 4: commoditisation and accountancy *Steve Hollis*	60

PART II SERVICES AND CORE BUSINESS PROCESSES

6	Green and sustainable innovation in a service economy *Faridah Djellal and Faïz Gallouj*	83
7	The three-stage model of service consumption *Rodoula H. Tsiotsou and Jochen Wirtz*	105
8	Creating and capturing value in the service economy: the crucial role of business services in driving innovation and growth *Michael Ehret and Jochen Wirtz*	129
9	Measuring business activity in the UK *Julian Frankish, Richard Roberts, David J. Storey and Alex Coad*	146
10	The growth of information-intensive services in the US economy *Uday Apte, Uday Karmarkar and Hiranya Nath*	170
11	Service and experience *Jon Sundbo*	205

PART III MANAGING SERVICE BUSINESSES

12 The organization of service business　225
    Andrew Jones

13 Managing experts and creative talent　242
    David J. Teece

14 Globalization of services　257
    Joanne Roberts

15 Internationalisation of services: modes and the particular case of KIBS　278
    Luis Rubalcaba and Marja Toivonen

16 In pursuit of creative compliance: innovation in professional service firms　301
    Timothy Morris, Michael Smets and Royston Greenwood

17 Business and professional service firms and the management and control of talent and reputations: retaining expert employees and client relationship management　316
    John R. Bryson

PART IV UNDERSTANDING SERVICE BUSINESS

18 How has logistics come to exert such a key role in the performance of economies, society and policy making in the 21st century?　333
    Andrew Potter and Robert Mason

19 Creative systems: a new integrated approach to understanding the complexity of cultural and creative industries in Eastern and Western countries　349
    Lauren Andres and Caroline Chapain

20 Tourism services: a sustainable service business?　371
    C. Michael Hall

21 Growth and spatial development of producer services in China　392
    Anthony G.O. Yeh and Fiona F. Yang

PART V CONCLUSION: A NEW RESEARCH AGENDA?

22 Developing the agenda for research on knowledge-intensive services: problems and opportunities　417
    John R. Bryson and Peter W. Daniels

Index　439

Contributors

Lauren Andres is Lecturer in Spatial Planning at the University of Birmingham, UK. Her fields of interest include: the policies and governance process of urban and economic regeneration and broadly the understanding of the role of (temporary) creative uses and intermediaries in shaping spaces and making places. She is also interested in assessing the forms of persistent resilience of groups of individuals and communities in a context of disturbances and pressures.

Uday Apte is Professor of Operations Management at the Graduate School of Business and Public Policy, Naval Postgraduate School (NPS), Monterey, CA, USA. Before joining NPS, Uday taught at the Wharton School, University of Pennsylvania, Philadelphia, and at the Cox School of Business, Southern Methodist University, Dallas, TX. His research interests include managing service operations, supply chain management, technology management, and globalisation of information-intensive services.

John R. Bryson is Professor of Enterprise and Competitiveness at Birmingham Business School and Adjunct Professor of Enterprise and Economic Geography in the School of Geography, Earth and Environmental Sciences, University of Birmingham, UK. His research interests include understanding the growth and dynamics of knowledge-intensive service firms, innovation and services, the interactions between services and manufacturing, the competitiveness of manufacturing in high cost locations, design and manufacturing, price and non-price sources of competitiveness, and infrastructure business models. He was elected a Fellow of the Academy of Social Sciences in 2009.

Caroline Chapain is a Lecturer in the Department of Management at Birmingham Business School, University of Birmingham, UK. Her research interests include: creative industries and creative cities and regions; economic restructuring, especially plant closures and redundancies; local and regional economic development; and quantitative and qualitative research methodologies.

Alex Coad is Senior Research Fellow at the Science Policy Research Unit (SPRU), University of Sussex, UK and External Associate Professor at Aalborg University (Denmark), Associate Fellow at the RATIO Institute (Stockholm, Sweden) and Associated Researcher at HUI (Stockholm). His research interests are focused on firm performance, entrepreneurship, business strategy and industrial dynamics.

Peter W. Daniels is Emeritus Professor of Geography at the University of Birmingham, UK. He has held a number of fellowships and visiting positions at universities in Australia, the United States, Hong Kong and Italy. A past-President of the European Research Network on Services and Space, his research interests include the growth of service industries and their role in economic and urban development. He has published journal papers as well as a number of single-authored and co-edited books on the rise of service industries, especially advanced services, in the UK, Europe, North America and the Asia-Pacific, and their role in globalisation and international trade.

Faridah Djellal is Professor of Economics and Dean of the Faculty of Economics and Social Sciences at the University Lille 1, France. Her research interests focus on performance, productivity in services, research and development, innovation in services, relationship innovation in employment and services, services innovation and economic geography, and innovation and public–private partnerships.

Michael Ehret is Reader in Technology Management at Nottingham Trent University, UK and Visiting Professor at Universitaet Rostock, Germany. His research focuses on business and technology marketing. He has published in leading academic journals, including the *Journal of Marketing* and *Industrial Marketing Management*.

Julian Frankish is Head of Business Economics and Research for Barclays Business Banking, UK. His role at Barclays involves detailed research into all aspects of the UK small business market.

Faïz Gallouj is Professor of Economics at University Lille 1, France. His research interests focus on the economics of innovation, the economics of services, and innovation and services. He has also worked extensively on productivity and performance in services. He is editor of many books on services, including *The Handbook of Innovation and Services* (Edward Elgar) and *Measuring and Improving Productivity in Services* (Edward Elgar).

Royston Greenwood is Telus Professor of Strategic Management at the Alberta School of Business, University of Alberta, Canada. His research interests include the dynamics of organisation change, managing professional service firms, and new business ventures.

C. Michael Hall is Professor in Marketing in the Department of Management, University of Canterbury, New Zealand. He has longstanding teaching, publication and research interests in tourism, regional development and social/green marketing, with particular emphasis on issues of place branding and marketing as well as conservation and environmental change, event management and marketing, and the use of tourism as an economic development and conservation mechanism. He is the author and editor of over 40 books as well as over 250 journal articles and book chapters.

Steve Hollis was former Midlands (UK) Chairman of KPMG. His early career focused on the audit of global companies before he qualified as a chartered tax practitioner and then specialised in corporate restructuring, along with all aspects of mergers and acquisitions. He has served on KPMG's boards for the UK, Europe and EMA and was appointed to chair the firm's Midlands practice in 2010. Steve has served on the Industrial Development Advisory Board, Sport England, and the Greater Birmingham and Solihull Local Enterprise Partnership.

Andrew Jones is Professor and Dean of the School of Arts and Social Sciences at City University London, UK. He is an inter-disciplinary social scientist with a background as an economic geographer. His research interests include: how the activities of organisations relate to the on-going development of a globalised economy; the nature of global knowledge management and the way in which local contexts and places shape competitiveness; and how the globalisation of financial and business services is caught up in the rise and development of emerging economies, especially in Asia.

Uday Karmarkar is Distinguished Professor in Decisions, Operations and Technology Management and L.A. Times Chair in Technology and Strategy in the School of Management, UCLA Anderson, USA. He is the founder and Director of UCLA Anderson's Business and Information Technologies Project (BIT), which studies the impact of new online information and communication technologies on business practices worldwide. His research interests are in information-intensive industries, competitive analysis, intelligent management systems, and operations and technology strategy for manufacturing and service firms. He has published over 70 articles and research papers.

Cheryl A. Kieliszewski is a research scientist at IBM Research, Almaden, San Jose, USA. She has over 15 years of research and applied human factors engineering experience, has been the technical assistant to the Almaden vice-president and lab director, managed a team focused on the design and implementation of advanced analytics for improved business intelligence and, most recently, is part of a team creating a platform to couple heterogeneous models through data exchange and is the client experience architect for the Accelerated Discovery Lab. Dr Kieliszewski is co-editor of the *Handbook of Service Science*.

Paul P. Maglio is a research scientist and manager at IBM Research, Almaden, San Jose, USA. He is currently working on a system to compose loosely coupled heterogeneous models and simulations to inform health and health policy decisions. Since joining IBM Research, he has worked on programmable Web intermediaries, attentive user interfaces, multimodal human–computer interaction, human aspects of autonomic computing, and service science. He is currently an Associate Adjunct Professor at UC Merced, USA, where he teaches cognitive science and service science.

Robert Mason is Senior Lecturer in Logistics at Cardiff Business School, UK. His research interests centre on the optimisation of supply chain system processes, which includes topics such as the integration of transport/logistics into national and international supply networks, the management of inter-organisational relationships (vertical and horizontal) and the organisation of enterprise to deliver customer value.

Timothy Morris is Professor of Management Studies at Saïd Business School, University of Oxford, UK. He is a founding member of the Novak Druce Centre for Professional Service Firms and a project director in Oxford University's Centre for Corporate Reputation. His research and teaching activities focus on the growing, and increasingly important, field of professional service firms.

Hiranya Nath is Professor of Economics at Sam Houston State University, USA. His primary research interests are in monetary economics and macroeconomic time series analysis, with specific interests in disaggregate price behaviour and information economics.

Mary O'Mahony is Professor of Applied Economics at King's College London, UK. Her research interests include measuring and explaining international differences in productivity, technology and growth; human capital formation and its impacts on productivity; and measuring performance in public services, including health and education. She joined King's in May 2013 and was previously Professor at Birmingham Business School, University of Birmingham, UK. She is currently a Visiting Fellow at the National Institute of Economic and Social Research, London, UK.

Andrew Potter is Reader in Transport and Logistics at Cardiff Business School, University of Cardiff, UK. His research interests lie in how transport can be more effectively integrated within supply chains, through process-, people- and technology-based approaches. This has often involved translating manufacturing-orientated frameworks into a logistics operating environment.

Joanne Roberts is Professor in Arts and Cultural Management at Winchester School of Art, University of Southampton, UK. She gained her doctorate at the Centre for Urban and Regional Development Studies, Newcastle University, UK, and she has held academic posts at Durham University, Newcastle University and Northumbria University, UK. Her areas of expertise include international business, innovation, and knowledge production and transfer.

Richard Roberts has worked for Barclays since 1984, first, briefly as an industrial economist but mainly as a specialist on small and medium-sized enterprises (SMEs). For many years he led the Business Economics Team within Barclays. This team considers all aspects of the SME customer base. Between 2001 and 2012, he was Chief Economist, Barclays UK Retail Banking activities.

Luis Rubalcaba is Professor in Economics in the Department of Applied Economics, University of Alcalá, Spain and Visiting Professor at VTT Technical Research Centre, Finland. His research interests focus on understanding the economics of service firms and employment, service trade and internationalisation, service innovation, business and professional service firms, public services, and regional policy and services.

Michael Smets is Associate Professor in Management and Organisation Studies and a Research Fellow at Green Templeton College, Oxford, UK. He is also a member of the Novak Druce Centre for Professional Service Firms, based at Saïd Business School, Oxford, UK. His research focuses on professional service firms, and especially their internationalisation, innovation and regulation.

David J. Storey is Professor of Enterprise at the Department of Business Management and Economics, University of Sussex, UK. He has two honorary Doctorates and has been Visiting Professor at the Universities of Manchester, Reading and Durham, UK, and was an International Fellow at Sydney University, Australia in 2009. He has made major contributions to research into entrepreneurship and to understanding the economics and management of small and medium-sized enterprises. In 1998 he received the International Award for Entrepreneurship and Small Business Research from the Swedish Council, and was awarded a Wilford White Fellowship in 2008.

Patrik Ström is Associate Professor in the School of Business, Economics and Law, University of Gothenburg, Sweden and President of the European Association of Research on Services. He is an economic geographer whose research focuses on understanding the internationalisation and competitiveness of knowledge-intensive services, with a particular focus on Japanese services in Europe and Asia.

Jon Sundbo is Professor in the Department of Communication, Business and Information Technologies, Roskilde University, Denmark. He has published widely on services and especially on innovation in services, services and the experience or performance economy, experience and management, and the management aspects of services.

David J. Teece is Director of the Tusher Center on Intellectual Capital, Professor of Business Administration and the Thomas W. Tusher Chair in Global Business at the Haas School of Business, University of California, Berkeley, USA. He is an authority on subjects including the theory of the firm and strategic management, the economics of technological change, knowledge management, technology transfer, and antitrust economics and innovation. He has held teaching and research positions at Stanford University, USA and Oxford University, UK, and has also received three honorary doctorates. According to Science Watch, he is the lead author of the most cited article in economics and business worldwide, 1995–2005. He is also one of the top 10 cited scholars in economics and business for the decade, and has been recognised by Accenture as one of the world's top 50 business intellectuals.

Marja Toivonen is Research Professor at VTT Technical Research Centre of Finland and Adjunct Professor at Aalto University, Finland. Her research interests include service innovation, producer services, service business models, service process modelling and the development of services.

Rodoula H. Tsiotsou obtained her PhD from Florida State University, USA and is currently Assistant Professor of Marketing in the Department of Marketing and Operations Management, University of Macedonia, Greece. Her research interests include services marketing, strategic marketing, relationship marketing, brand management, nonprofit marketing, sports marketing, tourism marketing and e-marketing.

Jochen Wirtz is Professor of Marketing at the National University of Singapore (NUS), Founding Director of the dual degree UCLA–NUS Executive MBA Program (ranked number 5 globally in the Financial Times EMBA rankings in 2012), Fellow of the NUS Teaching Academy (the NUS think-tank on education matters), Associate Fellow of Executive Education at the University of Oxford, UK's Saïd Business School, and International Fellow of the Service Research Centre at Karlstad University, Sweden. His research focuses on services marketing and his books include *Services Marketing: People, Technology, Strategy* (co-authored with Christopher Lovelock, 2011, 7th edition, Prentice Hall), *Essentials of Services Marketing* (co-authored with Chew and Lovelock, 2012, 2nd edition, Prentice Hall), and *Flying High in a Competitive Industry: Secrets of the World's Leading Airline* (co-authored with Heracleous and Pangarkar, 2009, McGraw Hill).

Fiona F. Yang is Research Assistant Professor in the Department of Urban Planning and Design, University of Hong Kong, China. She holds a PhD in economic geography from the University of Hong Kong, and an MSc in human geography and a BSc in economic geography and urban planning from Sun Yat-Sen University, China. From 2007 to 2008, she was an exchange Research Fellow at the Department of Geography, Hunter College, City University of New York, USA. Her academic interests focus on the service economy, China's urban and regional development, and Hong Kong–Guangdong economic integration.

Anthony G.O. Yeh is Head of the Department of Urban Planning and Design, Director of the Geographic Information Systems Research Centre and the Deputy Convenor of the Contemporary China Studies Strategic Research Area, University of Hong Kong, China. His areas of research specialisation are in urban planning and development in Hong Kong, China and South East Asia, and the applications of geographic information systems in urban and regional planning. He was elected an Academician of the Chinese Academy of Sciences in 2003, Fellow of The Academy of Sciences for the Developing World in 2010, and Fellow of the Academy of Social Sciences in the UK in 2013.

Preface: the structure of the *Handbook*

This *Handbook* brings together a collection of chapters that explores various aspects of service business with a focus on management, marketing, geography, planning, innovation and internationalisation. The *Handbook* deliberately brings together academics from across the social sciences and includes economists, geographers, planners, sociologists and academics from across the business school disciplines (entrepreneurship, marketing, operations). In 2007, we edited another *Handbook* on services that was also published by Edward Elgar (Bryson and Daniels, 2007). This brought together a very different collection of scholars and had a very different structure and title from this new *Handbook*. It was entitled *The Handbook of Service Industries* and was divided into five parts: conceptual perspectives on services, the development of service economies, trading services: from local to global production, services, technology and innovation, and service employment: embodied and emotional labour.

This new *Handbook* is not intended to replace our 2007 book, but to provide an alternative account of the shift towards service-led or service-dominated societies. The difference from our earlier *Handbook* is the emphasis that is placed in the new *Handbook* on service business or the activities and processes that lead to the production and consumption of services. It was not our intention to cover all aspects of services, as this would have led to the production of a very large volume. We were also aware that two specialist *Handbooks* that focused on services had been published since our 2007 collection. In 2010, Maglio et al. produced their *Handbook of Service Science* that provides an interdisciplinary approach to studying, improving, creating and innovating services. This collection was a call for the development of a new science of services that would lead to the development of a new set of methodologies and theories. In 2010 Gallouj and Djellal produced *The Handbook of Innovation and Services: A Multi-disciplinary Approach*. The Gallouj and Djellal *Handbook* focuses on understanding the role of innovation in all types of services.

It is interesting that all four *Handbooks* that focus on services are multi-disciplinary, but each has a very different objective. Service research is to be found across the social sciences and even in the discipline of engineering (Maglio et al., 2010). Our new *Handbook* is divided into four parts. The first part provides four disciplinary perspectives on services. This includes accounts from economics (Chapter 2), economic geography (Chapter 3) and service science (Chapter 4). Unusually, it also contains a chapter written by a practitioner (Chapter 5). In this chapter, Steve Hollis, a former partner at KPMG, provides a fascinating insider account of the challenges facing the Big 4 accountancy firms as they respond to new forms of competition. In the second part the focus is on exploring services and core business processes. This includes six chapters that explore sustainability and services (Chapter 6), service consumption (Chapter 7), the creation of value and its capture through the provision of services (Chapter 8), measuring new firm formation (Chapter 9), the growth of information-intensive services in the US (Chapter 10) and a detailed account of services as experiences (Chapter 11). In part III the focus is on understanding some of the management aspects of service businesses. This includes an analysis of the organisation of service businesses (Chapter 12), the management of creative employees (Chapter 13), the globalisation and internationalisation of services (Chapters 14 and 15), innovation in professional service firms (Chapter 16) and client relationship management (Chapter 17). The final part provides a series of industry case studies. It was not our intention to provide case studies of all service sectors. This part contains four chapters that explore logistics (Chapter 18), creative and cultural industries (Chapter 19), tourism (Chapter 20) and the growth of producer services in China (Chapter 21). In this part there is a mix of sector and country case studies.

Acknowledgements

A book project like this represents, and requires, the work of many participants. We particularly thank the individual chapter authors who contributed to this *Handbook*: Lauren Andres, Uday Apte, Caroline Chapain, Alex Coad, Faridah Djellal, Michael Ehret, Julian Frankish, Faïz Gallouj, Royston Greenwood, C. Michael Hall, Steve Hollis, Andrew Jones, Uday Karmarkar, Cheryl Kieliszewski, Paul P. Maglio, Robert Mason, Tim Morris, Hiranya Nath, Mary O'Mahony, Andrew Potter, Joanne Roberts, Richard Roberts, Luis Rubalcaba, Michael Smets, David Storey, Patrik Ström, Jon Sundbo, David Teece, Marja Toivonen, Rodoula Tsiotsou, Jochen Wirtz, Fiona Yang and Anthony Yeh. We appreciate their time, patience, expertise and commitment to the broader project of understanding service business and the global economy and the contribution services make to economic development and innovation.

Producing an edited collection is a partnership between many different individuals – the editors, chapter authors and their families, but also those service firms and employees who were willing to allow academics access and provided the empirical evidence that made many of the chapters in this *Handbook* possible. We should also not forget all the policy-makers who have given their time to discuss services. Finally, we appreciate the work of our editor at Edward Elgar, Matthew Pitman, for encouraging us and seeing the project through to completion.

John R. Bryson and Peter W. Daniels
July 2014
Birmingham

1. Service business: growth, innovation, competitiveness
John R. Bryson and Peter W. Daniels

INTRODUCTION

All capitalist economies are service economies and all manufacturing-dominated economies have been simultaneously service economies (Bryson et al., 2004). It is difficult to design, manufacture and sell a product without providing service functions. There are many different types of services, from those that directly support the production of goods and other services – the producer or business and professional services – to personal services including tourism, health care and education. A walk through a supermarket highlights the role that marketing, advertising and packaging design services play in our society. Simple products incorporate quantities of visible and often invisible service expertise; the former is seen in advertising campaigns and the latter is found in logistics services that ensure that raw materials and completed manufactured goods are transported and distributed efficiently. Much of the service expertise hidden within products arises from the constant search for product differentiation.

The process of corporate and product differentiation through blending service expertise with the manufacture of products is reflected extremely effectively in the development of the concept of the *unique selling proposition* (USP). After the Second World War, Rosser Reeves, an American advertising executive who ran the Ted Bates Advertising Agency, used to stand before a potential client, reach into his pocket and extract some loose change. He would select two quarters and announce that his job was to persuade consumers that the quarter in his right hand was worth more than the one in his left hand (Twitchell, 2004: 5–6). This process he labelled as the creation of a USP which involved, for Reeves, developing stories and associations that were attached to goods; this contributed to the development of strong brands. Reeves's brother-in-law, David Ogilvy (1911–1999), developed the 'proposition' part of the USP through placing stories on products (Ogilvy, 1983). He was the first Scottish sales representative for Aga, the manufacturer of cast iron range cookers, and during his time there Ogilvy created the company's sales manual; a book that defined a sales process built on understanding consumer behaviour and aspirations and the development of stories.

The economies and societies of developed market economies are saturated with service jobs, activities, firms and functions. In many developed market economies there has been a continual growth in service employment and a decline in employment in manufacturing. Nevertheless, the economic downturn that commenced in 2008 has seen a renewed interest by governments in stimulating manufacturing production. Thus, between 2012 and 2013 there was growth in employment in both manufacturing and services in Chile, Hungary, Ireland, Japan, Korea, New Zealand, Poland, the

Table 1.1 Employment in manufacturing and service, 2012 and 2013 (all persons)

	Service Employment 2012	Service Employment 2013	Manufacturing Employment 2012	Manufacturing Employment 2013
Australia	8,685,688	8,819,650(+)	948,107	921,999(−)
Austria	2,884,525	2,896,000(+)	660,075	651,125(−)
Canada	13,635,730	13,847,730(+)	1,785,517	1,734,217(−)
Chile	5,047,329	5,189,401(+)	881,428	882,185(+)
Czech Republic	2,876,650	2,935,525(+)	1,299,075	1,285,275(−)
Denmark	2,089,800	2,097,375(+)	333,825	325,000(−)
Finland	1,817,000	1,796,775(−)	356,725	350,300(−)
Germany	28,145,600	28,613,100(+)	7,917,050	7,839,925(−)
Hungary	2,520,425	2,567,350(+)	803,100	823,175(+)
Ireland	1,416,250	1,431,950(+)	208,775	213,125(+)
Italy	15,687,580	15,496,130(−)	4,207,650	4,128,775(−)
Japan	44,771,670	45,230,830(+)	10,317,500	10,390,000(+)
Korea	17,184,430	17,502,510(+)	4,104,900	4,184,017(+)
Netherlands	6,917,800	6,941,775(+)	771,700	768,225(−)
New Zealand	1,624,250	1,667,150(+)	245,875	247,675(+)
Norway	2,005,625	2,017,400(+)	239,125	228,775(−)
Poland	8,890,500	8,949,050(+)	2,905,800	2,968,675(+)
Slovak,Republic	1,379,500	1,417,825(+)	570,325	539,475(−)
Slovenia	562,100	549,550(−)	206,475	203,200(−)
Spain	13,244,230	13,017,480(−)	2,223,900	2,118,675(−)
Sweden	3,651,500	3,712,900(+)	537,925	524,900(−)
Switzerland	3,360,250	3,402,425(+)	576,275	581,325(+)
United Kingdom	23,511,900	23,917,420(+)	2,886,775	2,913,900(+)
United States	115,675,200	116,593,800(+)	14,686,420	14,869,080(+)

Source: OECD, 2014.

UK and the US (Table 1.1). This highlights the complexity of economic restructuring. Thus, the replacement of manufacturing employment with service jobs has been occurring, but this is not a simple process. Manufacturing remains an important activity in most developed market economies, but most people are employed in some form of service work. In an analysis of the new economy of American jobs, Moretti noted that:

> If you take a walk in one of America's cities, most of the people you see on the streets will be store clerks and hairstylists, lawyers and waiters, not innovators. About a third of Americans work either for government or in the education or health services sectors, which include teachers, doctors and nurses. Another quarter are in retail, leisure, and hospitality, which includes people working in stores, restaurants, movie theatres, and hotels. An additional 14 percent are employed in professional and business services, which includes employees of law, architecture and management firms. In total, two-thirds of American jobs are in the local services sector, and that number has been quietly growing for the past fifty years. (2013: 12)

For Moretti, the growth of service work is the effect and not the cause of economic development. This argument is based on productivity differentials between manufacturing and services. These tend not to alter dramatically for the delivery of local services; it takes the same time to teach a child to play the piano or for a hairdresser to cut hair as it did 40 years ago. Nevertheless, there have been major improvements in productivity in manufacturing firms; every year they improve their productivity through process and product innovations and often this involves replacing people with machines.

Technological developments provide new opportunities for the creation and delivery of new services. This includes all types of Apps for smart phones and many types of Web-enabled service delivery model. The development of new forms of e-commerce continues to transform high streets and shopping centres. It is interesting to note that the rise and decline of the high street shop has actually occurred over a relatively short space of time; they only only emerged as we know them in the late 17th century. Prior to this time goods were purchased from temporary stalls set up in market places or from 'shops' without glass windows. Instead, they were protected outside working hours by window-shutters that were let down and when open supported on posts for the display of goods (Gray, 1921: 110). The word 'shop' only emerged as a verb with the meaning of going to purchase goods in the mid-18th century (1764); prior to this the term meant to expose goods for sale or a workshop (Oxford English Dictionary, 1991).

The replacement of high street shops with out-of-town shopping centres or malls and the emergence of e-commerce reflect ongoing innovations in retail services. The latest trend has been the transformation of the high street and shopping centre from a place to shop for goods into a place of entertainment and experiences. A good example is the Bluewater Shopping Centre, Kent, UK. A 30 per cent stake in Bluewater was recently sold by Australia's Lend Lease to Land Securities (White, 2013: B5). Other publicly listed property developers competed with Land Securities to purchase this stake, requiring it to pay £656 million for its share, which provides the company with an overall net initial yield of 4.1 per cent. Bluewater attracts over 27 million visitors a year and the acquisition of the 30 per cent stake in the development is part of Land Securities's strategy to focus on developing a retail investment portfolio based on dominance, experience and convenience. Retail habits are changing; visits to shopping centres are less frequent but for a longer period of time and over greater distances. The retail mix within large shopping centres is changing; ten years ago food and drink outlets would account for 5 per cent of outlets but today developers aim for at least 20 per cent. Land Securities wants to increase Bluewater's leisure outlets to at least 24 per cent. This highlights the on-going transformation of shopping centres into leisure spaces that are saturated with service experiences.

For many years services were ignored by academics, they devoted most of their energy towards understanding manufacturing firms and even agricultural activities. The relatively recent discovery of services is surprising given their pivotal contribution to the first industrial revolution. Sociologists (Wright Mills, 1953; Urry, 1987, 1990; Lash and Urry, 1994) and geographers were amongst the first social scientists to discover service employment and activities. The early attention to services by geographers (Daniels, 1982, 1985) had its roots in a 1970s research agenda that focussed on understanding the growth and location of offices (Daniels, 1979; Goddard, 1986). It was soon recognised that this topic would be enriched by looking more closely at the business activities that were major users of office space, the rapidly growing business and professional services (Beyers and Alvine, 1985; Beyers and Lindahl, 1996; Bryson, 1996, 1997a, b; Bryson et al., 1993a, b, 1994, 1997, 2004).

In this chapter we provide some conceptual tools for understanding the world of service business. This includes a definition of services, service histories, services and the division of labour, services and productivity, and services and manufacturing. This reflects the broad objective for this *Handbook*, which is to provide a systematic account of service businesses. This can be sub-divided into two sub-objectives: first, to identify a series of building blocks that can be used to understand the operations and activities of service businesses, and, second, to provide a collection of chapters that can be used to support an advanced undergraduate or graduate teaching module that is focussed on service business.

DEFINING SERVICES

In 1986, Riddle argued that the 'service sector is one of the least understood portions of our global economy', even though 'no economy can survive without a service sector' (Riddle, 1986: 6). During the 1970s and early 1980s, very few academics were interested in exploring services. Economics was still heavily influenced by classical political economists, such as Adam Smith and David Ricardo, who drew a distinction between productive and unproductive labour. According to Smith (1997: 429–430), 'there is one sort of labour which adds to the value of the subject upon which it is bestowed: there is another which has no such effect. The former, as it produced a value, may be called productive; the latter, unproductive labour.' He used this simple bipolar division of labour to argue that a whole range of service activities was essentially unproductive. The traditional meaning of the term 'service' heavily influenced Smith's understanding of services.

The word 'service' is very problematic as it has too many meanings and associations (Fellowes, 1954: 719). It comes from the Latin *servitium* or 'slavery'. The meaning of the term has altered so that the act of serving, for example, is no longer associated with slavery. There are many types of 'service', and these include the occupation of a servant, a public or civil servant, or religious associations based on church service, public worship or 'Divine Service' (Fellowes, 1954: 719) and serving God. The various meanings of the term are all based on the concept of the 'act of serving' and on a relationship between master (*sic*) and servant. As a noun a 'service' is the action or process of servicing and as a verb it describes the performance of routine maintenance or repair work or the provision of a service. The use of the word 'service' as a verb is a very recent development. In this context, Gowers noted that the verb 'service' 'is a useful newcomer in an age when almost everyone keeps a machine of some sorts that needs periodical attention' (Gowers, 1982: 46). It is also worth noting that in 1925 the term 'service' was first applied to describe 'expert advice or assistance given to customers after sale by manufacturers or vendors' (Oxford English Dictionary, 1991: 1950).

At best, then, the word 'service' is a confusing term. The association between service, servants and unproductive labour partly explains the absence of a significant body of research on services prior to the 1980s. This is not to argue that no academic work on services was undertaken. A number of important monographs made major contributions to understanding service work as well as beginning to trace some of the dimensions of the emerging service economy. Some of these early studies avoided the term 'service', preferring to focus on exploring white-collar work (Wright Mills, 1953) or professional people (Lewis and Maude, 1952). One of the most important of the early studies is Greenfield's research on producer services (1966). This was followed by Bell's classic account of post-industrial society (1973) that influenced Gershuny's analysis of the new service economy (Gershuny, 1978; Gershuny and Miles, 1983) and the geographical analyses by Daniels (1982, 1985) and Illeris (1996).

Scholars have found it extremely difficult to construct a rigorous definition of 'services'. Originally, the category of 'services' was a 'residual' that embraced everything that was not included in the primary (extractive) or secondary (manufacturing) sectors of the economy. This has encouraged attempts to define and identify services on the basis of their primary characteristics, two of which are supposed to delimit them from other forms of economic activity. First, the output of a service is ephemeral or non-material (for example, a lecture or theatrical performance) and, second, the production and consumption of a service occur simultaneously. There are obvious difficulties with both of these characteristics; some services are not ephemeral and do have materiality (software, or even a haircut) and some can be stored (software, servicing of a machine).

All attempts to classify services also recognise the complexity and diversity of the business activities that need to be included. A simple classificatory schema of service activities identifies five different types:

1. Consumer services that provide services for final end-users.
2. Producer and business services that provide intermediate inputs into the activities of private- and public-sector organisations.
3. Public services provided directly by the state or indirectly by the private sector and not-for-profit organisations.
4. Not-for-profit organisations working beyond the confines of the state.
5. Informal services or unpaid service work, which is usually predominantly undertaken by women, and which is a vital element of people's daily lives.

Each type describes a heterogeneous collection of service functions. They comprise several distinctive sectors (law, accountancy, market research, technical consultancy, etc.) as well as a very broad size-range of firms. Producer services in particular make an important contribution to economic development as they contribute directly to the creation of added value; they contribute to a national economy's balance of payments through exports and have also given rise to dramatic growth rates both in employment and in new firm formation.

It is usually assumed that producer service firms perform an important role in knowledge creation, as well as in shaping regional competitiveness. However, it is easier to assert this relationship than it is to successfully test its existence because it is difficult and perhaps impossible to develop objective measures of the impacts of producer service firms on client companies or regional competitiveness.

SERVICE HISTORIES

It is worth noting at the outset the common mistake of assuming that transformation of economies towards services is a phenomenon of the 20th century. The social sciences have paid too little attention to the role services played during the industrial revolution and earlier. The considerable rise in the importance of services in the latter half of the 18th century has been largely ignored. A case example is the United Kingdom. Using fire office registers to undertake an investigation into the structure of London's economy between the years 1775 and 1825, Barnett, for example, provides an unusually detailed historical investigation from which it is concluded that:

> service industries made no less contribution to the British economy during the Industrial Revolution than manufacturing, and that nowhere was this more true than in London. Its service economy was on a very large scale, serving the nation as a whole as well as the capital ... London's service industries underpinned both its own and the national manufacturing and commercial infrastructure and at the same time contributed to the new 'commercialisation of leisure'. (Barnett, 1998: 183)

During the late 19th century London was already being transformed into a key world city, a process that has continued up to the present to ensure its status as a global city. Deane and Cole (1962: 166, 175) calculated that in 1851 some 45.3 per cent of the UK's national income was derived from service activities (trade, transport, housing, the professions, civil service, etc.). The structure of employment changed dramatically during the 19th century as a result of technological innovation and the increasing maturity and extension of the capitalist system. Employment growth occurred in occupations that facilitated the exchange of goods and services between producers and consumers. Between 1881 and 1901 the number of business clerks increased from 175,000 to 308,000, bank officials from 16,000 to 30,000 and insurance officials and clerks from 15,000 to 55,000 (Marsh, 1977: 124). During the 19th century international trade was continuously impeded by financial problems until it was enabled and supported by the establishment of bill markets and banking facilities of the kind associated with the flow of tea and silk from China to Europe between 1860 and 1890 (Hyde, 1973). As the UK was becoming an industrialised society it was simultaneously being transformed into a service economy; the growth of manufacturing employment went hand in hand with the growth of service employment. This was an on-going transformation and it is evident in many different indicators of social and economic transformation.

The development of a market economy based around agricultural products and manufactured goods required the development of supporting services. During the 18th century market towns developed to 'service and supply a thinly populated countryside, but also to gather the products of rural industries such as handloom weaving, glove making and stocking knitting for onward transmission to the regional and national markets' (Pinches, 2009: 44). Nevertheless, all market-based transactions, however primitive, required supporting services including logistics and accounting.

Circulation, trade and exchange are essential pre-conditions for the emergence of capitalism. These activities are closely related to service functions. The development of international trade is associated with the creation of a supporting international financial system. During the 18th century London developed as a centre of a 'wide intricate, multilateral network of world trade ... with its wide sheltered anchorages, its vast wharves and warehouses, its rich banks, its specialists in marine insurance and its world-wide mercantile contacts ... It drew to itself a cosmopolitan concentration of wealth and expertise' (Deane, 1969: 56).

Business and professional service (BPS) functions emerged as market-based activities became ever more complex. The history of capitalism is one in which important qualitative changes in the management and organisation of business emerge. These new processes and BPS functions gradually became commonplace. Thus, at the beginning of the 17th century a stock market was opened in Amsterdam; 'Government stocks and the prestigious shares of the Dutch East India Company had become objects of speculation in a totally modern fashion' (Braudel, 1988: 101). This was not the first stock market as State loan stock had been negotiable in Florence before 1328. The difference between the Florence and the Amsterdam stock markets was in the volume of transactions, fluidity and publicity (Braudel, 1988: 101). Thus, the difference between Amsterdam's first stock market and that of the London stock market in 2014 is one of scale and complexity. To Braudel, during the 17th century:

> speculation on the Amsterdam Stock Exchange had reached a degree of sophistication and abstraction which made it for many years a very special trading-centre of Europe, a place where people were not content simply to buy and sell shares, speculating on their possible rise or fall, but where one could by means of various ingenious combinations speculate without having any money or shares at all. (Braudel, 1988: 101)

The Amsterdam stock market was intertwined with the development of trade and a series of related supporting and facilitating services.

THE DIVISION OF LABOUR

The division of labour is central to the history of capitalism; new tasks, functions or activities emerge that reflect specialisation within labour markets. The onset of mass production with the development of the assembly lines in the early years of the last century is associated with the emergence of mass consumption. This requires the development of sophisticated forms of marketing, advertising, branding and packaging and this led to the emergence of another set of supporting service functions. It is impossible to separate service functions from manufacturing as each supports the other. Technological developments in the design and manufacture of office technology (the telephone, photocopiers, typewriter, telex machines, fax machines, computers, e-mail) enabled major innovations to occur in office-based services. Similarly, innovations in services, from advertising to logistics, played a critical role in the emergence of mass production systems.

The outsourcing of service functions and the creation of new types of service occupations may represent an increase or extension of the division of labour. An increasing division of labour reflects both increasing specialisation of activity with a resultant increase in the complexity of production, and alterations in the way in which production is organised. Here the important point is the *extended labour process* (Sayer and Walker,

1992; Walker, 1985), which is work that occurs before and after goods and services are physically produced. Thus, research and development, design, market research, trial production, product testing, marketing, customer care and sales are all essential parts of the production process. The fact that they can be separated in both time and space from the actual production process does not necessarily imply that they are not an integral part of the manufacturing sector of the economy. Ultimately, this means that the dramatic growth in business service employment reflects alterations in the way in which manufacturing production is organised, rather than the development of a service or knowledge economy.

SERVICES AND PRODUCTIVITY

In his economic theory of services, Baumol (1967, see also Baumol et al., 1989) distinguishes 'progressive' services (those oriented towards the application of technology in production and which can therefore achieve improved rates of output per capita) from 'non-progressive' services (substitution of technology for labour is not possible). In relation to the latter, the nature of the production process determines that the work done (such as a ballet or an opera, a consultancy) cannot be speeded up or abbreviated in the interests of improved productivity (by reducing the number of dancers or performers, devoting less time to researching and preparing a consultancy report). This would be unacceptable to those watching or listening to the performance or those paying to obtain the best advice from a consultant. There is therefore very little scope for productivity improvements of the kind possible in the 'progressive' services, where innovation, economies of scale or developments in information and communication technologies (ICTs), for example, can be adopted to achieve increases in productivity. The overall implication is that over time services become more costly relative to goods. If it is assumed that the demand for services is inelastic to price, but that demand will continue to increase as living standards rise, there will be a steady transfer of employment from the progressive to the non-progressive parts of the economy. The result is not only a general shift of employment from manufacturing to services but also a shift from the progressive to the non-progressive sectors within services.

The Baumol model has the attraction of simplicity but it does assume that measures of output and productivity used for progressive activities are readily transferable to non-progressive activities. Recent advances in ICTs indicate that many service activities are much more open to the substitution of new technology and economies of scale for production, thus diminishing the credibility of the 'productivity gap' thesis that is central to Baumol's case. But at least Baumol and fellow economists such as Fuchs (1965, 1969) recognised the new and expanding role for services in the economy. Meanwhile, others such as Galbraith (1967) effectively ignored services in their analyses of the growth and diversification of industrial economies and societies, even though in Europe and in the US some 40–50 per cent of the labour force was employed in service industries by the end of the 1950s.

MANUFACTURING AND SERVICES – HYBRID PRODUCTION PROCESSES AND HYBRID PRODUCTS

Manufacturing has changed but our understanding of it has not. The production process of both goods and services has become blurred; service firms increasingly provide service functions that are combined within goods whilst manufacturers of goods are stretching their value chains to profit from the provision of services. It has become difficult to differentiate between goods and services, and in many instances the attempt to apply such a distinction distorts understanding of production processes that are inherently complex. Four processes have been at work that require detailed attention: the shift towards service employment, changes to manufacturing, the blurring of services and manufacturing functions and the rise of hybrid products and production systems, and, finally, alterations in the control systems required to manage complex hybrid production systems.

First, the shift in employment towards services in all developed market economies reflects alterations in consumer behaviour (both business-to-business and business-to-consumer) and in the structure of production systems (Bryson et al., 2004). The growth of service functions also reflects changes in the skill sets required. In some accounts this employment shift is considered to indicate the rise of a service economy in which service functions and outputs become increasingly dominant (Illeris, 1996; OECD, 2000; Bryson et al., 2004; Bryson and Daniels, 2007). Such accounts have a tendency to over-emphasise the importance of the rise of service employment and under-emphasise productivity differentials that exist between manufacturing and services. Manufacturing employment has declined in many national economies but productivity improvements have meant that output has risen (Bryson, 2008a). To complicate matters further, many service jobs directly or indirectly support goods production (MacPherson, 1997; Bryson, 2009a). In some cases, service functions that were previously undertaken inside manufacturing firms have been externalised to specialist service suppliers (Goe, 1991; Beyers and Lindahl, 1996; Bryson et al., 2004: 83–85). This type of restructuring produces a statistical anomaly in which service employment appears to grow at the expense of manufacturing jobs (Bryson, 2009a).

Second, manufacturing has changed, or perhaps our understanding of it has changed during this century. Academics and policy-makers have begun to adopt a more complex definition of manufacturing that has been informed by relatively sophisticated accounts of commodity or value chains (Gereffi, 2001; Gereffi et al., 2005; Bryson, 2008b). In this context, the concept of value chain fragmentation has become an important conceptual tool for understanding the evolving structure and new geographies of production systems. The concept of 'fragmentation' highlights the fact that manufacturing involves more than just fabrication, but also includes service functions that are integral to the production of physical goods. This implies that academics and policy-makers have begun to shift their attention away from a narrow fabrication view of manufacturing to one in which manufacturing includes research and development, design functions, marketing and advertising, services that support production processes and a set of services that have been created to support customers' experiences of a product. Manufactured goods should now be conceptualised as products that contain different quantities of service inputs; some of these service inputs are wrapped into a good during the production process and some are wrapped around a completed product (Daniels and Bryson, 2002; Livesey, 2006; Department for Business, Enterprise and Regulatory Reform, 2008).

Third, the blurring that has occurred and continues to occur between goods and services implies that these terms are beginning to distort the ways in which academics and policy-makers conceptualise production systems (Howells, 2000; Daniels and Bryson, 2002). This has important implications for academic theory but also impacts on the ways in which national statistical agencies classify economic activity. It also means that regional policy can no longer be targeted at services or manufacturing sectors, but must become increasingly holistic and be designed to enhance the competitiveness of complex production systems consisting of hybridised manufacturing and service processes and ultimately hybrid production systems that produce hybrid products (Bryson, 2008a, b; Bryson et al., 2008b). A hybrid product contains a complex combination of services and manufactured inputs, with both sets of inputs required for the product to function. All production systems are hybrid systems but not all hybrid production systems produce hybrid products.

There is an interesting literature that argues that manufacturing is undergoing a process of 'servitisation' (Neely, 2007). But the term 'servitisation' suggests that services are just being applied to manufacturing as an additional element. This overlooks the ways in which the application of services to manufacturing is more than a process of application but one of transformation. With this in mind, our argument is that a blend of manufacturing and services has led to a process of hybridisation that has led to the development of hybrid production systems and hybrid products. The use of the term 'servitisation' reflects the application of an old language to a new process.

Fourth, the development of hybrid production systems has important implications for the science of management cybernetics that have yet to be explored to any great extent. Simple price-based competition has been replaced or supplemented with other forms of competition founded upon the exploitation of new forms of expertise supported by new control systems (Bryson et al., 2008b). The implication is that management systems must be designed to capture profits that come from services that are woven into and wrapped around products. This means that profit not only occurs at the point of sale of some goods, but also continues as part of a revenue stream that comes from the provision of a range of related services (Giarini, 2002). Management and control within hybrid production systems is a new area that will require considerable further research.

These four processes have important implications for the ways in which firms position their products and services in the marketplace and also for regional and national economic policy. Further detailed research, however, is required in order to explore many of the management and policy implications that come from a shift away from the production of goods towards the creation of hybrid products.

The transformation that is ongoing within manufacturing has its origins in the 19th century (Bryson, 2009b). The provision of services was an important element for many manufacturing firms during the Industrial Revolution. The difference between now and then is the extent and complexity of services that have been incorporated into manufacturing production systems and into physical goods. In 1993, Sandra Vandermerwe was one of the first academics to identify the ways in which firms in all sectors can create added value through the provision of services. The title of her book – *From Tin Soldiers to Russian Dolls: Creating Added Value Through Services* – highlights the extent of this transformation. According to Vandermerwe, 'by the 1990s, for one reason or another, most manufacturing and service firms had concluded that products without services, or services without products, was an untenable position. Both were patently necessary in order to compete. Besides, were products and services so different from each other?' (1993: 47). In this book she poses the question: 'Could a manufacturer really be considered part of the industrial sector anymore if such a large portion of its revenue depended on services?' (Vandermerwe, 1993: 47). This highlights the difficulties of classifying activities using languages and classification systems that reflect the economy as it was structured in the 18th century. It also calls for a new language to describe and understand the structure of economic activity (Daniels and Bryson, 2002). In 2002 we published a paper calling for the development of a new language for a new economic geography. At that time we had no answer to the difficulty of classifying activities into goods and services, but subsequently a solution to this difficulty has emerged (Bryson, 2009b) – namely, to use the term 'production'; all production is a blend of service and manufacturing tasks (Bryson and Rusten, 2011).

There are important control, management and organisational issues to be considered between BPS functions that can be delivered remotely and those that are still dependent upon social relationships that are constructed around face-to-face encounters. In this context it is important to remember that a production process consists of a set of processes and operations (Bryson, 2008b). A process refers to the collection of activities that combined together produce a product, whilst an operation refers to an activity that is performed at a particular point in a production process (Blackstone et al., 1997: 602). Production systems are increasingly founded upon complex combinations of manufacturing and service knowledge. The production of products and services should be conceptualised as a process that consists of a complex and evolving blending of manufacturing and service processes or, perhaps more correctly, producer service processes. This can be conceptualised as a simple equation in which:

Production (P) = Manufacturing Processes (M) + Service Processes (S).

It is impossible to manufacture without services and services cannot be created or delivered without manufactured products (Bryson et al., 2008a). It is therefore important for academics and policy-makers to begin to identify and conceptualise the complex interrelationships that occur between different elements of production processes that together create value. This 'coming together' to create a product (physical product or a service) can take place within the same company or can be part of a coordinated value chain of independent companies that are managed by a company or even an individual. The control of a process should take precedence over operations and all operations must be subordinate to the requirements of the process. Operations may be geographically distributed, since a production process may be designed around the benefits that can accrue from an international division of labour. The production process must be controlled so that the constraints or limitations on the process can be identified and taken into consideration. This also means that individual operations may have to be controlled using different systems to take into consideration the nature of the activity, for example service or product inputs into the overall process, and complexity that is associated with a geographically distributed production system.

A Case Example: Rolls-Royce and BAE Systems

Rolls-Royce is one of the most frequently cited examples of a firm that has shifted towards a service-based business model. In the financial year 2006–2007, 55 per cent of its revenues were derived from the delivery of services (Rolls-Royce, 2007: 15). In 1987 Rolls-Royce 'supported our engines in service by offering repair and overhaul arrangements which often failed to align our interests with those of our customers' (Rolls-Royce: 2007: 14). At this time, services were considered as a supporting set of functions rather than as an integral element within the firm's business model. Since 1987, Rolls-Royce has transformed itself into a provider of power rather than a provider of engines. This transformation has occurred in all the firm's core market segments, ranging from civil aviation to defence aerospace and marine engines. The transformation has involved the company in developing:

> comprehensive through-life service arrangements in each of our business sectors. These align our interests with those of our customers and enable us to add value through the application of our skills and knowledge of the product. In 2007, underlying aftermarket service revenues grew by nine per cent and represented 55 per cent of Group sales. This growth has been achieved partly as a result of the introduction of new products, but also because our ownership of intellectual property enables us to turn data into information that adds value to our customer. (Rolls-Royce, 2007: 14)

A good example of this shift is the mission ready management solutions (MRMS ®) package supplied by Rolls-Royce. MRMS provides the military with customised solutions that include total support packages and 'Power by the Hour'®. With the latter package, major airline and defence customers pay a fixed warranty and operation fee for the hours that an engine runs. Contract performance is measured against the performance of the fleet and in terms of ready for issue engine availability.

Rolls-Royce offers three types of service solutions. First, TotalCare is based upon an agreed rate per engine flying hour and this enables customers to engage in accurate financial forecasting. This package is designed for airline fleets and it transfers the technical and financial aspects of fleet maintenance from the customer to the service supplier. At the same time it converts Rolls-Royce into a service provider or, more precisely, a provider of hybrid products. Second, CorporateCare is intended for corporate and business jet customers and is designed to ensure that the aircraft is available when required; it may also result in increased residual value. Third, MRMS is targeted at defence customers and provides them with engine management and maintenance to ensure operational capability 24 hours per day and seven days per week. These types of hybrid products have transformed Rolls-Royce from a company that designs and manufactures engines to a provider of turnkey engine power (Table 1.2). To maximise profitability, Rolls-Royce must now focus on the effective management of an extended manufacturing value chain or its hybrid production system. This includes the design and development of engines, installation, after-sales maintenance, repair and overall services, and parts availability and management.

Many other manufacturing companies sell products that include services. Another excellent example is BAE Systems, the British manufacturer of defence equipment. Currently, BAE is trying to diversify from an over-dependence on defence-related

Table 1.2 *The transformation of Rolls-Royce from a provider of a good to a provider of hybrid products*

Good delivery of engine			→	Service delivery of power
Traditional Support	Enhanced Support	Advanced Support	Total Support	Extended support
Spare parts Repair Overhaul	Data forecasting services Technical logistics support Customer training	Comprehensive package integrating elements of basic and enhanced support Spares inclusive Repair Overall contracts	Complete, availability-based services Can cover all of an aircraft, or some element's of an aircraft's activity Configuration management and reliability enhancements covered	Partnered capability Turnkey service Non-propulsion-related support solutions
Customer responsibility			→	Service provider responsibility

Source: After Rolls-Royce, 2006.

products to commercial products that include engine controls, flight controls and cabin management systems. The company is heavily involved in the development of the F-35 Joint Strike Fighter. This new fighter has cost an estimated $1 trillion to develop (Clancy, 2014: B5). Lockheed Martin is developing the F-35 for the US government, but BAE is the largest partner and is responsible for producing 15 per cent of the aircraft. BAE is making parts for the fuselage and tail for every F-35 at its Samlesbury factory located in the north-west of England and the components are shipped to the final assembly line in Fort Worth, Texas. This British element accounts for 10 per cent of the plane, with another 5 per cent manufactured by BAE in the US. The F-35 is anticipated to have a life expectancy of 50 years. For BAE the F-35 programme offers two opportunities to generate profit – first, through the design and manufacture of components and, second, via the provision of a constant stream of services for all the planes. The manufacturing element represents a single moment to create value or profit, but the service element will provide BAE with service-generated profits for up to 50 years.

Both these examples of hybrid products are from heavy engineering, but the hybridisation between manufacturing and services is occurring across all manufacturing sectors.

CONCLUSIONS

The development of hybrid production systems and hybrid products represents a radical alteration to the ways in which products are produced and consumed. At one level, this represents a breakdown of the long-standing and simple bi-polar distinction that is made between services and manufacturing. Increasingly, this division is no longer helpful as manufactured products acquire many of the characteristics of services. This has three important implications that need to be considered.

First, economic theory needs to be refined to take into consideration the demise of the manufacturing/service divide. This has important implications for the ways in which the creation of profit is theorised. Profit should not be conceptualised as something that is created by a single transaction involving the transfer of the ownership of a product from a producer to a consumer. Instead, profit occurs through multiple moments of exchange. The physical good may have limited value whilst profit is created and acquired through the sale of embedded services, spare parts or consumables.

Second, the way in which governments collect statistical data about the economy will have to be revised. Too many service functions are hidden within manufacturing firms and many manufacturing firms are really providers of hybrid products. The long-established Standard Industrial Classification (SIC) codes may have to move towards a system in which firms are categorised on the basis of an analysis of jobs, functions and tasks (rather than just their principal product). Alternatively, new SIC codes could be developed to identify firms in particular sectors that have developed hybrid production systems and products.

Third, economic policy will have to alter to take into consideration the shift towards hybrid production systems. This means that policies must embrace complete value chains rather than being targeted at manufacturing or service elements. Manufacturing firms should be encouraged to develop service expertise and explore ways in which services can add value to their existing business models. Services and their hybridisation with manufacturing processes provide an important source of inimitability. This process of hybridisation requires further detailed study.

The shift in emphasis towards hybrid production systems is also a shift towards a production system that might be more resilient during periods of economic downturn or recession. A firm that only sells goods has great difficulties during an economic downturn unless it sells goods that are essential rather than a discretionary purchase. A firm that sells hybrid products may have difficulties in selling the manufactured product during an economic downturn, but should have developed an extensive network of existing captive 'service' customers. These may have signed service contracts that represent a continual income stream for the provider of a hybrid product or represent opportunities for the sale of new product-supporting services. Further research is required to compare different business models and their ability to create wealth during recessionary periods. We are currently living in extremely uncertain times and perhaps the only certainty is that economic competitiveness rests on the ability of companies to design and manage effective hybrid production systems and to produce hybrid products that blend services and manufacturing elements together in increasingly novel ways.

BIBLIOGRAPHY

Barnett, D. (1998), *London: Hub of the Industrial Revolution: A Revisionary History: 1775–1825*, Tauris Academic Studies: London.
Baumol, W. (1967), 'Macroeconomics of unbalanced growth: the anatomy of an urban crisis', *American Economic Review*, 57, 415–426.
Baumol, W., Blackman, S. and Wolf, E. (1989), *Productivity and American Leadership*, MIT Press: Cambridge, MA.
Bell, D. (1973), *The Coming of Post-Industrial Society: A Venture in Social Forecasting*, Basic Books: New York.
Beyers, W.B. and Alvine, M.J. (1985), 'Export services in post-industrial society', *Papers of the Regional Science Association*, 57, 33–45.
Beyers, W.B. and Lindahl, D.P. (1996), 'Explaining the demand for producer services', *Papers in Regional Science*, 75, 351–374.
Blackstone, J.H., Gardiner, L.R. and Gardiner, S.C. (1997), 'A framework for the systemic control of organizations', *International Journal of Production Research*, 35, 3, 597–609.
Braudel, F. (1988), *Civilization and Capitalism 15th–18th Century: The Wheel of Commerce*, Collins: London.
Bryson, J.R. (1996), 'Small business service firms and the 1990s recession in the United Kingdom: implications for local economic development', *Local Economy Journal*, 11, 221–236.
Bryson, J.R. (1997a) 'Business service firms, service space and the management of change', *Entrepreneurship and Regional Development*, 9, 93–111.
Bryson, J.R. (1997b), 'Small and medium-sized enterprises, Business Link and the new knowledge workers', *Policy Studies*, 18, 1, 67–80.
Bryson, J.R. (2008a), 'Service economies, spatial divisions of expertise and the second global shift', in P.W. Daniels et al. (eds), *Human Geography: Issues for the 21st Century*, Prentice Hall: London, third edition, pp. 339–357.
Bryson, J.R. (2008b), 'Value chains or commodity chains as production projects and tasks: towards a simple theory of production', in Dieter Spath and Walter Ganz (eds), *The Future of Services: Trends and Perspectives*, Carl Hanser Verlag: Munich, pp. 265–284.
Bryson, J.R. (2009a) 'Economic geography: business services', in R. Kitchin and N. Thrift (eds), *International Encyclopaedia of Human Geography*, Elsevier: Oxford, pp. 368–374.
Bryson, J.R. (2009b) 'Service innovation and manufacturing innovation: bundling and blending services and products in hybrid production systems to produce hybrid products', in F. Gallouj (ed.), *Handbook on Innovation in Services*, Edward Elgar: Cheltenham and Northampton, MA, pp. 679–700.
Bryson, J.R. (2009c), *Hybrid Manufacturing Systems and Hybrid Products: Services, Production and Industrialisation*, University of Aachen: Aachen.
Bryson, J.R. and Daniels, P.W. (2007), 'Service worlds', in J.R. Bryson and P.W. Daniels (eds), *The Handbook of Service Industries in the Global Economy*, Edward Elgar: Cheltenham and Northampton, MA.
Bryson, J.R. and Rusten, G. (2011), *Design Economies and the Changing World Economy: Innovation, Production and Competitiveness*, Routledge: London.
Bryson, J.R., Daniels, P.W. and Warf, B. (2004), *Service Worlds: People, Organisations, Technologies*, Routledge: London.
Bryson, J.R., Keeble, D. and Wood, P. (1993a), 'The creation, location and growth of small business service firms in the United Kingdom', *Service Industries Journal*, 13, 2, 118–131.
Bryson, J.R., Keeble, D. and Wood, P. (1993b), 'Business networks, small firm flexibility and regional development in UK business services', *Entrepreneurship and Regional Development*, 5, 265–277.
Bryson, J.R., Keeble, D. and Wood, P. (1994), *Enterprising Researchers: The Growth of Small Market Research Firms in Britain*, Business Services Research Monograph, Series 1, No. 2, Small Business Research Trust.
Bryson, J.R., Keeble, D. and Wood, P. (1997), 'The creation and growth of small business service firms in post-industrial Britain', *Small Business Economics*, 9, 4, 345–360.
Bryson, J.R., Taylor, M. and Cooper, R. (2008a), 'Competing by design, specialization and customization: manufacturing locks in the West Midlands (UK)', *Geografiska Annaler: Series B, Human Geography*, 90, 2, 173–186.
Bryson, J.R., Taylor, M. and Daniels, P.W. (2008b), 'Commercializing "creative" expertise: business and professional services and regional economic development in the West Midlands, UK', *Politics and Policy*, 36, 2, 306–328.
Clancy, R. (2014), 'There's going to be consolidation in defence', *Daily Telegraph*, Monday July 14, B5.
Daniels, P.W. (ed.) (1979), *Spatial Patterns of Office Growth and Location*, John Wiley & Sons: Chichester.
Daniels, P.W. (1982), *Service Industries: Growth and Location*, Cambridge University Press: Cambridge.
Daniels, P.W. (1985), *Service Industries: A Geographical Appraisal*, Methuen: London.

Daniels, P.W. and Bryson, J.R. (2002), 'Manufacturing services and servicing manufacturing: changing forms of production in advanced capitalist economies', *Urban Studies*, 39, 5–6, 977–991.
Deane, P. (1969), *The First Industrial Revolution*, Cambridge University Press: Cambridge.
Deane, P. and Cole, W. (1962), *British Economic Growth 1688–1959*, Cambridge University Press: Cambridge.
Department for Business, Enterprise and Regulatory Reform (2008), *Manufacturing: New Challenges, New Opportunities*, BERR: London.
Fellowes, E.H. (1954), 'Service', in E. Blow (ed.), *Grove's Dictionary of Music and Musicians*, Macmillan & Co. Ltd: London, pp. 719–723.
Fuchs, V.R. (1965), 'The growing importance of the service industries', *Journal of Business of the University of Chicago*, 38, 360–362.
Fuchs, V.R. (1968), *The Service Economy*, Columbia University Press: New York.
Fuchs, V.R. (1969), *Production and Productivity in the Service Industries*, National Bureau of Economic Research: New York.
Galbraith, J.K. (1967), *The New Industrial State*, Hamish Hamilton: London.
Gallouj, F. and Djellal, F. (eds) (2010), *The Handbook of Innovation and Services: A Multi-disciplinary Perspective*, Edward Elgar: Cheltenham and Northampton, MA.
Gereffi, G. (2001), 'Shifting governance structures in global commodity chains, with special reference to the internet', *American Behavioural Scientist*, 44, 1616–1637.
Gereffi, G., Humphrey, J. and Sturgeon, T. (2005), 'The governance of global value chains', *Review of International Political Economy*, 12, 78–104.
Gershuny, J. (1978), *After Industrial Society? The Emerging Self-Service Economy*, Macmillan: London.
Gershuny, J. and Miles, I. (1983), *The New Service Economy: The Transformation of Employment in Industrial Societies*, Pinter Publishers: London.
Giarini, O. (2002), 'The globalization of services in economic theory and economic practice: some conceptual issues', in J.R. Cuadrado, L. Rubalcaba and J.R. Bryson (eds), *Trading Services in the Global Economy*, Edward Elgar: Cheltenham and Northampton, MA, pp. 58–77.
Goddard, J.B. (1986), 'Advanced telecommunications and regional economic development', *The Geographical Journal*, 152, 383–397.
Goe, R. (1991), 'The growth of producer services industries: sorting through the externalization debate', *Growth and Change*, 22, 118–141.
Gowers, E. (1982), *The Complete Plan Words*, Penguin: Harmondsworth.
Gray, A.B. (1921), *Cambridge Revisited*, W. Heffer & Sons: Cambridge.
Greenfield, H.I. (1966), *Manpower and the Growth of Producer Services*, Columbia University Press: New York.
Howells, J. (2000). *The Nature of Innovation in Services*, report presented to the OECD Innovation and Productivity Workshop, Sydney, Australia, October.
Hyde, F.E. (1973), *Far Eastern Trade: 1860–1914*, Adam & Charles Black: London.
Illeris, S. (1996), *The Service Economy: A Geographical Approach*, Wiley: Chichester.
Lash, S. and Urry, J. (1994), *Economies of Signs and Spaces*, Sage: London.
Lewis, R. and Maude, A. (1952), *Professional People*, Phoenix House: London.
Livesey, F. (2006), *Defining High Value Manufacturing*, Institute for Manufacturing: Cambridge.
MacPherson, A. (1997), 'The role of producer services outsourcing in the innovation performance of New York state manufacturing firms', *Annals of the Association of American Geographers*, 87, 1, 52–71.
Maglio, P.P., Kieliszewski, C.A. and Spohrer, J.C. (eds) (2010), *Handbook of Service Science*, Springer: New York.
Marsh, D.C. (1977), *The Changing Structure of England and Wales: 1871–1961*, Routledge & Kegan Paul: London.
Moretti, E. (2013), *The New Geography of Jobs*, Mariner Books: Boston, MA.
Neely, A. (2007), *The Servitization of Manufacturing: An Analysis of Global Trends*, paper presented at the 14th European Operations Management Association Conference, Ankara, Turkey, June.
OECD (2000), *The Service Economy*, Business and Industry Policy Forum, OECD: Paris.
OECD (2014), *Short Term Labour Market Statistics*, available at http://stats.oecd.org/index.aspx?queryid=38899#, accessed 25 July 2014.
Ogilvy, D. (1983), *Ogilvy on Advertising*, Pan: London.
Oxford English Dictionary (1991), *The Shorter Oxford English Dictionary*, Clarendon Press: Oxford.
Pinches, S. (2009), *Ledbury: A Market Town and its Tudor Heritage*, Phillimore: Chichester.
Riddle, D.I. (1986), *Service-Led Growth: The Role of the Service Sector in World Development*, Praeger: New York.
Rolls-Royce (2006), *MRMS® Mission Ready Management Solutions*, Rolls-Royce: London.
Rolls-Royce (2007), *Annual Report 2007: A Global Business*, Rolls-Royce: London.
Sayer, A. and Walker, R. (1992), *The New Social Economy: Reworking the Division of Labour*, Blackwell: Oxford.

Smith, A. (1977 [1776]), *The Wealth of Nations*, Penguin Books: Harmondsworth.
Twitchell, J.B. (2004), *Branded Nation: The Marketing of Megachurch, College Inc., and Museumworld*, Simon & Schuster: New York.
Urry, J. (1987), 'Some social and spatial impacts of services', *Environment and Planning D: Society and Space*, 5, 5–26.
Urry, J. (1990), 'Work, production and social relations', *Work, Employment and Society*, 4, 271–280.
Vandermerwe, S. (1993), *From Tin Soldiers to Russian Dolls: Creating Added Value Through Services*, Butterworth-Heinemann: Oxford.
Walker, R. (1985), 'Is there a service economy? The changing capitalist division of labour', *Science and Society*, 49, 1, 42–83.
White, A. (2013), 'Land Secs goes shopping at Bluewater in £656m deal', *The Daily Telegraph*, 26 June, B5.
Wright Mills, C. (1953), *White Collar Work: The American Middle Classes*, Oxford University Press: New York.

PART I

UNDERSTANDING SERVICE BUSINESS: DISCIPLINARY PERSPECTIVES

2. Growth and productivity in EU service sectors[1]
Mary O'Mahony

1. INTRODUCTION

In recent years the performance of service industries has come to the forefront of research and policy debate on Europe's comparative economic performance. The much publicised acceleration in US output and productivity growth and Europe's failure to match this performance are primarily down to developments in market service sectors (Timmer et al., 2010). The focus on service sectors was facilitated by developments in the measurement of outputs and inputs, which for the first time allowed for quantitative comparisons across sectors, countries and time. The purpose of this chapter is to highlight some of the main findings from the existing quantitative analysis and present new estimates on service sector performance following the global financial crisis since late 2007.

In the past much quantitative research on growth and productivity relied on data for manufacturing sectors. This was because the physical nature of the output made it easier to measure but there was also an underlying belief that most of the interesting productivity developments occurred in these sectors, which were relatively capital intensive and invested heavily in research and development (R&D). Services were seen as labour intensive, often employing relatively unskilled workers, with little physical capital employed in production and few avenues for innovation. During the past decade or so economic analysts, especially in the United States, started to notice that output and labour productivity growth were accelerating in service sectors (Triplett and Bosworth, 2006). A vast literature emerged that tried to explain these trends, linked to adoption and diffusion of information and communications technology (ICT), as services were seen as intensive users of the new technology. This dispelled the low capital intensity low innovation myth. By distinguishing different types of assets it became apparent that the rapidly growing stocks of ICT capital had a much larger impact on growth than previously thought (Jorgenson, Ho and Stiroh, 2005). Another strand of the literature homed in on the need for organisational changes to reap the benefits from the new technology (Bresnahan, Brynjolfsson and Hitt, 2002), both enhanced skills of the workforce and managerial skills. This in turn led to a focus on the role of investments in intangible capital in generating growth (Corrado, Hulten and Sichel, 2005). Process innovations in service sectors are mostly organisational changes which complement ICT while product innovations in services are frequently facilitated by this technology.

Much of the research on the emergence of services as the generator of growth originated in the United States, with some research results also providing confirmation of aspects of this process for other industrial countries, primarily for the UK. Research for other large EU countries, however, appeared to suggest that an acceleration of growth arising from ICT was not widespread in Europe (see O'Mahony and Van Ark, 2003 for a summary). It was only following the construction of the EU KLEMS database (see section 2),[2] with its harmonised methodology, that research could verify that relatively slow growth in the EU was in fact driven mostly by developments in service sectors.

This chapter reviews recent developments in service sector performance since the mid 1990s, drawing from both the EU KLEMS database and recent updates. The main analysis focuses on two time periods. The first is the boom period from 1995 to 2007, contrasting with earlier decades. This summarises the evidence from EU KLEMS on sources of growth. The second time period covers the period of the financial crisis since 2008. This presents an overview of trends in output and labour productivity in EU market services as a whole, contrasting with performance in production sectors, followed by additional detail within service sectors for some countries. Before examining these two periods it is necessary first to sketch out the main features of the EU KLEMS database and the growth accounting methodology employed in the analysis – this is the subject of the next section. The main analysis by time period follows in sections 3 and 4. Finally, a concluding section draws out some implications from recent literature, concentrating on two possible explanations for the US growth lead: intangible capital linked to organisational changes and the regulatory environment.

2. THE EU KLEMS DATABASE

This section sketches the main features of the EU KLEMS database that are used in the analysis in subsequent sections – more detail is available in O'Mahony and Timmer (2009). The EU KLEMS database has been constructed on the basis of data from national statistical institutes, employing series from national accounts combined with census and survey data when required. The underlying rationale is that data should be processed according to harmonised procedures to ensure international comparability, including a common industrial classification, the use of similar price concepts for inputs and outputs, and consistent definitions of various labour and capital types. Nominal and constant price series for output (value added) at the industry level are taken directly from the National Accounts. As these series are often short (since revisions are not always taken back in time) different vintages of the National Accounts were bridged according to a common link-methodology. In cases where industry detail was missing additional statistics from censuses and surveys were used to fill the gaps.

Labour input is based on series of hours worked and wages of various types of labour collected from employment and labour force surveys. The database cross-classifies hours worked into 18 labour categories by educational attainment (high, medium and low skilled), gender and age (15–29, 30–49 and 50+). Total labour input is calculated as a weighted average of the growth of hours in each category, with weights equal to their wage bill shares.

In EU KLEMS, capital input is measured as capital services which are calculated as the weighted growth of stocks of eight types of assets. These comprise three types of ICT assets (information technology equipment, communication technology equipment and software) and five types of non-ICT assets (residential structures, non-residential structures, transport equipment, other machinery and equipment, and other fixed capital assets). For each individual asset, stocks have been estimated on the basis of investment series using the perpetual inventory method with geometric depreciation profiles.

Depreciation rates differ by asset and industry but have been assumed identical across countries. The basic investment series by industry and asset have been derived from capital flow matrices and benchmarked to the aggregate investment series from the National Accounts. Additional information was required to obtain investment series to break out ICT assets. When the deflator for computers did not contain an adjustment for quality change, a harmonised deflator based on the United States deflator has been used, as suggested by Schreyer (2002). The weights are based on the rental price of each asset, which consists of a nominal rate of return plus depreciation minus capital gains. This weighting is crucial in the analysis as ICT assets typically have very high depreciation rates and experience high capital losses, and so their contribution to overall input is much greater using this approach than would result from a simple (unweighted) summation across asset types. The main EU KLEMS database reports capital services estimates for two groups: ICT and non-ICT assets.

The series described above can be used in statistical analysis of the relationship between inputs, outputs and productivity growth at the industry level. However, the database also combines series to estimate sources of output growth, including productivity, using the method of growth accounting which was pioneered by Solow (1957) and further developed by Jorgenson and associates (Jorgenson and Griliches, 1967; Jorgenson, Gollop and Fraumeni, 1987). This decomposes the growth in real value added into the growth in labour and capital inputs, each weighted by their compensation shares of value added, and a residual termed multi-factor productivity (MFP) growth.[3] Labour compensation is adjusted to include the earnings of the self-employed. Capital compensation is calculated as value added minus labour compensation, which can be justified using an assumption of constant returns to scale. Underlying this method are assumptions of competitive output and input markets so that all inputs are paid their marginal products.

Growth accounting is a useful tool for describing sources of growth, dividing output growth into contributions from inputs and MFP. Contributions of aggregate capital and labour inputs in turn can be divided into volume growth and composition or 'quality' growth. The EU KLEMS database shows that there is a general shift towards more skilled and more experienced workers in the labour force. As such, labour input grows faster than suggested by a crude measure of hours worked, unadjusted for changes in labour composition. Similarly, especially since the mid 1990s the importance of short-lived ICT assets relative to non-ICT assets has increased. Consequently, capital service input growth rates are higher than capital stock growth rates as ICT assets deliver more services per unit of capital stock. Not accounting for this shift in the composition of capital biases input growth downwards, and consequently MFP growth upwards. Growth accounting is a useful first step in benchmarking sources of growth but it must be emphasised that it in no way suggests any causal relationship.

3. GROWTH AND PRODUCTIVITY IN SERVICES DURING THE INFORMATION TECHNOLOGY ERA

Performance in the Market Economy

The EU KLEMS database makes it possible for the first time to compare and analyse the role of high-skilled labour and ICT capital for productivity growth at an industry level between countries. The focus here is on the market economy, excluding health and education services, public administration and real estate, where output measurement poses particular difficulties. Also, in this section the aggregate EU estimates only cover ten countries, excluding Greece, Ireland, Luxembourg, Portugal and Sweden, where some key data series were unavailable.

Table 2.1 provides a summary picture of the contributions of factor inputs and MFP to output growth in the market economy, comparing the European Union with the United States for the periods 1980–1995 and 1995–2007. When comparing the period before and after 1995, the annual growth rate of output in the European Union and the United States shows a slight increase. Hours worked in the European Union grew after 1995 following a period of decline before then. In contrast, the growth in hours worked slowed down in the United States, even though the average growth rate in hours was comparable with that of the EU between 1995 and 2007. Labour productivity growth (output growth minus growth in hours worked) increased in the US market economy compared with a slowdown in Europe after 1995.

Table 2.1 shows that, even though contributions from labour composition were small, its positive sign implies that the process of transformation of the labour force to higher skills has proceeded in both regions. However, this process slowed down somewhat in the European Union after 1995. Concerning the contribution of capital deepening to labour

Table 2.1 Decomposition of labour productivity growth, market economy, EU and USA, 1980–2007

	European Union		United States	
	1980–1995	1995–2007	1980–1995	1995–2007
Real output	2.1	2.5	3.3	3.5
Hours worked	−0.5	0.8	1.3	0.9
Labour productivity	2.5	1.6	2.0	2.6
Contributions from:				
Labour composition	0.4	0.2	0.2	0.3
ICT capital services per hour	0.4	0.5	0.7	0.9
Non-ICT capital services per hour	0.8	0.4	0.3	0.3
Multi-factor productivity	1.0	0.6	0.8	1.2

Notes: Contributions to growth in labour productivity (annual average growth rates, in percentage points) are derived by multiplying the growth in each component by its respective compensation shares in value added. Data for European Union refer to ten countries: Austria, Belgium, Denmark, Finland, France, Germany, Italy, the Netherlands, Spain and the United Kingdom.

Source: Calculations based on EU KLEMS database.

productivity growth, measured by capital services per hour, Table 2.1 shows increasing contributions from information and communications technology in both regions but the absolute contribution in the United States is much larger than in the European Union. This in turn reflects higher levels of ICT capital in the USA, due to earlier adoption there of the technology. The contributions from non-ICT capital deepening declined in the EU while remaining static in the United States.

The largest difference in the source of growth between the European Union and the United States shown in Table 2.1 is in the contribution of multi-factor productivity growth. Whereas in the United States it accelerated from 0.8 percentage points from 1980–1995 to 1.2 from 1995–2007, the same measure declined from 1.0 to 0.6 percentage points between these two periods in the European Union. MFP includes the effects of technological change, along with non-constant returns to scale. But as a residual measure it also includes measurement errors and the effects from unmeasured output and inputs, such as R&D and other intangible investments, including organisational improvements. Broadly, it indicates the efficiency with which inputs are used in the production process, and its reduced growth rate is therefore a major source of concern across Europe.

Market Services and the European Slowdown

Both Europe and the United States have experienced a major shift of production and employment from manufacturing and other goods-producing industries towards services. Table 2.2 presents the shares in total hours worked by major sectors in 1995 and 2007. Data are provided for the major European countries and the USA, as before. Over the period 1980–2007, the share of labour input going to goods production declined, driven by a decrease in manufacturing share. On the other hand, the share of hours worked in

Table 2.2 Share of hours worked by sector

	European Union		United States	
	1995	2007	1995	2007
Goods production	0.43	0.37	0.35	0.30
Manufacturing	0.25	0.20	0.23	0.16
Other goods production[1]	0.18	0.17	0.12	0.14
Market Services	0.57	0.63	0.65	0.70
Distribution	0.20	0.19	0.23	0.22
Transport	0.06	0.06	0.04	0.05
Communications	0.02	0.02	0.03	0.02
Financial services	0.04	0.04	0.06	0.06
Business services[2]	0.11	0.17	0.15	0.19
Other services[3]	0.13	0.15	0.14	0.16

Notes:
1. Agriculture, mining, utilities and construction;
2. Excluding real estate;
3. Hotels and restaurants, recreation services, other community and personal services.

Source: EU KLEMS.

market services increased. While there are differences across European countries, even in Germany – a country in which manufacturing traditionally plays an important role – market services are now almost three times as big as manufacturing. Market services have grown fastest in the UK and employ almost five times the number of manufacturing workers. Within market services the main growth areas in terms of hours worked have been business services and other services. The latter includes hotels, restaurants and personal services.

The shift from manufacturing to services has important implications for productivity growth. Traditionally, manufacturing activities have been regarded as the main locus of innovation and technological change, and thus the central source of productivity growth. They were the key to post-World War II growth in Europe through a combination of economies of scale, capital intensification and incremental innovation (van Ark, O'Mahony and Timmer, 2008). Data from EU KLEMS can show to what extent this perception remains correct for the period from the mid 1990s. Table 2.3 shows the equivalent information to that contained in Table 2.1 for aggregate production and aggregate market services.

Looking first at productivity, in terms of both labour and multi-factor productivity it is clear from Table 2.3 that the main difference between the two regions occurs in market services, with productivity either declining or flat in the EU but accelerating in the USA. In contrast the two productivity measures show deceleration in production sectors in

Table 2.3 Decomposition of labour productivity growth, production and market services, EU and USA, 1980–2007

	Production		Market Services	
	1980–1995	1995–2007	1980–1995	1995–2007
European Union				
Real output	1.3	1.4	2.9	3.3
Hours worked	−2.2	−0.6	1.3	1.8
Labour productivity	3.5	2.0	1.6	1.5
Contributions from:				
Labour composition	0.4	0.2	0.3	0.2
ICT capital services per hour	0.3	0.3	0.6	0.7
Non-ICT capital services per hour	1.1	0.6	0.5	0.3
Multi-factor productivity	1.7	0.9	0.2	0.3
United States				
Real output	2.7	1.6	3.7	4.6
Hours worked	−0.1	−0.5	2.2	1.5
Labour productivity	2.8	2.1	1.5	3.1
Contributions from:				
Labour composition	0.3	0.1	0.2	0.3
ICT capital services per hour	0.6	0.6	1.2	1.2
Non-ICT capital services per hour	0.4	0.5	0.3	0.3
Multi-factor productivity	1.6	0.9	−0.1	1.3

Notes: As for Table 2.1.

both regions. In market services ICT capital deepening increased slightly in the EU, more than compensated by a decline in non-ICT capital per hour. In the United States the contribution of labour composition to market services' labour productivity growth showed a slight increase in the later period, in contrast to a decline in the EU. Although these contributions are small, as stated above, increasing the average skill levels of the workforce is important in adopting and using new technologies, and Europe appears to be falling behind the United States in the extent of improving the skills of its workforce.

Finally in this section it is useful to consider trends in the major sectors that make up market services. Table 2.4 shows the change in growth rates and contributions, comparing the period 1995–2007 with 1980–1995 in the two regions. This shows considerable diversity within the market services sector, with both communications and financial services showing accelerations in output and productivity in both the EU and the USA.

Table 2.4 Decomposition of labour productivity growth, market services, EU and USA, 1995–2007 minus 1980–1995

	Distribution	Transport	Communications	Financial services	Business services	Other services
European Union						
Market economy output	−0.04	0.08	2.68	1.03	−0.15	0.01
Hours worked	0.59	1.20	0.02	−0.64	−0.07	−0.14
Labour productivity	−0.63	−1.12	2.66	1.67	−0.08	0.15
Contributions from:						
Labour composition	−0.06	−0.13	−0.02	0.02	−0.26	−0.28
ICT capital services per hour	0.09	0.09	0.20	0.00	0.21	0.04
Non-ICT capital services per hour	−0.01	0.01	−0.07	−0.22	−0.36	0.07
Multi-factor productivity	−0.65	−1.09	2.55	1.87	0.34	0.32
United States						
Market economy output	0.14	−0.17	2.16	1.66	1.20	−0.25
Hours worked	−0.57	0.36	−0.86	−0.44	−1.96	−0.43
Labour productivity	0.71	−0.54	3.02	2.10	3.16	0.18
Contributions from:						
Labour composition	0.09	−0.01	−0.21	0.00	0.41	0.09
ICT capital services per hour	0.11	0.44	1.01	−0.68	0.45	−0.02
Non-ICT capital services per hour	0.06	0.04	0.09	−1.20	0.66	0.03
Multi-factor productivity	0.45	−1.00	2.14	3.99	1.63	0.09

Notes: As for Table 2.1.

The main differences are in distribution and business services where, in the USA, productivity growth and contributions from labour composition accelerate compared with decelerations in the EU.

Market services as the source of the productivity revival in the United States is confirmed in studies by Jorgenson, Ho and Stiroh (2005) and Triplett and Bosworth (2006). As differences in multi-factor productivity growth are important in accounting for the observed labour productivity differences between the two regions, it is useful to look at what lies behind MFP growth. While these factors may differ across sectors, the example of the retail sector may serve as an illustration of the complex interactions between productivity and investment. Over the past 25 years, the retail sector has undergone a substantial transformation due to benefits from the increased use of ICT. Commonly referred to as the 'lean retailing system', this has turned the retail industry from a low-technology sector that changed the products it sold at infrequent intervals into one that trades information by continuously matching goods and services to customer demands. Various studies, including McKinsey (2002) and McGuckin, Spiegelman and van Ark (2005), have discussed the reasons for superior performance in the USA retail industry relative to Europe.

Greater investment in ICT capital in the retail sector, consisting of the use of barcode scanners, inventory control tracking devices, communication equipment, transaction processing software, and so on, has a direct impact on the higher productivity growth achieved in the USA relative to the EU. However, the greater impact arguably stems from indirect benefits from ICT through increasing the potential for innovation as measured by MFP growth. These 'softer' innovations, such as the invention of new retail formats, service protocols, labour scheduling systems and optimised marketing campaigns (McKinsey 2002), have been more prevalent in the US than in Europe. These differences in turn are likely to be influenced by the regulatory environment in Europe, especially land zoning and labour market regulations which constrain new large-scale retail formats from achieving increases in market share. There is evidence that large-scale retailing has been a main driver of growth in the US, both because of increased competitive pressure on incumbent firms and higher productivity levels of new entrants (Foster, Haltiwanger and Krizan, 2006).

4. OUTPUTS AND LABOUR PRODUCTIVITY IN SERVICE SECTORS: THE IMPACT OF THE FINANCIAL CRISIS

This section focuses on performance in the most recent two years for which data are available, 2008 and 2009, relative to the boom years 1995–2007. It updates EU KLEMS using a combination of data from the OECD STAN database, National Accounts from EUROSTAT, and the US Bureau of Economic Analysis and US Bureau of Labour Statistics. As with any such exercise the extent of industry detail in the recent data is uneven across countries, so that the analysis here is confined to the two broad aggregates, production industries and market services. It examines growth in the period 1995 to 2009, divided into the boom period up to 2007 and the final two years. Attention is confined to labour productivity as the investment and labour composition data are not available from 2007. Given this, it is possible to cover the entire EU15 countries, so the results are not directly comparable with those in the previous section.

Table 2.5 Growth in gross value added, hours worked and labour productivity in the EU and US, 1995–2009 (per cent per annum)

	EU15		USA	
	1995–2007	2007–2009	1995–2007	2007–2009
Real gross value added				
Production	1.8	−6.0	1.6	−2.8
Market services	3.4	−1.4	4.6	−1.2
Hours worked				
Production	−0.6	−4.6	−0.5	−8.0
Market services	1.8	−0.9	1.5	−4.4
Labour productivity				
Production	2.4	−1.4	2.1	5.2
Market services	1.6	−0.5	3.1	3.2

Source: For sources see text.

Table 2.5 shows that the downturn in real output growth in the final two years hit the EU15 region harder than the USA, although the difference was primarily in production sectors. The second panel of the table shows trends in hours worked. Compared with production industries, hours worked declined by much less in market services in the EU15. In this region the trends in the final two years in hours worked mostly reflect declines in numbers employed in production industries. However, in market services there were significant reductions in both the numbers employed and hours worked per person, suggesting a greater proportion of people working for shorter hours, either through increased part-time work or greater absence. Hours worked in the USA declined by more in production industries than in market services, mirroring the pattern found in the EU, but with much deeper cuts. Finally, in terms of labour productivity growth, the table shows a stark contrast between the EU and the USA, with the latter accelerating productivity across the years of the crisis while in the EU labour productivity decelerated. Thus the financial crisis, if anything, seems to have exacerbated the difference between the two regions in productivity growth in market services.

An examination of data by sector within market services for a few large EU countries for which data are available suggests the downturn in output growth was most noticeable in distribution, with large declines also in business services, especially in the UK and France. Most countries in contrast experienced little change in output in the financial services sector. This raises some questions regarding the methods used in the National Accounts to measure output of this sector. In terms of labour productivity growth, most countries experienced negative growth in distribution although Spain and the UK fared better than most. In contrast labour productivity growth during 2007–2009 was positive in financial services, so that labour shedding more than compensated for downturns in output growth, and labour productivity growth even accelerated in Germany and the Netherlands. Italy and Spain registered the weakest productivity performance in financial services. Finally, labour productivity performance in business services showed the greatest diversity across countries, accelerating in Spain, Italy and the Netherlands, decelerating in France and Germany, and showing a pronounced downturn in the UK.

In summary, this section highlights the serious consequences of the recent financial crisis on output, labour input and labour productivity growth in the EU. It suggests that there are differences in its impact on production and services sectors, with the former most adversely affected. The impact also varies within market services, with the financial intermediation sector appearing least affected by the downturn, while distribution and business services have shown large impacts in some countries. However, even in the most adversely affected market services, real output declines were generally lower than in production industries. This in turn reflects a greater ability or willingness of firms to reduce labour input in production than in market services. These sectoral differences suggest significant differences in the underlying modes of production and the factors affecting these.

5. CONCLUSIONS

This chapter highlights the differences in output and productivity growth in market services, comparing the EU with the USA, both over the long run and the years immediately following the onset of the financial crisis. This begs the question of what are the driving forces behind this diverging performance? In this concluding section we focus on two areas that might help in explaining differences in performance in Europe relative to the USA: intangible capital and the regulatory environment.

Recent literature following the growth accounting tradition focuses on intangible capital which (apart from software) is not included as an input in EU KLEMS. The pioneering work in measuring intangible capital is by Corrado, Hulten and Sichel (2005, 2009), who investigated investments in intangible assets for the United States. These authors defined a number of types of intangible investments, including software, scientific and non-scientific R&D, brand equity and firm-specific expenditures such as on-the-job training and managing organisational changes. Their estimates suggest that these investments combined account for about 11 per cent of US GDP and have been growing rapidly. Similar studies for the UK (Giorgio Marrano, Haskel and Wallis, 2009), Finland (Javala, Aulin-Amhavarra and Alanen, 2007), the Netherlands (van Rooijen-Horsten, van den Bergen and Tanriseven, 2008), Japan (Fukao et al., 2009) and for EU countries (Jona-Lasinio, Iommi and Roth, 2009) present estimates that show that intangibles are also sizeable in other developed economies. Estimates of the impact of intangible capital on labour productivity growth suggest that these assets account for around 27 per cent in the US, only slightly less than this in the UK, France and Germany, and about 16 per cent in the EU15 as a whole. Thus intangible investments, many of which are intrinsically linked to organisational changes within workplaces, have been a neglected source of growth in the past.

Estimates of the impact of intangibles at the sector level are not yet readily available but the literature to date, primarily for the UK and US, indicates that these assets are more important in market services than in production industries. Overall, the research on intangible assets suggests that while the lower investments in these assets in Europe might explain some of the productivity growth gaps with the US, these are unlikely by themselves to provide a full explanation.

The strategies that firms adopt to increase productivity or reduce costs, through investment in tangible and intangible assets and adoption of new technologies, are linked to the market environment in which they operate. This will be affected by a range of institutional factors, many outside the control of firms. Recent literature has highlighted part of this institutional environment, the regulatory system, as likely to be an important part of the explanation for differences in productivity performance across countries.

According to economic theory, product market deregulation brings about substantial direct productivity gains. An increase in competition is likely to reduce slack, increase the incentives to organise work more efficiently, and foster the introduction of new and/ or improved production processes. The indirect impact of regulation through its effect on innovation is less clear. Some theories consider that competition is detrimental for productivity growth because it reduces the monopoly rents that reward new innovations, while others argue that competition has a positive influence on incentives to innovate in firms seeking to escape increased levels of competition. Recently the hypothesis that the relationship between the intensity of competition and innovation is non-linear has been favoured (Aghion et al., 2005). In order to reconcile the predictions of different models it is crucial to understand the interaction between levels of competition and to what extent changes in competitive conditions may impact on innovative behaviour of firms and on the markets as a whole, which may depend on technological characteristics of industries and/or distance to the technological frontier. But overall there seems to be consensus in the literature that, at least in the long run, competition is good for innovation.

Many service industries in Europe could benefit from a truly single market across Europe, in which competition can be strengthened and scale advantages may be realised. The European 'single market' programme has since the 1980s aimed at removing the barriers to free movement of capital, labour and goods, but the impact on the services industry is generally seen as limited. The present drive in Europe towards a greater openness of service product markets, for example through the adoption of a Services Directive in 2006 specifically aimed at creating a common market for services across the European Union, may hold the potential to increase productivity growth across Europe in the coming decade.

The importance of deregulating services increases for those that are key providers of intermediate inputs. In evaluating regulation it is necessary to take into account the negative productivity effects that the lack of competition has on downstream sectors, in particular for firms close to the technological frontier (Bourles et al., 2010). This is the case for professional services and retail trade, which are sectors that still need reform. In professional services, in particular, removing barriers to trade and foreign direct investment, easing market access along a number of dimensions including nationality and residency requirements, easing zoning laws and limiting the degree of self-regulation by professional associations should strengthen the competitive environment in European service sectors. The EU Services Directive adopted by EU countries in December 2006 aimed to open up the European internal market to cross-border trade in services by making it easier for service providers to set up business or offer their services in other EU countries. The objective is to promote high-quality service provision and enhance the rights of consumers for cross-border services within the EU while avoiding imposing unnecessary burdens on small and medium sized enterprises.

Product market regulations are only one side of the regulatory environment facing firms; the other being barriers to efficient operations of labour markets. The majority of studies at the economy level find evidence that strict labour market regulation reduces technological progress and productivity *growth*, although productivity *levels* tend to be higher due to greater use of skilled labour. Bassanini, Nunziata and Venn (2009) show that the effect of employment protection legislation is likely to vary across industries, with greater impacts on productivity in industries with higher lay-off propensities. Apart from this evidence the impact on individual sectors or firms remains largely unexplored. An alternative view supports the idea that employment security is considered essential for productivity (Storm and Naastepad, 2007, 2009), since it has a positive impact on the incidence of training and so raises human capital.

While there is a vast literature on the impacts of various types of regulation at the aggregate economy level there are few sectoral studies, especially ones targeted at service sectors. Further research requires indicators that are better capable of mapping competition within a sector, such as profit elasticity or knock-on effects from regulation of professional services. In view of the new mechanisms and channels through which market structure can affect stability, as revealed in the current crisis, the existing evidence on the interactions between regulations, competition and stability in the financial services industry also needs to be re-assessed.

This chapter ends with a few remarks on productivity post the financial crisis. The variations noted above across country and sector in real output growth and productivity following the financial crisis will be affected by changes in investment in physical, human and other forms of intangible capital as well as a host of institutional and environmental factors. Looking first at inputs, the most direct impact of the financial crisis is to raise the user cost of capital following reductions in banks' willingness to take risks and finance new investment. Since production industries are typically more capital intensive than services industries, a proportionally greater impact of the crisis in the former might be expected. Intangible investments such as R&D and organisational changes might also be more costly during the crisis if they depend on bank finance. There is some suggestion in the literature that innovation in services requires less external financing than in production industries, as the latter is more dependent on large-scale and costly R&D labs. For example, Dahlstrand and Cetindamer (2000) suggest that services firms are more capable of self-financing and use fewer bank loans than firms in production sectors.

In terms of market environment, the degree of international exposure is also likely to have an impact. Traditionally, international trade and investment have been concentrated in production industries but there is some evidence that the extent of international activity is increasing in services. Institutional factors, especially the degree of regulation in markets, are also likely to have major impacts, although the reverse is also likely; that is, that the crisis will have impacts on regulation, and in particular in financial services.

It is difficult at this stage to go beyond the simple descriptive analysis in section 4 above to investigate the sectoral impacts of the current economic crisis due to lack of up-to-date information on investment. However, some information can be gleaned from looking at impacts of past financial crises. In fact Barrell and Davis (2008) argue that systemic banking crises are the norm in developed economies rather than unusual events, although few are as severe as the crisis that began at the end of 2007. Efthyvoulou (2012) investigates the impacts of financial stress on productivity in previous periods, comparing production industries with market services. This controls for changes in capital services, skill composition of the workforce and other factors that impact on technical change, such as R&D expenditure, openness and foreign direct investment. Although the direct impact of financial stress is not found to be significantly different across the two broad sectors, the stress term interacted with an index of international activity is large and highly significant only for production industries. Therefore the overall impact of financial crises tends to be higher in production industries than in market services, consistent with the descriptive analysis above. However, it will be some time before the costs of the crisis, and the extent to which these vary by sector, are known.

NOTES

1. This chapter arose from research for the SERVICEGAP project funded by the European Commission, Research Directorate General as part of the 7th Framework Programme, Theme 8: Socio-economic Sciences and Humanities, Grant Agreement no. 244 552.
2. Available to download from www.euklems.net (accessed 28 November 2014).
3. The terms 'total factor productivity' (TFP) or the 'Solow Residual' are also frequently used in the literature.

REFERENCES

Aghion, P., N. Bloom, R. Blundell, R. Griffith and P. Howitt (2005), 'Competition and Innovation: An Inverted-U Relationship', *Quarterly Journal of Economics*, **120** (2), 701–728.
Barrell, R. and P. Davis (2008), 'The Evolution of the Financial Market Crisis in 2008', *National Institute Economic Review*, October, **206**, 5–14.
Bassanini, A., L. Nunziata and D. Venn (2009), 'Job Protection Legislation and Productivity Growth in OECD Countries', *Economic Policy*, **24** (58), 349–402.
Bourles, R., G. Cette, J. Lopez, J. Mairesse and G. Nicoletti (2010), 'Do Product Market Regulations in Upstream Sectors Curb Productivity Growth? Panel Data Evidence for OECD Countries', OECD Economics Department Working Paper No. 791.
Bresnahan, T., E. Brynjolfsson and L. Hitt (2002), 'Information Technology, Workplace Organization, and the Demand for Skilled Labor: Firm-Level Evidence', *Quarterly Journal of Economics*, **117** (1), 339–376.
Corrado, C., C. Hulten and D. Sichel (2005), 'Measuring Capital and Technology: An Expanded Framework', in C. Corrado, C. Hulten and D. Sichel (eds), *Measuring Capital in the New Economy: Studies in Income and Wealth Vol. 65*. Chicago, IL: The University of Chicago Press, pp. 11–46.
Corrado, C., C. Hulten and D. Sichel (2009), 'Intangible Capital and U.S. Economic Growth', *Review of Income and Wealth*, **55** (3), 661–685.
Dahlstrand, A.L. and D. Cetindamer (2000), 'The Dynamics of Innovation Financing in Sweden', *Venture Capital*, **2**, 203–221.
Efthyvoulou, G. (2012), 'The Impact of Financial Stress on Sectoral Productivity', *Economic Letters*, **116** (2), 240–243.
Foster, L., J. Haltiwanger and C.J. Krizan (2006), 'Market Selection, Reallocation, and Restructuring in the U.S. Retail Trade Sector in the 1990s', *Review of Economics and Statistics*, **88** (4), 748–758.
Fukao, K., H. Sumio, M. Tsutomu and T. Konomi (2009), 'Intangible Investment in Japan: Measurement and Contribution to Economic Growth', *Review of Income and Wealth*, **55** (3), 717–736.
Giorgio Marrano, M., J. Haskel and G. Wallis (2009), 'What Happened to the Knowledge Economy? ICT, Intangible Investment and Britain's Productivity Record Revisited', *Review of Income and Wealth*, **55** (3), 686–716.
Jalava, J., P. Aulin-Amhavarra and A. Alanen (2007), 'Intangible Capital in the Finnish Business Sector, 1975–2005', ETLA – The Research Institute of the Finnish Economy, Discussion Paper No. 1103.
Jona-Lasinio, C., M. Iommi and F. Roth (2009), 'Report on Gathering Information and Estimations for the Innodrive Project – Macro Approach', mimeo, Innodrive project, Brussels.
Jorgenson, D.W. and Z. Griliches (1967), 'The Explanation of Productivity Change', *Review of Economic Studies*, **34** (3), 249–283.

Jorgenson, D.W., F.M. Gollop and B. Fraumeni (1987), *Productivity and U.S. Economic Growth*. Cambridge, MA: Harvard Economic Studies.
Jorgenson, D.W., M. Ho and K.J. Stiroh (2005), 'Growth of U.S. Industries and Investments in Information Technology and Higher Education', in C.A. Corrado, J. Haltiwanger and D.E. Sichel (eds), *Measuring Capital in the New Economy*. Chicago, IL: University of Chicago Press, pp. 403–478.
McGuckin, R.H., M. Spiegelman and B. van Ark (2005), *The Retail Revolution: Can Europe Match U.S. Productivity Performance?* Perspectives on a Global Economy, Research Report R-1358-05-RR, The Conference Board, New York.
McKinsey Global Institute (2002), *Reaching Higher Productivity Growth in France and Germany – Retail Trade Sector*, Düsseldorf.
O'Mahony, M. and B. van Ark (eds) (2003), *EU Productivity and Competitiveness: A Sectoral Perspective. Can Europe Resume the Catching-Up Process?* Brussels: The European Commission, November.
O'Mahony, M. and M.P. Timmer (2009), 'Output, Input and Productivity Measures at the Industry Level: The EU KLEMS Database', *Economic Journal*, **119** (538), F374–F403.
Schreyer, P. (2002), 'Computer Price Indices and International Growth and Productivity Comparisons', *Review of Income and Wealth*, **48** (1), 15–31.
Solow, R. (1957), 'Technical Change and the Aggregate Production Function', *Review of Economics and Statistics*, **39** (3), 312–320.
Storm, S. and C.W.M. Naastepad (2007), 'Why Labor Market Regulation May Pay Off: Worker Motivation, Co-ordination and Productivity Growth', Economic and Labour Market Paper 2007/4, Employment and Labour Market Analysis Department, International Labour Organization, Geneva.
Storm, S. and C.W.M. Naastepad (2009), 'Labour Market Regulation and Productivity Growth: Evidence for Twenty OECD Countries (1984–2004)', *Industrial Relations: A Journal of Economy and Society*, **48** (4), 629–654.
Timmer, M.P., R.C. Inklaar, M. O'Mahony and B. Van Ark (2010), *Economic Growth in Europe*, Cambridge: Cambridge University Press.
Triplett, J.E. and B.P. Bosworth (2006), 'Baumol's Disease has been Cured: IT and Multifactor Productivity in U.S. Service Industries', in D.W. Jansen (ed.), *The New Economy and Beyond: Past, Present, and Future*, Cheltenham and Northampton, MA: Edward Elgar, pp. 34–71.
van Ark, B., M. O'Mahony and M.P. Timmer (2008), 'The Productivity Gap between Europe and the U.S.: Trends and Causes', *Journal of Economic Perspectives*, **22** (1), 25–44.
van Rooijen-Horsten, M., D. van den Bergen and M. Tanriseven (2008), 'Intangible Capital in the Netherlands: A Benchmark', Statistics Netherlands, Discussion Paper No. 08001, 2008.

3. Service research and economic geography
Patrik Ström

INTRODUCTION

Economic geographers identified the shift from manufacturing to service employment and the growth of the service sector at an early stage. These initial geographical studies of services often used a more descriptive and explorative approach compared with, for example, management and marketing theory (Marshall, 1988; Takeuchi, 1992; Daniels, 1993; Daniels et al., 1993; Illeris, 1996; Alvstam, 1998), where the normative approach to service delivery or strategy was the focus. Research in economic geography has been active on many levels in the study of the service economy; macro-oriented studies have been complemented by micro-oriented studies of sub-sectors or individual firms. Studies within geography have dealt not only with the geographical aspects of the development of service economies but to a large extent have also focused on conceptual and classification issues. Studies in the area of corporate economic geography are often aligned with international business research (Jones, 2005; Rusten and Bryson, 2010; Ström and Wahlqvist, 2010). Economic geographers were also actively involved as founders of the cross-disciplinary research organization, the European Association for Research on Services, during the early 1990s (www.reser.net).

Geographers made major contributions to identifying and charting the shift towards service employment and the growth of service firms (Bryson et al., 2004). This provides the context for this chapter, in which the role and contribution of economic geographers to service industry research will be explored. Since the service sector embraces a large and diverse set of sub-sectors, this contribution will focus on business services and, in particular, the more advanced, so-called professional business services or knowledge-intensive services. Both of these activities also share similarities with advanced services within the creative industries, many of which are at the cutting edge of service development, technological capability, creativity and user interaction (Johns, 2006; Grabher et al., 2008; Power, 2010). An additional underlying reason for this focus is the interconnectedness of these services with manufacturing and other service operations where value creation is at this interface and innovation has been strong (Bryson and Rusten, 2011). These services have also demonstrated substantial productivity gains, including highly globalized services but also controversial financial services (Daniels, 2012). The chapter begins with a macro-oriented and conceptual perspective before turning to the geographical analyses of service location strategies and internationalization. The final part of the chapter examines aspects of service research in Asia and in emerging markets.

CONCEPTS AND CLASSIFICATIONS

Economic geographers were involved in exploring services on a global scale using a spatial perspective at a relatively early stage (Daniels, 1993; Illeris, 1996). Factors that have pushed development forward are the evolutionary tendencies of the global economy, the increased expansion of information technology and the fact that the demand for various services has steadily increased, linked to urbanization and export trade in goods and services. In terms of location related to services, proximity to customers, the need for information-intensive surroundings, transportation and recruitment are essential requirements. All these aspects are important for service location but are highly related to the kind of service that is offered. The role of proximity in terms of customers and transportation possibilities varies between consumer and producer services, as well as the importance of being able to attract university graduates (Illeris, 1996). Access to information and being part of the cosmopolitan atmosphere in major cities, such as London or New York, is vital and very much part of the strategic choice made by international service firms wanting to supply their customers on a global scale (Daniels, 1993). The issue of internalization of service activities is also part of these studies; it has an impact on the locational strategies pursued by individual firms but it is also related to the cultural impacts on services. Marshall and Wood (1995) attempt to provide a complete picture of the service sector in wider terms, focusing on 'urban and regional development'. As well as discussing some explanations for the growth of service industries, they also include a social perspective. They discuss the difficulties associated with new forms of relatively low-skilled service employment, the restructuring of public services and the possibility that the service sector can act as a creator of welfare. This broader approach to describing and analysing the shifts in service industries has subsequently been further elaborated by Bryson et al. (2004) in a comprehensive overview that takes into account the emergence of the internet and rapid technological transformation since the mid-1990s.

The problems of classification and conceptualization and the service sector are well known. Broadly, a distinction is made between producer services and consumer services (e.g. Marshall and Wood, 1995; Bryson et al., 2004) but, in order to obtain analytical sharpness, sub-classifications are necessary for useful empirical research. The real problem is not to fit services within these classifications, but rather to be able to distinguish rather unclear distinctions in terms of sub-sectors to undertake spatial analyses. Economic geographers have refined the conceptualization but, more importantly, they have tried to narrow the field of research by identifying and labelling the specific sub-sectors within producer services (Sidaway and Bryson, 2002; Bryson and Daniels, 2007). The other possibility is to use a relatively broad term and then work within a number of sub-sectors, which, in the end, might not be related (Bryson and Rusten, 2011).

Service internalization or externalization plays an important role in the classification and measurement challenge. The process of externalization can transform or transfer service functions that were previously produced in-house to independent private service firms. On the one hand, it is possible that the growth of services reflects the transfer of employment from inside firms to private sector firms rather than the growth of new service jobs (Ström, 2005). Nevertheless, externalization only occurs at a single point in time and results in new service jobs as externalized service functions expand and develop new service products. On the other hand, it is also possible to argue that, in the modern globalized economy, more and more research and development and other types of service functions contribute to the final product, and therefore that the value added of service might be under-valued (Daniels and Bryson, 2002). Service firms operate through co-production with their clients and the supplier firm's employees are highly skilled and often have years of specific working experience. This has also been noted in various empirical studies (Illeris, 2002; Aoyama and Izushi, 2003; Grabher et al., 2008).

The localization of services is discussed in terms of time and physical infrastructure, especially its relationship with the increase in the specialization of the division of labour and outsourcing; further research is needed to attempt to capture the magnitude and characteristics of service operations that are being outsourced (Alvstam, 1998; Bryson, 2007b). The issue of geography in terms of both production and consumption is important. Some services might be very locally bounded while others are rather ubiquitous, flexible and moveable. Such a conceptualization touches upon the difficulty associated with the agglomeration of services and then, in particular, various service sub-sectors. Some services are clustered because they want to be seen to be related to the right places, the 'image factor' that generates a feeling of importance in terms of, for example, being located in a centre noted for creativity or a strong research cluster such as Silicon Valley. Such considerations can be perceived from either an objective or a subjective perspective. This shows the difficulty of discussing services in general; different types of services have their own suite of characteristics and needs.

Alvstam (1998) has suggested that geographical inertia, whereby the national border is a boundary, should be considered; is there a limit to the extent to which buying behaviour can be globalized? State borders impose a form of containment as national regulatory frameworks apply to services and their tradability. In conceptual terms this can work in the opposite direction for some service sub-sectors. Differences in national regulations have created a demand for specialist advice that is often provided by American management consultancy firms and much of this advice has focused on advising foreign service firms how to adapt to different national service regulations. Service research suffers from too many studies that adopt a 'snap shot' approach based around the analysis of a single time period. This reflects methodological and funding constraints. The failure of service researchers to adopt a longitudinal perspective is unfortunate as there is much to learn from understanding the changing nature of the interactions between client firms and their service providers (Bryson et al., 1999).

LOCATION AND EMBEDDEDNESS

One of the most valuable theoretical contributions made by economic geographers is to the understanding of the location of services. The geographical proximity and the embeddedness of firms within regions endowed with good infrastructure, solid institutional surroundings and the possibility for knowledge to flow easily are very important for locational decisions. The early work of Weber (1909) and Marshall (1920) is to some extent still valid but needs revisions that recognize alterations in the functioning of regional and national economies. One approach is Porter's diamond framework (1990), which has been developed from its original scope to better incorporate services and the creative industries (Power, 2010). Additional concepts that try to catch the dynamics of regions and countries have been put forward (e.g. Lundvall, 1992; Storper, 2000; Asheim and Isaksen, 2002). Lundvall (1992), for example, discusses what he calls 'National Innovation Systems', where the institutional environment affects the ability of firms to expand and prosper, and a debate has appeared regarding learning regions and whether knowledge spillovers are created through the co-location of firms in a limited geographical area (Asheim and Coenen, 2005). A number of different analytical approaches have been put forward, and there is some overlap and confusion regarding both conceptual scope and empirical testing (Malmberg and Maskell, 2002; Martin and Sunley, 2003).

One aspect of location is the role of proximity to client organizations. Juleff-Tranter (1996) has studied the impact of proximity on service demand and not only shows that firms tend to use co-located service firms, but also demonstrates that the service sector is a very important source of demand for producer services. This strengthens the hypothesis that service firms can be a vital part of economic development without primarily being dependent on manufacturing clients (Bryson et al., 2004). The regional supply and proximity of services has been discussed by many, both in terms of mega-cities (Daniels, 1998) and less densely populated areas (Bennett and Smith, 2002). Many kinds of business services need face-to-face interactions between suppliers and clients. The proximity of services has also been explored in studies of advanced manufacturing (Gertler, 2003).

People including skilled experts are the most important factor in the production of business services. This is especially the case for professional business services where both the firm and fee-earning employees determine proximity issues. Professional employees working for business service firms have been shown to favour environments rich in culture, alternative job opportunities and related factors essential for their career plans such as 'quality of life' (Florida, 2002; Pratt, 2004; Ström and Wahlqvist, 2010; Borggren, 2011). These can often be one-person companies with strong prior connections to clients, through earlier work experience (Beyers and Lindahl, 1996a; Beyers, 2002a; Keeble and Nachum, 2002) so that these high flyers and lone eagles (Beyers and Lindahl, 1996a) are much more geographically dispersed.

The locational dimension for producer services has been discussed both from a narrow regional perspective (e.g. Sjøholt, 1993; Ishimaru, 1994; Beyers and Lindahl, 1996b; Keeble and Nachum, 2002) and from a more outward perspective (O'Farrell et al., 1996, 1998; Bryson, 2007b; Faulconbridge et al., 2008) that includes firms' internationalization processes. Keeble and Nachum (2002) reveal that professional business service firms are highly dependent on integrated clusters, with high levels of collaboration, knowledge transfer and labour mobility. The increased need for knowledge diffusion – that is, the availability of expertise and information – has been conceptualized as *relationship capitalism* (Bryson, 2000). In combination with reputation and image, this creates much of the foundation for the localization of business services (Amin, 1997; Amin and Roberts, 2008). Rusten and Bryson (2010) also discuss the complexity of global service production; networks of expertise constantly form and reform into project teams that may be localized or dispersed, often with a focus on prosperous business regions.

Beyers (2012) has elaborated on what he calls the service imperative for future research agendas. The first imperative is micro-considerations, where services play an important part in regional economic development. The second relates to trade and sourcing and the macro-perspective. Beyers argues that services require additional study to completely understand their role within international business. This has also been advocated by Ström and Wahlqvist (2010). Regarding the problem of measuring international trade and foreign direct investment (FDI) within services, Beyers concludes that the only remedy is well-structured surveys and suitable case studies following the operations of a specific company. More case studies are also required to understand the geographical impact on firms. The last imperative discusses the framing of research and the measurements used. Clusters and complex agglomerations of service activity need to be explained in more detail. This might be achieved through additional case study research or by using more of a social network approach. The evolutionary tendency for new industries to grow and become more important in regional and national economies is also a question that deserves more geographic research of the kind undertaken by Beyers (2002b, 2012) on cultural industries. Such evolutionary research connects very well with the discussion on differences in sub-sector development between mature industrialized countries.

'Place' for Advanced Business Services?

A starting-point when studying the location of business services is the fact that they are heavily concentrated in a primary city or in a number of larger cities. The larger the city, the more diverse is the supply of various professional business service firms (Jones, 2002; Hermelin and Rusten, 2007). The degree of spatial concentration varies according to specific sub-sectors. Empirical research (e.g. Keeble and Nachum, 2002; Hermelin, 2007) has shown that accounting and data processing or technical consulting seem to be more dispersed than, for example, advertising, management consulting or IT-related services. Larger firms tend to locate their head offices in metropolitan regions and large cities, and may be more scattered than smaller firms in terms of the geographical extent of their physical premises. It is worth noting that much producer service work is undertaken in clients' offices. Other shifts that are helping to create a specific structure for business service location are the emergence of increased international activities through trade and investments in connection with technological development (Daniels, 1995, 2000) and the formation of 'concentrated decentralization' based around suburban regions or second-tier cities (Hermelin, 1997; Daniels and Bryson, 2005; Jones, 2009). It is suggested that the concepts of 'urbanization' and 'localization' economies should be used separately from 'agglomeration' economies to explain service-specific locations (Daniels, 1993). On the one hand, a location in a major urban area provides access to a well-developed infrastructure. Localization, on the other hand, is a concept that highlights local knowledge spillover and proximity to supply and demand. Marshall and Wood (1995: 24) present a list of factors that are important localization factors for service activities (Table 3.1), but that also act as forces driving the different sub-sectors towards spatial centralization.

All these issues work together to form the local environment for professional business services. Harrington and Daniels (2006) suggest a categorization to explain the location of services with different value added. The more specific or niche oriented the service provider, and the more novel the service encounter, the more these firms are to be found in global cities or high-order urban centres. Over time, many of these factors have become available in areas other than metropolitan districts. One of the most important factors making this possible is the rapid development of various forms of information

Table 3.1 Localization factors for understanding the geography of service activities

Accessibility and proximity to each other
Good physical access to customers and large range of local business activities
Readily accessible transport facilities
A competitive market environment supporting high quality
High-quality telecommunications infrastructure
High-quality labour force
Ample numbers of clerical and administrative staff
An ample supply of offices of suitable quality
A high-quality urban environment, including cultural and social facilities

Source: Marshall and Wood (1995).

and telecommunications technologies. Other factors that might push services out of metropolitan areas are the salary costs of employees, office space and congestion. Additionally, the issue of quality of life becomes more important but conditions such as a pleasant residential environment may actually encourage the dispersion of producer services.

Research has also demonstrated the impact of proximity on service demand. Not only do client firms tend to use co-located service firms, but the service sector is itself a very important source of demand for producer services (Juleff-Tranter, 1996; Bryson et al., 2004). This strengthens the hypothesis that service firms are a vital part of economic development without necessarily being primarily dependent upon manufacturing clients. Many kinds of business services need a face-to-face interaction between suppliers and clients (Faulconbridge, 2008). The importance of proximity to services has also been noted in studies more oriented towards manufacturing firms. In a discussion of the conditions conducive to 'stickiness' in a globalized economy, Markusen (1996) identifies proximity to services and knowledge flows as relevant to all the various forms of industrial district in her typology. Higher knowledge content and a dependency upon externalities surrounding firms have become even more important when 'ubiquitification' makes knowledge more transferable, thus reducing traditional locational competitive advantages (Gertler, 2003; Ström and Schweizer, 2011).

Business Service Strategy and Location

A number of studies have explored the relationship between producer services and the creative industries, with the purpose of applying or testing Porter's locational and competitive strategy approach. O'Farrell et al. (1993), Beyers and Lindahl (1996b, 1999) and Power (2010) have in various ways tried to apply Porter's framework to the services and content industries. Differentiation is highly important and costs seem to be less important than expected. Differences in diversification are apparent between rural and urban regions. In knowledge-intensive sectors and the content industry market positionality is more important than traditional concerns about the protection of intellectual property rights. Market appeal creates differentiation from competitors (Power, 2010). Differences can often be traced back to the surrounding business environment and a debate about whether services might be an engine of economic growth in more dispersed, peripheral or even economically distressed areas. For example, Wernerheim and Sharpe (2003) examine the assumption that supporting services in rural areas will create new engines of economic growth. Professional business services firms tend to locate where they complement the underlying industrial structure. This implies that firms in metropolitan areas are unlikely to be footloose enough to respond to policy interventions. It is necessary to build basic requirements rather than trying to develop completely new types of services that will spur regional service growth. Perhaps there is something 'in the air' at specific locations (Wernerheim and Sharpe, 2003) which means that a certain geographical location can create an ownership-specific advantage for a firm active in a specific industry (Dunning, 1998).

Over the last decade there has been a surge in economic geography research on matters connected to the perceived importance of intangible locational advantages. A discussion about evolutionary paths and relational aspects of agglomeration has received most attention. This has largely taken place within a broad industry perspective in an effort to establish an understanding of relational aspects of the knowledge economy (Bathelt and Glückler, 2003; Boschma and Martin, 2010), but has also become increasingly important for the explanation of service industry location (Ström and Schweizer, 2011; Glückler and Ries, 2012). The complexity of global production networks in combination with the growth of advanced services in metropolitan areas increases the importance of relationships amongst decision makers in firms. These networks have widened the scope of many of the advanced global service providers and have also increased the complexity of how strategy is developed. Even if global cities are of great importance within these networks, diffusion of strategy decision making has been accompanied by the emergence of more complex spatial distributions (Jones, 2002).

INTERNATIONALIZATION OF SERVICES

Because of classification, conceptualization and measurement difficulties, research on service internationalization is complicated. Traditional international trade theories such as the Heckscher–Ohlin approach have been used as a frame of reference for concluding that trade theory does not help very much in explaining trade in services (Daniels, 1993; Bryson et al., 2004; Dicken, 2011). Instead, approaches based on technological development, government regulations or cultural differences are of more importance in understanding the patterns and structure of service trade (Beyers, 2012). It is also increasingly recognized that service sector FDI is important for analysing the internationalization of services, since it is often the preferred way for companies to enter overseas markets and is specifically determined by locational factors (Rusten and Bryson, 2010). Studies of internationalization within the service industries have focused on more specific sub-sectors and often incorporated consideration of internationalization strategies (e.g. O'Farrell et al., 1998; Jones, 2002; Schulz, 2005; Ström and Mattsson, 2006; Bryson, 2007b; Faulconbridge et al., 2008). In general, the results show that firms located in major metropolitan areas are more likely to be aware of, and to follow up, various internationalization opportunities. Firms in more peripheral regions also show an interest in working internationally, but the opportunities are more limited. In the case of trade and FDI-oriented research, this tends to be focused on understanding either the specific characteristics of trade and FDI in general or upon understanding flows of FDI between places.

There are major differences between trade in goods and trade in services (Daniels, 1993; Bryson et al., 2004; Dicken, 2011). First, even though information technology has helped to bring parts of the world closer together, many of the activities in which service industries are engaged must take place simultaneously and may also need to be co-located. It is sometimes important to have direct contact between producers and buyers. Secondly, service industries are still heavily regulated in many countries, which affects opportunities for FDI and trade. In some countries the service sector has been almost completely removed from global competition as local providers are preferred over foreign firms. For example, India and China restrict service FDI, requiring either local provision or foreign firms to enter strategic partnerships with local providers. Finally, there are many obstacles to service trade, notably non-tariff barriers such as a requirement for legal advisors, medical doctors or accountants to be licensed locally. It is often very difficult to calculate the cost disadvantage imposed through these various types of actions, especially taking account of cultural and language barriers together with the risky nature of international trade (Illeris, 1996). These are less visible but no less important barriers to international service trade. Cultural problems are evident for any kind of international transaction but in the case of services they may be an absolute barrier (Jones, 2002; Ström, 2005). These barriers have their greatest impact on verbal- and media-based services, with standardized and highly technological services less susceptible to this general rule. Another factor is risk in general, such as that arising from political situations, insurance, finance or adaptation to new legal systems. It is argued that these barriers and/or risks may be greater for service firms than for firms exporting physical goods. The complexity of managing international service operations spanning countries or continents also poses new challenges (Jones, 2005).

Another important issue for the internationalization of service production is the complexity of value added in relation to manufactured goods (Daniels, 2000; Daniels and Bryson, 2002). The question of value added is valid both in relation to where services are produced and how services are traded in relation to the indirect contribution made by services to the export of goods. In relation to the classification and conceptual difficulties associated with service trade, empirical findings show that the direct contribution of services to national exports is growing slowly and the pattern of trade is highly concentrated. Instead it might be useful to acknowledge and nurture the indirect contribution of services to overall national export activity. The support provided by service inputs to many manufacturing sectors enhances total export competitiveness. Specialized services are embodied in the goods and services exported elsewhere within a specific country or beyond into the world market (Daniels and Bryson, 2002). One conclusion is that FDI in the form of mergers and acquisitions seems to be more important than greenfield investments because of the intangible character of services and the need to locate at specific places to be recognized as a serious service provider (Bryson et al., 2004). This is most likely an outcome of a need for new companies to rapidly acquire the intangible assets that lie behind the competitiveness of many service functions in mature economies (Rubalcaba-Bermejo and Cuadrado-Roura, 2002: 48).

SERVICE RESEARCH IN ASIA AND EMERGING MARKETS

A number of economic geographers have drawn attention to the fact that limited research has been undertaken on services activities and functions located in Asia and other emerging markets such as Eastern Europe (O'Connor and Hutton, 1998; Harrington and Daniels, 2006; Ström and Yoshino, 2009; Daniels et al., 2011, 2012; Asian Development Bank (ADB), 2012; Di Meglio et al., 2012). The interconnectedness of mature economies such as Japan and emerging markets in East and Southeast Asia has created a complex economic network of production and knowledge networks (Alvstam et al., 2009). It resembles to some degree the economic geography of Europe, where the new Member States of the European Union have been connected through market widening and increased FDI, of which one expression is the economic integration of the service economy. This has yet to materialize in Asia (Ström and Yoshino, 2009; ADB, 2012).

In Asia the high level of employment in manufacturing throughout the region does show signs of being reduced in favour of the service sector (O'Connor and Hutton, 1998; Daniels and Harrington, 2007). There seems to be a different sectoral context for the development of services in the Asia Pacific; a significant proportion takes the form of internalized service operations and service bundling in larger firms (Ström and Mattsson, 2005). Economic development is closely connected to the evolution of a small number of mega-cities in which new service clusters give rise to further economic development or evolution (Lai, 2011, 2012; Daniels et al., 2012). The political and economic environment in Asia is different from that in Europe or North America. It has been more common for governments in Asia to take a more active role in economic development. China, Korea and Japan are probably the most well-known examples (Edgington and Haga, 1998; Ernkvist and Ström, 2008). This raises a number of interesting issues, such as the role of regulation and the cultural context as explanatory variables in service sector development in Asia. The region has become more integrated through the development of information and communication technologies and this has had an impact on the possibilities for bringing in external service expertise and for managing companies over larger distances. The push forward of intra-regional economic co-operations, such as APEC and ASEAN, could create a better framework for service sector development and internationalization in Asia but progress has been slow.

Based on employment data, Daniels (1998) has revealed that producer services are less prominent in Asia than in Europe and North America. This has also been confirmed in later studies (ADB, 2012). However, this might be an outcome of different ways of managing and organizing firms, rather than a real under-development of services (Ishimaru, 1994). The externalization of services that has been a driving force behind development in the West does not seem to exist to the same extent in Asia (Bramklev and Ström, 2011). Further research in this area is needed, including more detailed data, in order to develop a more complete picture of particular services and locations in Asia. This should take more account of internal corporate structures, ethnic ties and relations within industrial groupings, such as the Japanese *keiretsu* or the Korean *chaebols*. Daniels (2001) explored the Asian service market by conducting a selected number of Asian case studies and concluded that Asian firms were experiencing problems arising from the impact of foreign service providers that capture both foreign and local clients. The characteristics of specific sub-sectors, in terms of regulations and capacity, will naturally influence the competitive situation (Daniels, 2001). O'Connor and Daniels (2001) used Australian–APEC trade within producer services as a case study for developing a framework for understanding service globalization. Their findings support the development of a framework where the characteristics of the home and host market significantly influence the possibility of trade in advanced producer services.

More research, of a case-oriented character, in fields such as management consultancy, advertising and other business services is needed to establish how these services are organized, compared with their Western counterparts. The empirical data also leave much to be desired as regards the operations of Western service producers in Asia. It is most likely that Asia can be a very attractive market for Western service firms which have experience, established reputations and strong competitive advantages (Jones, 2005).

Yeung and Lin (2003) try to develop Asian theorizing in economic geography further by stressing the importance of critical reasoning when applying Western theory in Asia. They argue that Asia is about more than just 'regional studies' on the fringe of conventional economic geographical research. Instead, the region should be used as a field laboratory for developing theory that will be incorporated into mainstream research (Yeung, 2009). One area that is acknowledged as successful is the discussion of transnationalization and transnationalism (Yeung and Lin, 2003: 118). The use of grounded theory in the economic geographies of Asia has generated valuable conceptual insights into networks, embeddedness and Asian internationalization that would have been difficult to identify and measure using conventional theory. With the rise of service industries in Asia and the future challenges to sustained regional competitiveness (ADB, 2012), this kind of grounded research is even more important. A longitudinal approach would enable an analysis of how national economies evolve into more mature service economies, as has been the case for many of the emerging markets in Asia and elsewhere. The key problem for this kind of research is how to assemble accurate disaggregated data.

The Asian context offers a number of interesting issues in relation to location and economic development. In the discussion of national embeddedness, the different theoretical perspectives put forward illuminate the relations between firms and their surroundings and what drives fruitful agglomeration; consideration of the role of the state is often secondary. In Asia, however, the state had, and continues to have, a crucial role in the formation of economic geographical patterns (O'Connor and Hutton, 1998). The influence of governments in upgrading economies and creating specific roles for large cities such as Shanghai in terms of service dominance is clearly apparent (Ström and Yoshino, 2009). The institutional setting, APEC and ASEAN for instance, is there, but limited progress has been made in liberalizing trade. Instead, regional development has been focused on a number of large cities acting as hubs for incoming investments and local business activity (Edgington and Haga, 1998; O'Connor and Hutton, 1998; Sassen, 2001). The dispersed service activities found outside the global cities in Western economies do not exist to the same extent. There seems to be a concentration of advanced service activities in specific regions where international access is good (Xu and Yeh, 2010). The existence of expatriate communities acting as contact bases functions as a source of influence on the national business environment (Jones, 2005; Faulconbridge, 2008).

CONCLUSIONS

This chapter has provided a general overview of the contribution of economic geographers to service research. Economic geographers have contributed to the conceptualization and understanding of how value added is produced in space and how this is related to both service production and the interconnectedness between services and manufacturing. The spatial dimension of economic transformation has proved to be highly important despite the technological advances that have helped to compress time and space (The Economist, 2012). Spatial characteristics have shaped how economic geographies have been developing and changing at both the micro and macro levels. Agglomeration has been shown to be vital for the development of advanced services and how clusters of these specialized, knowledge-intensive activities are interconnected on the global scale. The chapter has endeavoured to show the difficult conceptual and empirical data challenges involved when undertaking research on services. It goes without saying that additional research on the formation of the service economy in space and place is vital for understanding the continuing evolution of local and national economies and the developing global economy. Such research is important for informing the development of new policies and management practices that will enhance the contribution service functions, activities and employment make to economic development.

REFERENCES

ADB (2012) *Asian Development Outlook 2012 Update: Services and Asia's Future Growth*, Manila: ADB.
Alvstam, C.G. (1998) Lokaliseringsteori för produktion av tjänster – 'Ueber den Mangel eines Standorts der Dienstleistungen', *Svensk Geografisk Årsbok (74)*, Lund: Sydsvenska Geografiska Sällskapet.
Alvstam, C., Ström, P. and Yoshino, N. (2009) On the economic interdependence between China and Japan – challenges and possibilities. *Asia Pacific Viewpoint*, 50(2): 198–221.
Amin, A. (1997) Placing globalization. *Theory, Culture and Society*, 14(2): 123–137.
Amin, A. and Roberts, J. (eds) (2008) *Community, Economic Creativity and Organization*, Oxford: Oxford University Press.
Aoyama, Y. and Izushi, H. (2003) Hardware gimmick or cultural innovation? Technological, cultural, and social foundations of the Japanese video game industry. *Research Policy*, 32(3): 423–444.
Asheim, B. and Coenen, L. (2005) Knowledge bases and regional innovation systems: comparing Nordic clusters. *Research Policy*, 34: 1173–1190.
Asheim, B.-T. and Isaksen, A. (2002) Regional innovation systems: the integration of local 'sticky' and global 'ubiquitous' knowledge. *Journal of Technology Transfer*, 27: 77–86.
Bathelt, H. and Glückler, J. (2003) Toward a relational economic geography. *Journal of Economic Geography*, 3: 117–144.
Bennett, R.J. and Smith, C. (2002) The influence of location and distance on the supply of business advice. *Environment and Planning A*, 34: 251–270.
Beyers, W.B. (2002a) Services and the New Economy: elements of a research agenda. *Journal of Economic Geography*, 2: 1–29.
Beyers, W.B. (2002b) Culture, services and regional development. *The Service Industries Journal*, 22(1): 4–34.
Beyers, W.B. (2012) The service industry research imperative. *The Service Industries Journal*, 32(3–4): 657–682.
Beyers, W. and Lindahl, D.P. (1996a) Lone eagles and high fliers in rural producer services. *Rural Development Perspectives*, 12: 2–10.
Beyers, W.B. and Lindahl, D.P. (1996b) Explaining the demand for producer services: is cost-driven externalization the major factor? *Papers in Regional Science*, 75(3): 351–374.
Beyers, W.B. and Lindahl, D.P. (1999) The creation of competitive advantage by producer service establishments. *Economic Geography*, 75(1): 1–20.
Borggren, J. (2011) *Kreativa individers bostadsområden och arbetsställen, belysta mot bakgrund av näringslivets*

omvandling och förändringar i bebyggelsestrukturen i Göteborg, Department of Geography, University of Gothenburg, Series B, no. 120.
Boschma, R. and Martin, R. (2010) *Handbook of Evolutionary Economic Geography*, Cheltenham and Northampton, MA: Edward Elgar.
Bramklev, C. and Ström, P. (2011) A conceptualization of the product/service interface: case of the packaging industry in Japan. *Journal of Service Science Research*, 3(1): 21–48.
Bryson, J.R. (2000) Spreading the message: management consultants and the shaping of economic geographies in time and space. In Bryson, J., Daniels, P.W., Henry, N. and Pollard, J. (eds) *Knowledge, Space, Economy*, London: Routledge.
Bryson, J.R. (2007a) Lone eagles and high fliers: rural-based business and professional service firms and information communication technology. In Rusten, G.and Skerratt, S. (eds) *Information and Communication in Technologies in Rural Society: Being Rural in a Digital Age*, London: Routledge.
Bryson, J.R. (2007b), The 'second' global shift: the offshoring or global sourcing of corporate services and the rise of distanciated emotional labour. *Geografiska Annaler*, 89B: 31–43.
Bryson, J.R. and Daniels, P.W. (2007) *The Handbook of Service Industries in the Global Economy*, Cheltenham and Northampton, MA: Edward Elgar.
Bryson, J.R. and Rusten, G. (2011) *Design Economies and the Changing World Economy: Innovation, Production and Competitiveness*, London: Routledge.
Bryson, J.R., Daniels, P.W. and Ingram, D.R. (1999) Methodological problems and economic geography: the case of business services. *The Service Industries Journal*, 19(4): 1–16.
Bryson, J.R., Daniels, P.W. and Warf, B. (2004) *Service Worlds: People, Organizations, Technologies*, London: Routledge.
Daniels, P.W. (1993) *Service Industries in the World Economy*, Oxford: Blackwell.
Daniels, P.W. (1995) Services in a shrinking world. *Geography*, 80(2): 97–110.
Daniels, P.W. (1998) Economic development and producer services growth: the APEC experience. *Asia Pacific Viewpoint*, 29(2): 145–159.
Daniels, P.W. (2000) Export of services or servicing exports? *Geografiska Annaler*, 82B(1): 1–15.
Daniels, P.W. (2001) Globalization, producer services, and the city: is Asia a special case? In Stern, R.M. (ed.) *Services in the International Economy*, Ann Arbor, MI: The University of Michigan Press.
Daniels, P.W. (2012) Service industries at a crossroads: some fragile assumptions and future challenges. *The Service Industries Journal*, 32(3–4): 619–639.
Daniels, P.W. and Bryson, J.R. (2002) Manufacturing services and servicing manufacturing: changing forms of production in advanced capitalist economies. *Urban Studies*, 39(5–6): 977–991.
Daniels, P.W. and Bryson, J.R. (2005) Sustaining business and professional services in a second city region. *The Service Industries Journal*, 25(4): 505–524.
Daniels, P.W. and Harrington, J.W. (2007) *Services and Economic Development in the Asia-Pacific*, Aldershot: Ashgate Publishing.
Daniels, P.W., Ho, K.C. and Hutton, T.A (2012) *New Economic Spaces in Asian Cities: From Industrial Restructuring to the Cultural Turn*, London: Routledge.
Daniels, P.W., Illeris, S., Bonamy, J. and Philippe, J. (1993) *The Geography of Services*, London: Frank Cass.
Daniels, P.W., Rubalcaba, L., Stare, M. and Bryson, J. (2011) The new member state (NMS) and the transformation of the European services landscape. *Tijdschrift voor economische en sociale geografie*, 102(2): 146–161.
Di Meglio, G., Stare, M. and Jaklič, A. (2012) Explanation for public and private service growth in the enlarged EU. *The Service Industries Journal*, 32(3–4): 503–514.
Dicken, P. (2011) *Global Shift: Mapping the Changing Contours of the World Economy*, New York: Guilford Press.
Dunning, J.H. (1998) Location and the multinational enterprise: a neglected factor? *Journal of International Business Studies*, 29: 45–66.
Edgington, D.W. and Haga, H. (1998) Japanese service sector multinationals and the hierarchy of Pacific Rim cities. *Asia Pacific Viewpoint*, 39(2): 161–178.
Ernkvist, M. and Ström, P. (2008) Enmeshed in games with the government: governmental policies and the development of the Chinese online games industry. *Games and Culture*, (3): 98–126.
Faulconbridge, J.R. (2008) Managing the transnational law firm: a relational analysis of professional systems, embedded actors, and time-space-sensitive governance. *Economic Geography*, 84(2): 185–210.
Faulconbridge, J.R., Hall, S.J.E. and Beaverstock, J.V. (2008) New insight into the internationalization of producer services: organizational strategies and spatial economies for global headhunting firms. *Environment and Planning A*, 40: 210–234.
Florida, R. (2002) *The Rise of the Creative Class; and How It's Transforming Work, Leisure Community, and Everyday Life*, New York: Basic Books.

Gertler, M. (2003) Tacit knowledge and the economic geography of context, or The undefinable tacitness of being (there). *Journal of Economic Geography*, 3: 75–99.

Glückler, J. and Ries, M. (2012) Why being there is not enough: organized proximity in place-based philanthropy. *The Service Industries Journal*, 32(3–4): 515–530.

Grabher, G., Ibert, O. and Flohr, S. (2008) The neglected king: the customer in the new knowledge ecology of innovation. *Economic Geography*, 84(3): 253–280.

Harrington, J.W. and Daniels, P.W. (2006) *Knowledge-Based Services, Internationalization and Regional Development*, Aldershot: Ashgate.

Hermelin, B. (1997) *Professional Business Services – Conceptual Framework and a Swedish Case Study*, Uppsala: Department of Social and Economic Geography, Uppsala University.

Hermelin, B. (2007) The urbanisation and suburbanisation of the service economy: producer services and specialisation in Stockholm. *Geografiska Annaler*, 89B: 59–74.

Hermelin, B. and Rusten, G. (2007) The organizational and territorial changes of services in a globalized world. *Geografiska Annaler*, 89B: 5–11.

Illeris, S. (1996) *The Service Economy: A Geographical Approach*, Chichester: Wiley.

Illeris, S. (2002) Are service jobs as bad as theory says? Some empirical findings from Denmark. *The Service Industries Journal*, 22(4): 1–18.

Ishimaru, T. (1994) The spatial characteristics of manufacturing cities and service cities in Japan. *Jimbun-Chiri/Human Geography (Kyoto)*, 44(2): 284–298.

Johns, J. (2006) Video games production networks: value capture, power relations and embeddedness. *Journal of Economic Geography*, 6(2): 151–180.

Jones, A. (2002) The 'global city' misconceived: the myth of 'global management' in transnational service-firms. *Geoforum*, 33: 335–350.

Jones, A. (2005) Truly global corporations? Theorizing 'organizational globalization' in advanced business-services. *Journal of Economic Geography*, 5(2): 177–200.

Jones, A. (2009) Theorizing global business spaces. *Geografiska Annaler*, 91B(3): 203–218.

Juleff-Tranter, L.E. (1996) Advanced producer services: just a service to manufacturing? *The Service Industries Journal*, 16(3): 389–400.

Keeble, D. and Nachum, L. (2002) Why do business service firms cluster? Small consultancies, clustering and decentralization in London and Southern England. *Transactions of the Institute of British Geographers*, 27(1): 67–90.

Lai, Karen P.Y. (2011) Marketisation through contestation: reconfiguring China's financial markets through knowledge networks. *Journal of Economic Geography*, 11(1): 87–117.

Lai, Karen P.Y. (2012) Differentiated markets: Shanghai, Beijing and Hong Kong in China's financial centre network. *Urban Studies*, 49(6): 1275–1296.

Lundvall, B.-Å. (1992) *National Innovation Systems: Towards a Theory of Innovation and Interactive Learning*, London: Pinter Publishers.

Malmberg, A. and Maskell, P. (2002) The elusive concept of localization economies: towards a knowledge-based theory of spatial clustering. *Environment and Planning A*, 34(3): 429–449.

Markusen, A. (1996) Sticky places in slippery space: a typology of industrial districts. *Economic Geography*, 72(3): 293–313.

Marshall, A. (1920) [1890] *Principles of Economics: An Introductory Volume*, London: Macmillan and Co.

Marshall, J.N. (1988) *Services and Uneven Development*, Oxford: Oxford University Press.

Marshall, J.N. and Wood, P.A. (1995) *Services and Space: Key Aspects of Urban and Regional Development*, Harlow: Longman.

Martin, R. and Sunley, P. (2003) Deconstructing clusters: chaotic concept or policy panacea? *Journal of Economic Geography*, 3: 5–35.

O'Connor, K. and Daniels, P.W. (2001) The geography of international trade in services: Australia and the APEC region. *Environment and Planning A*, 33: 281–296.

O'Connor, K. and Hutton, A.T. (1998) Producer services in the Asia Pacific region: an overview of research issues. *Asia Pacific Viewpoint*, 39(2): 139–143.

O'Farrell, P., Hitchens, D.M. and Moffat, L.A.R. (1993) The competitive advantage of business service firms: a matched pairs analysis of the relationship between generic strategy and performance. *The Service Industries Journal*, 13(1): 40–64.

O'Farrell, P., Wood, P.A. and Zheng, J. (1996) Internationalization of business services: an interregional analysis. *Regional Studies*, 32(2): 101–118.

O'Farrell, P., Wood, P.A. and Zheng, J. (1998) Regional influences on foreign market development by business service companies: elements of a strategic context explanation. *Regional Studies*, 32(1): 31–48.

Porter, M.E. (1990) *The Competitive Advantage of Nations*, London: Macmillan.

Power, D. (2010) The difference principle? Shaping competitive advantage in the cultural product industries. *Geografiska Annaler, Series B*, 92(2): 145–158.

Pratt, A.C. (2004) The cultural economy: a call for spatialized 'production of culture' perspectives. *International Journal of Cultural Studies*, 7(1): 117–128.
Rubalcaba-Bermejo, L. and Cuadrado-Roura, J.R. (2002) A comparative approach to the internationalisation of service industries. In Cuadrado, J.R., Rubalcaba- Bermejo, L. and Bryson, J.R. (eds) *Trading Services in the Global Economy*, Cheltenham and Northampton, MA: Edward Elgar.
Rusten, G. and Bryson, J.R. (2010) Pacing and spacing services: towards a balanced economic geography of firms, clusters, social networks, contracts and the geographies of enterprise. *Tidschrift voor Economische en Sociale Geographie*, 101(3): 248–261.
Sassen, S. (2001) *The Global City: New York, London, Tokyo*, 2nd ed., Princeton, NJ: Princeton University Press.
Schulz, C. (2005) Foreign environments: the internationalisation of environmental producer services. *The Service Industries Journal*, 25(3): 337–354.
Sidaway, J.D. and Bryson, J.R. (2002) Constructing knowledges of 'emerging markets': UK-based investment managers and their overseas connections. *Environment and Planning A*, 34: 401–416.
Sjøholt, P. (1993) The dynamics of services as an agent of regional change and development. *The Service Industries Journal*, 12(2): 36–59.
Storper, M. (2000) Globalization, localization, and trade. In Clark, C.L., Feldman, M.P. and Gertler, M.S. (eds) *The Oxford Handbook of Economic Geography*, Oxford: Oxford University Press.
Ström, P. (2005) The Japanese service industry: an international comparison. *Social Science Japan Journal*, 8(2): 253–266.
Ström, P. and Mattsson, J. (2005) Japanese professional business services: a proposed analytical typology. *Asia Pacific Business Review*, 11(1): 49–68.
Ström, P. and Mattsson, J. (2006) Internationalisation of Japanese professional business service firms: client relations and business performance in the UK. *The Service Industries Journal*, 26(3): 249–265.
Ström, P. and Schweizer, R. (2011) Space oddity – on managerial decision making and space. In Schlunze, R., Agola, N. and Baber, W. (eds) *New Perspectives on Managerial Geography: Launching New Perspectives on Management and Geography*, Houndmills: Palgrave Macmillan.
Ström, P. and Wahlqvist, E. (2010) Regional and firm competitiveness in the service based economy – combining economic geography and international business theory. *Tijdschrift voor economische en sociale geografie*, 101(3): 287–304.
Ström, P. and Yoshino, N. (2009) Japanese financial service firms in East and Southeast Asia: location pattern and strategic response in changing economic conditions. *Asian Business and Management*, 8(1): 33–58.
Takeuchi, A. (1992) Strategic management of Japanese steel manufacturing in the changing international environment. *International Review of Strategic Management*, 3: 189–203.
The Economist (2012) A Sense of Place: Special Report, Technology and Geography, 27 October.
Weber, A. (1909) *Theory of the Location of Industries*, Chicago: University of Chicago Press.
Wernerheim, M.C. and Sharpe, C.A. (2003) 'High order' producer services in metropolitan Canada: how footloose are they? *Regional Studies*, 37(5): 469–490.
Xu, J. and Yeh, A.G.O. (2010), Planning mega-city regions in China: rationales and policies. *Progress in Planning*, 73: 17–22.
Yeung, H.W.C. (2009) Regional development and the competitive dynamics of global production networks: an East Asian perspective. *Regional Studies*, 43(3): 325–351.
Yeung, H.W.C. and Lin, G.C.S. (2003) Theorizing economic geographies of Asia. *Economic Geography*, 79(2): 107–128.

4. The new scientific study of service
Paul P. Maglio and Cheryl A. Kieliszewski

THE NEW SCIENTIFIC STUDY OF SERVICE

Over the last ten years, we and others at IBM Research have begun to consider seriously the nature of service businesses and the opportunities for improvement and innovation in service (Horn, 2005; IBM Research, 2005; Spohrer and Maglio, 2008). It should be no surprise that scientists at IBM would focus on what has become IBM's biggest and fastest-growing set of businesses: outsourcing and consulting services (IBM, 2012; Spohrer and Maglio, 2008). We found a rich and diverse set of disciplinary research on service, including economics, marketing, operations, industrial engineering, computer science, design and more (for reviews, see Fisk and Grove, 2010 and Spohrer and Maglio, 2010b). But we also found fragmentation and a lack of awareness among researchers and scientists in these various disciplines (Rust, 2004; Spohrer and Maglio, 2010b). *Service science* is the term we used to try to draw the various disciplinary threads together into a single, coherent study of service phenomena (Maglio, Srinivasan, Kreulen and Spohrer, 2006; Spohrer, Maglio, Bailey and Gruhl, 2007). It was a bold and perhaps foolish idea to try to create a new science of service, and ten years later it is not yet clear whether we have succeeded. Nevertheless, we would argue that there has certainly been good progress and there are still further avenues to explore and develop. In this chapter, we will review some of this progress and some of these prospects.

The chapter is organized in three parts. First, we define some terms, describing our fundamental view that core to service is *value cocreation*. Second, given that service requires multiple parties to cocreate value, we argue that *the study of service systems requires multiple methods*. Third, to put this new study of service systems on solid scientific ground, we suggest that *the future of service science rests with the use of computational models*.

SERVICE IS VALUE COCREATION

The traditional view holds that services constitute the third sector of the economy: service activities are those economic activities that are left over after accounting for agriculture and manufacturing (Fitzsimmons and Fitzsimmons, 2010). Over time, disciplines including economics, marketing, operations, management, engineering and more have all focused some attention on service activities, primarily from the perspective of this traditional view (for some history, see Brown, Fisk and Bitner, 1994; Maglio, Kieliszewski and Spohrer, 2010a; and Vargo and Lusch, 2004). Yet, even given this broadly agreed upon view of service, different disciplines have not used the same basic definition of service. For instance, economics defines service as a distinct type of exchange (other than exchange of goods) – a category for counting and analysing jobs, businesses and exports (e.g., Triplett and Bosworth, 2004). Traditionally, marketing defines service as a distinct type of exchange (Shostack, 1977), delivered by a distinct type of process (Bitner and Brown, 2006), and often characterized by customized human interactions or 'moments of truth' with customers (Carlzon, 1987). In the field of operations, service is usually defined as a process that is dependent on customer inputs (Chase, 1981; Sampson and Froehle, 2006). The disciplines of engineering and operations research have defined service by the distinct type of modelling and optimization problems that result from customer variability (Dietrich and Harrison, 2006; Mandelbaum and Zeltyn, 2008; Riordan, 1962). In computer science, service is defined by a particular kind of abstraction for network-accessible capabilities with unique discovery, composition and modelling challenges (Sheth, Verma and Gomadam, 2006; Zhang, 2007). These are just a few examples; there are many more disciplinary views of service (see also Spohrer and Maglio, 2010b).

The past ten years have witnessed an explosion of activity aimed at knitting together multiple disciplinary views of service into a single, unified whole (e.g., Demirkan, Spohrer and Krishna, 2011; Salvendy and Karwowski, 2010). We were even a little bit responsible for some of this. For example, we helped edit a volume on service science that drew together many disciplinary perspectives in one place, providing some context for discussion among related fields (Maglio, Kieliszewski and Spohrer, 2010a). Years earlier, we helped organize the first service science conference, which was held in New York, which brought together a broad set of international academics and practitioners from many disciplines (see Hefley and Murphy, 2008). In addition, we have been involved in special issues of journals devoted to service science, all with contributions drawn from multiple disciplines and perspectives (Maglio, Spohrer, Seidman and Ritsko, 2008; Spohrer and Riecken, 2006). We suggest that, while there has been a lot of support and enthusiasm for these and similar activities, significant cross-disciplinary dialogues have actually remained quite limited.

Given the traditional view that service activities are separate and left over after agriculture and manufacturing, academic disciplines have developed multiple characterizations of services, with each offering unique perspectives. This has encouraged each discipline to stay apart from every other. Service science aims to bring these perspectives together, and to do this we must develop a unified definition. In our view, *capabilities, interaction, change* and *value* are fundamental to service. At its most basic, service results in change in one entity brought about as a result of interaction with another entity (Hill, 1977). To be effective, this change must be preferable or leave the entities better off than they were before they interacted (Vargo, Maglio and Akaka, 2008). Entities interact because each may have specialized knowledge and capabilities (Bastiat,1850/1979), and a group composed of specialized capabilities can achieve more than individuals with multiple capabilities (Ricardo, 1817/2004). Simply put, service is the application of competences for the benefit of one another, making all economic activity an exchange of service for service (Vargo and Lusch, 2004). More precisely, we define service as *value cocreation* – value as change that people prefer – and value cocreation as a change or set of related changes that people prefer that is realized as a result of communication, planning or other purposeful actions (Maglio and Spohrer, 2013; Spohrer and Maglio, 2010a).

Thus, in our view, all economic activity depends on service because all value is cocreated (Vargo and Lusch, 2004). This service-dominant logic effectively flips the usual 'goods dominant' worldview on its head and takes service to be the primary category of economic activity. Service-dominant logic can be difficult to understand, or even perhaps to accept, because a goods-dominant logic has served us so well for so long. When all economic activities are seen as service-for-service exchange, goods become just one way of transmitting service, that is, for applying human competence. Increasing automation in manufacturing shifts value-cocreation opportunities from production-assembly activity to design, distribution and marketing; representing a shift from vertically integrated companies toward orchestrated value networks or service systems (see also Normann and Ramirez, 1993; Quinn, 1992). Because of increasing use of technology for routine manual and routine cognitive activities, value-cocreation opportunities migrate over time toward expert thinking and complex communication skills (Levy and Murnane, 2004). It is not so much that there has been a broad increase in economic activity in the service sector over time, but that increasing knowledge and technology have enabled new ways to create effective value-creating interactions.

Knowledge and informational activities are core to service. Their impact on industrial productivity became particularly noticeable as the evolution of information and communication technology (ICT) improved and provided an environment for service interactions on a global scale (e.g., Bryson, Daniels and Warf, 2004; De Bandt and Dibiaggio, 2002; Herzenberg, Alic and Wial, 1998). Though obviously relevant to business-to-business services, ICT is central to all types of modern service arrangements, as economic circumstances have shifted from industrial to informational. For example, the personal use of mobile phone applications allows one to determine, within a matter of moments, if one should take a train, hail a taxi or drive in a private vehicle, given current traffic conditions, perceived reliability and availability of each mode of transportation. Technology-enabled and technology-enhanced activities provide one key context for service (Davis, Spohrer and Maglio, 2011; Glushko, 2010; Lyons, 2012) and knowledge-intensive activities provide another (den Hertog, 2002; Sheth, 2010).

In summary, service is value cocreation (Maglio, Kieliszewski and Spohrer, 2010b; Spohrer and Maglio, 2010b) and this occurs naturally as entities with distinct capabilities interact to realize mutually beneficial outcomes. Service phenomena arise in a real-world ecology of entities, their interactions and their capacity for finding mutually beneficial outcomes. Service science is the study of these value-cocreation phenomena.

THE STUDY OF SERVICE SYSTEMS REQUIRES MANY METHODS

Our own multi-disciplinary framework for service science puts together approaches from marketing, economics, operations, computer science and more (Maglio and Spohrer, 2013). In our view, the service system is the fundamental abstraction needed to understand service phenomena and build the scientific study of service (IfM and IBM, 2008; Maglio and Spohrer, 2008; Spohrer and Maglio, 2010a; Vargo, Lusch and Akaka, 2010). Within this abstraction, each entity in a service system is a collection of resources, including people, technologies, organizations and information (Spohrer, Maglio, Bailey and Gruhl, 2007) that interact by granting one another access rights (Spohrer and Maglio, 2010b). Value emerges when entities work together for mutual benefit (Vargo, Maglio and Akaka, 2008), and to work together effectively entities must coordinate joint actions and compute the potential future value of joint action (Maglio and Spohrer, 2013). This general theory of service system operations implies that service systems are complex systems that incorporate interactions among resources that may have unexpected and unanticipated consequences.

Our theory is rooted firmly in a service-dominant worldview (see also Maglio and Spohrer, 2008). That is not to say that other theories, based on other worldviews, are not equally reasonable. For example, Unified Service Theory is based on an operational view in which service is defined by operations that require customer input or involvement (Sampson, 2010a,b; Sampson and Froehle, 2006). One big implication is that providers and customers are bound together primarily by co-production activities, focusing attention on production processes rather than on value creation. The co-production activities may be realized far from the interactions of provider and customer. (See also Ng, 2012, who distinguishes between *worth*, which is calculated to make a decision at a certain point, and *value-in-context*, which is benefit realized, perhaps far from the time of decision.) Another approach to understanding service is based on the notion that it fundamentally involves the exchange of access to resources rather than the exchange of resources, the so-called 'rental paradigm' (Lovelock and Gummesson, 2004). Though similar in some ways to the service-dominant view (e.g., both depend on service as a kind of exchange of capabilities among entities), the rental paradigm distinguishes between goods-based activities, which involve a transfer of ownership, and service-based activities, which do not. But whatever the perspective, it seems the object of any theory of service is to understand how interactions between entities create mutual value.

Key priority areas for the study of service or value cocreation include growth, culture, innovation, design, optimization, branding, experience, measurement and technology (see Chesbrough and Spohrer, 2006; Ostrom et al., 2010). Such a broad range of topics requires a broad range of methods. Moreover, because service results from interactions, methods must consider not only production-side or provider-side processes, but customer-side processes as well (see also Lillrank, 2010; Sampson, 2010a,b). Understanding service means understanding the relationship between value cocreation and provider–customer co-production (Ordanini, Zarantonello and Parasuraman, 2010). Traditional studies of service innovation follow a traditional view of service activities by industry to classify innovation types and innovation processes (see, for instance, Miles, 2010). In these sorts of studies, service innovation processes reside wholly inside service organizations. More recently, service innovation has been distinguished from other sorts of innovation by the need to incorporate customers in the innovation process, for instance through co-development (Edvardsson, Gustafsson, Kristensson and Witell, 2010). In this case, the customer is seen not only as an input to production but as an input to new product development as well. Involving multiple stakeholders outside a firm service innovation is taken even further by the 'open innovation' view, in which any external party may be a source of new ideas (Chesbrough, 2011). The customer co-development and open innovation perspectives reflect an inherently service-dominant worldview (see also Ordanini and Parasuraman, 2011).

The interactive nature of service is particularly clear in the area of service experience and design. For instance, service blueprinting couples service experience and design, providing a process-oriented visualization of where, when, who, how and with what a service is enacted (Shostack, 1984). Central to service is value cocreated with customers, whether the value is realized as reliable shipments from one point to another, or personalized marketing and ticket sales for major league baseball games (Bitner, Ostrom and Morgan, 2008). Service blueprinting can be extended with techniques from new service development, interaction design and service design, integrating the firm's service concept with the customer value constellation of service offerings, the firm's internal service system and customer service encounters (Patricio, Fisk, Falcão e Cunha and Constantine, 2011).

Going one step further, design may be seen as a service itself in which, at the highest level of abstraction, meta-design captures the outline of potential experiences developed through cocreation and informs the implementation of a service architecture, such as a physical departure area in an airport or a personalized healthcare experience (Evenson and Dubberly, 2010). Designing a service in this way places people at the centre of the experience to discover what technical or informational elements are required to support the customer experience and identify new services or improved efficiencies in the system (Evenson and Dubberly, 2010). Many service systems are information-intensive, and even those services that are experience- or interaction-intensive still require a degree of information exchange (Glushko, 2010). The identification of systematic relationships or patterns of information exchange can create powerful contexts for service design and options for fulfilment of a cocreated service experience. Service design contexts range from a person-to-person service encounter to technology-enhanced, location-based and context-aware services, all of which can be mixed and matched to assemble a complex service system (Glushko, 2010).

There is a clear need to link information technology research with service science (Davis, Spohrer and Maglio, 2011; Raj and Sambamurthy, 2006). At the infrastructure level, service-oriented architecture promotes loose coupling of software components for interoperability across platforms and dynamic choreography of business processes, but even here multiple perspectives are required to transform architectural decisions into business value (Demirkan et al., 2008). In general, information technologies embody capabilities that can be traded against human capabilities to create efficiencies in service systems (Campbell, Maglio and Davis, 2011; see also Zysman, 2006). In some cases, technology (or information) can substitute for interactions with service personnel (Glushko and Nomorosa, 2013), and in others it can entirely reconfigure service processes and responsibilities (Campbell, Maglio and Davis, 2011). In general, distribution of effort (and of roles and responsibilities) across a distributed service system, including customers, suppliers and firm, is a design parameter that determines a system's effectiveness (Roels, Karmarkar and Carr, 2010). Technology does not by itself determine value cocreation; rather, value cocreation depends in large part on social connections among human participants in a service system that can be enhanced through technology (Breidbach, Kolb and Srinivasan, 2013).

Our approach to service science follows from the service-dominant worldview, which sees all value as cocreated. Such a general framework requires an equally general and broad set of tools and methods to study and improve value cocreation activities. Here, we have described just a few of these methods, particularly as they relate to service experience, service innovation and technology-enabled service. We now discuss why computational modelling is a fundamental method of service science and for underpinning and providing coherence to the development of a multi-disciplinary service science.

COMPUTATIONAL MODELS ARE THE FUTURE OF SERVICE SCIENCE

There is a long history of analytical modelling to understand and improve service systems (e.g., Riordan, 1962). On the analytical side, many optimization models of service systems focus on shortening process wait times and streamlining tasks (e.g., Paul and MacDonald, 2013), and recently, optimization models of service have focused more on human resources and skill-matching, among other related issues (e.g., Mojsilović and Connors, 2010). Service domains rife with optimization models include transportation and logistics (e.g., Saltzman, 2012) and healthcare processes (e.g., Rouse, 2008).

Complex service systems pose particular challenges for modelling, often requiring computational simulation rather than analytical approaches. One reason for this is that services often include interactions among multiple entities with emergent or aggregate properties (see also Lazer et al., 2009). Another is that any given service exists in two states: a potential state and an actual state (Shostack, 1982). These characteristics make it difficult to understand the behaviour of service systems and to know how to design them to best create value. A service can be experienced and described in many different ways, depending on background and perspective; for instance, students asked to identify and categorize a local subway system often created descriptions that used multiple aspects of multiple frameworks to highlight the various complexities of the system (Glushko, 2012).

Key to modelling service systems is modelling interactions among technologies, people and organizations (Spohrer, Maglio, Bailey and Gruhl, 2007), both in space and in time (Ferrario and Guarino, 2009). Actions and events at the time of co-production and value cocreation must also be considered in the context of actions that serve to set up the system in the first place (Ferrario and Guarino, 2009; Stucky, Cefkin, Rankin, Shaw and Thomas, 2011). When provider and customer work together, realization of value anticipated in contracts and agreements depends on what transpires at the time of service delivery (Stucky, Cefkin, Rankin, Shaw and Thomas, 2011). This line of thought suggests that a service system ought to be examined as two loosely coupled systems – the governing system (in which there is the potential for value creation) and the actualizing system (in which value is ultimately created) – to detect misalignments that can derail value realization or to identify opportunities for improving value propositions (Stucky, Cefkin, Rankin, Shaw and Thomas, 2011).

Another way to model service systems is by their structure or business architecture, that is, by representing each of the entities found in an enterprise and then identifying relationships and linkages between them (Leung and Bockstedt, 2009). Business architecture frameworks differ in how they address the four business domains of strategy and structure, business networks, operations, and performance and revenue (Glissmann and Sanz, 2010). As a technique for analysing the design of an enterprise, business architecture frameworks have often been applied to problems of business transformation, but the techniques and methodologies have also explicitly incorporated service components within the conceptual models. Such integration of service components can render all four business domains, serving as an integration model for the overall business architecture (Glissmann and Sanz, 2010).

Computational approaches to modelling service systems include agent-based models and system dynamics models, among others. For instance, agent-based modelling can be used to simulate behaviour or outcomes that result from complex interactions among many parts in scientific domains (Bonabeau, 2002) and in business domains (Rand and Rust, 2011). For service systems, this might help us understand how different physical layouts of a servicescape (shopping mall, airport terminal, office environment) may affect customer satisfaction (Rand and Healy, 2010). System dynamics models can also be used to examine unexpected and emergent consequences of interactions between human behaviour and physical aspects of a production system, such as the impact of efficiency gains and cost containment initiatives on service delivery and quality (Oliva and Sterman, 2001, 2010). In this case, the computational model can account for the dynamics of people (e.g., total numbers, experience, quit rates) and the inclusion and exclusion of particular operational structures (e.g., desired service capacity, work week, training, overtime) to examine elements, such as decision-making processes and customer and employee expectations.

A single, uniform approach to modelling may not always be sufficient for service systems, which may require composite or hybrid approaches, such as hybrids that include business process modelling and agent-based modelling (see Apte, Karmarkar, Kieliszewski and Leung, 2012). One promising direction is composite modelling of complex service systems as value constellations (Kieliszewski, Maglio and Cefkin, 2012). By coupling component-level or entity-level models, composite models can help uncover opportunities for reconfiguring roles and relationships to unlock value (cf. Normann and Ramirez, 1993; Ordanini and Parasuraman, 2012). Combining heterogeneous simulation models and heterogeneous datasets to create composite simulation models of complex systems can also foster and facilitate cross-disciplinary models and collaborations (Maglio, 2011; Tan et al., 2012). For instance, a composite model may combine real-world component models such as agent-based, stochastic, deterministic, system dynamics models to emulate a complex business ecosystem and gauge the impact of changes on it. There are major research challenges in developing such a composite modelling technology and in gaining adoption and use by a scientific and business community (Cefkin et al., 2010). The goal is to combine expert models of constituent real-world systems related to a particular set of issues, such as healthcare, natural resources or urban planning, to create an interoperating complex composite system model with which policy-makers and stakeholders can work to try out alternative solutions in a systematic way.

CONCLUSION

The new field of service science aims to draw various disciplinary threads together into a single, coherent study of service. In this chapter, we have outlined several aspects of this emerging science. Specifically, we described our view of service as *value cocreation*, which implies that service depends fundamentally on multiple parties working together in complex arrangements called *service systems*. In turn, we argued that understanding and improving service systems requires deploying multiple scientific methods, including computational modelling, to characterize the behaviour of complex systems under various conditions. Because complex service systems may have nonlinear and emergent properties, particular kinds of simulation models, such as system dynamics models, agent-based models and hybrid models, may in fact be the most appropriate computational techniques for service scientists.

BIBLIOGRAPHY

Apte, U., Karmarkar, U., Kieliszewski, C. and Leung, Y.T. (2012). Exploring the representation of complex processes in information-intensive services. *International Journal of Services Operations and Informatics*, 7(1), 52–78.
Bastiat, F. (1850/1979). *Economic Harmonies: The Foundation for Economics Education*. Irvington-on-Hudson, NY: Foundation for Economic Education.
Bitner, M.J. and Brown, S.W. (2006). The evolution and discovery of services science in business schools. *Communications of the ACM*, 49(7), 73–78.
Bitner, M.J., Ostrom, A.L. and Meuter, M.L. (2002). Implementing successful self-service technologies. *Academy of Management Executive*, 16(4), 96–108.
Bitner, M.J., Ostrom, A. and Morgan, F. (2008). Service blueprinting: A practical technique for service innovation. *California Management Review*, 50(3), 66–94.
Blomberg, J. (2008). Negotiating meaning of shared information in service system encounters. *European Management Journal*, 26, 213–222.
Bonabeau, E. (2002). Agent-based modeling: Methods and techniques for simulating human systems. *Proceedings of the National Academy of Sciences*, 99(Suppl 3), 7280–7287.
Breidbach, C.F., Kolb, D.G. and Srinivasan, A. (2013). Connectivity in service systems: Does technology-enablement impact the ability of a service system to co-create value? *Journal of Service Research* (DOI, 1094670512470869, published on-line).
Brown, S.W., Fisk, R.P. and Bitner, M.J. (1994). The development and emergence of services marketing thought. *International Journal of Service Industry Management*, 5(1), 21–48.
Bryson, J.R., Daniels, P.W. and Warf, B. (2004). *Service Worlds: People, Organizations, Technologies*. New York, NY: Routledge.
Campbell, C.S., Maglio, P.P. and Davis, M.M. (2011). From self-service to super-service: How to shift the boundary between customer and provider. *Information Systems and eBusiness Management*, 9(2), 173–191.
Carlzon, J. (1987). *Moments of Truth*. Cambridge, MA: Ballinger.
Cefkin, M., Glissman, S., Haas, P.J., Jalali, L., Maglio, P.P., Selinger, P. and Tan, W.C. (2010). Splash: A progress report on building a platform for a 360 degree view of health. In D. Sundaramoorthi, M. Lavieri and H. Zhao (eds), *Proceedings of the 5th INFORMS Workshop on Data Mining and Health Informatics*, DM-HI 2010 (splashvision20100728.pdf).
Chase, R.B. (1978). Where does the customer fit in a service operation? *Harvard Business Review*, 56, 137–142.
Chase, R.B. (1981). The customer contact approach to services: theoretical bases and practical extensions. *Operations Research*, 29(4), 698–706.
Chesbrough, H. (2011). *Open Services Innovation: Rethinking Your Business to Grow and Compete in a New Era*. San Francisco, CA: Wiley.
Chesbrough, H. and Spohrer, J. (2006). A research manifesto for services science. *Communications of the ACM*, 49(7), 35–40.
Davis, M.M., Spohrer, J.C. and Maglio, P.P. (2011). Guest editorial: How technology is changing the design and delivery of services. *Operations Management Research*, 4(1–2), 1–5.
De Bandt, J. and Dibiaggio, L. (2002). Informational activities as co-production of knowledge and values. In J. Gadrey, and F. Gallouj, (eds), *Productivity, Innovation and Knowledge in Services: New Economic and Socio-economic Approaches*. Cheltenham and Northampton, MA: Edward Elgar.
Demirkan, H., Kauffman, R.J., Vayghan, J.A., Fill, H.G., Karagiannis, D. and Maglio, P.P. (2008). Service-oriented technology and management: Perspectives on research and practice for the coming decade. *Electronic Commerce Research Applications*, 7, 356–376.
Demirkan, H., Spohrer, J.C. and V. Krishna (eds) (2011). *The Science of Service Systems*. New York, NY: Springer.
den Hertog, P. (2002). Co-producers of innovation: On the role of knowledge-intensive business services in innovation. In J. Gadrey, and F. Gallouj, (eds), *Productivity, Innovation and Knowledge in Services: New Economic and Socio-economic Approaches*. Cheltenham and Northampton, MA: Edward Elgar.
Dietrich, B. and Harrison, T. (2006). Serving the services industry. *OR/MS Today*, 33(3), 42–49.
Edvardsson, B., Gustafsson, A., Kristensson, P. and Witell, L. (2010). Service innovation and customer co-development. In P.P. Maglio, C.A. Kieliszewski and J.C. Spohrer (eds), *Handbook of Service Science*. New York: Springer.

Evenson, S. and Dubberly, H. (2010). Designing for service: Creating an experience advantage. In G. Salvendy and W. Karwowski (eds), *Introduction to Service Engineering*. Hoboken, NJ: John Wiley & Sons.
Ferrario, R. and Guarino, N. (2009). Towards an ontological foundation for services science. *First Future Internet Symposium, FIS 2008*, Vienna, Austria, 29–30 September 2008. Revised Selected Papers (DOI: 10.1007/978-3-642-00985-3_13).
Fisk, R.P. and Grove, S.J. (2010). Service science: Research and innovations in the service economy. In P.P Maglio, C.A. Kieliszewski and J.C. Spohrer (eds), *Handbook of Service Science*. New York, NY: Springer.
Fisk, R.P., Brown, S.W. and Bitner, M. (1993). Tracking the evolution of the services marketing literature. *Journal of Retailing*, 69, 61–103.
Fitzsimmons, J.A. and Fitzsimmons, M. (2010). *Service Management: Operations, Strategy, and Information Technology* (seventh edition). New York, NY: McGraw Hill.
Glissmann, S. and Sanz, J. (2010). Business architecture for the design of enterprise service systems. In P.P. Maglio, C.A. Kieliszewski and J.C. Spohrer (eds), *Handbook of Service Science*. New York, NY: Springer.
Glushko, R.J. (2010). Seven contexts for service system design. In P.P. Maglio, C.A. Kieliszewski and J.C. Spohrer (eds), *Handbook of Service Science*. New York, NY: Springer.
Glushko R.J. (2012). Describing service systems. *Human Factors in Ergonomics and Manufacturing*, 23(1), 11–18 (DOI: 10.1002/hfm.20514).
Glushko, R.J. and Nomorosa, K.J. (2013). Substituting information for interaction: A framework for personalization in service encounters and service systems. *Journal of Service Research*, 16(1), 21–38.
Hefley, B. and Murphy, W. (2008). *Service Science, Management and Engineering: Education for the 21st Century*. New York, NY: Springer.
Herzenberg, S.A., Alic, J.A. and Wial, H. (1998). *New Rules for a New Economy: Employment and Opportunity in Postindustrial America*. Ithaca, NY: Cornell University Press.
Hill, T.P. (1977). On goods and services. *Review of Income and Wealth*, 23(4), 314–339.
Horn, P. (2005). The new discipline of services science. *BusinessWeek*, 21 January (http://www.businessweek.com/technology/ content/jan2005/tc20050121_8020.htm).
IfM and IBM (2008). *Succeeding through Service Innovation: A Service Perspective for Education, Research, Business and Government*. Cambridge, UK: University of Cambridge Institute for Manufacturing.
IBM (2012). *2012 IBM Annual Report* (available at: http://www.ibm.com/annualreport/2012/bin/assets/2012 ibmannual.pdf, accessed 28 November 2014).
IBM Research (2005). *Services Science: A New Academic Discipline?* 17–18 May, San Jose, CA: IBM Research.
Johnson, B.C., Manyika, J.M. and Yee, L.A. (2005). The next revolution in interactions. *McKinsey Quarterly*, 2005/4, 20–33.
Kandogan, E., Maglio, P.P., Haber, E. and Bailey, J. (2012). *Taming Information Technology: Lessons from Studies of System Administrators*. New York, NY: Oxford University Press.
Kieliszewski, C.A., Maglio, P.P. and Cefkin, M. (2012). On modeling value constellations to understand complex service system interactions. *European Management Journal*, 3(5), 438–450 (http://dx.doi.org/10.1016/j.emj.2012.05.003, accessed 28 November 2014).
Lazer, D., Pentland, A., Adamic, L., Aral, S., Barabasi, A.L., Brewer, D., Christakis, N., Contractor, N., Fowler, J., Gutmann, M., Jebara, T., King, G., Macy, M., Roy, D. and Van Alstyne, M. (2009). Computational social science. *Science*, 323, 721–723.
Lemey, E. and Poels, G. (2011). Towards a service system ontology for service science. *Proceedings of the 9th International Conference on Service-Oriented Computing* (ICSOC 2011), 250–264.
Leung, Y.T. and Bockstedt, J. (2009). Structural analysis of a business enterprise. *Service Science*, 1(3), 169–188.
Levy, F. and Murnane, R.J. (2004). *The New Division of Labor: How Computers are Creating the Next Job Market*. Princeton, NJ: Princeton University Press.
Lillrank, P. (2010). Service processes. In G. Salvendy and W. Karwowski (eds), *Introduction to Service Engineering*. Hoboken, NJ: John Wiley & Sons.
Lovelock, C. and Gummesson, E. (2004). Whither services marketing? In search of a new paradigm and fresh perspectives. *Journal of Service Research*, 7, 20–41.
Lyons, K. (2011). A framework that situates technology research in the field of service science. In H. Demirkan, J.C. Spohrer and V. Krishna (eds), *The Science of Service Systems*. New York, NY: Springer.
Lyons, M. (2012). Introduction to technology innovation cases. In L.A. Macaulay, I. Miles, J. Wilby, Y.L.Tan, L. Zhao and B. Theodoulidis (eds), *Case Studies in Service Innovation*. New York, NY: Springer.
Maglio, P.P. (2011). Modeling complex service systems. *Service Science*, 3(4), i–ii.
Maglio, P.P. and Mabry, P.L. (2011). Commentary: Agent-based models and systems science approaches to public health. *American Journal of Preventive Medicine*, 40(3), 392–394.
Maglio, P.P. and Spohrer, J. (2013). A service science perspective on business model innovation. *Industrial Marketing Management*, 42, 665–670.

Maglio, P.P. and Spohrer, J. (2008). Fundamentals of service science. *Journal of the Academy of Marketing Science*, 36, 18–20.
Maglio, P.P., Kieliszewski, C.A. and Spohrer, J.C. (2010a). *Handbook of Service Science*. New York, NY: Springer.
Maglio, P.P., Kieliszewski, C.A. and Spohrer, J.C. (2010b). Introduction: Why a handbook? In P.P. Maglio, C.A. Kieliszewski and J.C. Spohrer (eds), *Handbook of Service Science*. New York, NY: Springer.
Maglio, P.P., Nusser, S. and Bishop, K. (2010). A service perspective on IBM's brand. *Marketing Review St. Gallen*, 6, 44–48.
Maglio, P.P., Spohrer, J., Seidman, D.I. and Ritsko, J.J. (2008). Special issue on SSME. *IBM Systems Journal*, 47.
Maglio, P.P., Srinivasan, S., Kreulen, J.T. and Spohrer, J. (2006). Service systems, service scientists, SSME, and innovation. *Communications of the ACM*, 49(7), 81–85.
Maglio, P.P., Vargo, S.L., Caswell, N. and Spohrer, J. (2009). The service system is the basic abstraction of service science. *Information Systems and e-business Management*, 7, 395–406.
Mandelbaum, A. and Zeltyn, S. (2008). Service engineering of call centers: Research, teaching, and practice. In B. Hefley and W. Murphy (eds), *Service Science, Management, and Engineering: Education for the 21st Century*. New York, NY: Springer.
Miles, I. (2010). Service innovation. In P.P. Maglio, C.A. Kieliszewski and J.C. Spohrer (eds), *Handbook of Service Science*. New York, NY: Springer.
Mojsilović, A. and D. Connors (2010). Workforce analytics for the services economy. In P.P Maglio, C. Kieliszewski and J.C. Spohrer (eds), *Handbook of Service Science*. New York, NY: Springer.
Ng, I. (2012). *Value and Worth: Creating New Markets in the Digital Economy*. Cambridge: Innovorsa Press.
Normann, R. (2001). *Reframing Business: When the Map Changes the Landscape*. Chichester, England: Wiley.
Normann, R. and Ramirez, R. (1993). From value chain to value constellation: Designing interactive strategy. *Harvard Business Review*, 71, 65–77.
Oliva, R. and Sterman, J.D. (2001). Cutting corners and working overtime: Quality erosion in the service industry. *Management Science*, 47(7), 894–914.
Oliva, R. and Sterman, J.D. (2010). Death spirals and virtuous cycles. In P.P. Maglio, C.A. Kieliszewski and J.C. Spohrer (eds), *Handbook of Service Science*. New York, NY: Springer.
Ordanini, A. and Parasuraman, A. (2011). Service innovation viewed through a service-dominant logic lens: A conceptual framework and empirical analysis. *Journal of Service Research*, 14, 3–23.
Ordanini, A. and Parasuraman, A. (2012). A conceptual framework for analyzing value-creating service ecosystems: An application to the recorded-music market. In Stephen L. Vargo and Robert F. Lusch (eds), *Review of Marketing Research, Special Issue – Toward a Better Understanding of the Role of Value in Markets and Marketing*, 9.
Ordanini, A., Zarantonello, L. and Parasuraman, A. (2010). The link between customer co-production and value co-creation: A theory-based framework. Presentation at the 19th Annual Frontiers in Services Conference, Karlstad, Sweden, June.
Ostrom, A., Bitner, M.J., Brown, S.W., Goul, M., Smith-Daniels, V., Demirkan, H. and Rabinovich, E. (2010). Moving forward and making a difference: Research priorities for the science of service. *Journal of Service Research*, 13, 4–36.
Patricio, L., Fisk, R.P., Falcão e Cunha, J. and Constantine, L. (2011). Multilevel service design: From customer value constellation to service experience blueprinting. *Journal of Service Research*, 14(2), 180–200.
Paul, J.A. and MacDonald, L. (2013). A process flow-based framework for nurse demand estimation. *Service Science*, 5(1), 17–28.
Quinn, J.B. (1992). *The Intelligent Enterprise: A Knowledge and Service Based Paradigm for Industry*. New York, NY: The Free Press.
Raj, A. and Sambamurthy, V. (2006). The growth of interest in services management: Opportunities for information systems scholars. *Information Systems Research*, 14(4), 327–331.
Rand, W. and Healy, J. (2010). Analyzing servicescapes using agent-based modeling. Presentation at Frontiers in Service, Karlstad, Sweden, June.
Rand, W. and Rust, R.T. (2011). Agent-based modeling in marketing: Guidelines for rigor. *International Journal of Research in Marketing*, 28(3), 181–193.
Ricardo, D. (1817/2004). *The Principles of Political Economy and Taxation*. Mineola, NY: Dover Publications.
Riordan, J. (1962). *Stochastic Service Systems*. New York, NY: Wiley.
Roels, G. Karmarkar, U.S. and Carr, S. (2010). Contracting for collaborative services. *Management Science*, 56(5), 849–863.
Rouse, W.B. (2008). Health care as a complex adaptive system: Implications for design and management. *The Bridge: Linking Engineering and Society*, 38(1), 17–25.
Rust, R.T. (2004). A call for a wider range of service research. *Journal of Service Research*, 6, 24–36.

Saltzman, R.M. (2012). Planning for an aging fleet of shuttle vehicles with simulation. *Service Science*, 4(3), 195–206.
Salvendy, G. and Karwowski, W. (eds) (2010). *Introduction to Service Engineering*. Hoboken, NJ: John Wiley & Sons.
Sampson, S.E. (2010a). The unified service theory: A paradigm for service science. In P.P. Maglio, C. Kieliszewski and J.C. Spohrer (eds), *Handbook of Service Science*. New York, NY:Springer.
Sampson, S.E. (2010b). A unified services theory. In G. Salvendy and W. Karwowski (eds), *Introduction to Service Engineering*. Hoboken, NJ: John Wiley & Sons.
Sampson, S.E. (2012). Visualizing service operations. *Journal of Service Research*, 15(2), 182–198.
Sampson, S.E. and Froehle, C.M. (2006). Foundations and implications of a proposed unified services theory. *Production and Operations Management*, 15(2), 329–343.
Sheth, A. (2010). Computing for human experience: Semantics-empowered sensors, services, and social computing on the ubiquitous web. *IEEE Internet Computing*, 14(1), 88–91.
Sheth, A., Verma, K. and Gomadam, K. (2006). Semantics to energize the full services spectrum. *Communications of the ACM*, 49(7), 55–61.
Shostack, G.L. (1977). Breaking free from product marketing. *Journal of Marketing*, 41, 73–80.
Shostack, G.L. (1982). How to design a service? *European Journal of Marketing*, 16, 49–63.
Shostack, G.L. (1984). Designing services that deliver. *Harvard Business Review*, 62(1), January–February, 133–139.
Smith, A. (1776/2000). *The Wealth of Nations*. New York, NY: The Modern Library.
Spohrer, J. and Maglio, P.P. (2008). The emergence of service science: Toward systematic service innovations to accelerate co-creation of value. *Production and Operations Management*, 17(3), 1–9.
Spohrer, J. and Maglio, P.P. (2010a). Service Science: toward a smarter planet. In G. Salvendy and W. Karwowski (eds), *Introduction to Service Engineering*. Hoboken, NJ: John Wiley & Sons.
Spohrer, J. and Maglio, P.P. (2010b). Toward a science of service systems: Value and symbols. In P.P. Maglio, C.A. Kieliszewski and J.C. Spohrer (eds), *Handbook of Service Science*. New York, NY: Springer.
Spohrer, J. and Riecken, D. (2006). Special Issue on Services Science, *Communications of the ACM*, 49(7).
Spohrer, J., Maglio, P.P., Bailey, J. and Gruhl, D. (2007). Steps toward a science of service systems. *Computer*, 40, 71–77.
Stucky, S.U., M. Cefkin, Y. Rankin, B. Shaw and J. Thomas (2011). Dynamics of value co-creation in complex IT service engagements. *Information Systems and E-Business Management*, 9(2), 267–281.
Tan, W.C., Haas, P.J., Mak, R.L., Kieliszewski, C.A., Selinger, P., Maglio, P.P., Glissmann, S., Cefkin, M. and Li, Y. (2012). Splash: A platform for analysis and simulation of health. *Proceedings of ACM SIGHIT International Health Informatics Symposium (IHI 2012)*, New York, NY: ACM Press.
Thorwarth, M. and Arisha, A. (2012). A simulation-based decision support system to model complex demand driven healthcare facilities. *Proceedings of the Winter Simulation Conference* (WSC '12), Berlin (DOI: 10.1109/WSC.2012.6465019).
Triplett, J.E. and Bosworth, B.P. (2004). *Productivity in the U.S. Services Sector: New Sources of Economic Growth*. Washington, DC: The Brookings Institution.
Vargo, S.L. and Lusch, R.F. (2004). Evolving to a new dominant logic for marketing. *Journal of Marketing*, 68, 1–17.
Vargo, S.L., Lusch, R.F. and Akaka, M.A. (2010). Advancing service science with service-dominant logic: Clarifications and conceptual development. In P.P. Maglio, C.A. Kieliszewski and J.C. Spohrer (eds), *Handbook of Service Science*. New York, NY: Springer.
Vargo, S.L., Maglio, P.P. and Akaka, M.A. (2008). On value and value co-creation: A service systems and service logic perspective. *European Management Journal*, 26(3), 145–152.
Zhang, L.J. (2007). *Modern Technologies in Web Services Research*. Hershey, PA: IGI Publishing.
Zysman, J. (2006). The 4th service transformation: The algorithmic revolution. *Communications of the ACM*, 49(7), 48.

5. The role of the Big 4: commoditisation and accountancy
Steve Hollis[1]

THE ROLE OF THE BIG 4

The Big 4 accounting firms as we know them now comprise four mega global professional services firms: PwC, Deloitte, Ernst & Young and KPMG. The combined annual fee income for these firms is over $100 billion and together they employ approaching 700,000 talented professionals. These four firms impact directly or indirectly on every business, government and organisation in just about every corner of the planet. Every year the Big 4 train over 100,000 new professionals, and the majority leave following the formal part of their training to develop their careers in industry and commerce. This global training programme has developed probably the largest body of alumni on the planet, with an increasing number of the 'old boys and girls' occupying senior positions in just about every large company in the world. From the outside, an impartial observer could be forgiven for concluding that this is a great example of world domination! The reality is that the competition among the Big 4 and with other service providers is intense. The Big 4 operate in a market that has become increasingly sophisticated in the procurement of professional services. Despite this competition (or perhaps because of it?), the Big 4 are a role model for how professional services firms can survive, and indeed prosper, in good times and bad.

So how have these firms evolved? What is their role? How have they transformed into the global giants we now find? The origins of PwC, Deloitte and KPMG date back to 19th-century London. Ernst & Young has its origins in America, with two brothers, Theodore and Alvin Ernst, who set up in practice together in Cleveland in 1903. Each of these very small firms expanded through mergers and acquisitions (M&A) during the course of the 19th and 20th centuries. In the case of PwC, for example, currently the biggest of the Big 4 (global revenues in FY 2012 $31.5 billion), the journey commenced with two accountants, Samuel Lowell Price and Edwin Waterhouse, who joined forces in 1894 to form Price Waterhouse & Co. The firm became respected as one of the finest in London and to meet the needs of its growing international clients, it opened an office in New York in 1890 and began to establish separate partnerships across the globe. William Cooper and his three brothers were good competition for Samuel Price and Edwin Waterhouse and in 1957 the British firm Coopers merged with the US firm formed by William Lybrand to create a firm under the name of Coopers & Lybrand. In 1998, Coopers & Lybrand merged with Price Waterhouse to form PricewaterhouseCoopers and rebranded this trading name to PwC in 2010. This journey from modest beginnings to global titan is replicated for each of the other three 'Big 4' firms and the more recent merger activity is shown in Figures 5.1 and 5.2.

Figure 5.1 The mergers that made the Big 4

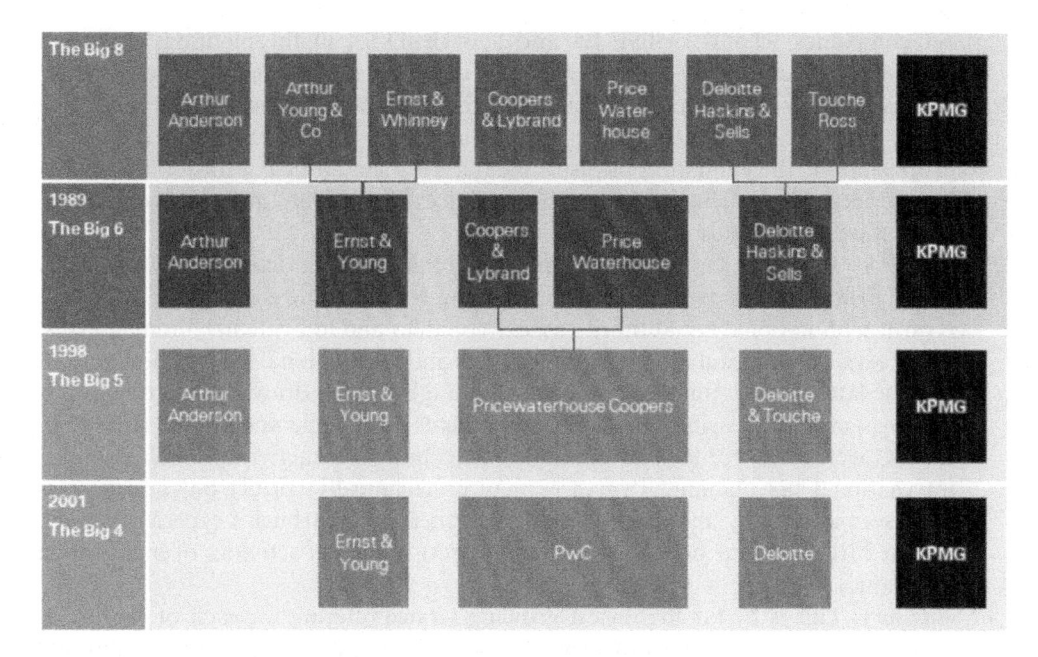

Note: The scale of the Big 4 is based on FY12 revenues as shown in Figure 5.2.

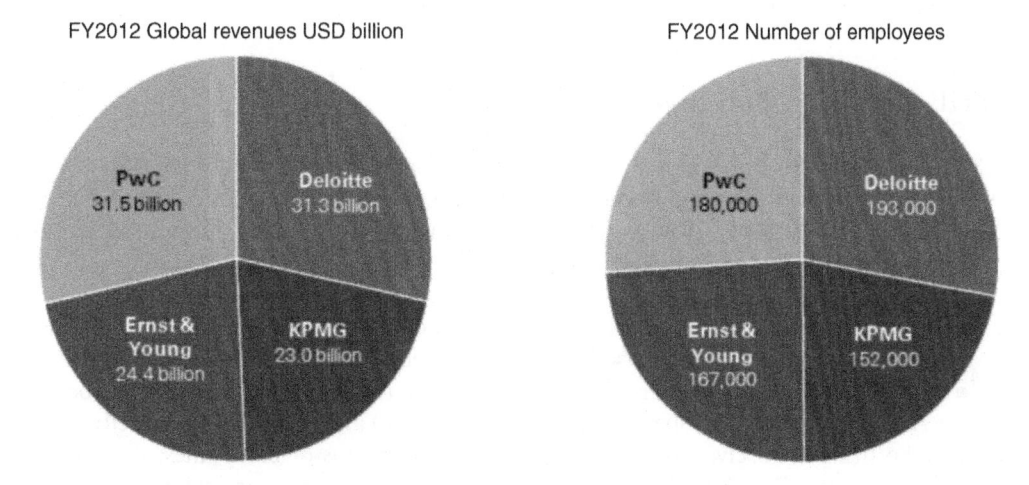

Figure 5.2 The scale of the Big 4

Each of the Big 4 provides a vast array of services to its clients which broadly fall into three areas:

1. *Assurance.* Statutory audits are the traditional mainstay for the Big 4 and, at the last count, between them the four firms are auditors to all but seven of the Fortune

500 companies, 99 of the FTSE 100 and 27 of the DAX 30; this picture is replicated across most parts of the globe. From this dominance of the global audit market space, the Big 4 have successfully extended their assurance services to cover a broad range of financial and non-financial activities. At the core of the assurance value proposition is the validation of historic data. This should be contrasted with advisory services (and particularly consulting services), where the emphasis is on recommendations for a future transaction or event.
2. *Taxation.* The Big 4 are the largest providers of taxation services globally, and each of the firms provides taxation services ranging from routine compliance (filing tax returns with national tax authorities for companies and individuals), indirect taxes, sales taxes, customs duties, employment taxes and international tax. In recent years, it is the latter service (international tax planning) that has drawn most attention to the Big 4 as an increasing number of their clients have come under intense scrutiny from Governments for their complex global tax mitigation activities. These complex international tax planning structures may be technically correct but are increasingly seen as morally abhorrent – look no further than Starbuck's experience in the UK in 2012 to see the heightened public interest in the tax activities of high profile companies.
3. *Advisory.* This is by far the fastest growing service offering for each of the Big 4, and encompasses a broad range of activities. These typically include: traditional consulting (encompassing strategy, performance improvement and related services); corporate finance (including financial and commercial due diligence); restructuring services; pensions and actuarial services; valuation services; and crisis management and sustainability services.

THE CHALLENGE FOR THE BIG 4

On the surface, the Big 4 portray a picture of globally integrated businesses working seamlessly across international boundaries delivering consistently high quality services to global and local clients. However, if you peel away the outer veneer, the actual picture is very different:

- The Big 4 operate a franchise model under a common brand (although there can be variations on the name used in some countries).
- The national practices are agglomerated into regional units – typically being Americas, EMA (Europe, Middle East and Africa) and ASPAC (Asia Pacific).
- The firms themselves are set up as either partnerships or companies, depending on the prevailing laws in each country. In some cases there are external owners/shareholders, but in most cases the partnership/company is owned by the 'partners' who work in the partnership/company.
- The 'power' typically vests in country leadership teams and not their global peers.
- Profits are shared very much as the country leadership determines for the partners in their country with a varying degree (which is typically very limited) of influence to reflect non-country activities.

- The basis for sharing profits differs from one country to the next, ranging from lockstep (equal profit sharing) to meritocracy (performance driven bonuses).
- Within country practices, profits are typically heavily influenced by personal performance and not team performance and, despite huge internal PR from country leadership on the importance of playing for the team, when the chips are down there is a strong culture of 'eat what you kill'.
- In most cases, national practices distribute the full amount of each year's profits – that is, there is no retention of profits in the business.

There is a very long list of similar such inconsistencies between the external perception and the internal reality. To a corporate reader, rightly you will observe that this structure is no way to run a business! It must be remembered, however, that despite their organisation structures, each of the Big 4 has enjoyed phenomenal success with only the occasional hiccup. Whilst this may have been the case for the past 100 years, is the tide turning?

During the 1980s, 1990s and 2000s, the global economy experienced unprecedented international activity and the Big 4 played a full part, working with companies and organisations of all sizes. Their growth was fuelled by their clients' growth. There was generally more demand for the Big 4's services than they were capable of supplying, particularly in what were hot areas such as M&As in the debt fuelled 2000s. These 'hot activities' fuelled extraordinary profitability and the Big 4 successfully used this to build their global brand strength across all their services. It is always dangerous to generalise, but one of the consequences of an extended demand driven part of the business cycle is that this understandably results in the professionals (who are the only real assets in any professional services firm) being focussed on being high quality producers. What's more, the strategies deployed by the firms at times of demand led growth focus on production efficiency and performance improvement. Indeed, in a demand led part of the business cycle, the leadership's challenge is a pleasure:

- Growth creates opportunities for career progression/promotion/recruitment.
- Profits and employee salaries increase.
- There is less fee resistance and greater client loyalty.

The leadership/management challenge is to keep the supply and development lines for new professional talent running smoothly. One thing there is no doubting is that each of the Big 4 is truly world class in this area. But then there are 'blips' that challenge their progress, and in the last ten years there have been two 'blips' in particular that have focussed the minds of the Big 4. It is interesting to see how the firms reacted to these 'blips' and, in the case of the latest interruption, what this may hold for the future.

Enron

The first of these 'macro' blips came with the Enron scandal, which was revealed in October 2001 when the full effect of the Houston based company's manipulation of accounting loopholes and policies was exposed. Enron was shortly followed by a number of other large companies that had transgressed in manipulating the financial information used by investors and suppliers alike. Per se there is nothing unusual about accounting scandals and there have been many predating Enron in relatively recent years from the 1970s to the 1990s (Barlow Clowes, Polly Peck, Cedant, Waste Management Inc. to name just a few). What distinguished Enron, however, in this rogues' gallery was not just the scale of the accounting manipulation, but the losses incurred by shareholders (over $11 billion) and most notably the consequent dissolution of Enron's audit firm, Arthur Anderson.

Prior to Enron, the Big 4 were actually the Big 5, with Arthur Anderson in this elite club. Anderson was a relatively young firm. It was founded in 1913 in Illinois and its founder, Arthur Anderson, headed the firm until his death in 1947. This firm had grown very successfully from its American roots into the firm that was, with good reason, always considered to be the most global of the Big 5.

The relationship between Enron and Arthur Anderson dated back to the early 1990s, and the Houston office (which serviced the Enron account) had over 1400 employees before the collapse (out of a total global head count of over 85,000 employees). In 2000, Enron paid Arthur Anderson $25 million for audit services (making this one of the biggest audit fees in the Fortune 500) and non-audit fees (principally for consultancy and taxation advice) of $27 million. This made Enron one of Arthur Anderson's largest and, at the time, most prestigious accounts. But as big as it was, the Enron account represented less than three-tenths of 1 per cent of Arthur Anderson's global fees, which exceeded $9 billion.

Whilst at first sight the Enron scandal appeared to be a relatively small problem in Houston, little did the Big 5 (soon to become the Big 4) foresee the full ramifications that would ensue. Prior to Enron, little attention was paid to the scale of audit and non-audit fees and, in particular, the multiple of non-audit fees to audit fees. The audit provided the best platform for the audit firm to really get under the skin of a company's business and this platform typically gave the audit firm a competitive advantage to provide non-audit services (such as taxation, consulting, etc.). So much so that it was not uncommon for some of the (then) Big 5 to generate over 80 per cent of their total fees (audit and non-audit) from their audit client base.

Unfortunately (or certainly so for what became the Big 4), immediately following Enron their clients switched off the non-audit fees tap, and this heralded the introduction of the Sarbanes–Oxley regulation that imposed stricter audit fee and non-audit fee governance. Politicians did what politicians do, and assumed the Enron behaviour was endemic and therefore built an expensive and cumbersome regulatory framework to safeguard the independence of auditor firms with their clients. How could an auditor be truly independent if his or her firm was receiving non-audit fees that could be jeopardised by the auditor's opinion on the financial statement?

In the UK, this resulted in overnight losses in excess of 30 per cent for each of the surviving Big 4 firms. On the positive side, however, Enron opened the door to the non-audit accounts. In theory this presented a bigger opportunity to deploy each firm's non-audit services to a previously untapped market; that is the other Big 4 firms' clients. The only impediment to the successful implementation of this theory was the skill base housed within each of the firms. Partners and professional staff had in the main been accustomed to dealing with a relatively easy 'sell' into existing audit clients. The scary world of developing new relationships with hitherto untouched Boards of Directors was quite a challenge.

Each of the firms 'right sized' their practices following Enron (particularly at the partner level) and embarked on Client Relationship Development Programmes, investing in Customer Relationship Management Programmes and sales training. Whilst each of the Big 4 would claim credit that it was enlightened leadership that led to the successful replacement (and more) of its non-audit fee income surrendered by its audit clients, the reality is that 2004 marked the start of the biggest M&A boom the world has ever seen. The events surrounding Enron had produced a little short term pain, but unknowingly at the time, it prepared each of the firms for significant midterm gains.

In the history of the Big 4 and the accounting profession, whilst Enron did not feel like a 'blip' at the time, it served to further strengthen the Big 4. Pre Enron, the then Big 5 were very strong audit firms developing a broad range of non-audit services, but typically more centred around their audit base. Post Enron, very few large audit client accounts changed hands (except obviously for the Arthur Anderson clients that were hotly contested by the remaining Big 4), and each of the firms made great strides in developing its 'sphere of influence' beyond the traditional audit client base. And the ultimate irony in this event in accounting history probably lies with Mr Arthur Anderson himself. He built his firm on the ruthless pursuit of honesty, integrity and an underlying commitment to the continuous improvements of accounting standards. Anderson's motto was 'think straight, talk straight'. Many lessons were learnt from the tragic loss of a great global firm, and in a quiet corner, be sure that the surviving Big 4 would concur 'but for the grace of God!'

Lehman's

The second and most recent macro event impacting on the Big 4 occurred in the autumn of 2008. The bountiful supply of cheap and apparently risk free credit abruptly dried up and the world economy entered what then turned out to be the biggest ever financial crisis that spawned the deepest recession since the great depression of the 1930s. The Big 4, unlike most of the banks and Governments they served, entered the start of this recession in reasonably good shape, buoyed by their good run in the biggest bull market boom the world has ever seen.

The Lehman's 'blip', however, was very different from Enron. The misdemeanours of Enron were laid by all (politicians, commentators, businesses) fairly and squarely at the door of the accounting profession. By way of contrast, the Lehman's crisis (and what followed) was not down to the accountants. Indeed many within the Big 4 commented that the 2008 financial crisis was a good crisis for accountants!

It is likely, however, that Lehman's will prove to be the single biggest defining moment for the Big 4. None of the Big 4 has, post Lehman's, faced an immediate burning platform that threatened its very survival. Contrast this with Enron, where overnight there was a 30 per cent reduction in fee income. Each of the firms has since 2008 implemented rightsizing programmes for partners and staff to reflect the reduced demand, particularly for advisory services. This in itself is not unusual and indeed the Big 4 and their forebears have successfully used recessions in the past to spring clean their ranks and to offload the dead wood that was hanging around the firm. This has presented an opportunity to remove professionals that are not 'fit for purpose' and restock the shelves with new stock or professional talent, ready for the next upward phase in the business cycle. The

66   *Handbook of service business*

danger for the Big 4 nearly five years on from Lehman's is that, with no sign of an economic upturn and continuing low levels of corporate activity, the rate of commoditisation across almost all of their services is accelerating apace and it is not clear what range of talent or expertise should now be deployed to restock their shelves in the face of this market challenge.

THE DIFFERENTIATION CONUNDRUM

The challenge facing the Big 4 is perhaps best depicted using the classic bell curve. Pre Lehman's, the picture was very much as shown in Figure 5.3. Because there was in the main plentiful demand, the Big 4 could occupy a middle ground where they could provide both low and high margin services to clients. Low margin services would typically include compliance related activities (e.g. tax returns, audit work, etc.) and high margin activities would include M&A related services (corporate finance, financial due diligence, etc.). The brand strength of the Big 4 was such that, unlike almost any other business, the market accepted the Big 4 occupying a middle space and able to play at either end (i.e. effectively commodity pricing at one extreme and premium pricing for value based services at the other extreme). The trading environment for the Big 4 changed rapidly post Lehman's and each firm saw the demand side contract significantly. What then happened was that the bell curve effectively inverted, as shown in Figure 5.4.

The Big 4's success has always been modelled on their ability to be close to the markets the firms served and thereby grow and learn with their clients. The rapid contraction in demand post Lehman's brought an added dimension where the market quickly differentiated low margin and high margin activities and the brands that provided those products and services. If you were a firm, such as the Big 4, that had happily been positioned in the middle ground in the bell curve, that middle ground had now been swiftly removed.

How has this translated in practice? Finance Directors only need whisper the words 'audit tender' and this produces typically a 20 per cent reduction in fees. Whilst previously, the Big 4 were accepted in the middle ground playing at both ends, the market hardened and swiftly sought to differentiate commodity and value providers. The once profitable services related to M&A became commoditised.[2] Loyalty from clients expe-

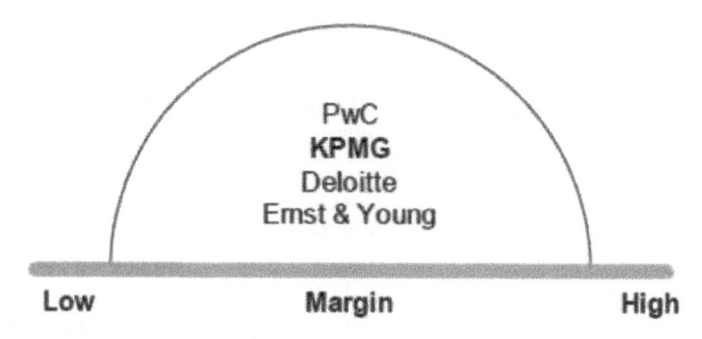

Figure 5.3   The Big 4 margin profile pre-Lehman's

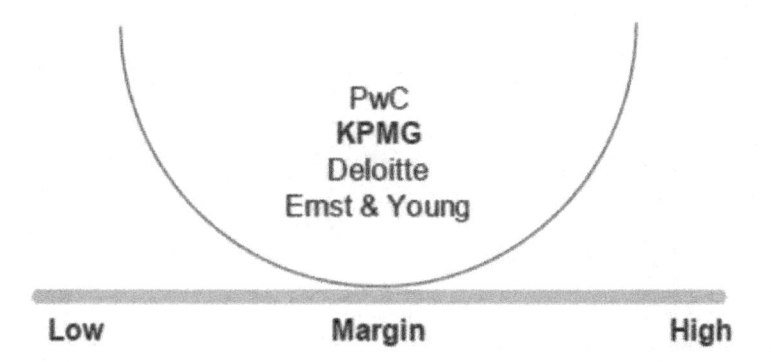

```
         PwC
         KPMG
         Deloitte
         Ernst & Young

  Low         Margin          High
```

Figure 5.4 The Big 4 margin profile post Lehmans

rienced in the good times became a rare phenomenon. Profit per partner has been held steady by reducing the number of partners. The post Lehman's era is proving to be a real challenge for the Big 4. So what happens next?

THE ELEPHANTS IN THE ROOM

The Big 4 are no different from any other professional service firm, whatever its size, in one key respect. The vast majority of professionals in each of the Big 4 are far more comfortable ploughing their particular field of expertise than they are developing a broad perspective on a client's business and the issues that business faces. Professionals all too often close their minds to anything which is not in their area of expertise and within their comfort zone. Each of the firms will run training programmes encouraging their professionals to:

- think like their clients;
- see things from their clients' perspective;
- put themselves in their clients' shoes.

In practice most of this well intentioned personal development falls on stony ground. This observation is not intended to be critical. In many ways, it is wholly understandable. Human nature is such that, if you have worked hard to develop a depth of knowledge in a particular specialist area, you will rightly be most comfortable demonstrating your expertise in that area.

Now for the shock and what every professional in the Big 4 hates to hear – clients take the firm's reservoir of expertise as a given. One of the main reasons the pricing model has been shot post Lehman's is the reality to clients buying services from the Big 4: they all look the same. This is the first elephant in the room: the brand 'Big 4' probably has more value than the constituent parts.

Shortly after the Enron crisis, one of the Big 4 commissioned a piece of in-house research to explore the difference in importance attached to technical expertise and relationship development, first asking their service professionals (partners, directors and

68   *Handbook of service business*

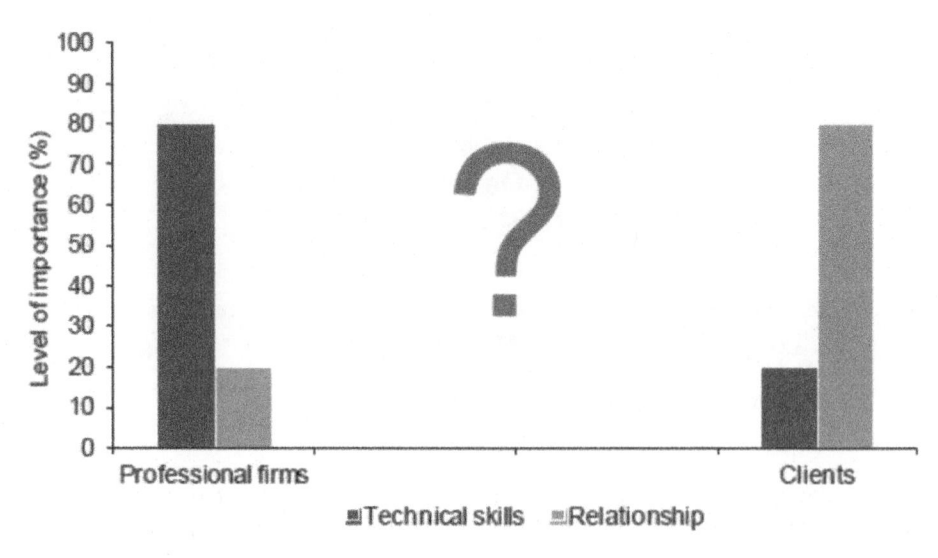

Source: KPMG Markets Research 2003.

Figure 5.5   What do clients want?

senior managers) and then asking the clients they served. The results were startling and are set out in Figure 5.5. Professionals weighted broadly 80 per cent importance to their technical knowledge and skills whilst clients only weighted this on average at around 20 per cent. Professionals found it hard to believe that clients would want to see them face to face to develop a relationship and yet this is exactly what clients were looking for. This elephant may have been thwarted had the world economy not seen the pre Lehman's boom, thereby forcing the Big 4 (and many other professional services firms) to focus minds on finding and developing new relationships. The more pressing issue for the Big 4 (and frankly any professional services firm) faced with the 80:20 technical skills/relationship divide is that any other business would recalibrate the investment/ development budgets to address (in this case) the relationship skills chasm. Again for very understandable reasons, the leadership in Big 4 firms (at all levels, from Global Boards to Office Management teams) is drawn from talented professionals who typically made their name through their particular specialist area. This is their comfort zone but unfortunately (in the absence of a new global boom in international activity) this will serve to see the gradual decline in their firms' fortunes.

The other big elephant, which is not unrelated to its cousin above, is rampant introspectionism. This is always prevalent in a down cycle as professional services firms run for safe ground, but this again simply fuels an even faster rate of commoditisation. This introspective behaviour can be seen at so many different levels, from the shop floor to the Boardroom:

- In the Boardroom, the Profit and Loss (P&L) account is a key driver of behaviour in any organisation and no more so than for large firms of accountants who have a passion for their numbers. As the P&L accounts for the Big 4 are in the main at the

country level, this can present inter-country (cost sharing, fee sharing, etc.) issues in working together.
- On the shop floor, a professional who has developed a good working relationship with a particular client (and with this a steady flow of fees) is most reluctant to introduce a colleague from a different specialist area lest he or she 'messes up' (client sharing).

Professionals naturally search out safety and this introspective behaviour is again wholly understandable. The challenge the Big 4 and professional service firms face, however, is how to break the mould of old and reinvigorate behaviours to better serve their firm's future. This introspective behaviour is simply making it easier for the clients to pick out the services (and professionals) they want and drive the pricing model for those services (down not up).

SO WHAT IS NEXT?

This chapter is not intended to be a treatise on the future of the Big 4 and the wider professional service firms market. The reality is that, for the Big 4 in particular, through their size and global reach, everything else being equal, they should outperform the market. There is little or no threat of new entrants that will challenge their global dominance (much to the dismay of regulators who quietly regret the loss of Anderson and the added perceived additional competition that came with five firms rather than four). Regulators would not accede to any further mergers between the Big 4 (the last attempt was in 1998 between KPMG and Ernst & Young, which was a defensive move against the subsequent merger of Price Waterhouse and Coopers & Lybrand). Despite public perception to the contrary, there is and will continue to be intense competition between the Big 4 and other, albeit in the main smaller, service providers. It is this competition that has underpinned the success each of the Big 4 has enjoyed to date.

So what is next? The reality for most professional service firms (including the Big 4) is that most recessions are relatively short and are followed by a much longer period of economic upside. Over the period covered by the high and low in any cycle, there is typically more gain than pain. Indeed, prior to the most recent global recession, for the Big 4 even the down cycles were good for business, as professionals were 'resprayed' or reallocated to work on corporate/government salvage operations through their variously named corporate restructuring divisions. Since 2008, with the exception of a small number of global casualties, including Lehman's, the demand for the Big 4's restructuring services has been much reduced as compared with the much milder recessions that have occurred in the recent past.

This in itself is symptomatic of what is happening across the Big 4's services. In the case of corporate restructuring, the Big 4's previous dominant position has been surrendered to a combination of smaller specialist firms (which in the main have either been formed by or are made up of professionals from the Big 4) and more sophisticated buying patterns from banks and institutions typically using more internal resources and limiting the input from the Big 4.

As the finishing touches were being made to this chapter, we are approaching the fifth anniversary of the Lehman's collapse. Sovereign debt is going into unchartered waters and everything looks set for a long period of economic consolidation which to professionals who experienced the thrills and spills of the market pre-Lehman's all look, well frankly, boring! The combination of significant growth in the number of professionals pre-Lehman's and a sustained period of calm waters as the world economy shrugs off the worst effects of the most recent recession is not a good mix for professional service firms and the Big 4 alike. There is relatively little scope for growth from the market. Any growth is therefore going to increasingly depend on being able to steal increasingly bigger pieces of your competitor's cake. In the case of the Big 4's competitors and for other professional services firms, this will mark a round of defensive mergers and firms collapsing (which is certainly the case for example for legal firms in most parts of the globe).

Coming back to the question, so what is next? A view on what the future actually holds for the Big 4 is deferred until the end of this chapter. Before getting there, it is hopefully informative to look at the strategies deployed by the Big 4 to try to address the relationship skills chasm referred to earlier (see Figure 5.5). The reality is that the Big 4 and most professional service firms will not lose market share because they do not have the skills to do the job. The skills are there in abundance.

THE OPTIONS

First-hand experience of the Big 4 is that too many clever people try to do too many well intentioned (but clever) things at the same time, with the result that the results are inevitably sub-optimal. This is often referred to as the 'tyranny of partnership' and it is this partnership ecosystem that frustrates corporate leaders. A leading and highly respected corporate baron once commented that in his global empire, if he told everyone in his employ to stand on a sixpence (this was a small British predecimalisation coin), the only question would be 'when?' Contrast this with a partnership culture where in addition to the question 'when' there would be:

- Why?
- What is the significance of a sixpence?
- How can you physically stand on a small coin?
- Isn't there an alternative?
- Etc., etc., etc.

Whilst for those reading this who can empathise with the partnership's *modus operandi*, this may seem mildly amusing, this is probably at the heart of the change needed for any professional services firm to prosper in the future. The reality is that the options (which follow) are probably less important than the ability to implement and align professionals to get behind a shared strategy. As a Big 4 global senior partner famously said to a gathering of country senior partners – 'if you are not on the bus, get off and get out'. Unfortunately, at the time of making this 'command' the economies of the member firms were enjoying good growth and the country senior partners did not take kindly to their global leader's rallying call.

The creation of an operating model capable of implementing the options that follow requires a hard look at a firm's values, appraisal and counselling, key performance indicators (KPIs), remuneration model, contracts with senior staff/partners, and so on. The operating model will reflect the issues that need to be addressed. The tough call for most professional services firms is that the operating model will need to be applied without exception and the tyranny of partnership will need to be expunged. This means tackling those sacred cows and, in no small measure, some of the firm's sacred partners!

All too often (in fact in just about every instance), professional service firms will agree to options that will drive positive change but then try and graft these on to an operating model that is not fit for purpose. This health warning is duly served! And on to the options. There are principally two key levers for change in any professional services firm:

1. The firm's products and services.
2. Route to market and service delivery.

Products and Services

Increasingly, clients will stipulate what they expect in the products and services they buy. Add to this a competitor's ability to copy and replicate new delivery techniques and good ideas and you are witnessing a high quality race to the bottom. An interesting case study is to look at how the audit market has twisted and turned over the last ten years or so to the extent that it is now a high quality commodity.

Most companies of any size in all corners of the planet are required by statute to have an audit. This is effectively an annual health check by an appropriately accredited accountant to report his or her findings to the investment community. The global audit market is somewhere in the order of $20 billion and individual contracts for transnational corporations can easily exceed $100 million. There is a lot to play for.

As we have already seen, prior to the collapse of Enron (and particularly the regulation that followed to restrict the provision of non-audit services), the audit service was the platform used to sell a range of non-audit services to a client. The audit product itself was keenly priced and it was not uncommon to price audits on the basis of an undertaking being given by the client to buy a minimum amount of other high margin services (typically tax and financial due diligence services). Without saying it, the accounting professionals basically accepted that the audit was a commodity and innovation was driven by the ability to sell more of the firm's so-called specialist services. This worked well in the Big 5 firms in the pre-Enron years for a number of reasons:

- Most firms generated their total fees from their audit clients.
- The role of the audit partner was pivotal to the firm's success – they held the key relationships in the client and were instrumental in positioning non-audit services.
- The client's relationships with their professional advisers (accountants, lawyers, PR, consultants, etc.) were seen as 'special' and were free from the grip of procurement. This made pricing discussions a much more personal affair.
- Changing auditor is very similar to changing your bank. Most clients would rather not do it.

This may not read as a particularly stimulating example of product innovation for the audit service. For what were the Big 5 prior to Enron's collapse, however, the audit was key to the development of the firm's other services. The big firms developed corporate finance teams to compete with the investment banks, consulting teams to compete with the strategy houses, pension teams to compete with actuarial firms, and so on. There were no non-audit professional service firms that were safe from the Big 5's onslaught to steal their cake. And in the race to build the Big 5's non-audit services, they had a unique competitive advantage. Whilst the audit was a stand-alone service offering, every company needs one and for those delivering the service, they get regular contact with the client's most senior management. The Big 5 didn't need marketing departments – their cost of sale was wrapped up in the audit service line. This is a great example of product innovation at its best, not in the core service itself (i.e. the audit), but how that core service can be used to spin new services.

To bring the Big 5 audit analysis up to date post Enron (when the Big 5 became the Big 4), regulators across the world basically redesigned the remaining Big 4's business model. The audit service had to stand on its own two feet. The independence of the audit partner became paramount. The audit partners and staff were increasingly restricted in their ability to be the conduit to introduce non-audit colleagues. Furthermore, the audit practices became increasingly isolated to the extent that the auditors and services were being rewarded from the fruits of their audit labours with a reduced ability to enjoy any form of cross subsidy from the more profitable non-audit services.

What followed was a period of intense process re-engineering for the delivery of audit services. This was the only option available in a market that had accepted the audit as a commodity for many years. Unfortunately for the audit, there is a world shortage of examples where commoditised products and services have successfully become de-commoditised. Recent years have seen the greater use of technology replacing its human masters and the audit profession being forced down an increasingly narrow field of expertise which is becoming less attractive to new recruits. Add to this that the audit service line was used to over-recruit new talent to enable it to develop resource with core skills that would transfer to fuel the growth in non-audit services, and you can see that the old, very prosperous business model is broken.

It is difficult to see what the future holds for the audit profession and would-be auditors out there. The reality, however, is that what the accounting profession has seen in the redesign of a core service is being replicated across all professional service firms. In the case of the Big 4, the new services that were so successfully spun from the audit base are now, just like their audit heritage, in the grip of commoditisation.

Route to Market and Service Delivery: The Service Delivery Model

If one accepts that arresting the ravages of commoditisation by focussing on a firm's products and services is doomed to fail (particularly in flat markets), the only other lever is to focus on the firm's route to market and how those services are delivered. This is at the heart of brand differentiation for any service based business. A cup of coffee is a cup of coffee, but why do hard pressed consumers flock to certain well known coffee shops? It is no different for the Big 4 and professional service firms generally.

The starting point for any firm that is serious about driving change in its approach to the market and its clients is to understand the firm's existing service delivery model.

For a client account of any size, there will be a team of professionals drawn from a cross section of different service lines whose job it is to work together to sell as many services as possible into a client. The client service team will have a team captain who will typically be a partner from one of the service lines (let's call this talented professional the 'Lead Partner'). As with any team, the captain (or in this case Lead Partner) has a pivotal role.

Heidi Gardner, who was at the time a Reader at the London Business School, conducted a research project looking at the team dynamics within a selection of 15 different client service teams at KPMG. The findings were somewhat surprising for the Big 4 firm concerned.

First, for the 15 different client service teams that Heidi studied, she found the Lead Partners had 15 different ways of organising the team's activities, developing the account strategy, communicating within the team, and so on. In all cases, the 15 Lead Partners were amongst the most successful partners in the firm, but there was no consistency in their approach to servicing the accounts for which they were responsible. In a global accounting firm that prides itself on the consistent delivery of high quality services across the globe, this was indeed a shock. There would be chaos if every auditor, for example, was allowed to carry out their audit using their own model and not the firm's.

The Heidi Gardner research prompted KPMG to examine the best parts of the 15 different client service operating models she had uncovered and use these to piece together a framework of best practice. The intention was that this client services framework could then be used by all Lead Partners in the hope that the consistent application of best practice would result in above average growth for the firm in the market.

The challenge was two-fold. First, how could hundreds of pages of research and feedback be condensed onto one page? Second, and equally vexing, how could the firm's professionals be aligned to buy into a common platform? The tyranny of partnership is a very strong force and professionals instinctively know what is best and are not willing to accept someone else's view as better than theirs – even an academic from a leading global business school.

In driving any change in a partnership environment, it is important that this change is driven from within the ranks. Leadership and management are always made to feel important, but the reality in most professional service firms is that they are there to keep the score. The obvious cohort to develop the client service framework was the 15 Lead Partners that had been at the hub of the research.

The framework itself is set out in Figure 5.6. In its simplest form, the Lead Partners identified two key areas that were present in the most successful client service teams:

- There are 'things' that simply need to be in place. Referred to in the client service framework as 'components', the Lead Partner team condensed these components into three keys areas:

    1. Understanding the client – the depth of understanding of the client's business and the quality of the relationships with the key players.
    2. Positioning – the client service team is matched to the client team; the client has a good understanding of the client service team's capability. There is a true 'partnership' between the client and the firm.
    3. Strategy – there is a plan with an account plan that drives accountability.

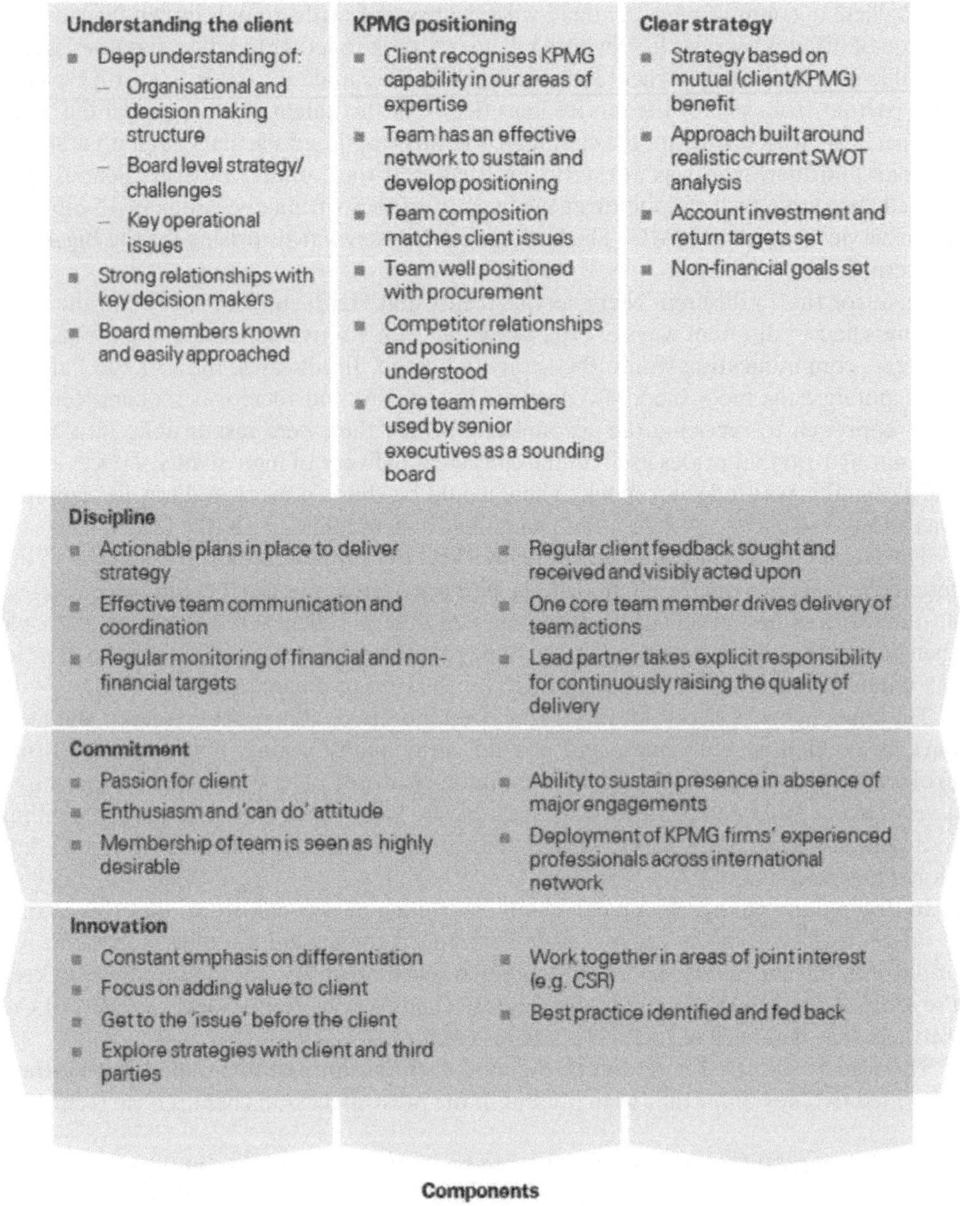

74 *Handbook of service business*

Understanding the client
- Deep understanding of:
  - Organisational and decision making structure
  - Board level strategy/challenges
  - Key operational issues
- Strong relationships with key decision makers
- Board members known and easily approached

KPMG positioning
- Client recognises KPMG capability in our areas of expertise
- Team has an effective network to sustain and develop positioning
- Team composition matches client issues
- Team well positioned with procurement
- Competitor relationships and positioning understood
- Core team members used by senior executives as a sounding board

Clear strategy
- Strategy based on mutual (client/KPMG) benefit
- Approach built around realistic current SWOT analysis
- Account investment and return targets set
- Non-financial goals set

Discipline
- Actionable plans in place to deliver strategy
- Effective team communication and coordination
- Regular monitoring of financial and non-financial targets
- Regular client feedback sought and received and visibly acted upon
- One core team member drives delivery of team actions
- Lead partner takes explicit responsibility for continuously raising the quality of delivery

Commitment
- Passion for client
- Enthusiasm and 'can do' attitude
- Membership of team is seen as highly desirable
- Ability to sustain presence in absence of major engagements
- Deployment of KPMG firms' experienced professionals across international network

Innovation
- Constant emphasis on differentiation
- Focus on adding value to client
- Get to the 'issue' before the client
- Explore strategies with client and third parties
- Work together in areas of joint interest (e.g. CSR)
- Best practice identified and fed back

Components

Note: CSR: Corporate Social Responsibility; SWOT: Strengths, Weaknesses, Opportunities and Threats.

Figure 5.6 The client service framework

- There is a way of doing things and behaviours that bring out the best in the client service team. The framework refers to these as 'mindsets' and again under three headings:

    1. Discipline – in a busy schedule, there are many competing demands on a professional's time and the natural tendency is to gravitate to the safe zone of just doing the job. Effective communication between the client service team members and with the client is key to putting the team first.
    2. Commitment – one thing each of the 15 Lead Partners had in common was a passion and enthusiasm for their client. This passion and enthusiasm was infectious and the Lead Partner's team put their modesty to one side and observed that there was no shortage of professionals who wanted to join them.
    3. Innovation – this was the most difficult area, but probably held the key to maintaining margins and building a pre-eminent relationship with the client. In the war against commoditisation, delivering the same service in the same way plays into the client's hands to reduce fees and commoditise. Getting different parts of the firm to work together and find new ways of improving existing services is key.

The client services framework is therefore very simple in concept but, as the Lead Partner team found, very difficult to institutionalise across the firm. The breakthrough came when the firm recognised that the KPIs and measures were not driving the behaviour that was needed to support the framework. Most professional services firms use measures that play to their core competence around high quality production. Measures such as:

- chargeable hours;
- utilisation;
- fee recovery;
- lock-up (the amount of time not billed);
- unpaid fees.

These measures have a common thread. They are production based and measure what has happened. For a firm that is keen to measure the rate of its own commoditisation, these are good measures. The focus therefore turned to designing measures that would drive the required change in behaviour. The team came up with new measures to support the client service framework:

1. Relationships – the team designed a scale to measure the quality of the relationship with their clients and targets.
2. Penetration – how many of the firm's different service lines had been engaged by the client.
3. Client feedback – again, using a numerically based scale across six different areas, the firm built a qualitative client feedback score card.
4. Proposal hit rate – probably one of the best measures to help judge how a firm's

76   *Handbook of service business*

> **1** Strong: Know client's personal agenda. You are a trusted adviser.
>
> **2** Comfortable: Review client's business agenda. You are used to providing a service.
>
> **3** Weak: Client knows you but does not use you.
>
> **4** Poor: Client does not know you.

Figure 5.7   The relationships matrix

offering stands in the market is by keeping an accurate score on the wins and losses in head to head competitions to win new work.

On the face of it, these measures look very sensible and in each case an improvement in some or all of these measures on a client account would result in an improvement in the traditional production measures. The early experience was however very different. To a proud professional these market facing measures are scary and potentially a source of some threat to their air of intellectual excellence. For example, in the case of measuring the quality of relationships, a simple four part matrix was introduced, as shown in Figure 5.7. The surprise was that most professionals rated the relationships with their clients as either strong or comfortable, and yet in the feedback from their client the rating was invariably lower.

In a professional's development from a raw graduate to being the King of the Jungle in his or her specialist field, the time devoted to understanding the social science part of their job could typically be measured in minutes. This mismatch in the measure of relationships was therefore understandable and again was consistent with the 80:20 research findings (technical skills: relationships) referred to earlier (see Figure 5.5). In the case of accountants, this challenge is heightened by their need for objectivity (this is rooted in the accountant's love for numbers). Fortunately, the now widely used trust equation which is reproduced in Figure 5.8 came to the rescue. Simply ascribing a measure on a ten part scale (10 = Outstanding and 1 = Very Poor) to the numerator components

$$T = \frac{C + R + I}{SI}$$

Where: T = Trust
C = Credibility
R = Reliability
I = Intimacy
SI = Self interest

Figure 5.8   The trust equation

(C, R and I), and for SI (self interest) a score of 1 for Outstanding Selflessness and 10 for Outright Selfish Behaviour, gave a framework to build greater integrity into the relationships measure. Scores of 10+ were rated as Strong, 8+ Comfortable, and so on.

There is nothing new in this trust equation and many academics and sales trainers have built their careers going through the practical application of this simple formula. What is new and different, however, is to take a professional service firm that has been built on the quality of its production and face up to the simple fact in a fast moving market that this technical excellence is a given and the relationship is key. This does not mean abandoning the technical training and development programmes. These are at the core of any firm's value proposition, but these alone are not enough.

THE QUEST FOR THE PERFECT RELATIONSHIP

As we have seen, the single biggest challenge that professional service firms face is being able to differentiate their products and services. But what exactly are the characteristics of a relationship that will drive this differentiation?

Research on this very subject by John DeVincentis that was later adapted by the global sales guru Professor Neil Rackham[3] proved to be most informative (Figure 5.9). Working with Neil Rackham and drawing on further research from the Integrated

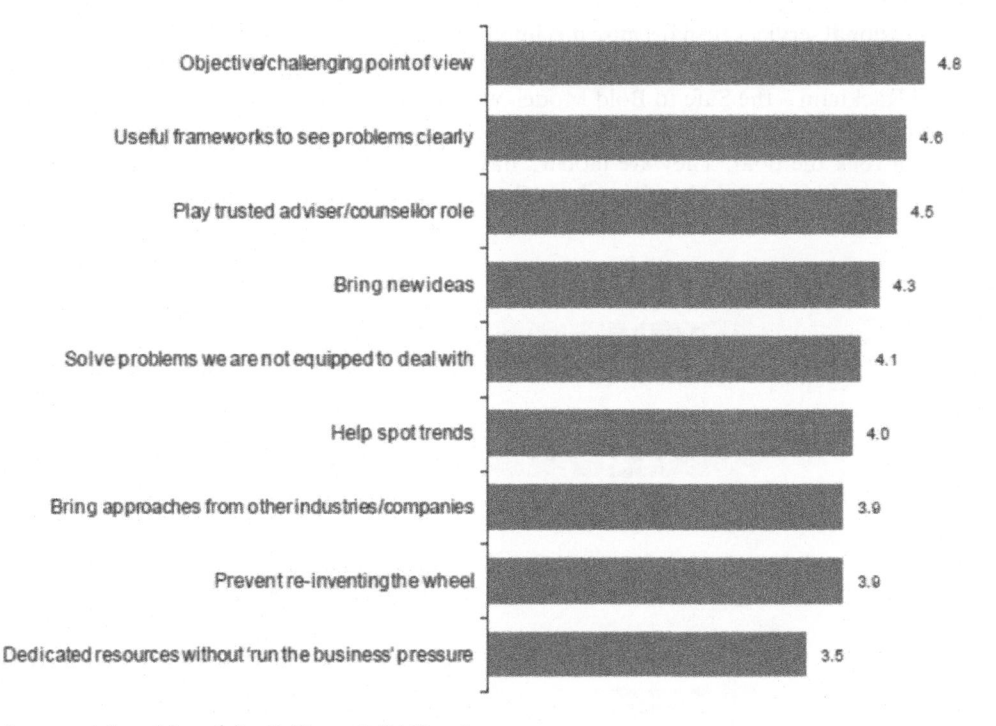

Characteristic	Score
Objective/challenging point of view	4.8
Useful frameworks to see problems clearly	4.6
Play trusted adviser/counsellor role	4.5
Bring new ideas	4.3
Solve problems we are not equipped to deal with	4.1
Help spot trends	4.0
Bring approaches from other industries/companies	3.9
Prevent re-inventing the wheel	3.9
Dedicated resources without 'run the business' pressure	3.5

Source: Adapted from John DeVincentis BAH study.

Figure 5.9   How do advisers create value?

78   *Handbook of service business*

Sales Executive Council, there emerged a model comparing 'high performers' and 'core performers' who had demonstrated good client relationship handling skills. The breakthrough was the recognition that there were two distinct categories of relationship skills:

- Relationship Builders – these are professionals who are good team players, help others and act very much as 'go to' members of the team, linking the client with the right source of expertise in the firm.
- Challengers – these are professionals who again have very good relationship handling skills, but they use this relationship to debate issues, push clients and challenge their thinking.

The power of understanding the differences between Relationship Builders and Challengers was realised when looking at the relative performance in low complexity and high complexity scenarios. Reverting back to the post Lehman's bell curve (Figure 5.4), the clear realisation is that to capture the high margin ground to the right hand side needed a new breed of professionals – Challengers.

FROM SAFE TO BOLD

The business need is for challengers, but most professionals are attracted to a career in a professional services firm because it is intellectually stimulating, interesting and above all relatively safe. The task is to be bold, and what follows is a simple model developed with Neil Rackham – the Safe to Bold Model, which is set out in Figure 5.10.

The starting point is to recognise that professionals are developed to be responsive and to not rock the boat. They are modern day servants to their demanding clients. At the core of the Safe to Bold Model is the realisation that to challenge clients, business service

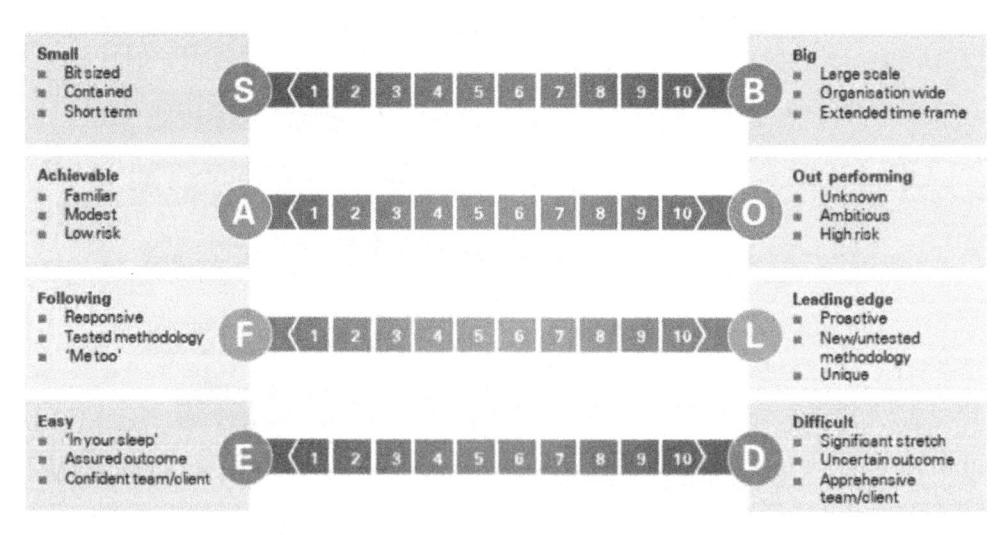

Figure 5.10   Safe or bold?

professionals must provide inputs that will need to be big, have an element of risk, be innovative and difficult.

In practice, the Safe to Bold Model was used by getting two different client service teams to critique each other and in particular the account plans, strategy and ideas that each team had for its account. Using the client service framework to improve the quality of client service (Figure 5.6), the trust equation to assess and address relationship shortfalls (Figure 5.8) and the new KPIs, client service teams were empowered to challenge. And what do the clients see – a more confident client team, hungry to help. This is most definitely a differentiation.

CONCLUSION: SO WHAT IS NEXT?

The world of professional service firms is changing, has changed and will continue to change. The only asset professional service firms have is people. These people are servants to the needs of their clients and inevitably, notwithstanding the excellence in a professional service firm's practice strategy, it is the market and the clients in that market who will have the biggest role in shaping the future of professional service firms.

The reality is that the Safe to Bold Model briefly outlined above has created competitive advantage for those client service teams brave enough to shrug off old habits and open their minds to new ways. As the market for services gets tougher, inevitably Safe to Bold will become 'business as usual' and so the next innovation will emerge.

It is anyone's guess what is going to happen to the global economy. If, however, one works on the basis that there is unlikely to be a speedy return to the days of a debt fuelled pre-Lehman's boom, the consequence is that the market will not ride to rescue professional service firms' good fortunes.

Hopefully the case is clear that if professional service firms compete on product improvement alone, this marks a race to the bottom. Like any other business, success will be driven by a professional service firm's ability to innovate, challenge the norm, re-engineer and, of course, to be BOLD!

NOTES

1. This chapter has been written by a practitioner rather than an academic. It provides an analysis of firm strategy in response to commoditisation.
2. Commoditisation occurs when differentiation by quality, innovation, reputation and other forms of distinctiveness in a market place is replaced with a focus on price. Thus, goods and services that have been commoditised are purchased primarily on price. This means profit margins are reduced as competition is increasingly based on price rather than other qualities of the product or service. Commoditisation may occur as a product or service matures and other companies enter the market to compete on price. The process of commoditisation is not to be confused with the conversion of a good or service in to a commodity that can be sold to realize exchange- and use-values. The process of commoditisation refers to existing commodities that are now predominately purchased on the basis of price rather than non-price factors (*The Editors*).
3. N. Rackham (1995), *SPIN-selling*, Gower Publishing: Farnham; J. DeVincentis and N. Rackham (1999), *Rethinking the Sales Force: Redefining Selling to Create and Capture Customer*, McGraw-Hill: New York.

PART II

SERVICES AND CORE BUSINESS PROCESSES

6. Green and sustainable innovation in a service economy
Faridah Djellal and Faïz Gallouj

INTRODUCTION

Contemporary developed economies can be described according to three basic perspectives that are extremely important research objects and economic policy issues. They are *service* economies, *innovation* economies, and they aspire to be *sustainable development* economies.

First, whether we like it or not, modern economies are undeniably service economies. The tertiary sector is the main source of wealth and job creation in all developed countries, and emerging economies provide another example of what can be called the Fisher–Clark–Fourastié law, reflecting a sectoral shift of the workforce from the primary to the secondary and then to the tertiary sector. Second, today's economies are innovation economies, focused on permanent innovation, quality and knowledge. The terms 'new economy' or 'net economy' are also frequently used. Unlike the first category, this second facet of contemporary economies has mostly positive connotations (at least in economic theory[1]). Innovation and knowledge are considered powerful drivers of socio-economic progress. While this argument is not new, the magnitude and rapidity of innovation and cognitive dynamics are greater than ever. Finally, today's economies are, or aspire to be, sustainable development economies, and green economies in particular. Thus, environmental issues are no longer considered only militant and utopian concepts but are now a major part of socio-economic and political discourses.

Considered separately, each of these perspectives has been the subject of extensive research, which we need not review here. Yet, the strength of these individual research trajectories contrasts with the weak interfaces between them when they are examined in pairs. For example, while many of today's economies are both service and innovation economies, paradoxically, despite a growing awareness, they are not sufficiently considered as economies of innovation *in* services, that is to say, as economies in which the innovation efforts of service firms are proportional to their contribution to economic aggregates. It is as if services and innovation are two parallel universes, insufficiently open to each other. Similarly, in a post-industrial service economy, the concept of sustainable development was essentially built in response to environmental damage and the subsequent socio-economic damage associated with an intensive agricultural and industrial economy (depletion of non-renewable resources, proliferation of waste, pollution, natural disasters, desertification, deforestation, global warming, social exclusion in rich countries, global financial crisis, and growing inequalities between North and South). Sustainable development continues to convey a strong industrial connotation, even though some services (tourism, transport, etc.) can also harm the environment, and the increased awareness of the social or socio-economic aspect of sustainable development is paving the way for a greater focus on services (Djellal and Gallouj, 2010). Thus, as Jean Gadrey (2010, p. 94) rightly notes, 'services are ignored by political environmentalism, while environmentalism is neglected by the service economy' or 'with a few exceptions, the economics of services, as currently constituted, takes little account of environmental or social considerations'. The impact of services on the environment and social inequalities is rarely a major concern.

The interface between the innovation economy and the sustainable development economy is less tenuous. But it is technologically biased. There is an extensive literature that analyses the link between technological innovation and sustainable development mainly from an environmental perspective (Kemp and Soete, 1990, 1992; Fussler and James, 1996; Rennings, 2000; Marinova et al., 2007). This bias is reinforced by the ambivalent status of technology, seen both as a source of the problem (e.g. pollution source) and as a solution (reparative, de-polluting or decontaminating technology).

The purpose of this chapter is to help to link these three essential facets of the modern economy: services, innovation and sustainable development, with a view to seeking an answer to the following questions: to what extent can innovation in services be considered green in that it addresses environmental concerns and, more generally, to what extent can it be considered sustainable innovation that promotes social and economic inclusion?

To address these questions, we first examine the more general hypothesis that the service economy is greener than the industrial economy. This hypothesis is based on a number of real or imagined characteristics of services (particularly their intangible and interactive nature). We then examine how the different theoretical perspectives used to understand innovation in services can address the issue of environmental and socio-economic sustainability.

IS THE SERVICE ECONOMY 'GREENER' AND MORE SUSTAINABLE THAN THE GOODS-PRODUCING ECONOMY?

The idea that, given some of their intrinsic characteristics, services are 'greener' and more 'sustainable' than industry is common in the literature. This appealing and optimistic assumption is based on the idea that in terms of environmental protection 'the office is better than the factory'.

The main theoretical argument that supports this hypothesis is that services are intangible and goods are tangible. It is the processing of tangible goods that gobbles up natural resources and harms the environment. The 'ecological footprint'[2] of services is generally considered to be lower than that of industry and agriculture (Gadrey, 2004). The environmental argument is also supported and reinforced by a socio-economic argument, within the broader perspective of sustainable development. Thus, the service sector appears to be naturally more 'sustainable' because it is the privileged domain of activities motivated by a civic and social purpose: service activities (public, private or non-profit) aimed at reducing unemployment and promoting personal development and social cohesion and inclusion.

Nonetheless, the assumption that services are green and sustainable is disputed, and the question as to the intangible nature of services is under debate. The ecological footprint of some services, such as transportation, tourism, distribution and advertising, is particularly high. The socio-economic damage caused by the recent financial crisis and the role played by financial services in particular argue against the socio-economic sustainability of services, at least in some forms. The most highly service-oriented contemporary economies are not responsible for less pollution but, on the contrary, pollution increases with the degree of tertiarization in these economies (Gadrey, 2010). The most service-based economies are thus the most polluting.

Though it may be true that services are generally more environmentally friendly than manufacturing, their potential to cause environmental damage is probably greatly underestimated (Fourcroy et al., 2012). The idea that services are green or appreciably greener than industry will not hold true in the future unless major innovations are made, as Gadrey (2010) points out. In what follows we do not claim to provide a definitive answer to the question, 'Is the service economy "greener"?', but we will suggest some lines of inquiry based on the characteristics of services discussed in the literature and on a number of recent evolutions of these characteristics.

The Intangibility of Services: A Myth?

Compared with goods, the main characteristic of services that supposedly makes them environmentally friendly is their intangibility. It is important, first, to understand what this concept means, then to determine if it is a factor conducive to environmental sustainability, and finally to consider whether services really are intangible.

a) The intangibility of services and their environmental sustainability
Adam Smith (1976) is generally credited with authorship of the idea that services are intangible. Focusing mainly on public servants, domestic servants, artists, lawyers and doctors, Smith wrote in *The Wealth of Nations* (p. 352), 'the work of all of them perishes in the very instant of their production'. This supposedly intrinsic technical characteristic of services was (and still is) a core issue in contemporary research on services, whether in economics or management. This means that contrary to a product, which is inseparable from the technical components that make it up, a service is intangible and evanescent. It does not have a 'tangible' form that can be accumulated and circulated economically, independent of its delivery medium (a good, an individual). It is not a given entity; it is a process, an act, a change of state. One cannot 'sell' or 'transfer' such a state. Nor can it be 'repaired', at least in the way that a tangible good can. The Lancasterian analysis in terms of characteristics (discussed below) reflects the intangibility of services (Gallouj and Weinstein, 1997).

The intangibility of services is often associated with the idea that services have a smaller environmental impact than tangible goods. After all, producing goods would appear to consume more energy and natural resources than providing a service. At the macroeconomic level, some research has pointed towards a shift to an increasingly intangible (dematerialized) economy, considered less harmful to the environment (Ettighoffer, 1992; Romm et al., 1999). The appealing idea that the inexorable tertiarization of our economies will naturally lead to greater environmental sustainability is present in the literature, including that emanating from international institutions (OECD, 2000).

Although the connection with tangibility is not always clear, some statistical findings apparently support the idea that services are 'greener', that is to say, more sustainable from an ecological perspective. For example, in 2005 services (excluding transport) represented 9 per cent of total final energy consumption and 12 per cent of CO_2 emissions worldwide (International Energy Agency, 2008). Of course, this average masks significant disparities between countries. If the analysis includes only the top 14 energy-consuming countries[3] representing 85 per cent of global final energy consumption, services account for 16 per cent of final energy use and 17 per cent of total CO_2 emissions (International Energy Agency, 2007). This more satisfactory energy and emissions performance is supposedly due to the fact that services are less likely to use energy-hungry heavy equipment. However, the statistics are in fact misleading because analysts looked only at the emissions associated with heating and lighting buildings and the energy consumed by technical systems. They ignored other types of consumption involved in overlooked tangible aspects, such as the embodied or grey energy corresponding to manufacturing and maintaining the technologies in question.

b) Services are less intangible than we think

In reality, services are more tangible than is generally assumed. This is true both of services that have always been tangible, such as transportation for example, and services that have become more tangible over time given the increasing use of technologies and the pervasiveness of information and communications technologies (ICTs).

Thus, the hypothesis that services are intangible is easily refuted in some cases. In fact, for many services, tangibility is a central characteristic. This is true of services whose principal medium is tangible objects.[4] These services are labeled logistics and material-processing services (Gadrey, 1991; Gallouj, 1999); their core business is to transport, process, repair and make available tangible objects. Such services include the transportation of goods, retailing, food services, supplying water, gas or electricity, rental, and maintenance and repair services. Their relation to tangible objects is obvious and essential while they are also among the most capital intensive and least green services.

In all services (admittedly to varying degrees), some of the tangibility was introduced by technological innovation. This is a consequence of the natural technological trajectory of increasing mechanization as described in the evolutionary theory of technical change (Nelson and Winter, 1982). The technologies in question are not limited to information technologies even though they do occupy a central place in services, where they are described as pervasive. Many other key technologies in services (transport, cooking, refrigeration, heating, biotechnology, etc.) often hybridize with IT. In the case of IT, although software is 'invisible technology', the same is not true for hardware, which is by definition tangible and known to consume large amounts of exhaustible natural resources (rare metals) and energy. These types of equipment are being developed at a rapid pace, with planned obsolescence and extremely short life cycles (Desmarchelier et al., 2011). They also pose serious waste treatment problems. In short, whether in terms of hardware or software, the assumption that ICTs are low MIPS[5] technologies is questionable.

In some companies and sectors services are also becoming industrialized, which can lead to greater tangibility. Such industrialization can take different forms. The first is when services are replaced by industrial goods used at home (self-service within the meaning of Gershuny (1978)): for example, not going to the laundromat but instead using one's own washing machine; not going to the cinema but instead watching a DVD at home. The second is when work processes are standardized, which (in the case of services) is synonymous with, or leads to, a standardization of the service itself. The product in this case is not a good but a quasi-product: for example, a standard insurance contract or financial product; a standard tourist package; a standard menu item that is identical throughout a fast food chain. In the latter case, industrialization means eliminating anything that is not a standard case. Note that the industrialization strategy (particularly in its second form) has sometimes been established as a strategic rule involving the systematic industrialization of services by using industrial methods of production of the kind advocated by Levitt (1972). Likewise, Shostack (1984) sees this industrialization strategy as a solution to the 'divergence' (degree of freedom) and the complexity of service delivery. Like other authors (Kingman-Brundage, 1992; Lovelock, 1992), she recommends the development of delivery models ('flowcharting', 'blueprinting') that are veritable 'manuals' of service delivery.

c) The tangibility/intangibility debate and the convergence between goods and services
As the convergence of goods and services continues apace, the question of intangibility as a characteristic specific to services becomes less relevant. Increasingly, many contemporary authors report a blurring of the boundaries between the sectors that is redefining the nature of 'products' (Vandermerwe and Rada, 1998; Barcet and Bonamy, 1999). There is also a dialectic of industrialization of services and 'servitization' of goods (increased importance of service and the service relationship as a mode of coordination between economic agents in the sectors producing industrial and agricultural goods). Some authors (Broussolle, 2001) also show that, as technical systems shared by industry and services, ICTs contribute to this 'blurring'.

The blurring of boundaries and the shift towards an 'all service' economy are manifest in different ways. For example, information and service are increasingly important components in the value of most industrial and agricultural goods. Whether for potatoes, perfume or calculators, services and information (R&D expenditures, transportation, distribution, marketing, etc.) are key components of the value produced. The widespread trend towards 'services around the product' (pre-sales services, sales services, after-sales services) (Furrer, 1997) also blurs the services/goods distinction. It is also driven by the transformation of iconic industrial corporations (e.g. IBM, Benetton) into service companies, to the extent that now most of their revenue comes from this type of activity. Another example is the new business model of certain industrial enterprises which now lease their goods rather than selling them (Rank Xerox copiers) or refurbish or recycle goods rather than manufacturing them. Fuji Xerox, the manufacturer of copy machines and cartridges, manufactures new machines and cartridges by reusing and recycling parts. This company creates goods by using resources rather than simply manufacturing new products.

Along with this blurring of boundaries (and the growing intangibility of manufacturing), a new message is appearing whereby certain businesses now no longer define themselves as goods manufacturers but as providers of solutions, functions or experiences. Clearly, when products are considered in this manner, the binary opposition between the tangible and the intangible loses much of its relevance.

We conclude that, from a systemic and historical (dynamic) perspective, the argument that services are intangible and therefore environmentally friendly is debatable. Even if one accepts the idea that services are less energy hungry and less polluting than industry, they do use more energy and pollute more than we think. The second essential characteristic of services, namely their interactivity, is itself a significant source of environmental degradation.

Interactivity: A Source of Tangibility

The service economy has tended to consider intangibility and interactivity as two separate intrinsic technical characteristics of services. In fact, in terms of environmental sustainability, these two characteristics are inseparable. Interactivity has a tangible, environmentally damaging dimension, which is expressed at two levels.

The first level (of the tangible dimension of interactivity) is direct. It is also the oldest. Interactivity actually means the co-production of the service by the provider and the consumer. Co-production is often synonymous with a face-to-face encounter, meaning the two parties must travel, which mobilizes all the tangible components involved in getting from one place to another. This may involve the customers/users/consumers going to the provider's location (retailing, food service, hotels, education or health), the service provider going to the customer's location (consulting, door-to-door sales, in-home services) or both parties travelling (passenger transport). Likewise, another factor is the daily commute to and from work. It is therefore important to measure the environmental impact of this tangibility not only by assessing the direct impact (CO_2 emissions, energy consumption) of the tangible goods used in providing the service (individual or collective means of transport), but also the indirect impact in terms of the energy and resources needed to produce the goods (at least for the portion used in services) and to produce the durables (machinery) used to manufacture the goods (Gadrey, 2010; Fourcroy et al., 2012). This is known as 'embodied' or 'grey' energy. The negative environmental impact is a positive function of the geographical distance between actors and the frequency of their interactions. Moreover, co-production and face-to-face encounters generally require premises (agencies, offices, shops, hospitals, universities, etc.) which are clearly tangible and which require material resources to build and maintain.

The second level (of the tangible dimension of interactivity) is indirect. It is related to the development of ICTs. For many services, ICT has been proposed as a remedy for the adverse ecological impact of the tangibility involved in co-production. Online services (home banking, e-commerce, e-government, video conferencing, telecommuting, telemedicine) reduce physical travel and the related impact. While this argument is not totally wrong, the ecological benefit of these technological developments must be looked at from different perspectives. First, that which is gained by reducing the tangibility of travel must be offset against what is lost by increasing the tangibility involved in the introduction of ICTs. Moreover, some forms of online services increase the physical flow of goods (online retailing). More generally, then, ICT has actually caused an increase in material flows and it has been shown that the growth of information flows is positively correlated with the growth of material flows.

Our conclusion is that the perceived environmental impact depends on where the perimeter of the tangibility of services is set. A restrictive view of the perimeter (or system) of services gives positive results but it seems more appropriate (or realistic) to adopt a more expanded perimeter (or system), whether in terms of function, space or time. Thus, the real or (socially) 'constructed' increase in the tangibility of services makes them less environmentally friendly.

The Baumol Model: Services and Pollution

There is little theoretical research focused on the relationship between the growth of services and the issue of pollution. The few exceptions are based on a revisited version of Baumol's (1967) model of unbalanced growth, including a paper by Baumol himself (Baumol, 2010) and a particularly interesting analysis by Gadrey (2010). In the original *unbalanced growth* model (Baumol, 1967) the economy is divided into two sectors: 1) a sector (called stagnant) in which labour productivity is constant, given its low level of technology; and 2) a sector (called progressive) in which labour productivity increases due to the introduction of technology.

According to Baumol, most services belong to the non-progressive or stagnant sector. Few productivity gains can be realized because the possibilities for mechanization are limited given that the final product is often identical to the labour factor itself. Baumol's reasoning is illustrated by particularly expressive and well-known examples, such as how could the productivity of a wind quintet be increased – certainly not by increasing the playing speed of the musicians.

In his 2010 paper, Baumol states that the 'cost disease' (chronic inflation) affecting services is a factor of pollution. After all, in such a context it can be more costly to repair an item than to throw it away and buy a new one. The service society is therefore a 'throw-away society'.

Gadrey (2010) bases his analysis of pollution in a service economy on Baumol's model. He shows, however, that the cost disease can to some extent be halted if we internalize environmental externalities (e.g. by introducing taxes). Indeed, in this case the cost of goods in the progressive sector will increase even if the sector achieves gains in labour productivity. According to Gadrey, it is as if a fixed proportion of stagnant service (a sort of service provided by nature) had been incorporated into the model. The consequence is that 'the relative price of industrial goods compared to stagnant services is no longer bound to tend toward zero' (Gadrey, 2010, p. 106). Gadrey adds that, given the scarcity of natural resources, we should see a gradual increase in the 'price' of nature's services. The conclusion of this analysis is that the share of industry in employment that was bound to decrease in Baumol's original model should increase, and consequently the share of services in employment could diminish in the future.

GREEN AND SUSTAINABLE INNOVATION IN SERVICES

Our goal here is not to review the literature on innovation in services (for recent surveys, see Howells, 2007; Droege et al., 2009; Djellal and Gallouj, 2010; Gallouj, 2010). Rather, we want to examine how the major research on innovation in services conducted during the past two decades explicitly addresses, or can be extrapolated to address, the issue of green innovation, and more generally sustainable innovation. We will also identify gaps in the literature and explore potential avenues of research.

The field of innovation in services is built upon three different theoretical perspectives (assimilation, differentiation and integration[6]), each reflecting different conceptions of the relationship vis-à-vis the dominant field of innovation in manufacturing. *Assimilation* considers innovation in services in the same way as innovation in industry, focusing on the relationships with technical systems. Insofar as it focuses on adopted innovation (from industry), the assimilationist perspective is also a subordinate perspective. *Differentiation* (or *demarcation*) is centred on the specific characteristics of services and can detect innovation activities where the traditional (technologist or assimilationist) perspective cannot. *Integration* uses the same analytical framework to look at goods and services, namely technological innovation and non-technological innovation. As we shall see, these three theoretical perspectives can also be useful for considering the connection between innovation in services and sustainable development.

Assimilation: The Pre-eminence of Technological Trajectories

There is a fair amount of literature that focuses on the connection between technological innovation in industry and sustainable development in environmental terms (Kemp and Soete, 1990, 1992; Fussler and James, 1996; Rennings, 2000; Marinova et al., 2007). The concept of sustainable development is often associated primarily with technological innovations that have a positive or negative impact on the environment. Similarly, as pointed out in many studies (Gallouj and Weinstein, 1997; Sundbo, 1998; Miles, 2002; Gallouj, 2002a; Hipp and Grupp, 2005; Tether, 2005; Windrum and García-Goñi, 2008), it is the technologist or assimilationist perspective that has long prevailed and continues to enjoy a large audience in services innovation studies and in public policies to support innovation in services (Rubalcaba, 2006). It is therefore not surprising that the specific issue of innovation in services and its relationship with sustainable development is no exception to this technologist bias. In other words, in services as elsewhere, technologies are often seen as sources of environmental problems (pollution from transportation), and technological innovations are seen as solutions (electric vehicles, various examples of e-services: telemedicine, home banking, online shopping, e-government, etc.).

The assimilationist perspective on innovation in services is strongly focused on the relationship to technology (the production or use of tangible technologies). But it can also reflect certain organizational and strategic choices (mentioned above) that aim to make services more similar to industry, or at least to a certain industry at a certain point in time (usually Fordist industry of the post-war period). From this perspective, assimilation strategies aim to make services 'goods like any other' by eliminating or attenuating their specific intangible and interactive characteristics. The idea is to make services more tangible, less interactive, less subjective (that is to say, less socially constructed) and more anonymous, so that they differ as little as possible from goods. This 'assimilation' of services, which is synonymous with industrialization, may rely on various well-known mechanisms: standardization, mechanization of production processes or mechanization of the service itself (replacing services with tangible goods). Assimilation or industrialization often equates to the application of the Fordist production model to services. The textbook example of this type of strategy is the fast food industry (and especially McDonald's). But there are many other examples in fields such as low-cost airlines, mass tourism and mass food retailing. The industrialization of service is often seen as having a negative impact on sustainable development. After all, it is a key driver of growth in the types of services that are particularly harmful to the environment or public health (air travel, mass tourism, mass retailing generating urban mobility and congestion, unhealthy fast food).

A consideration of the assimilationist perspective, in its technological aspect, provokes a number of comments. The issue of innovation from the technological perspective may be considered in two ways: production or adoption. From this perspective, services innovation studies that promote an assimilationist perspective are essentially studies of the diffusion of innovation and its adoption by services. This is the case with the reverse cycle model of Barras (1986). Therefore services appear dependent on industry when it comes to innovation, especially green or sustainable innovation. As Pavitt (1984) puts it, services are subordinate to industry, or supplier dominated. Thus, a city that introduces natural gas or electric vehicles for public transport (clean, quiet, low maintenance) is not an innovator per se but simply an adopter of innovation.

However, the view that services are subordinate with respect to goods is losing ground. Some service firms (especially larger ones; for example, in retailing) may be the main drivers behind sustainable or green technological innovation projects. Similarly, some service organizations are able to exert pressure to guide technology towards greener trajectories. In this regard, some service firms (e.g. mass retailing; see Gallouj, 2007) have a decisive influence on the industrial suppliers which are dependent on them, so much so that we should introduce a 'consumer dominated' innovation trajectory as opposed to Pavitt's 'supplier dominated' trajectory.

The technologies used in services are varied, although some theoretical models (Barras's reverse cycle, for example) focus solely on ICTs. Indeed, as we noted earlier, other types of technologies play a fundamental role in services, such as those involved in cooking, refrigeration, transportation, biotechnology, and so on. However, ICTs are undeniably central to the dynamics of innovation in services. The numerous studies featuring the assimilative perspective of innovation have always stressed the pervasive nature of ICTs in services. As we have seen in the first part of this chapter, ICTs are not necessarily green technologies. The processes involved in manufacturing and using ICT equipment consume scarce resources and energy (their MIPS is high), and from this point of view the idea of a dematerialized society is an illusion. However, in services the use of ICT can be a vehicle for sustainability and inclusion. Examples include video conferencing instead of travelling to meetings (business trips); new working arrangements (telecommuting); using ICT for measuring, checking and monitoring certain sustainable development indicators; and social networks that challenge governments and enable citizens to mobilize quickly.

The assimilationist perspective essentially focuses on green innovation, that is to say, technological innovation aimed at solving ecological and environmental problems. However, in a context of broader sustainable development, it is important to take into account not only environmental technologies but also social technologies. In contemporary service firms there is a major technological innovation trajectory focused on solving social problems, be they problems of the disabled, the elderly or people with socio-economic difficulties (Djellal and Gallouj, 2006). These innovations include domestic robots, smart homes, remote monitoring technologies and more generally technologies for the disabled.

Differentiation: An Intangible and Social Innovation Trajectory

In services innovation studies, the differentiation perspective is more recent. There are fewer studies on differentiation than on assimilation but the number is growing. All of the studies denounce the 'myopic' nature of technologist approaches, which underestimate the dynamics of innovation in services by focusing only on technological systems (which are more visible and more spectacular). The research in terms of differentiation seeks, sometimes theoretically, but more often empirically, to identify innovations that are invisible to traditional indicators by conducting qualitative or quantitative surveys (for a review, see Gallouj, 1994; Gallouj and Djellal, 2010). These hidden or invisible innovations are non-technological innovations, which are more difficult to identify and quantify than traditional technological product and process innovations. These include organizational and social innovations but also non-technological product or process innovations. Examples would include a new insurance contract, a new financial product, a new field of expertise in consulting, a new service offered by a government, new tourism, hotel or restaurant concepts, a new consulting methodology, a new care protocol in a hospital, or a new cleaning protocol in a cleaning company.

The research into sustainable or green innovation in services is also eager to identify non-technological forms of innovation (Seyfang and Smith, 2006). Sustainable non-technological innovation in services intersects a vast and prolific field, but one that has so far been little explored by economic theory: the field of social innovation (Djellal and Gallouj, 2012). It is invisible because it is intangible. These innovations are not necessarily spectacular. They are bottom-up or 'grassroots' innovations (as opposed to 'mainstream green business innovations' (Seyfang and Smith, 2006)), developed by individuals or organizations in response to local issues, and are in keeping with the interests and values of the communities concerned. In order to provide examples of sustainable innovations in services from a differentiation perspective, Table 6.1 distinguishes several types of services according to the main medium mobilized (materials, individual, information, knowledge) and sustainability according to environmental and socio-economic issues.

Materials-processing services are those whose main medium is a tangible object, which is transported, transferred or repaired. Examples include the transport of goods, automotive repair, and the supply of water, gas and electricity.[7] The waterless cleaning systems used in some garages, materials recycling, selling the use of a product rather than the product itself (photocopiers, vehicles), providing additional services around the product, and various forms of sharing are examples of (mostly) non-technological green innovations. The innovation has less to do with technical solutions than behaviours within firms and organizations (business models). There are many examples of non-technological innovations that pursue socio-economic goals. These include the multiple forms of fair trade, direct farm outlets, community-supported agriculture programmes and continued supply of water, gas or electricity for low-income populations. Harrisson et al. (2012) provide an interesting illustration of how a utility company (Hydro-Quebec, a Canadian electric company) and a consortium of consumer protection associations formed an alliance to jointly develop a number of social innovations allowing access to electricity defined as an 'essential service'.

Table 6.1 *Examples of sustainable innovation from a differentiation perspective*

Type of service	Examples of innovations according to sustainable development aspects	
	Environmental	Socio-economic
Materials processing *Goods transport, water, gas and electricity distribution*	Waterless cleaning, materials recycling, selling the use of a product rather than the product itself (photocopiers, vehicles), providing additional services around the product, and various forms of sharing	No gas, water or electricity cut-offs, fair trade, producer outlets, community-supported agriculture schemes
Processing of individuals *Transport, personal services, health, education*	Car sharing, work integration enterprises, sustainable tourism (agro-tourism, cycling, industrial tourism)	Work integration enterprises, sustainable tourism (linked to local social fabrics), care of the elderly, services for individuals living in hardship, cooperative child care centres
Information processing *Banking, insurance, family allowance offices, local governments*	Information on the environmental and social situation, loans at preferential rates	Microloans, multiservice information and mediation centres, public service centres, public service and advice centres
Processing of organizational knowledge *Consultancy services*	New area of expertise (environmental law, sustainable development consultancy services), ad hoc innovation, methodological innovations (MIPS, environmental standards and certifications)	New area of expertise (social law, sustainable development consultancy services), ad hoc innovation, methodological innovations

Individuals-processing services are those whose main medium is the individual himself, whose location, appearance or emotional, intellectual or physical health, and so on, are being changed or processed. Examples include health services, passenger transport, education, recreation, and so on. Environmentally speaking, car sharing, sustainable tourism and innovative initiatives in the field of elderly care, whether it be new types of housing situations or tailored services (Djellal and Gallouj, 2006), are examples of non-technological innovations in services. Again, most of these examples pursue socio-economic goals in that they seek not only to preserve the environment but also to promote economic development and enhance and preserve the local socio-economic fabric. They are not green innovations in the strict sense but more broadly sustainable innovations.

Information-processing services are those whose main medium is codified information, which is produced, entered, transported, and so on. Banking, insurance and government services fall into this category. Examples of the many invisible innovations in this area include microloans in response to the problem of exclusion from banking services, ethical finance initiatives that develop during financial crises, and loans at preferential rates in order to encourage firms to install environmentally friendly machinery. Worthy of mention in the public services is the development by local authorities (possibly in partnership with private companies, particularly in areas where services to individuals are inadequate) of facilities ('one-stop shops') providing services for people in hardship, such as multiservice information and mediation centres, public service centres and advice centres.

Organizational knowledge-processing services mainly target the knowledge of the organization which is produced, maintained, capitalized, and so on. These are knowledge-intensive services in which knowledge is both the input and the principal output (Miles et al., 1994; Gallouj, 2002b; Toivonen, 2006). The main examples of such activities are consulting, engineering, and research and development.

The innovations generated by knowledge-intensive services would seem, by definition, to be environmentally friendly. After all, knowledge-intensive services are among the most intangible of service activities. They produce cognitive solutions that do not seem to have a direct adverse impact on environmental sustainability. Consulting services have often been seen as the purest services based on the three traditional intrinsic technical criteria of services: intangibility, interactivity and non-stockability. It is therefore not surprising that they were among the first to be studied empirically to identify non-technological forms of innovation. With a view to identifying hidden innovation, breaking with the traditional distinction between product innovation and process innovation, Gallouj (1994) instead distinguishes three types of innovation: *ad hoc* innovation (jointly developing, with the client, a novel solution to a problem); *new field of expertise* innovation (detecting an emerging field of knowledge and providing advice in that field) and *formalization* innovation (implementing methods to make the service less ill-defined).

This is part of a trajectory of cognitive rationalization (as opposed to industrial rationalization mentioned above) described by Gadrey (1996) as follows: standardization (typification) of cases, formalization of problem-solving procedures (methods), and use of individual or organizational routines. This typology of innovation can readily be applied to the field of sustainable development. After all, many ad hoc solutions are co-produced by consultants and their clients in response to environmental or social problems. The increased awareness of sustainable development issues has led to the emergence of many new fields of expertise innovations and numerous specialist firms, for example in environmental law, social law and sustainable development consulting. Likewise, there have been many methodological innovations (formalization innovations) in the field of sustainable development. The MIPS indicator mentioned above and the introduction of new environmental standards can be cited by way of example. Thus, Nicolas (2004) analyses the way in which the introduction of standards and eco-labels such as the organic farming standard has given rise to an organizational learning process for firms, which is based on the use of external knowledge-intensive services (e.g. training services). In fact, some cognitive solutions provided by consultants can have a negative impact on social sustainability, particularly when they involve company closures or job cuts. Likewise, as we have noted earlier in this chapter, consultants also use technical systems (especially ICTs), offices and means of travel that have a negative impact on the environment.

Integration: Incorporating the Ecological and Sustainable Components of Innovation

In services innovation studies, the integration perspective reflects recent studies aiming to develop theoretical models able to account for technological and non-technological innovation in goods and in services. As discussed earlier, integration is based on the observation that the boundary between goods and services is becoming blurred, which means abandoning the goods/services dichotomy and focusing (in production, consumption and exchange) on solutions, systems, functions or experiences. The theoretical research prospects that support these approaches include: the functional economy (Stahel, 1997), the experience economy (Pine and Gilmore, 1999), the service-dominant logic (Lusch and Vargo, 2006) and the characteristics-based approaches (Gallouj and Weinstein, 1997).

The issue of sustainable development provides a new argument in favour of integration. After all, as defined in the Brundlandt Report (World Commission on Environment and Development, 1987), the concept of sustainable development has economic, environmental and social aspects. A satisfactory representation of sustainable innovation should be able to reconcile these different aspects.

To some extent, the integration of goods and services (the transition from an economy based on the production and consumption of goods to one based on the production and consumption of hybrid solutions or packages) can be envisaged as a factor of sustainability. This is the view supported by the functional economy. After all, by adding services to their product or by increasing the service content of their goods, and selling the use of goods rather than the goods themselves, firms are optimizing the period of use and reducing the relative share of materials-processing activities which are the causes of environmental damage. But here again, the potentially tangible nature of the 'add-on service' can cast doubt on this reasoning.

In the field of economics, the main integrative analytical framework for goods and services and their multiple forms of innovation is based on an approach to the product in terms of characteristics, in line with the Lancasterian model and the new consumer theory. Drawing from Saviotti and Metcalfe (1984), Gallouj and Weinstein (1997) (see also Gallouj, 2002a) adapt to services a model originally built for goods. Taking into account the specific characteristics of the product in services (in particular its potential intangibility and frequent interactivity), these authors define any product (whether a good or a service) as the conjunction of internal competences (those of the provider) [C] and external competences (those of the client) [C'] and/or internal technical characteristics [T] and external technical characteristics [T'][8] to produce the service characteristics [Y], that is to say, the use values. Figure 6.1 illustrates this general representation and a number of specific cases, including pure services, self-service arrangements, along with hybrid solutions (goods and services), for example, a car and the various upstream and downstream services that go with it: insurance, maintenance, financing, warranty, etc.

This representation makes it fairly easy to include sustainability issues by including sustainable service characteristics (Y_{iD}), that is to say, environmental and socio-civic use values, as well as the corresponding competences and technical characteristics.

As potential producers of service characteristics, technical characteristics are not neutral. They can be considered to be based on different value systems and viewpoints.

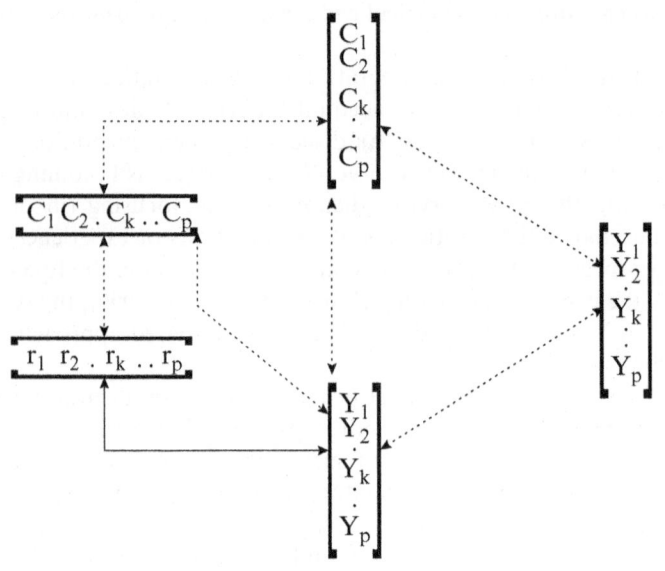

96 *Handbook of service business*

The general representation of a product (good or service) (after Gallouj and Weinstein, 1997)

The car in Saviotti and Metcalfe's representation

Service characteristics addition

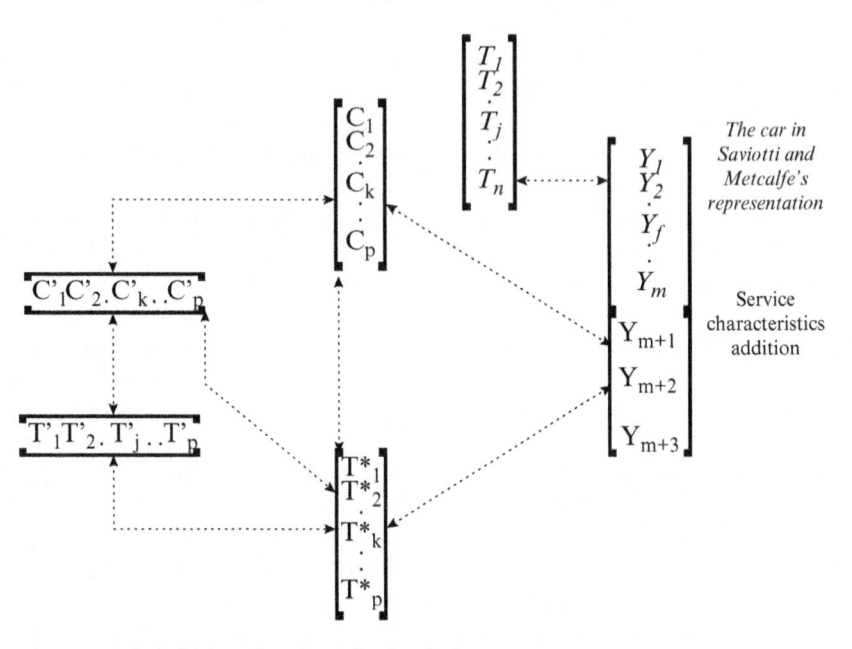

A car as a hybrid (good and services) solution

Notes: Y: service characteristics; T and T*: service provider's technical characteristics; C: service provider's competences; T': customer's technical characteristics; C': customer's competences.

Figure 6.1   The representation of the product in terms of vectors of characteristics and competences (after Gallouj and Weinstein, 1997; Gallouj, 2002a)

Green and sustainable innovation 97

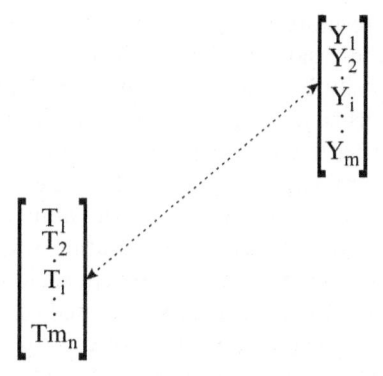

$$\begin{bmatrix} Y_1 \\ Y_2 \\ \cdot \\ Y_i \\ \cdot \\ Y_m \end{bmatrix}$$

$$\begin{bmatrix} T_1 \\ T_2 \\ \cdot \\ T_i \\ \cdot \\ Tm_n \end{bmatrix}$$

The representation of a good according to Saviotti and Metcalfe (1984)

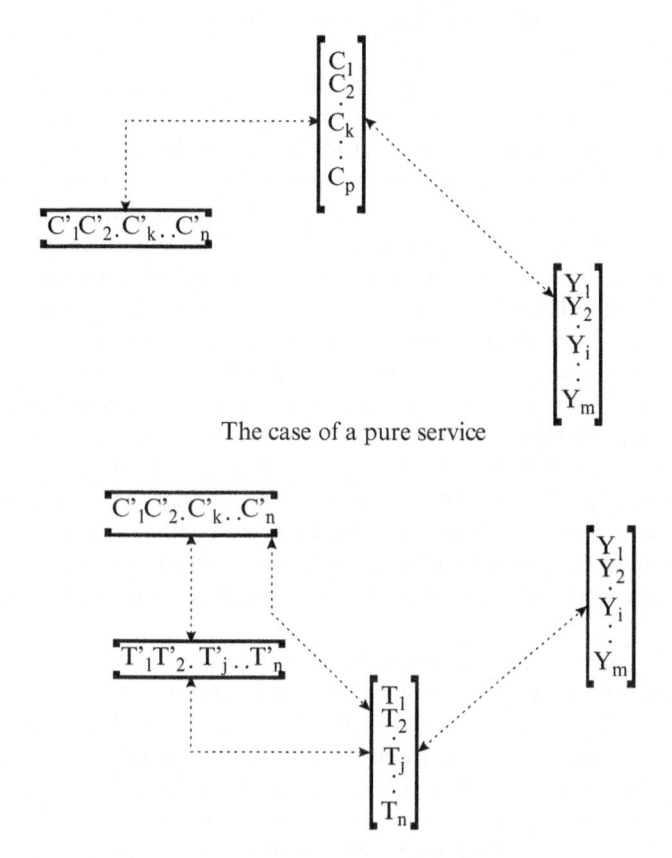

$$\begin{bmatrix} C_1 \\ C_2 \\ \cdot \\ C_k \\ \cdot \\ C_p \end{bmatrix}$$

$\boxed{C'_1 C'_2 . C'_k . . C'_n}$

$$\begin{bmatrix} Y_1 \\ Y_2 \\ \cdot \\ Y_i \\ \cdot \\ Y_m \end{bmatrix}$$

The case of a pure service

$\boxed{C'_1 C'_2 . C'_k . . C'_n}$

$\boxed{T'_1 T'_2 . T'_j . . T'_n}$

$$\begin{bmatrix} T_1 \\ T_2 \\ \cdot \\ T_j \\ \cdot \\ T_n \end{bmatrix}$$

$$\begin{bmatrix} Y_1 \\ Y_2 \\ \cdot \\ Y_i \\ \cdot \\ Y_m \end{bmatrix}$$

Self-service and e-services

Figure 6.1 (continued)

Some technical solutions are more fair (or deemed to be more fair) than others (e.g. technologies adapted for the disabled). Intangible technical characteristics (methods, organization modes) can easily be seen as organizational arrangements falling within the domestic and civic worlds (arrangements to ensure anonymity, confidentiality, discretion, fairness in the order in which requests are processed).

Competences are not neutral either. It is possible to identify *sustainable competences* (environmental, social and civic), that is to say, the ability to provide a service concerned with preserving the environment or promoting social inclusion (maintaining relationships with clients (users) in sometimes severe socio-economic hardship). Such ecological, social and civic competences can be accepted or promoted by the organization or they can be repressed. The competences of certain customers may be particularly weak (socio-economically disadvantaged customers, customers suffering from cultural and cognitive disabilities). This weakness can or should be compensated by the social and civic competences of empathy and 'translation' on the part of customer service agents.

In the case of automobiles, Saviotti and Metcalfe (1984) include some negative externalities (pollution, congestion) in the service characteristics vector. But we can also include sustainability as a positive factor. For example, the catalytic converter (technical characteristic T_{iD}) results in lower pollution emissions (service characteristic Y_{iD}). It is in the public services sector that it is easiest to find examples of socio-civic characteristics. The principles of public service (continuity of service, fairness, equal treatment) do indeed promote social relations based on equal treatment, fairness and justice. In the case of the French postal system, examples include the fair treatment of users (at the counters, along delivery routes), fair access, non-discrimination (of young people, foreigners), assistance to marginalized populations, social tariffs and social banking services (accounts for low-income earners, reasonable penalties, advice for individuals living in hardship) (see Gallouj et al., 1999). But of course, the sustainable environmental and socio-civic innovation trajectory is also at work in market services, as evidenced by the growing importance of corporate social responsibility. For example, private insurance companies can include sustainable socio-civic characteristics by imposing limits on searching for private background information on customers or adopting a premium structure that evens out differences between generations or social classes (Gadrey, 1996).

In product analyses in terms of characteristics, innovation is considered in terms of changes in characteristics. An innovation is born when certain characteristics are modified, intentionally in most cases. The changes involve either the addition, subtraction or formatting of characteristics. Addition consists of adding one or more characteristics to an existing product. One of its most complex variants is when several products are combined to develop a new product, as described by Henderson and Clark (1990). In contrast, subtraction consists of removing one or more characteristics from an existing product. Again the most complex variant of this method is dissociation, in which an innovative product is developed by splitting up (dissociating) an existing product. Finally, formatting includes all strategies that aim to make the service characteristics less ill-defined and to clarify the relationships between different terms of the service vectors. It may include developing methodologies, introducing technologies or management toolkits, incorporating the service into an organization, and so on.

Table 6.2 The various models of innovation and the dynamic of characteristics

Innovation model	Nature of the action on the characteristics	Examples of sustainable innovations
Radical	– *Narrow definition*: creation of a new set of characteristics {[C'*], [C*], [T*], [Y*]} – *Broad definition*: creation of a new set of characteristics {[C'*], [C*], [T*]} even though [Y] remains unchanged	Wind turbines Electric vehicles
Ameliorative	– No change in the general structure of the system – Increase in the weight (quality) of characteristics	Increased energy efficiency, less pollution, increased solidarity with disadvantaged populations
Incremental	– No change in the general structure of the system – Addition (or elimination) of characteristics	Add-on technology Addition of socio-civic characteristics (e.g. the French Postal System) Service around the product
Recombinative	Combining or splitting of groups of characteristics	Sustainable tourism Sustainable trade
Formalization	– Formatting and standardization of characteristics – Clarifying the correspondences (mapping)	Sustainable methodologies, MIPS, eco-labels

These general principles of innovation can easily be used to develop sustainable environmental or socio-civic innovations in services. After all, one can add or remove sustainable competences or sustainable (service or technical) characteristics. One can also combine, separate or format these elements to achieve sustainable innovation trajectories. The implementation of these general principles makes it possible to elaborate a typology of innovations (summarized in Table 6.2), illustrated with examples of sustainable service innovations.

Radical innovation is simple to define, since in theory it corresponds to the creation of a new set of characteristics and competences. Examples include wind turbines and electric vehicles. The benefit of the representation in terms of characteristics is the ability to highlight different forms of minor innovations as opposed to major innovations. After all, one way to produce minor innovations is to increase the prominence of certain sustainable green or socio-civic characteristics without changing the main specifications of the service (the {C, C', T, T', Y} system). These include for example increasing energy efficiency, reducing pollution levels and improving assistance for disadvantaged groups. This method is called 'ameliorative innovation'. Another method, which we call 'incremental innovation', denotes the addition (or elimination) of characteristics, again without fundamentally modifying the {C, C', T, T', Y} system defining the product. This includes 'add on' technologies that are introduced into products or production processes to meet certain environmental concerns. But socio-civic concerns may also result in incremental innovations, for example when social or civic characteristics are added to a service (see the examples provided above in the case of postal services). Finally, the practice of adding services to an existing product (Furrer, 1997) may also be considered as a form of sustainable incremental innovation, since it contributes to the 'dematerialization' of a firm's activities, which, in turn, if it is real (see our first part), enhances environmental sustainability.

Recombinative innovation is a form of innovation that relies on the basic principles of dissociation and association (i.e. the splitting or combining) of final and technical characteristics. It leads to new products either by combining or splitting existing products. Many forms of sustainable tourism or trade are examples of recombinative innovation.

Finally, formalization innovation aims to make the service more tangible. Unlike the other forms of innovation already mentioned, it does not involve adding, subtracting, combining or splitting characteristics; rather it involves formatting and standardizing them. Examples include the development of certifications, standards and methodologies aimed at increasing sustainability.

CONCLUSION

Contemporary developed economies are service and innovation economies that aspire to sustainable development. The issue of sustainability, however (especially in its environmental form), is often primarily associated with manufacturing industry and material technologies. Services are often, on average, considered greener than industry although there are notable counter-examples, including tourism and transport. 'The office pollutes less than the factory' is the image that sums up this optimistic assumption, reinforced by the essential socio-economic role of services which create most of the jobs in modern economies.

In this chapter we have endeavoured to show that the reality is more complex. Thus, both analytically and empirically, the notion of 'environment friendly' services is debatable. Analytically, the idea that services are by definition intangible and that post-industrial economies have become dematerialized is called into question. Thus, a systemic view of services, which includes service delivery, travel and conditioning service delivery premises, reveals many direct and indirect sources of tangibility (energy consumption and pollution) that can be measured (Fourcroy et al., 2012).

Just as services cannot be considered sustainable or green (by definition), neither can innovation in services. Whatever the analytical perspective adopted (assimilation, differentiation or integration), the dynamics of innovation in services incorporates contradictory relations with the issue of sustainable development. The assimilationist perspective, focusing primarily on the adoption of technical systems, points to certain trajectories that are environmentally friendly and others that harm the environment. The differentiation perspective, which focuses on identifying intangible forms of innovations (non-technological innovations, social innovations), suggests that innovation in services is inherently more sustainable than industrial innovation. Yet, a number of arguments refute or at least diminish the validity of this conclusion. First, certain intangible innovations involve tangible aspects that are often forgotten, such as travel on the part of users and providers, and the processes involved in producing the technical systems used in the service provision (embodied or grey energy). In addition, certain intangible innovations can fundamentally undermine the socio-economic aspects of sustainability. Such is the case with the financial innovations that resulted in the recent financial crisis, and with certain strategic or organizational solutions (in terms of business relocations or reorganizations) promoted by consultants. The intellectually appealing integration perspective also uses a unique model to look at service characteristics, including sustainable ones. Consequently, the issue of sustainability and innovation in services requires more research and improved measurement and evaluation in recognition of the fact that the future of sustainable development will be played out in the service sector, whether positively or negatively. In services, as elsewhere, innovation will play a fundamental role in guiding economies towards sustainable development (Desmarchelier et al., 2013).

NOTES

1. Economic history provides many illustrations of social resistance to technological progress.
2. The ecological footprint of a population is a simple indicator that estimates the surface area of the planet on which that population depends in order to sustain its economic activities.
3. The 14 IEA countries in question are: Austria, Canada, Denmark, Finland, France, Germany, Italy, Japan, the Netherlands, New Zealand, Norway, Sweden, the United Kingdom and the United States.
4. In the definition of Gadrey (1991), a service may have three other mediums: the individual himself, codified information and knowledge of organizations.
5. The MIPS indicator (Material Intensity Per Service Unit) measures the non-renewable natural resources consumed to produce a good or service.
6. This distinction was established by Gallouj in the early 1990s (see Gallouj, 1994; Gallouj and Weinstein, 1997). It has been taken up and reiterated by Coombs and Miles (2000). A fourth perspective is considered in the literature (Gallouj, 2010), but we will not discuss the details of it here. It is the inversion perspective, which reflects the active role of knowledge-intensive services in promoting innovation in other organizations.
7. These services have a certain material dimension, though, especially in the case of electricity, they are not tangible objects.
8. The inclusion of clients' technical characteristics was suggested by De Vries (2006) in order to take account of the new channels of consumption and delivery (e.g. when consumers use their own technologies to access a service on the web).

BIBLIOGRAPHY

Amin, A., Cameron, A. and Hudson, R. (2002), *Placing the Social Economy*, Routledge, London.
Barcet, A. and Bonamy, J. (1999), Eléments pour une théorie de l'intégration biens/services. *Économies et Sociétés*, série EGS, 1(5), 197–220.
Barras, R. (1986), Towards a theory of innovation in services. *Research Policy*, 15, 161–173.
Baumol, W.J. (1967), Macroeconomics of unbalanced growth: the anatomy of urban crisis. *American Economic Review*, 57, 415–426.
Baumol, W. (2010), The two-sided cost disease and its frightening consequences. In Gallouj, F. and Djellal, F. (eds), *The Handbook of Innovation and Services: A Multidisciplinary Perspective*, Edward Elgar Publishing, Cheltenham and Northampton, MA, pp. 84–92.
Broussolle, D. (2001), *Les NTIC et l'innovation dans la production de biens et services: des frontières qui se déplacent*, 11th RESER International Conference, Grenoble, October.
Coombs, R. and Miles, I. (2000), Innovation measurement and services: the new problematique. In Metcalfe,

S. and Miles, I. (eds), *Innovation Systems in the Service Economy: Measurement and Case Study Analysis*, Kluwer, Boston, pp. 85–103.
De Vries, E. (2006), Innovation in services in networks of organizations and in the distribution of services. *Research Policy*, 35(7), 1037–1051.
Desmarchelier, B., Djellal, F. and Gallouj, F. (2011), Economic growth by waste generation: the dynamics of a vicious circle. *Lecture Notes in Economics and Mathematical Systems*, 652, 129–138.
Desmarchelier, B., Djellal, F. and Gallouj, F. (2013), Environmental policies and eco-innovations by service firms: a multi-agent adaptation model. *Technological Change and Social Forecasting*, 80(7), September, 1395–1408.
Djellal, F. and Gallouj, F. (2006), Innovation in care services for the elderly. *Service Industries Journal*, 26(3), 303–327.
Djellal, F. and Gallouj, F. (2010), Innovation in services and sustainable development. In Maglio, P.P., Kieliszewski, C.A. and Spohrer, J.C. (eds), *The Handbook of Service Science*, Springer, New York, pp. 533–557.
Djellal, F. and Gallouj, F. (2012), Social innovation and service innovation. In Franz, H.-W., Hochgerner, J. and Howaldt, J. (eds), *Challenge Social Innovation Potentials for Business, Social Entrepreneurship, Welfare and Civil Society*, Springer, Berlin, pp. 119–137.
Droege, H., Hildebrand, D. and Heras Forcada, M. (2009), Innovation in services: present findings, and future pathways. *Journal of Service Management*, 20(2), 131–155.
Ettighoffer, D. (1992), *L'entreprise virtuelle. Ou les nouveaux modes de travail*. Editions Odile Jacob, Paris.
Fourcroy, C., Gallouj, F. and Decellas, F. (2012), Energy consumption in services industries: challenging the myth of non-materiality. *Ecological Economics*, 81(September), 155–164.
Furrer, O. (1997), Le rôle stratégique des services autour des produits. *Revue Française de Gestion*, March–May, 98–108.
Fussler, C. and James, P. (1996), *Driving Eco-Innovation: A Breakthrough Discipline for Innovation and Sustainability*, Pitman Publishing, London.
Gadrey, J. (1991), Le service n'est pas un produit. Quelques implications pour l'analyse économique et pour la gestion. *Politiques et management public*, 9(1), March, 1–24.
Gadrey, J. (1996), *Services: La productivité en question*, Desclée de Brouwer, Paris.
Gadrey, J. (2004), Services, croissance, décroissance. *Alternatives économiques*, (228), September. http://www.alternatives-economiques.fr/services-2c-croissance-2c-decroissa_fr_art_183_20542.html.
Gadrey, J. (2010), The environmental crisis and the economics of services: the need for revolution. In Gallouj, F. and Djellal, F. (eds), *The Handbook of Innovation and Services*, Edward Elgar Publishing, Cheltenham and Northampton, MA, pp. 93–125.
Gallouj, C. (2007), *Innover dans la grande distribution*. De Boeck, Bruxelles.
Gallouj, F. (1994), *Economie de l'innovation dans les services*. L'Harmattan, Paris.
Gallouj, F. (1999), Les trajectoire de l'innovation dans les services: vers un enrichissement des taxonomies évolutionnistes. *Économies et Sociétés, Série EGS*, 1(5), 143–169.
Gallouj, F. (2002a), *Innovation in the Service Economy: The New Wealth of Nations*, Edward Elgar Publishing, Cheltenham and Northampton, MA.
Gallouj, F. (2002b), Knowledge intensive business services: processing knowledge and producing innovation. In Gadrey, J. and Gallouj, F. (eds), *Productivity, Innovation and Knowledge in Services*. Edward Elgar Publishing, pp. 256–284.
Gallouj, F. (2010), Services innovation: assimilation, differentiation, inversion and integration. In Bidgoli, H. (ed.), *The Handbook of Technology Management*, John Wiley and Sons, Hoboken, NJ, pp. 989–1000.
Gallouj, F. and Djellal, F. (eds) (2010), *The Handbook of Innovation and Services: A Multidisciplinary Perspective*, Edward Elgar Publishing, Cheltenham and Northampton, MA.
Gallouj, F. and Weinstein, O. (1997), Innovation in services. *Research Policy*, 26, 537–556.
Gallouj, F., Gadrey, J. and Ghillebaert, E. (1999), La construction sociale du produit financier postal. *Annals of Public and Cooperative Economics*, 70(3), 417–445.
Gershuny, J. (1978), *After Industrial Society? The Emerging Self-Service Economy*, Macmillan, New York.
Harrisson, D., Chaari, N. and Comeau-Vallée, M. (2012), Intersectoral alliance and social innovation: when corporations meet civil society. *Annals of Public and Cooperative Economics*, 83(1), 1–24.
Henderson, R.M. and Clark, K.B. (1990), Architectural innovation: the reconfiguration of existing product technologies and the failure of established firms. *Administrative Science Quarterly*, 35(1), March, 9–30.
Hipp, C. and Grupp, H. (2005), Innovation in the service sector: the demand for service-specific innovation measurement concepts and typologies. *Research Policy*, 34(4), 517–535.

Howells, J. (2007), Services and innovation: conceptual and theoretical perspectives. In Bryson, J.R. and Daniels, P.W. (eds), *The Handbook of Service Industries*, Edward Elgar Publishing, Cheltenham and Northampton, MA, pp. 34–44.
International Energy Agency (2007), *Energy Use in the New Millennium: Trends in IEA Countries*, OECD, IEA, Paris.
International Energy Agency (2008), *Worldwide Trends in Energy Use and Efficiency: Key Insights from IEA Indicator Analysis: In Support of the G8 Plan of Action*, OECD, IEA, Paris.
Kemp, R. and Soete, L. (1990), Inside the 'green box': on the economics of technological change and the environment. In Freeman, C. and Soete, L. (eds), *New Explorations in the Economics of Technological Change*, Pinter Publishers, London, pp. 245–257.
Kemp, R. and Soete, L. (1992), The greening of technological progress: an evolutionary perspective. *Futures*, 24(5), 437–457.
Kingman-Brundage, J. (1992), The ABCs of service system blueprinting. In Lovelock, C. (ed.), *Managing Services*, Prentice-Hall International Editions, Englewood Cliffs, NJ, pp. 96–102.
Levitt, T. (1972), Production line approach to service. *Harvard Business Review*, 50(September–October), 41–52.
Lovelock, C. (1992), A basic toolkit for service managers. In Lovelock, C. (ed.), *Managing Services*, Prentice-Hall International Editions, Englewood Cliffs, NJ, pp. 17–30.
Lusch, R. and Vargo, S. (2006), Service-dominant logic: reactions, reflections and refinements. *Marketing Theory*, 6(3), 281–288.
Marinova, D., Annandale, D. and Phillimore, J. (eds) (2007), *The International Handbook on Environmental Technology Management*, Edward Elgar Publishing, Cheltenham and Northampton, MA.
Miles, I. (2002), Services innovation: towards a tertiarization of innovation studies. In Gadrey, J. and Gallouj, F. (eds), *Productivity, Innovation and Knowledge in Services*, Edward Elgar Publishing, Cheltenham and Northampton, MA, pp. 164–196.
Miles, I., Kastrinos, N., Flanagan, K., Bilderbek, R., den Hertog, P., Huntink, W. and Bouman, M. (1994), *Knowledge-Intensive Business Services: Their Role as Users, Carriers and Sources of Innovation*, PREST, University of Manchester, Manchester.
Nelson, R. and Winter, S. (1982), *An Evolutionary Theory of Economic Change*, Belknap Harvard, Cambridge, MA and London.
Nicolas, E. (2004), Apprentissage organisationnel et développement durable. La norme AB. *Revue française de gestion*, 2(149), 153–172.
OECD (2000), *The Service Economy*, OECD Publications, Paris.
Pavitt, K. (1984), Sectoral patterns of technical change: towards a taxonomy and a theory. *Research Policy*, 13, 343–373.
Pine, J. and Gilmore, J. (1999), *The Experience Economy*, Harvard Business School Press, Boston, MA.
Rennings, K. (2000), Redefining innovation – eco-innovation research and the contribution from ecological economics. *Ecological Economics*, 32, 319–322.
Romm, J., Rosenfeld, A. and Herrman, S. (1999), *The Internet Economy and Global Warming*, Center for Energy and Climate Solutions, Washington, DC.
Rubalcaba, L. (2006), Which policy for innovation in services? *Science and Public Policy*, 33(10), 745–756.
Saviotti, P.P. and Metcalfe, J.S. (1984), A theoretical approach to the construction of technological output indicators. *Research Policy*, 13, 141–151.
Seyfang, G. and Smith, A. (2006), Community action: a neglected site of innovation for sustainable development? CSERGE Working Paper, EDM 06-10.
Shostack, G.L. (1984), Service design in the operating environment. In George, W. and Marshall, C. (eds), *Developing New Services*, American Marketing Association, Chicago, IL, Proceedings Series, pp. 27–43.
Smith, A. (1976) (1st edition, 1776), *The Wealth of Nations*, University of Chicago Press, Chicago, IL.
Stahel, W. (1997), The functional economy: cultural and organizational change. In Richards, D.J. (ed.), *The Industrial Green Game: Implications for Environmental Design and Management*, National Academy Press, Washington, DC, pp. 91–100.
Sundbo, J. (1998), *The Organisation of Innovation in Services*, Roskilde University Press, Copenhagen.
Sundbo, J. and Toivonen, M. (2011), *User-Based Innovation in Services*, Edward Elgar Publishing, Cheltenham and Northampton, MA.
Tether, B. (2005), Do services innovate (differently)? Insights from the European Innobarometer Survey. *Industry and Innovation*, 12, 153–184.
Toivonen, M. (2004), Expertise as business: long-term development and future prospects of knowledge-intensive business services. PhD, Helsinki University of Technology.
Toivonen, M. (2006), Future prospects of knowledge-intensive business services (KIBS) and implications to regional economies. *ICFAI Journal of Knowledge Management*, 4(3), 18–39.

Vandermerwe, S. and Rada, J. (1988), Servitization of business: adding value by adding services. *European Management Journal*, 6(4), 314–324.
Windrum, P. and García-Goñi, M. (2008), A neo-Schumpeterian model of health services innovation. *Research Policy*, 37(4), 649–672.
World Commission on Environment and Development (1987), *Our Common Future*, Oxford University Press, Oxford.

7. The three-stage model of service consumption[1]
Rodoula H. Tsiotsou and Jochen Wirtz

INTRODUCTION

In addition to simultaneous production and consumption and the customer's participation in the service production, process is one of the main characteristics of services (Grönroos 2000a).

> Services are produced in a process wherein consumers interact with the production resources of the service firm . . . the crucial part of the service process takes place in interaction with customers and their presence. What the customer consumes in a service context is therefore fundamentally different from what traditionally has been the focus of consumption in the context of physical goods. (Grönroos 2000b, p. 15)

The consumption of services has been considered as 'process consumption' (Grönroos 1994) because production is part of service consumption and is not simply viewed as the outcome of a production process, as is the case in the traditional marketing of physical goods. The service-dominant logic also supports that service should be defined as a process (rather than a unit of output) and refers to the application of competencies (knowledge and skills) for the benefit of the consumer. Here, the primary goal of a business is value co-creation as 'perceived and determined by the customer on the basis of value-in-use' (Vargo and Lusch 2004, p. 7).

Consistent with this reasoning, academics gradually shifted from an output focus adapted from the goods literature to a process focus. Several models describing the various stages of the service consumption process have been proposed in the literature. This chapter adopts the three-stage perspective (comprising the pre-purchase, encounter and post-encounter stages) of consumer behaviour (Lovelock and Wirtz 2011; Tsiotsou and Wirtz 2012) and discusses relevant extant and emerging research on each stage.

The chapter is organized as follows. First, the three-stage model of service consumption is presented, followed by important new research developments concerning each stage. The chapter concludes by outlining emerging research topics and directions for future research.

THE THREE-STAGE MODEL OF SERVICE CONSUMPTION

According to the three-stage model of service consumption, consumers go through three major stages when they consume services: the pre-purchase stage, the service encounter stage and the post-encounter stage (Lovelock and Wirtz 2011, pp. 36–37; Tsiotsou and Wirtz 2012). This approach is helpful because it assists academics in developing a clear research focus and direction, and managers in setting objectives and shaping consumer behaviour in a targeted manner, and therefore facilitates efficient resources allocation (Blackwell, Miniard and Engel 2003; Hensley and Sulek 2007). Research has been conducted on all three stages to examine their major determinants, influences (direct and indirect), processes and outcomes (Figure 7.1).

PRE-PURCHASE STAGE

CONSUMER BEHAVIOUR	KEY CONCEPTS
- Need Awareness - Information Search - Evaluation of Alternatives - Make Decision on Service Purchase	- Need Arousal - Information Sources, Perceived Risk - Multi-attribute Model and Search, Experience and Credence Attributes

SERVICE ENCOUNTER STAGE

CONSUMER BEHAVIOUR	KEY CONCEPTS
- Request Service from Chosen Supplier or Initiation of Self-Service - Service Delivery Interactions	- An Integrative Model of Service Encounters, Servicescapes, Service Scripts - Low-Contact Service Encounters (Voice to Voice and Self-Service Encounters)

POST-ENCOUNTER STAGE

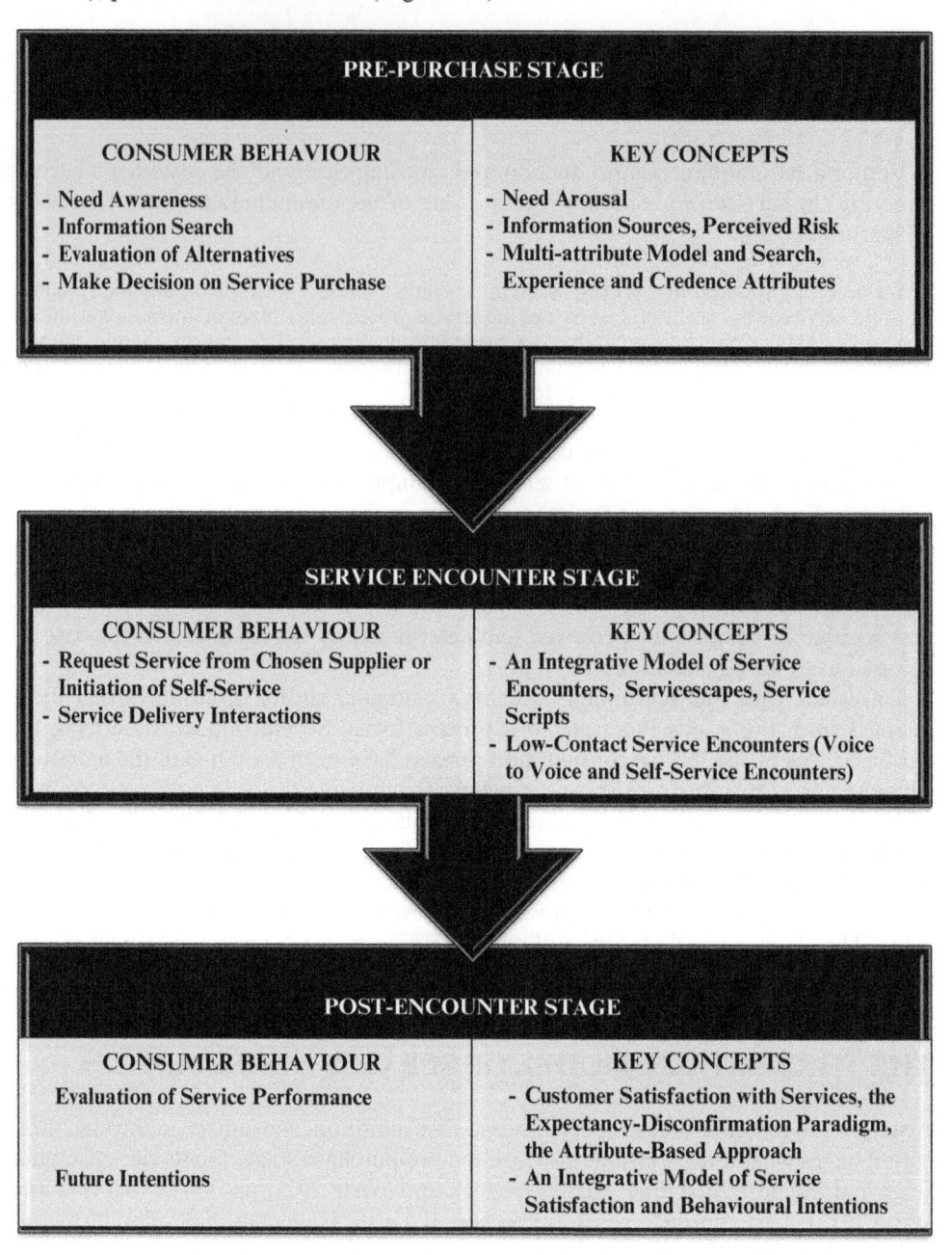

CONSUMER BEHAVIOUR	KEY CONCEPTS
Evaluation of Service Performance Future Intentions	- Customer Satisfaction with Services, the Expectancy-Disconfirmation Paradigm, the Attribute-Based Approach - An Integrative Model of Service Satisfaction and Behavioural Intentions

Figure 7.1 The three-stage model of service consumption

The Pre-purchase Stage

The pre-purchase stage of the decision-making process for services is more complex in comparison with that for goods as it involves a composite set of factors and activities (Fisk 1981). Because consumers participate in the service production process, the decision-making process takes more time and is more complicated than in the case of goods. Consumer expertise, knowledge (Byrne 2005) and perceived risk (Diacon and Ennew 2001) play important roles in this pre-purchase phase.

In the pre-purchase stage, a need arousal triggers consumers to start searching for information and evaluate alternatives before they make a purchase decision. There are various sources that could trigger needs: the unconscious mind (e.g., impulse buying), internal conditions (e.g., hunger) or external sources (e.g., marketing mix), to name a few. Consumers can engage in impulse buying or 'unplanned behaviour'. Impulse buying occurs less frequently in services than in goods due to the higher perceived risk and variability associated with services (Murray and Schlacter 1990; Sharma, Sivakumaran and Marshall 2009). However, service research has neglected the role of impulse buying although it is an important phenomenon extensively studied in the goods context (Kacen and Lee 2002; Mattila and Wirtz 2008). As such, the information search process described in the next section focuses on conscious consumer decision-making processes.

According to the notion of planned purchase behaviour, once consumers recognize a need or problem they are motivated to search for solutions to satisfy that need or resolve that problem (Figure 7.2). The information obtained in the pre-purchase stage

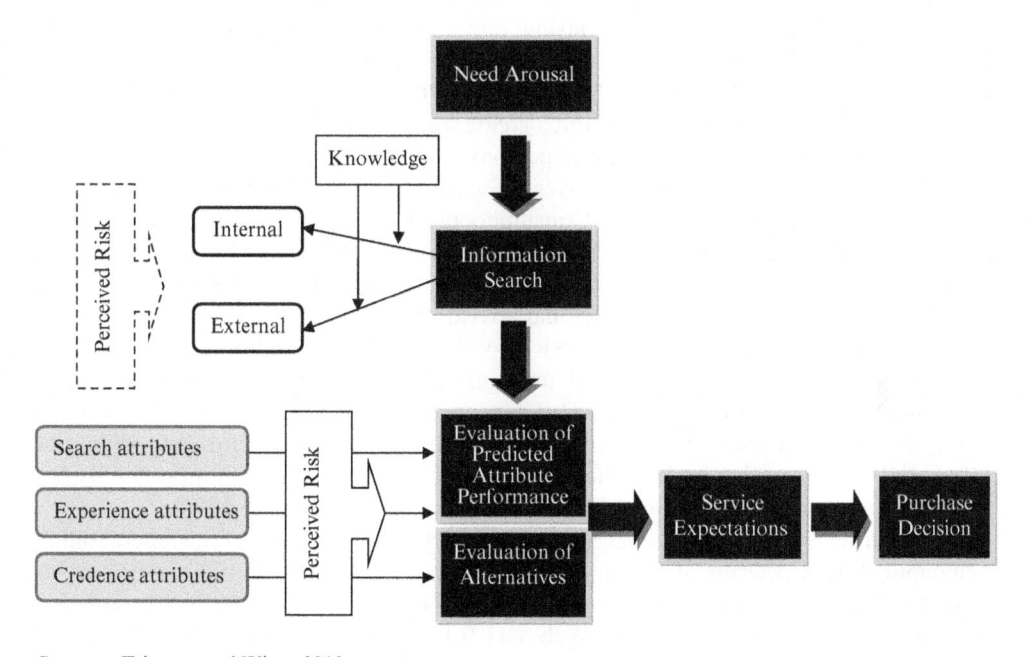

Source: Tsiotsou and Wirtz, 2012.

Figure 7.2 The pre-purchase process of consumers in services

has a significant impact on consumers' purchase decisions (Alba and Hutchinson 2000; Konus, Verhoef and Neslin 2008; Mattila and Wirtz 2002).

Information search

Consumer information search in services is more extensive than in goods (Alba and Hutchinson 2000; Mattila and Wirtz 2002) due to the uncertainty and perceived risk associated with a purchase decision. Both uncertainty and perceived risk are considered to be higher in services due to their intangible nature and variability (Bansal and Voyer 2000; Murray and Schlacter 1990) and because of the high degree of price uncertainty due to service firms' revenue management strategies (Kimes and Wirtz 2003; Wirtz and Kimes 2007).

Because of the above, service consumers typically do not limit themselves to a single source of information, but employ multiple sources of information depending on their orientation (multichannel orientation), their tendency to innovate and the perceived pleasure of the shopping experience. They search for information from multiple sources to explore and evaluate alternative service offerings, develop performance expectations of offers in the consideration set, save money and reduce risk (Konus, Verhoef and Neslin 2008).

In addition, service consumers acquire information not only from multiple sources but from different types of sources. Thus, they seek information from trusted and respected personal sources such as family, friends and peers; they use the Internet to compare service offerings and search for independent reviews and ratings; they rely on firms with a good reputation; they look for guarantees and warranties; they visit service facilities or try aspects of the service before purchasing; they examine tangible cues and other physical evidence and ask knowledgeable employees about competing services (Boshoff 2002; Lovelock and Wirtz 2011, pp. 41–42; Zeithaml and Bitner 2003).

In general, consumers not only exhibit a greater propensity to search for more information, but they also tend to explore more personal sources of information such as friends, family and co-workers (Bansal and Voyer 2000; Murray and Schlacter 1990; Wirtz et al. 2012; Xiao, Tang and Wirtz 2011). Consumers use these personal sources of information because they trust them more than any other source. For example, recent research evidence supports that family is a predominantly trustworthy source of information, considered more reliable than professional advisors (e.g., accountants or financial planners) when buying retirement services (Rickwood and White 2009). Moreover, consumer expertise, perceived risk and perceived acquaintances' expertise contribute to the active search for word of mouth (Alba and Hutchinson 2000; Mattila and Wirtz 2002). Thus, word of mouth as a source of consumer information has become a more important and influential concept within services than in the goods context due to their intangibility and higher perceived risk (Bansal and Voyer 2000; Murray and Schlacter 1990).

The Internet constitutes another source of information, although consumers' online behaviour differs in terms of the amount of search time spent on goods versus services websites. A study conducted in an online retailing context found that the average time consumers spent searching on the Web was 9.17 minutes on automotive sites, 9.26 minutes on telecom/Internet sites, 10.44 minutes on travel sites and 25.08 minutes on financial sites (Bhatnagar and Ghose 2004). Demographic characteristics, such as gender, education, age and Internet experience, influence the time consumers spend searching for information (Bhatnagar and Ghose 2004; Ratchford, Lee and Talukdar 2003). The more time consumers devote to searching for information via the Internet and the more often they do so, the more such online gathered information influences the purchase decision (Bhatnagar and Ghose 2004).

In order to assist consumers in their search, online services have developed electronic recommendation agents, also known as 'smart agents', as an element of their services (Aksoy et al. 2006; Diehl, Kornish and Lynch 2003; Haubl and Murray 2003; Haubl and Trifts 2000). Through the use of recommendation agents and avatars as an entertainment and informational tool, online services aim to fulfil consumers' desire for a more interpersonal shopping experience (Holzwarth, Janiszewski and Neumann 2006). Electronic recommendation agents provide consumers with information about products and their attributes after searching for a large amount of data using consumer-specified selection criteria in order to assist them in their purchase decisions (Aksoy et al. 2006; Diehl, Kornish and Lynch 2003). Research findings support that this practice can lead to desirable outcomes. For example, avatars or virtual salespeople acting as sales agents have been found to increase purchase intentions, enhance positive attitudes toward products and increase consumer satisfaction with products (Holzwarth, Janiszewski and Neumann 2006). Moreover, electronic recommendation agents can reduce the prices paid by consumers (Diehl, Kornish and Lynch 2003) and improve the quality of their decisions (Ariely, Lynch and Aparicio 2004; Haubl and Trifts 2000). However, to be effective, recommendation agents' attribute weightings and decision strategies need to be congruent with those of their target consumers to achieve high-quality purchase decisions, reduced search time and increased website loyalty and satisfaction (Aksoy et al. 2006).

Evaluation of alternative service offers
During the search process, consumers form their consideration set, learn about the service attributes they should consider and form expectations of how firms in the consideration set perform on those attributes (Lovelock and Wirtz 2011, p. 42). Multi-attribute models have been widely used to simulate consumer decision making. According to these models, consumers use service attributes (e.g., quality, price and convenience) that are important to them to evaluate and compare alternative offerings of firms in their consideration set. Each attribute is weighted according to its importance.

An example of a multi-attribute model applied to restaurant services is presented in Table 7.1. To make a purchase decision, consumers might use either the very simple linear compensatory rule (in which case the consumer would choose 'New Restaurant' in the example in Table 7.1) or the more complex, but also more realistic, conjunctive rule (e.g., if price should have a minimum rating of '8', then 'Current Restaurant' would be chosen). Consumers using the same information can ultimately choose different alternatives if they use different decision rules.

Multi-attribute models are based on the assumption that consumers can evaluate all important attributes before making a purchase decision. However, this is often not the case in services because some attributes are more difficult to evaluate than others. According to Zeithaml (1981), there are three types of attributes: search attributes, experience attributes and credence attributes. *Search attributes* refer to tangible characteristics consumers can evaluate before purchase (Paswan et al. 2004; Wright and Lynch

Table 7.1   Application of a multi-attribute model to restaurant services

SERVICE ATTRIBUTES	Current Restaurant	'Mom's' Restaurant	New Restaurant	Importance Weight
Quality of Food	8	9	10	30%
Convenience of Location	8	10	9	25%
Price	8	7	6	20%
Opening Hours	9	8	9	5%
Friendliness of Staff	8	9	9	15%
Restaurant Design	6	9	10	5%
MEAN SCORE	7.8	8.7	8.8	100%

Note: A high performance score on price means a low (i.e., attractive) price from the consumer's perspective.

1995). These attributes (e.g., price, brand name, transaction costs) help consumers to better understand and evaluate a service before making a purchase and therefore reduce the sense of uncertainty or risk associated with a purchase decision (Paswan et al. 2004). *Experience attributes*, on the other hand, cannot be reliably evaluated before purchase (Galetzka, Verhoeven and Pruyn 2006). Consumers must 'experience' the service before they can assess attributes like reliability, ease of use and consumer support. *Credence attributes* are characteristics that consumers find hard to evaluate even after making a purchase and consuming the service (Darby and Karni 1973). This can be due to a lack of technical experience or means to make a reliable evaluation, or because a claim can be verified only a long time after consumption, if at all (Galetzka, Verhoeven and Pruyn 2006). Here, the consumer is forced to believe or trust that certain tasks have been performed at the promised level of quality. Because most services tend to be ranked highly on experience and credence attributes, consumers find them more difficult to evaluate before making a purchase (Mattila and Wirtz 2002; Zeithaml 1981).

After consumers have evaluated the possible alternatives, they are ready to make a decision and move on to the service encounter stage. This next step may take place immediately, or may involve an advance reservation or membership subscription.

The Service Encounter Stage

The service encounter stage involves consumer interactions with the service firm. In this stage, consumers co-create experiences and value, and co-produce a service while evaluating the service experience.

Nowadays, customers are empowered and engaged in the service delivery process. *Consumer engagement* has recently attracted research attention in the branding and services literature (Brodie et al. 2011). Consumer engagement has been considered the emotional tie that binds the consumer to the service provider (Goldsmith 2011) and can be used as a proxy for the strength of a firm's consumer relationships based on both emotional and rational bonds consumers have developed with a brand (McEwen 2004). Bowden (2009) supports the view that engagement is a construct particularly applicable to services because services usually involve a certain degree of interactivity such as that seen between consumers and frontline personnel, and therefore implies a reciprocal relationship. Engagement might include feelings of confidence, integrity, pride and passion in a firm/brand (McEwen 2004). In addition to these affective elements, consumer engagement with service brands has been considered a behavioural manifestation toward a brand or firm that goes beyond a purchase and includes positive word of mouth, recommendations, helping other consumers, blogging, writing reviews and even engaging in legal action (van Doorn et al. 2010). Recent works recognize that consumer engagement involves cognitive (e.g., absorption), emotional (e.g., dedication) and behavioural (e.g., vigour and interaction) elements (Brodie et al. 2011; Patterson, Yu and de Ruyter 2006). Brodie et al. (2011, p. 260) define customer engagement as 'a psychological state that occurs by virtue of interactive, cocreative customer experiences with a focal agent/object (e.g., a brand) in focal service relationships'. Thus, service encounters could provide the context in which customers can create, express and enhance their engagement (positive or negative) with a service firm.

However, in order for customers to become engaged in the co-production of a service or in co-creation of value during the service encounter stage, they have to be motivated, and must have the ability and knowledge to provide and integrate various resources (e.g., information, effort and time) (Schneider and Bowen 1995; Lusch and Vargo 2006). According to the service-dominant logic, all social and economic actors are resource integrators (Vargo 2008) who co-create value. Co-creation of value could be distinguished into co-creation for use (for the benefit of the customer) and co-creation for others (for the benefit of other customers) (Humphreys and Grayson 2008). In contact, research has shown that consumers of health care services could not only provide information but also ideas for new service development during the service encounter stage (Elg et al. 2012).

Service encounters are complex processes where consumer interactions and surrounding environmental factors form consumers' expectations (Coye 2004), satisfaction, loyalty, repurchase intentions and word-of-mouth behaviour (Bitner, Brown and Meuter 2000). The service encounter is generally considered a service delivery process, often involving a sequence of related events occurring at different points in time. When consumers visit the service delivery facility, they enter a service 'factory' (e.g., a motel is a lodging factory and a hospital is a health treatment factory) (Noone and Mattila 2009). However, service providers focus on 'processing' people rather than the inanimate objects found in traditional goods factories. Consumers are exposed to many physical clues about the firm during the service delivery process. These include the exterior and interior of its buildings, equipment and furnishings, as well as the appearance and behaviour of service personnel and other customers. The performance along these dimensions constitutes a significant predictor of consumer satisfaction (Verhoef, Antonides and de Hoog 2004).

An integrative model of service encounters
The proposed service encounter model is an integration of the servuction (combining the terms 'service' and 'production') model and the servicescape/environmental model (Figure 7.3). The servuction model focuses on the various types of interactions that take place in a service encounter and together create the consumer's service experience. The servuction system consists of a technical core invisible to the customer and the service

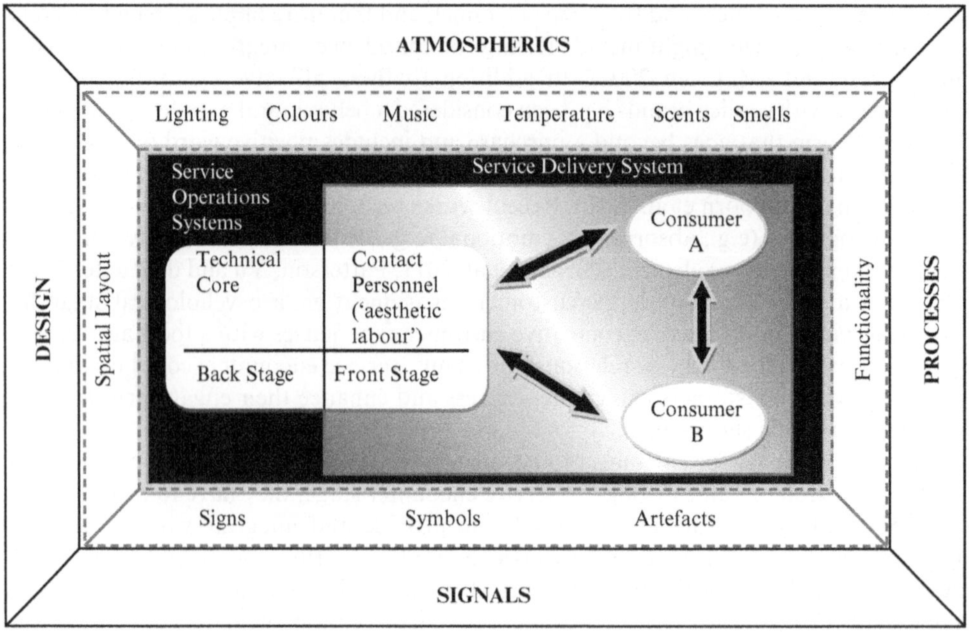

Figure 7.3 — An integrative model of service encounters

Outer frame labels: DESIGN · PROCESSES · ATMOSPHERICS · SIGNALS

Atmospherics: Lighting, Colours, Music, Temperature, Scents, Smells

Signals: Signs, Symbols, Artefacts

Design side: Spatial Layout
Processes side: Functionality

Service Operations Systems / Service Delivery System:
- Technical Core
- Contact Personnel ('aesthetic labour')
- Back Stage | Front Stage
- Consumer A
- Consumer B

Source: Tsiotsou and Wirtz (2012).

Figure 7.3 An integrative model of service encounters

delivery system visible to and experienced by the consumer (Eiglier and Langeard 1977; Langeard et al. 1981). As in the theatre, the visible components can be termed 'front stage' or 'front office', while the invisible components can be termed 'back stage' or 'back office' (Chase 1978; Grove, Fisk and John 2000).

The servuction system includes all the interactions that together make up a typical consumer experience in a high-contact service. Consumers interact with the service environment, service employees and even other consumers present during the service encounter. Each type of interaction can either create value (e.g., a pleasant environment, friendly and competent employees, other consumers who are interesting to observe) or destroy value (e.g., another consumer blocking your view in a movie theatre). Firms have to coordinate all interactions to ensure their consumers have the service experience for which they came.

Servicescapes

The Servicescape perspective considers all the experiential elements consumers encounter in a service context. The physical service environment consumers experience plays a significant role in shaping the service experience and enhancing (or undermining) consumer satisfaction, especially in high-contact people-processing services. Service environments, also called servicescapes, relate to the style and appearance of the physical surroundings and other experiential elements encountered by consumers at service delivery sites (Bitner 1992).

According to Lovelock and Wirtz (2011, p. 255), servicescapes serve four purposes: (1) they engineer the consumer experience and shape consumer behaviour; (2) they convey the planned image of the firm and support its positioning and differentiation strategy; (3) they are part of the value proposition; and (4) they facilitate the service encounter and enhance both service quality and productivity.

Bitner (1992) identified several dimensions of service environments, including ambient conditions, spatial layout/functionality, and signs, symbols and artefacts. *Ambient conditions* refer to environmental characteristics that pertain to the five senses. Ambient conditions are perceived both separately and holistically and include lighting and colour schemes, size and shape perceptions, sounds such as noise and music, temperature, and scents or smells. *Spatial layout* refers to environmental design and includes the floor plan, the size and shape of furnishings, counters, and potential machinery and equipment, and the ways in which they are arranged. *Functionality* refers to the ability of such items to facilitate the performance of service transactions and, therefore, the *process* of delivering the core service. Spatial layout and functionality create the visual and functional servicescape in which delivery and consumption take place. *Signs, symbols and artefacts* communicate the firm's image, help consumers find their way and convey the service script (the scenario consumers and employees should enact). Signals are aimed at guiding consumers clearly through the service delivery process and teaching the service script in an intuitive manner. Because individuals tend to perceive these dimensions holistically, the key to effective design is how well each individual dimension fits together with everything else (Bitner 1992).

Building on Bitner's (1992) servicescape model and theoretical perspectives on behavioural settings, approach-avoidance models and social facilitation theory, Tombs and McColl-Kennedy (2003) propose the social-servicescape model to conceptualize human elements and provide an account of how they influence consumption experiences. The social-servicescape model recognizes three separate aspects of the overall service experience: elements of the social-servicescape (including the purchase occasion as context and social interaction aspects), consumers' affective responses and consumers' cognitive responses. The social-servicescape model explains the influence of social interaction on consumer affect through social density, the displayed emotions of others, the susceptibility of the consumer to emotional contagion, and consumer awareness of the emotions of others (Tombs and McColl-Kennedy 2003).

In order to assist customers in satisfying their social and physical motives, service firms design servicescapes that create a communal atmosphere and facilitate personalization and ownership. However, often customers exhibit territorial behaviour, which has both positive and negative outcomes. On the one hand, territorial behaviour exhibited by regular customers such as occupying a whole table in a cafe or smoking in a restaurant is an indication of customer comfort and relaxation which might increase their loyalty. On the other hand, such behaviours might affect negatively the service operations as well as other customers' service experience. A recent study by Griffiths and Gilly (2012) has shown that certain servicescape designs encourage approach and territorial behaviours, which in turn positively affect territorial customers' loyalty, and avoidance behaviour by other customers.

Service scripts

Service scripts could assist in all interactions that take place within a service encounter by specifying the behavioural sequences employees and consumers are expected to learn and follow during the service delivery process. Employees receive formal training (cf., Grandey et al. 2010), whereas consumers learn scripts through experience, observation, communication with others, and designed communications and education (Harris, Harris and Baron 2003). 'Customers are not only capable of detecting the presence or absence of a script but can also detect the degree of scripting' (Victorino et al. 2012, p. 397). Moreover, customers' capability to recognize the script of a service encounter does not differ between standardized and customized services (Victorino et al. 2012). The more experience a consumer has with a service company, the more familiar that particular script becomes. Any deviation from this known script may frustrate both consumers and employees and can lead to dissatisfaction. If a company decides to change a service script (e.g., by using technology to transform a high-contact service into a low-contact one), service personnel and consumers need to be educated about the new approach and the benefits it provides. In addition, unwillingness to learn a new script can give customers a reason not to switch to a competing service provider.

Many service dramas are tightly scripted (such as flight attendants' scripts for economy class), thus reducing variability and ensuring uniform quality. However, not all services involve tightly scripted performances. Scripts tend to be more flexible for providers of highly customized services – designers, educators, consultants – and may vary by situation and by consumer.

The remainder of this section on the service encounter presents low-contact service encounters and specifically voice-to-voice encounters and self-service encounters.

Low-contact service encounters

Low-contact services involve little, if any, physical contact between consumers and service providers. Instead, contact takes place at arm's length through electronic or physical distribution channels. In practice, many high-contact and medium-contact services are becoming low-contact services as part of a fast-growing trend whereby convenience plays an increasingly important role in consumer choice (Lovelock and Wirtz 2011). Voice-to-voice and self-service encounters have become increasingly common and have recently attracted research interest.

Voice-to-voice encounters Voice-to-voice encounters have, until recently, been an under-investigated topic in the service literature. Service encounters with a telephone-based customer service representative are often moments of truth that influence consumers' perceptions of a firm. Voice-to-voice encounters can be important because the telephone is frequently the initial contact medium for the consumer (e.g., price checking) with a firm (Unzicker 1999), they can lead to purchase or non-purchase decisions, they are increasingly used as the platform through which transactions are conducted (e.g., making a booking or placing an order), and are used as a channel for after-sales service and service recovery processes (Whiting and Donthu 2006).

Voice-to-voice encounters play a significant role in developing, sustaining and managing consumer relationships (Anton 2000) and enhancing satisfaction (Feinberg et al. 2002). Customers expect that employees responding to their calls will exhibit 'adaptive-ness', 'assurance', 'empathy' and 'authority'. That is, it is expected that the call centre representative will adjust his or her behaviour to the customer, provide clear information to the customer about the procedures, will empathise with the customer's emotions/ situation, and has the authority to solve problems and answer questions (Burgers et al. 2000). Voice-to-voice encounters typically involve waiting time, music and information. Music and information have become two common tools firms use to keep consumers occupied while they wait and thereby reduce their perceptions of waiting time. However, recent research has shown that it is only when the customer likes the music that it reduces the perceived waiting time and increases satisfaction (Whiting and Donthu 2006).

Self-service encounters Self-service technology-enabled encounters allow for the production and consumption of services without relying on service personnel (e.g., automated teller machines, self-scanning checkouts and Internet banking). Self-service technologies (SSTs) allow consumers to 'produce a service independent of direct service employee involvement' (Curran, Meuter and Surprenant 2003, p. 209). For consumers, SSTs often require the co-production of services, increased cognitive involvement and new forms of service behaviour, while they can offer greater customization and more satisfying experiences (Meuter et al. 2000; Prahalad and Ramaswamy 2004). However, self-service encounters not only benefit consumers but also frequently benefit service providers by providing them with direct and immediate feedback from their consumers (Voorhees and Brady 2005), so improving service design, developing consumer loyalty (Voss et al. 2004) and reducing costs (Heracleous and Wirtz 2006).

Research on the application of SSTs has focused on factors that either facilitate or inhibit their adoption and usage by customers. Perceived usefulness, ease of use, reliability and fun have been identified as key drivers of consumer attitudes toward SSTs (Weijters et al. 2007). Dabholkar, Bobbit and Lee (2003) consider self-scanning checkouts in retail stores and find that control, reliability, ease of use and enjoyment are important usage determinants of this kind of SST. Consumer characteristics such as a lack of confidence, anxiety, technology-related attitudes and self-efficacy might inhibit the use of SSTs and successful co-production, especially in complex services (Boyle, Clark and Burns 2006; Dabholkar and Bagozzi 2002; Meuter et al. 2000).

Consumers are often dissatisfied with SSTs if they deliver poor service (Meuter et al. 2000) or the technology fails (Holloway and Beatty 2003; Meuter et al. 2000), and if they cause frustration they might engender poor service delivery and technological failure (Harris et al. 2006). Due to these reasons and because SSTs might deter consumers from voicing their complaints (Forbes, Kelley and Hoffman 2005), consumers might avoid engaging in SST-enabled encounters (Bitner, Ostrom and Meuter 2002) and even switch service providers (Forbes, Kelley and Hoffman 2005). SST-enabled service encounters also reduce the opportunity for service providers to get in touch with consumers, determine their emotional state (Freidman and Currall 2003) and detect service failures (Pujari 2004). This research shows that SSTs have enormous potential but need to be designed with great care and attention to consumer needs and behaviours.

The Post-encounter Stage

The last stage of service consumption is the post-encounter stage and involves consumers' behavioural and attitudinal responses to the service experience. Consumer satisfaction and perceived service quality have dominated the research agenda at this stage of the service consumption process due to their association with business performance (Brady and Robertson 2001). However, consumers who are satisfied and have high perceptions of service quality do not necessarily return to the same service provider or buy their services again (cf., Keiningham and Vavra 2001). As a result, there has recently been a shift in the consumer research agenda toward other important post-purchase outcomes, such as perceived service value, consumer delight, consumer reactions to service failures (e.g., complaining and switching behaviour) and consumer responses to service recovery.

Customer satisfaction with services

Consumer satisfaction with services has been explained by several conceptual models such as the expectancy–disconfirmation paradigm (Oliver 1980) and the perceived performance model (Churchill and Surprenant 1982), as well as attribution models (Folkes 1984), affective models (Mattila and Wirtz 2000; Westbrook 1987; Wirtz and Bateson 1999) and equity models (Oliver and DeSarbo 1988).

The following section describes two prevailing approaches – the expectancy–disconfirmation paradigm and the attribution model of satisfaction – and reviews current research supporting these approaches.

The expectancy–disconfirmation paradigm Most customer satisfaction research is based on the expectancy–disconfirmation model of satisfaction (Oliver 1980), where confirmation or disconfirmation of consumers' expectations is the key determinant of satisfaction (Oliver 1980; Wirtz and Mattila 2001). According to the expectancy–disconfirmation paradigm, consumers evaluate the service performance they have experienced and compare it with their prior expectations (Figure 7.4).

Consumers will be reasonably satisfied as long as perceived performance falls within the zone of tolerance, that is, above the adequate service level. When performance perceptions approach or exceed desired levels, consumers will be very pleased. Consumers with such perceptions are more likely to make repeat purchases, remain loyal to the service provider and spread positive word of mouth (Liang, Wang and Farquhar 2009; Wirtz and Chew 2002). Thus, satisfaction is related to important post-purchase attitudes and behaviours such as consumer loyalty (Vazquez-Carrasco and Foxall 2006; Yang and Peterson 2004), frequency of service use (Bolton and Lemon 1999), repurchase intentions (Cronin, Brady and Hult 2000), service recommendations to acquaintances (Zeithaml, Berry and Parasuraman 1996) and compliments to service providers (Goetzinger, Park and Widdows 2006).

When service performance is well above the expected level, consumers might be delighted. Consumer delight is a function of three components: (1) unexpectedly high levels of performance; (2) arousal (e.g., surprise, excitement); and (3) positive affect (e.g., pleasure, joy or happiness) (Oliver, Rust and Varki 1997). Consumer delight is distinct from consumer satisfaction and has its own responses to a service experience. Consumer delight has a threshold above which each increase has a greater impact on behavioural

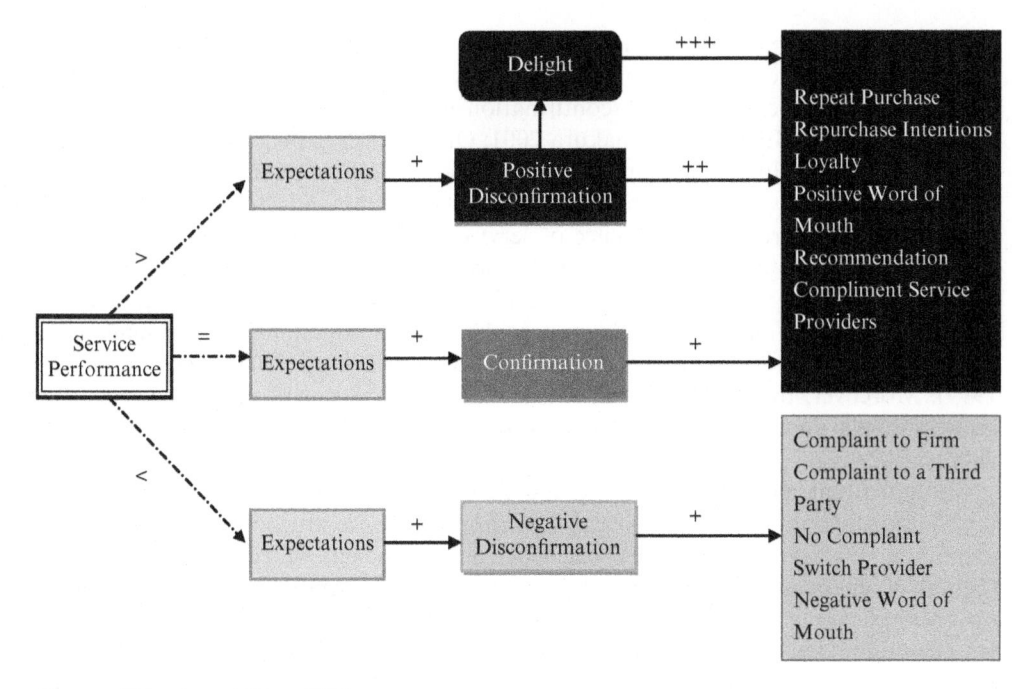

Figure 7.4 Consumer satisfaction (expectancy/disconfirmation) and its outcomes in services

Source: Tsiotsou and Wirtz (2012).

intentions (Finn 2012). However, delight might not always act in favour of the service firm, because it raises consumers' expectations (Santos and Boote 2003). This can lead to consumers becoming dissatisfied if service levels return to the previously lower levels, and it will probably take more effort to delight them in the future (Rust and Oliver 2000). Some firms are therefore strategically focusing delighting customers on soft factors (e.g., personalization) rather than hard process factors (e.g., give a free birthday cake on a customer's birthday). The latter creates hard expectations – the birthday cake soon becomes 'as expected' and loses its power to delight. For the former, customers are less likely to develop hard and raised expectations and customers can continuously be wowed by the firm's excellent service delivery (Heracleous and Wirtz 2010).

The expectancy–disconfirmation framework generally works well when consumers have sufficient information and experience to purposefully choose a service from the consideration set expected to best meet their needs and wants (Wirtz and Mattila 2001). However, this may not always be the case for services. For example, the expectancy–disconfirmation model seems to work very well for services with search and experience attributes, but less so for those with credence attributes. Consumers cannot assess the latter type of attributes directly and rely on tangible cues and expectations to form their views on satisfaction. If no tangible evidence contradicts their expectations, customers tend to evaluate credence attributes as meeting their expectations and will be satisfied (Wirtz and Mattila 2001).

The attribute-based approach to satisfaction Attribute-based perspectives are frequently used for explaining consumer satisfaction because they complement the multi-attribute choice models and expectancy–disconfirmation paradigm (Busacca and Padula 2005; Kano et al. 1984; Mittal and Kamakura 2001; Oliver 2000, p. 247). Based on the study of Weiner (2000), Oliver (2009, pp. 302–303) proposed that expectancy–disconfirmation precedes attribute evaluations, which in turn affect consumer satisfaction. Recent empirical evidence supports the significance of service attributes in influencing overall satisfaction (Akhter 2010; Mittal, Kumar and Tsiros 1999). The attribute-based approach argues that both cognitive (expectations) and affective (desires-motives associated with personal objectives) elements should be considered when examining the consumer satisfaction formation process (Bassi and Guido 2006; Oliver 2000, p. 250; Wirtz and Bateson 1999). Moreover, the affective component of satisfaction is expected to be greater in services than in goods due to the interactive and experiential nature of the former (Oliver 2000, p. 252).

Multi-attribute models provide several benefits to theory and practice in understanding the satisfaction formation process. Focusing on service attributes: (a) is useful for identifying the specific attributes which act as antecedents of customer satisfaction (Mittal, Kumar and Tsiros 1999); (b) facilitates the conceptualization of commonly observed phenomena such as mixed feelings toward a service (consumers are satisfied with certain attributes and dissatisfied with others) (Mittal, Ross and Baldasare 1998); (c) allows customers to render evaluations of their post-purchase experiences at an attribute level rather than only at the product level (Gardial et al. 1994); and (d) helps firms identify and manage attributes that have a strong impact on satisfaction and dissatisfaction (Mittal, Ross and Baldasare 1998).

The attribute-based approach considers the evaluation of different attributes of a service as an antecedent of overall satisfaction (Oliver 1993). Singh (1991) supports that there is sufficient and compelling evidence to suggest consumer satisfaction can be considered a collection of multiple satisfactions with various attributes of the service experience. Satisfaction with service attributes thus results from the observation of attribute-specific performance and strongly influences consumers' overall satisfaction (Oliver 1993).

Although these satisfaction approaches offer a framework with which to examine and understand consumer behaviour, they encourage the adoption of a 'zero defects' service paradigm (Bowden 2009). In other words, in their effort to maximize satisfaction, these models treat all consumers within the consumer base as homogeneous. For example, they regard newly acquired consumers as the same as loyal consumers, although the two groups might differ in the importance they place on each attribute (Mittal and Kamakura 2001). Furthermore, service consumers cannot always freely choose the service that best fits their needs, wants and desires. Services are time and location specific, both of which restrict consumer choice, and consumers are frequently locked into a specific provider. For example, in situations where switching costs are high, needs congruency would be a better comparison standard for modelling satisfaction than would expectations (Wirtz and Mattila 2001). Consumers use multiple standards in the satisfaction process (e.g., expectations as well as needs), and because needs-congruence explains satisfaction better than do expectations, it should be incorporated into the modelling of satisfaction in reduced consumer choice situations.

Due to the above deficiencies in existing perspectives modelling satisfaction, we propose an integrative model that combines the abovementioned perspectives to provide a more comprehensive framework for explaining the formation of service satisfaction and its outcomes.

AN INTEGRATIVE MODEL OF SERVICE SATISFACTION AND BEHAVIOURAL INTENTIONS

Our model supports that when consumers use a service, they rate its transaction quality (e.g., the quality of food, the friendliness of the server and the ambiance of a restaurant), which when combined with the satisfaction derived from key attributes (i.e., attribute satisfaction) and the perceived value of the specific transaction then lead to a judgment of the level of overall satisfaction with a particular service experience. Over time and over many satisfaction judgements, customers then form a belief about the overall service quality a firm offers. This in turn influences behavioural intentions (e.g., purchase intentions, remaining loyal to the firm and positive word of mouth) (see Figure 7.5).

Using the general living systems theory, Mittal, Kumar and Tsiros (1999) propose that a consumption system consists of attribute-level evaluations, satisfaction and behavioural intentions, and several subsystems. Their study shows that evaluations of a number of attributes lead to an overall level of satisfaction, which in turn influences customers' behavioural intentions. A service encounter is a multi-attribute experience

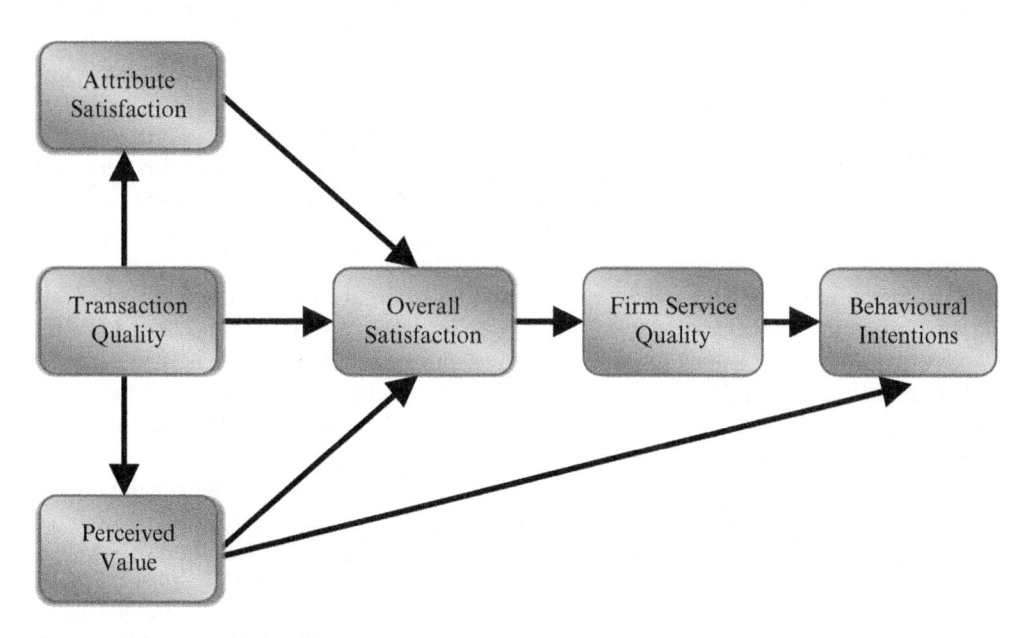

Source: Tsiotsou and Wirtz (2012).

Figure 7.5 Consumer satisfaction (expectancy/disconfirmation) and its outcomes in services

comprising satisfaction with service attributes, such as the provider, the offering, the location, information and facilitation, which together form overall satisfaction (Akhter 2010). Overall satisfaction reflects the level of satisfaction with the overall service experience, and is a global evaluation of a specific service consumption experience.

The attribute-based model has also been used in an online context to explain the link between SST attributes and quality satisfaction. Efficiency, ease of use, performance, perceived control and convenience have been identified as the main Internet-based self-service technology (ISST) attributes determining consumer satisfaction with service quality (Yen 2005). Thus, consumer satisfaction with ISST is not only a function of the benefits associated with its usage (e.g., convenience) and the attributes related to reduced barriers to use (e.g., ease of use), but also of its ability to perform the expected functions properly.

However, the relationship between attribute-level performance and overall satisfaction is more complex than it may seem. Evidence has shown that there is a nonlinear and asymmetric relationship between service attribute importance and attribute-level performance evaluations, a relationship that can be unstable over time (Busacca and Padula 2005; Kano et al. 1984; Mittal and Kamakura 2001). Research on consumer delight suggests that there is a nonlinear relationship in attribute-based judgements, probably due to the role affect plays in consumer satisfaction judgements as opposed to the weighting or importance consumers assign to a particular attribute only (Bowden 2009). Moreover, the phenomenon of 'fundamental attribution error' has been observed in the literature (Oliver 2000, p. 252). According to this 'error', negative attribute performance has a greater effect on overall satisfaction than does positive attribute performance (Mittal, Ross and Baldasare 1998). Furthermore, research shows that attribute weights do not remain stable but change over time due to modified consumer goals (Mittal, Kumar and Tsiros 1999). One possible explanation for these findings might be found in the dimensions of attributions proposed by Weiner (2000) and adopted in the marketing field by Oliver (2010, pp. 295–296). The locus of causality (internal-self vs. external-others), the stability of service attribute performance (stable vs. variable) and the degree to which an attribute is under the control of the service provider might influence the relationship between attribute satisfaction and overall satisfaction.

Furthermore, it has been shown that the halo effect can threaten the interpretability of such attribute-specific satisfaction data. For example, a long waiting time not only lowers the attribute rating of speed of service, but research has shown that all other attribute ratings are likely to be reduced as well (Wirtz and Bateson 1995). Halo is particularly acute in satisfaction measurement of services with a high degree of ambiguous and credence attributes (Wirtz 2003).

Finally, factors other than attribute-level evaluations might also influence the formation of consumer satisfaction. Spreng, MacKenzie and Olshavsky (1996, p. 17) stated that 'attribute-specific satisfaction is not the only antecedent of overall satisfaction, which is based on the overall experience, not just the individual attributes'. Lages and Fernandes (2005) suggest that any evaluation of a service provider is made at four abstract levels of a hierarchy, comprising simple attributes of the service offering, transactional service quality, value and more complex personal values. The present model proposes that, in addition to attribute satisfaction, transaction quality and service values are further antecedents of overall satisfaction with services.

The Role of Service Quality and Service Value

Before further proceeding to explain the model in Figure 7.5, it is necessary to distinguish between the transaction-specific and service-related aspects of service quality. Transaction-specific quality refers to consumers' perceptions of a specific service encounter experience, whereas a firm's service quality reflects evaluations of quality based on cumulative experience that are developed over time. Inconsistencies in the literature regarding the role of service quality in relation to satisfaction and purchase intentions can be attributed to interchangeable use of the above types of service quality, which are often not distinguished from each other. We thus posit that transaction quality precedes overall consumer satisfaction, which in turn influences the formation of perceptions of a product's or firm's overall service quality.

At a transaction level, it has been proposed that perceptions of the quality of service attributes are antecedents of satisfaction with the service experience (Otto and Ritchie 1995). Wilson et al. (2008, pp. 78–79) have proposed that satisfaction results from service quality evaluations (in addition to product quality and price) that mirror consumers' perceptions of its five dimensions: reliability, responsiveness, assurance, empathy and tangibles. It should be noted here that the early service literature considered these dimensions to be components of the perceived service quality of the firm (Boulding et al. 1993) and not as transaction specific. Brady and Cronin (2001) proposed that service quality is a multifaceted concept comprising three dimensions and nine sub-dimensions (in parentheses): interaction quality (attitudes, behaviour and expertise), physical environment quality (ambient conditions, design and social factors) and outcome quality (waiting time, tangibles and valence). Consumers evaluate service quality based on these three dimensions assessed via each of their three corresponding sub-dimensions. Additional empirical evidence has also demonstrated that the quality of the service delivery personnel (Johnson and Zinkham 1991) and physical environment (Bitner 1992) attributes have an impact on satisfaction with the service experience.

Another construct gaining increasing research attention is service value. Service value is the 'utility of a product based on perceptions of what is received and what is given' (Zeithaml 1988, p. 14). Empirical evidence shows that transaction service quality is a significant determinant of service value (Cronin, Brady and Hult 2000; Hu, Kandampully and Juwaheer 2009). Perceived service value is considered highly personal, idiosyncratic and variable among consumers (Holbrook 1994). It also seems reasonable to suggest that consumers evaluate transaction-specific attributes first before evaluating the service value of the service encounter experience. Transaction quality-related attributes may therefore represent most of the positive benefit drivers of consumer service value (Hu, Kandampully and Juwaheer 2009). Moreover, service value has been shown to have a direct effect on both consumer satisfaction with the service experience (Cronin, Brady and Hult 2000; Hu, Kandampully and Juwaheer 2009; Lin, Sher and Shih 2005; Varki and Colgate 2001) and behavioural intentions (Cronin, Brady and Hult 2000; Hu, Kandampully and Juwaheer 2009).

Service quality at the firm level has been linked to consumers' behavioural intentions. Boulding et al. (1993) conducted two studies in a service context and found that consumers' perceptions of a firm's overall service quality will influence their behaviour intentions expressed as positive word of mouth and recommendation of the service. Perceptions of a firm's overall service quality are relatively stable but will change over time in the same direction as transaction satisfaction ratings (Boulding et al. 1993; Palmer and O'Neill 2003). Consumers' repurchase intentions are influenced by their perceptions of overall service quality at the time of repurchase (i.e., consumers try to predict how good the next service transaction will be), and not by the individual transaction satisfaction formed immediately after a consumption experience (Boulding et al. 1993; Palmer and O'Neill 2003). For example, consumers might return to a hair stylist if they think the stylist is generally fantastic, even if they were unhappy the last time they went there because they believe the poor experience was an exception. However, a second or even third dissatisfaction evaluation will reduce the overall service quality perception of the firm more dramatically and jeopardize repeat purchases.

The strength of the relationship between satisfaction and consumers' behavioural intentions is often influenced, moderated or mediated by other factors. For example, consumers' adjusted expectations (Yi and La 2004) and characteristics such as personality traits (e.g., the need for social affiliation and relationship proneness) may act as mediators (Vazquez-Carrasco and Foxall 2006), whereas switching costs and consumer demographics (e.g., age and income) may act as moderators (Homburg and Giering 2001; Wirtz et al. 2014) in the relationship between satisfaction and behavioural outcomes.

DISCUSSION

Consumer behaviour in the services context has increasingly attracted research attention across all three stages of the consumption process. However, post-purchase behaviour seems to dominate consumer behaviour research in the services field, with the other two stages – the pre-purchase and service encounter stages – being under-investigated and requiring further research attention. Moreover, the consumer behaviour literature in services has gradually become delinked from the goods perspective and has moved on from merely adapting models developed in the goods literature and trying to apply and contrast them to a service context. New models and approaches (e.g., the servuction model, the servicescape/environmental approach and relationship marketing) have increasingly been developed from a service perspective.

In addition to presenting new developments in the consumer behaviour literature in services, this chapter also identifies several research gaps that warrant further attention. The first has emerged from the realization that the influence of the service environment on consumers' emotional reactions, evaluations and behaviour is more complicated than generally assumed. New research developments indicate that the effects of environmental elements depend on the service setting (e.g., private vs. public), the congruency between these elements, and consumers' individual characteristics. Research is needed to further clarify the complexities involved in the influence of the service environment, not only on consumers, but also on employees and the social interactions taking place in a servicescape.

The application of new technologies and their impact throughout the three stages of service consumption is another important area for further research. The growth of new technologies, ranging from smart-phone apps to biometrics, and their use in services are giving rise to questions about their acceptance. The Internet has brought about several changes in consumer expectations, as well as true interactivity, consumer-specific, situational personalization, and the opportunity for real-time adjustments to a firm's offerings (Rust and Lemon 2001). One can expect the advent of smart phones and tablet computers (e.g., the iPhone and iPad) with their many applications being created by individual service firms (e.g., Singapore taxi firms have created applications to make booking taxis easier) to further revolutionize self-service applications.

Furthermore, the Internet has changed the role of consumers from being simply receivers of services to becoming actively involved in the production and delivery processes (Xue and Harker 2002). These new consumer roles and determinants of the co-creation of value in e-services need further examination. In an online context, e-service quality dominates the literature and is followed in importance by e-service value (Parasuraman, Zeithaml and Malhotra 2005; Santos 2003). However, all recent e-service quality and value models are based on traditional service models. Because consumer evaluations of e-services and mobile services differ from those of traditional offline services (Rust and Lemon 2001), there is a need to develop and test new models of e-service quality and value (Parasuraman, Zeithaml and Malhotra 2005).

The role of avatars in enhancing the consumer experience, increasing trust and loyalty and developing consumer relationships with service providers has not been investigated in much detail in the service literature. There is also limited research related to the use of recommendation agents and mobile services and their effect in improving the quality of consumer decisions (Haubl and Trifts 2000).

Finally, more research on consumer behaviour in a service context is needed to shed light on various aspects of the purchase decision process and the development of consumer–firm relationships. In sum, this chapter provides an overview of key developments in the consumer behaviour literature in the services field and highlights relevant issues warranting further research attention.

NOTE

1. This chapter is based on Lovelock and Wirtz (2011) and Tsiotsou and Wirtz (2012). It is an adapted and updated version of these earlier publications.

REFERENCES

Akhter, S.H. (2010), 'Service attributes satisfaction and actual repurchase behaviour: The mediating influence of overall satisfaction and intention', *Journal of Satisfaction and Dissatisfaction and Complaining Behavior*, **23**, 52–64.
Aksoy, L., P.N. Bloom, N.H. Lurie and B. Cooil (2006), 'Should recommendation agents think like people?', *Journal of Service Research*, **8** (4), 297–315.
Alba, J.W. and J.W. Hutchinson (2000), 'Knowledge calibration: What consumers know and what they think they know', *Journal of Consumer Research*, **27** (2), 123–156.
Anton, J. (2000), 'The past, present, and future of customer access centers', *International Journal of Service Industry Management*, **11** (2), 120–130.
Ariely, D., J.G. Lynch Jr. and M. Aparicio IV (2004), 'Learning by collaborative and individual-based recommendation agents', *Journal of Consumer Psychology*, **14** (1/2), 81–95.
Bansal, H.S. and P.A. Voyer (2000), 'Word-of-mouth processes within a services purchase decision context', *Journal of Service Research*, **3** (2), 166–177.

Bassi, F. and G. Guido (2006), 'Measuring customer satisfaction: From product performance to consumption experience', *Journal of Consumer Satisfaction, Dissatisfaction, and Complaining Behavior*, **19**, 76–89.
Bhatnagar, A. and S. Ghose (2004), 'Online information search termination patterns across product categories and consumer demographics', *Journal of Retailing*, **80** (3), 221–228.
Bitner, M.J. (1992), 'Servicescapes: The impact of physical surroundings on customers and employees', *Journal of Marketing*, **56**, 57–71.
Bitner, M.J., S. Brown and M. Meuter (2000), 'Technology infusion in service encounters', *Journal of the Academy of Marketing Science*, **28** (1), 138–149.
Bitner, M.J., A.L. Ostrom and M.L. Meuter (2002), 'Implementing successful self-service technologies', *Academy of Management Executive*, **16** (4), 96–109.
Blackwell, Roger D., Paul W. Miniard and James F. Engel (eds) (2003), *Consumer Behavior*, Orlando, FL: Harcourt College Publishers.
Bolton, R.N. and K.N. Lemon (1999), 'A dynamic model of customers' usage of services: Usage as an antecedent and consequence of satisfaction', *Journal of Marketing Research*, **36** (2), 171–186.
Boshoff, C. (2002), 'Service advertising: An exploratory study of risk perceptions', *Journal of Service Research*, **4** (4), 290–298.
Boulding, W., A. Kalia, R. Staelin and V.A. Zeithaml (1993), 'A dynamic process model of service quality: From expectations to behavioural intentions', *Journal of Marketing Research*, **30** (1), 7–27.
Bowden, J.LH. (2009), 'The process of customer engagement: A conceptual framework', *Journal of Marketing Theory and Practice*, **17** (1), 63–74.
Boyle, D., S. Clark and S. Burns (2006), *Hidden Work: Co-production by People Outside Paid Employment*, York: Joseph Rowntree Foundation.
Brady, M.K. and J.J. Cronin Jr. (2001), 'Some new thoughts on conceptualizing perceived service quality: A hierarchical approach', *Journal of Marketing*, **65** (3), 34–49.
Brady, M.K. and C.J. Robertson (2001), 'Searching for a consensus on the antecedent role of service quality and satisfaction: An exploratory cross-national study', *Journal of Business Research*, **51** (1), 53–60.
Brodie, R.J., L.D. Hollebeek, B. Juric and A. Ilic (2011), 'Customer engagement: Conceptual domain, fundamental propositions and implications for research', *Journal of Service Research*, **14** (3), 252–271.
Burgers, A., K. de Ruyter, C. Keen and S. Streukens (2000), 'Customer expectation dimensions of voice-to-voice service encounters: A scale-development study', *International Journal of Service Industry Management*, **11** (2), 142–161.
Busacca, B. and G. Padula (2005), 'Understanding the relationship between attribute performance and overall satisfaction: Theory, measurement and implications', *Marketing Intelligence & Planning*, **23** (6), 543–561.
Byrne, K. (2005), 'How do consumers evaluate risk in financial products?', *Journal of Financial Services Marketing*, **10** (1), 21–36.
Chase, R.B. (1978), 'Where does the customer fit in a service organization?', *Harvard Business Review*, **56** (November/December), 137–142.
Churchill, G.A. Jr. and C. Surprenant (1982), 'An investigation into the determinants of customer satisfaction', *Journal of Marketing Research*, **19** (4), 491–504.
Coye, R.W. (2004), 'Managing customer expectations in the service encounter', *International Journal of Service Industry Management*, **15** (1), 54–71.
Cronin, J.J. Jr., M.K. Brady and G.T.M. Hult (2000), 'Assessing the effects of quality, value, and customer satisfaction on consumer behavioural intentions in service environments', *Journal of Retailing*, **76** (2), 193–217.
Curran, J.M., M.L. Meuter and C.F. Surprenant (2003), 'Intentions to use self-service technologies: A confluence of multiple attitudes', *Journal of Service Research*, **5** (3), 209–224.
Dabholkar, P.A. and R.P. Bagozzi (2002), 'An attitudinal model of technology-based self-service: Moderating effects of consumer traits and situational factors', *Journal of the Academy of Marketing Science*, **30** (3), 184–201.
Dabholkar, P.A., L.M. Bobbit and E.-J. Lee (2003), 'Understanding consumer motivation and behaviour related to self-scanning in retailing: Implications for strategy and research on technology-based self-service', *International Journal of Service Industry Management*, **14** (1), 59–95.
Darby, M.R. and E. Karni (1973), 'Free competition and the optimal amount of fraud', *Journal of Law and Economics*, **16** (1), 67–86.
Diacon, S. and C. Ennew (2001), 'Consumer perceptions of financial risk', *Geneva Papers on Risk and Insurance: Issues and Practice*, **26** (3), 389–409.
Diehl, K., L.J. Kornish and J.G. Lynch Jr. (2003), 'Smart agents: When lower search costs for quality information increase price sensitivity', *Journal of Consumer Research*, **30** (1), 56–71.
Eiglier, P. and E. Langeard (1977), 'A new approach to service marketing', *Marketing Consumer Services: New Insights*, 31–58.
Elg, M., J. Engström, L. Witell and B. Poksinska (2012), 'Co-creation and learning in health-care service development', *Journal of Service Management*, **23** (3), 328–343.

Feinberg, R., L. Hokama, R. Kadam and I. Kim (2002), 'Operational determinants of caller satisfaction in the banking/financial service call center', *International Journal of Bank Marketing*, **20**, 174–180.
Finn, A. (2012), 'Customer delight: Distinct construct or zone of nonlinear response to customer satisfaction?', *Journal of Service Research*, **15** (1), 99–110.
Fisk, Raymond P. (1981), 'Toward a consumption/evaluation process model for services', in J.H. Donnelly and W.R. George (eds), *Marketing of Services*, Chicago, IL: American Marketing Association, pp. 191–195.
Folkes, V.S. (1984), 'Consumer reactions to product failure: An attributional approach', *Journal of Consumer Research*, **10** (4), 398–409.
Forbes, L.P., S.W. Kelley and K.D. Hoffman (2005), 'Typologies of e-commerce retail failures and recovery strategies', *Journal of Service Marketing*, **19** (5), 280–292.
Friedman, R.A. and S.C. Currall (2003), 'E-mail escalation: Dispute exacerbating elements of e-mail communication', *Human Relations*, **56**, 1325–1348.
Galetzka, M., J.W.M. Verhoeven and T.H. Pruyn (2006), 'Service validity and service reliability of search, experience and credence services', *International Journal of Service Industry Management*, **17** (3), 271–283.
Gardial, S.F., D.S. Clemons, R.B. Woodruff, D.W. Schumann and M.J. Burns (1994), 'Comparing consumers' recall of prepurchase and postpurchase product evaluation experiences', *Journal of Consumer Research*, **20**, 548–560.
Goetzinger, L., J.K. Park and R. Widdows (2006), 'E-customers' third party complaining and complimenting behaviour', *International Journal of Service Industry Management*, **17** (2), 193–206.
Goldsmith, Ronald E. (2011), 'Brand engagement and brand loyalty', in A. Kapoor and C. Kulshrestha (eds), *Branding and Sustainable Competitive Advantage: Building Virtual Presence*, Hershey, PA: IGI Global, pp. 122–135.
Grandey, A., A. Rafaeli, S. Ravid, J. Wirtz and D.D. Steiner (2010), 'Emotion display rules at work in the global service economy: The special case of the customer', *Journal of Service Management*, **21** (3), 388–412.
Griffiths, M.A. and M.C. Gilly (2012), 'Dibs! Customer territorial behaviors', *Journal of Service Research*, **15** (2), 131–149.
Grönroos, C. (1994), 'From marketing mix to relationship marketing: Towards a paradigm shift in marketing', *Management Decision*, **32** (2), 4–20.
Grönroos, Christian (2000a), 'Service reflections: Service marketing comes of age', in T.A. Swartz and D. Iacobucci (eds), *Handbook of Services Marketing and Management*, London: Sage Publications, Inc., pp. 13–20.
Grönroos, C. (2000b), 'Christian Grönroos: Hanken Swedish School of Economics, Finland', in R.P. Fisk, S.F. Grove and J. Joby (eds), *Services Marketing Self-Portraits: Introspections, Reflections, and Glimpses from the Experts*, Chicago, IL: American Marketing Association, pp. 71–108.
Grove, S.J., R.P. Fisk and J. John (2000), 'Services as theater: Guidelines and implications', in T. Swartz and D. Iacobucci (eds), *Handbook of Services Marketing and Management*, Thousand Oaks, CA: Sage Publications, pp. 21–35.
Harris, K.E., D. Grewal, L.A. Mohr and K.L. Bernhardt (2006), 'Consumer responses to service recovery strategies: The moderating role of online versus offline environments', *Journal of Business Research*, **59** (4), 425–431.
Harris, R., K. Harris and S. Baron (2003), 'Theatrical service experiences: Dramatic script development with employees', *International Journal of Service Industry Management*, **14** (2), 184–199.
Haubl, G. and K.B. Murray (2003), 'Preference construction and persistence in digital marketplaces: The role of electronic recommendation agents', *Journal of Consumer Psychology*, **13** (1/2), 75–91.
Haubl, G. and V. Trifts (2000), 'Consumer decision making in online shopping environments: The effects of interactive decision aids', *Marketing Science*, **19** (1), 4–21.
Hensley, R.L. and J. Sulek (2007), 'Customer satisfaction with waits in multi-stage services', *Managing Service Quality*, **17** (2), 152–173.
Heracleous, L. and J. Wirtz (2006), 'Biometrics: The next frontier in service excellence, productivity and security in the service sector', *Managing Service Quality*, **16** (1), 12–22.
Heracleous, L. and J. Wirtz (2010), 'Singapore Airlines' balancing act – Asia's premier carrier successfully executes a dual strategy: It offers world-class service and is a cost leader', *Harvard Business Review*, **88** (7/8), 145–149.
Holbrook, Morris (1994), 'The nature of customer value, an axiology of services in the consumption experience', in R.T. Rust and R.L. Oliver (eds), *Service Quality: New Directions in Theory and Practice*, Thousand Oaks, CA: Sage Publications, pp. 21–71.
Holloway, B.B. and S.E. Beatty (2003), 'Service failure in online retailing: A recovery opportunity', *Journal of Service Research*, **6** (1), 92–106.
Holzwarth, M., C. Janiszewski and M.M. Neumann (2006), 'The influence of avatars on online consumer shopping behaviour', *Journal of Marketing*, **70** (4), 19–36.

Homburg, C. and A. Giering (2001), 'Personal characteristics as moderators of the relationship between customer satisfaction and loyalty – an empirical analysis', *Psychology & Marketing*, **18** (1), 43–66.
Hu, H., J. Kandampully and T. Juwaheer (2009), 'Relationships and impacts of service quality, perceived value, customer satisfaction, and image: An empirical study', *Service Industries Journal*, **29** (2), 111–125.
Humphreys, A. and K. Grayson (2008), 'The intersecting roles of consumer and producer: A critical perspective on co-production, co-creation and prosumption', *Sociology Compass*, **2**, 1–18.
Johnson, M. and G.M. Zinkham (1991), 'Emotional responses to a professional service encounter', *Journal of Services Marketing*, **5** (2), 5–16.
Kacen, J.J. and J.A. Lee (2002), 'The influence of culture on consumer impulsive buying behaviour', *Journal of Consumer Psychology*, **12** (2), 163–176.
Kano, N., N. Seraku, F. Takahashi and S. Tsuji (1984), 'Attractive quality and must-be quality', *Hinshitsu (Quality, The Journal of the Japanese Society for Quality Control)*, **14** (2), 39–48.
Keiningham, T.L. and T.G. Vavra (eds) (2001), *The Customer Delight Principle: Exceeding Customers' Expectations for Bottom-Line Success*, New York: McGraw-Hill.
Kimes, S.E. and J. Wirtz (2003), 'Has revenue management become acceptable? Findings from an international study on the perceived fairness of rate fences', *Journal of Service Research*, **6** (2), 125–135.
Konus, U., P.C. Verhoef and S.A. Neslin (2008), 'Multichannel shopper segments and their covariates', *Journal of Retailing*, **84** (4), 398–413.
Lages, L.F. and J.C. Fernandes (2005), 'The SERPVAL scale: A multi-item measurement instrument for measuring service personal values', *Journal of Business Research*, **58** (11), 1562–1572.
Langeard, E., J.E. Bateson, C.H. Lovelock and P. Eiglier (1981), *Services Marketing: New Insights from Consumers and Managers*, Marketing Science Institute, Report # 81–104 (August).
Liang, C.-J., W.-H. Wang and J.D. Farquhar (2009), 'The influence of customer perceptions on financial performance in financial services', *International Journal of Bank Marketing*, **27** (2), 129–149.
Lin, C.-H., P.J. Sher and H.-Yu Shih (2005), 'Past progress and future directions in conceptualizing customer perceived value', *International Journal of Service Industry Management*, **16** (4), 318–336.
Lovelock, C. and J. Wirtz (2011), *Services Marketing: People, Technology, Strategy* (7th edn), Upper Saddle River, NJ: Prentice Hall.
Lusch, R.F. and S.L. Vargo (2006), 'Service-dominant logic: Reactions, reflections, and refinements', *Marketing Theory*, **6** (September), 281–288.
Mattila, A.S. and J. Wirtz (2000), 'The role of pre-consumption affect in post-purchase evaluation of services', *Psychology & Marketing*, **17** (7), 587–605.
Mattila, A.S. and J. Wirtz (2002), 'The impact of knowledge types on the consumer search process: An investigation in the context of credence services', *International Journal of Service Industry Management*, **13** (3), 214–230.
Mattila, A.S. and J. Wirtz (2008), 'The role of environmental stimulation and social factors on impulse purchasing', *Journal of Services Marketing*, **22** (7), 562–567.
McEwen, W. (2004), 'Why satisfaction isn't satisfying', *Gallup Management Journal Online*, November (1–4). Available at: http://businessjournal.gallup.com/content/14023/why-satisfaction-isnt-satisfying.aspx (accessed 28 November 2014).
Meuter, M.L., A.L. Ostrom, R.I. Roundtree and M.J. Bitner (2000), 'Self-service technologies: understanding customer satisfaction with technology-based service encounter', *Journal of Marketing*, **64** (3), 50–64.
Mittal, V. and W.A. Kamakura (2001), 'Satisfaction and repurchase behaviour: The moderating influence of customer and market characteristics', *Journal of Marketing Research*, **38** (1), 131–142.
Mittal, V., P. Kumar and M. Tsiros (1999), 'Attribute-level performance, satisfaction, and behavioural intentions over time: A consumption-system approach', *Journal of Marketing*, **63** (2), 88–101.
Mittal, V., W.T. Ross and P.M. Baldasare (1998),'The asymmetric impact of negative and positive attribute-level performance on overall satisfaction and repurchase intentions', *Journal of Marketing*, **62** (1), 33–47.
Murray, K.B. and J.L. Schlacter (1990), 'The impact of services versus goods on consumers' assessment of perceived risk and risk variability', *Journal of the Academy of Marketing Science*, **18** (1), 51–65.
Noone, B.M. and A.S. Mattila (2009), 'Consumer reaction to crowding for extended service encounters', *Managing Service Quality*, **19** (1), 31–41.
Oliver, R.L. (1980), 'A cognitive model of the antecedence and consequences of customer satisfaction decisions', *Journal of Marketing Research*, **17** (September), 460–469.
Oliver, R.L. (1993), 'A conceptual model of service quality and service satisfaction: Compatible goals, different concepts', in T.A. Swartz, D.E. Bowen and S.W. Brown (eds), *Advances in Services Marketing and Management*, Vol. 2, Greenwich, CT: JAI, pp. 65–85.
Oliver, R.L. (2000),'Customer satisfaction with service', in T.A. Swartz and D. Iacobucci (eds), *Handbook of Services Marketing and Management*, Thousand Oaks, CA: Sage, pp. 247–254.
Oliver, R.L. (2009), *Satisfaction: A Behavioural Perspective on the Consumer*, New York: ME Sharpe Inc.

Oliver, R.L. (2010), *Satisfaction: A Behavioral Perspective on the Consumer* (2nd edn), New York: M.E. Sharpe, Inc.
Oliver, R.L. and W.S. DeSarbo (1988), 'Response determinants in satisfaction judgments', *Journal of Consumer Research*, **14** (4), 495–507.
Oliver, R.L., R.T. Rust and S. Varki (1997), 'Customer delight: Foundations, findings, and managerial insight', *Journal of Retailing*, **73** (3), 311–336.
Otto, J.E. and J.R.B. Ritchie (1995), 'Exploring the quality of the service experience: A theoretical and empirical analysis', *Advances in Services Marketing and Management*, **4**, 37–61.
Palmer, A. and M. O'Neill (2003), 'The effects of perceptual processes on the measurement of service quality', *Journal of Services Marketing*, **17** (3), 254–274.
Parasuraman, A., V.A. Zeithaml and A. Malhotra (2005), 'E-S-QUAL: A multiple-item scale for assessing electronic service quality', *Journal of Service Research*, **7** (3), 213–233.
Paswan, A.K., N. Spears, R. Hasty and G. Ganesh (2004), 'Search quality in the financial services industry: A contingency perspective', *Journal of Services Marketing*, **18** (5), 324–338.
Patterson, P., T. Yu and K. de Ruyter (2006), 'Understanding customer engagement in services', *Advancing Theory, Maintaining Relevance: Proceedings of ANZMAC 2006 Conference*, Brisbane, 4–6 December.
Prahalad, C.K. and V. Ramaswamy (2004), 'Co-creation experiences: The next practice in value creation', *Journal of Interactive Marketing*, **18** (3), 5–14.
Pujari, D. (2004), 'Self-service with a smile? Self-service technology (STT) encounters among Canadian business-to-business', *International Journal of Service Industry Management*, **15** (2), 200–219.
Ratchford, B.T., M.S. Lee and D. Talukdar (2003), 'The impact of the Internet on information search for automobiles', *Journal of Marketing Research*, **40** (2), 193–209.
Rickwood, C. and L. White (2009), 'Pre-purchase decision-making for a complex service: Retirement planning', *Journal of Services Marketing*, **23** (3), 145–153.
Rust, R.T. and K.N. Lemon (2001), 'E-service and the consumer', *International Journal of Electronic Commerce*, **5** (3), 85.
Rust, R.T. and R.L. Oliver (2000), 'Should we delight the customer?', *Journal of the Academy of Marketing Science*, **28** (1), 86–94.
Santos, J. (2003), 'E-service quality: A model of virtual service quality dimensions', *Managing Service Quality*, **13** (3), 233–246.
Santos, J. and J. Boote (2003), 'A theoretical exploration and model of consumer expectations, post-purchase affective states and affective behaviour', *Journal of Consumer Behavior*, **3** (2), 142–156.
Schneider, B. and D.E. Bowen (1995), *Winning the Service Game*, Boston, MA: Harvard Business School Press.
Sharma, P., B. Sivakumaran, and R. Marshall (2009), 'Exploring impulse buying in services vs. products – Towards a common conceptual framework', *Advances in Consumer Research – Asia-Pacific Conference Proceedings*, **8**, 195–196.
Singh, J. (1991), 'Understanding the structure of consumer satisfaction evaluation of service delivery', *Journal of the Academy of Marketing Science*, **19** (3), 223–224.
Spreng, R.A., S.B. MacKenzie and R.W. Olshavsky (1996), 'A reexamination of the determinants of consumer satisfaction', *Journal of Marketing*, **60** (July), 15–32.
Tombs, A. and J.R. McColl-Kennedy (2003), 'Social-servicescape conceptual model', *Marketing Theory*, **3** (4), 37–65.
Tsiotsou, Rodoula H. and Jochen Wirtz (2012), 'Consumer behavior in a service context', in V. Wells and G. Foxall (eds), *Handbook of Developments in Consumer Behavior*, Cheltenham and Northampton, MA: Edward Elgar, pp.147–201.
Unzicker, D. (1999), 'The psychology of being put on hold: An exploratory study of service quality', *Psychology and Marketing*, **16** (4), 327–350.
van Doorn, J., K.N. Lemon, V. Mittal, S. Nass, D. Pick, P. Pimer and P.C. Verhoef (2010), 'Customer engagement behavior: Theoretical foundations and research directions', *Journal of Service Research*, **13** (3), 253–266.
Vargo, S.L. (2008), 'Customer integration and value creation', *Journal of Service Research*, **11** (2), 211–215.
Vargo, S.L. and R.F. Lusch (2004), 'Evolving to a new dominant logic for marketing', *Journal of Marketing*, **68** (January), 1–17.
Varki, S. and M. Colgate (2001), 'The role of price perceptions in an integrated model of behavioural intentions', *Journal of Service Research*, **3** (3), 232–240.
Vazquez-Carrasco, R. and G.R. Foxall (2006), 'Influence of personality traits on satisfaction, perception of relational benefits, and loyalty in a personal service context', *Journal of Retailing and Consumer Services*, **13**, 205–219.
Verhoef, P.C., G. Antonides and A.N. de Hoog (2004), 'Service encounters as a sequence of events: The importance of peak experience', *Journal of Service Research*, **7** (1), 53–64.
Victorino, L., R. Verma, B.L. Bonner and D.G. Wardell (2012), 'Can customers detect script usage in service encounters? An experimental video analysis', *Journal of Service Research*, **15** (4), 390–400.

Voorhees, C.M. and M.K. Brady (2005), 'A service perspective on the drivers of complaint intentions', *Journal of Service Research*, **8** (2), 192–204.
Voss, C., A.V. Roth, E.D. Rosenzweig, K. Blackmon and R.B. Chase (2004), 'A tale of two countries' conservatism, service quality, and feedback on customer satisfaction', *Journal of Service Research*, **6** (3), 212–223.
Weijters, B., D. Rangarajan, T. Falk and N. Schillewaert (2007), 'Determinants and outcomes of customers' use of self-service technology in a retail setting', *Journal of Service Research*, **10** (1), 3–21.
Weiner, B. (2000), 'Attributional thoughts about consumer behavior', *Journal of Consumer Research*, **27** (3), 382–387.
Westbrook, R.A. (1987), 'Product/consumption-based affective responses and postpurchase processes', *Journal of Marketing Research*, **24** (3), 258–270.
Whiting, A. and N. Donthu (2006), 'Managing voice-to-voice encounters: Reducing the agony of being put on hold', *Journal of Service Research*, **8** (3), 234–244.
Wilson, Alan, V.A. Zeithaml, Mary-Jo Bitner and Dwayne Gremler (2008), *Service Marketing – Integrating Customer Focus Across the Firm*, Berkshire: McGraw-Hill.
Wirtz, J. (2003), 'Halo in customer satisfaction measures – the role of purpose of rating, number of attributes, and customer involvement', *International Journal of Service Industry Management*, **14** (1), 96–119.
Wirtz, J. and J.E.G. Bateson (1995), 'An experimental investigation of halo effects in satisfaction measures of service attributes', *International Journal of Service Industry Management*, **6** (3), 84–102.
Wirtz, J. and J.E.G. Bateson (1999), 'Consumer satisfaction with services: Integrating the environmental perspective in services marketing into the traditional disconfirmation paradigm', *Journal of Business Research*, **44** (1), 55–66.
Wirtz, J. and P. Chew (2002), 'The effects of incentives, deal proneness, satisfaction and tie strength on word-of-mouth behaviour', *International Journal of Service Industry Management*, **13** (2), 141–162.
Wirtz, J. and S.E. Kimes (2007), 'The moderating effects of familiarity on the perceived fairness of revenue management pricing', *Journal of Service Research*, 9 (3), 229–240.
Wirtz, J. and A.S. Mattila (2001), 'Exploring the role of alternative perceived performance measures and needs-congruency in the consumer satisfaction process', *Journal of Consumer Psychology*, **11** (3), 181–192.
Wirtz, J., P. Chew and C. Lovelock (2012), *Essentials of Services Marketing*, Singapore: Prentice Hall.
Wirtz, J. P. Xiao, J. Chiang and N. Malhotra (2014), 'Contrasting switching intent and switching behavior in contractual service settings', *Journal of Retailing*, **90** (4), 463–480.
Wright, A.A. and J.G. Lynch Jr. (1995), 'Communication effects of advertising versus direct experience when both search and experience attributes are present', *Journal of Consumer Research*, **21** (4), 708–718.
Xiao, P., C. Tang and J. Wirtz (2011), 'Optimizing referral reward programs under impression management consideration', *European Journal of Operational Research*, **215**, 730–739.
Xue, M. and P.T. Harker (2002), 'Customer efficiency: Concept and its impact on e-business management', *Journal of Service Research*, **4** (4), 253–267.
Yang, Z. and R.T. Peterson (2004), 'Customer perceived value, satisfaction, and loyalty: The role of switching costs', *Psychology & Marketing*, **21** (10), 799–822.
Yen, H.R. (2005), 'An attribute-based model of quality satisfaction for Internet self service technology', *Service Industries Journal*, **25** (5), 641–659.
Yi, Y. and S. La (2004), 'What influences the relationship between customer satisfaction and repurchase intentions?', *Psychology & Marketing*, **21** (5), 351–373.
Zeithaml, V.A. (1981), 'How consumer evaluation processes differ between goods and services', in J.A. Donnelly and W.R. George (eds), *Marketing of Services*, Chicago, IL: American Marketing Association, pp. 186–190.
Zeithaml, V.A. (1988), 'Consumer perceptions of price, quality and value: A means–end model and synthesis of evidence', *Journal of Marketing*, **52** (3), 2–22.
Zeithaml, V.A. and M.J. Bitner (eds) (2003), *Services Marketing: Integrating Customer Focus across the Firm*, New York: McGraw-Hill.
Zeithaml, V.A., L.L. Berry and A. Parasuraman (1996), 'The behavioural consequences of service quality', *Journal of Marketing*, **60** (2), 31–46.

8. Creating and capturing value in the service economy: the crucial role of business services in driving innovation and growth[1]
Michael Ehret and Jochen Wirtz

INTRODUCTION

Developed economies are service economies. Managers and researchers have yet to pay attention to one of the major drivers of this phenomenon: the rise of business services through the opening up of new business models. Contrary to popular opinion, the share of consumer services in economic output has remained rather stable over time, while the share of business services has been rising continuously (Ehret and Wirtz 2010). Business services constitute the backbone of the networked enterprise. While the vertically integrated firm used to dominate the age of manufacturing, service economies build on networks of specialized firms where businesses can hire almost any conceivable business activity or resource as a service.

In the domain of strategy, the rise of business services goes hand in hand with the strategic shift from product-centric strategies to service-based business models. Companies use business models as a strategic response to the increasing competitive pressure which shows in reduced margins and shorter time frames to capture the value from product innovation. Chances are shrinking that products with cost or differentiation advantages today will still be leaders tomorrow. As a response, firms organize around opportunities in order to craft value propositions that are difficult for others to copy. This requires companies to focus on domains where they are unique and have competencies that provide a sustainable competitive advantage, while using business services to integrate underlying resources.

By pursuing stronger focus, firms have been shifting a substantial share of business activity beyond the domain of their legal boundaries to the advantage of specialized providers of business services. Consumer goods firms outsource the management of complex manufacturing technology and their supply chains. Increasingly, they even team up with external suppliers and inventors in order to speed up time to market and boost the scale of innovation. Procter & Gamble set the pace in consumer goods with its landmark 'Connect and Develop' programme (Huston and Sakkab, 2006). For mobile systems like IOS, Android or Palm OS, attracting a sufficient number of external software development companies implies the make or break of a platform (Harper and Endres, 2010).

Business services are at the centre of this fundamental transformation taking place in developed economies. Arguably, the IT industry pioneered the move from manufacturing-centred to service-based business models. Having served on boards of major consumer goods companies, then IBM CEO Lou Gerstner felt that the company could do more than selling boxes and should focus on managing complex IT systems and processes for major client corporations (Ploetner, 2008). In manufacturing technologies, industrial suppliers are increasingly using machines and technical modules as platforms for the delivery of services. Providers of paper manufacturing machines like Germany-based Voith now operate entire factories for their clients as a service. Manufacturers of aircraft engines are joining their customers in operating their planes, tracking engine performance 24/7, and managing the service and maintenance networks (Ehret and Wirtz, 2010; Lay et al., 2010; Ploetner, 2012).

Even firms which began as consumer brands are starting to spot their own opportunities in business services. Take Amazon, which established itself as an online-retailing household name, but now drives the major share of its growth from offering IT services from its server farms, and from channel management by opening its fulfilment and retail infrastructure to third-party retailers (Chesbrough, 2011; Levy, 2011).

In this chapter we will elucidate key features of business services. First we show macroeconomic evidence that the rise of business services is the driving force transforming industrial into service economies. Next, we discuss three economic theories that explain the value provided by business services. Finally, we discuss key elements of service-based business models for capturing and delivering value in the service economy.

BUSINESS SERVICES AND THE RISE OF THE SERVICE ECONOMY

Macroeconomic research has started to note the role of business services as economic growth engines. OECD studies show a continuously rising share of business services in adding value to the output of manufacturing (Woelfl, 2005; OECD, 2008). In the case of the USA, businesses have increased the use of service inputs from around 30 per cent in 1992 to 40 per cent in 2008. If public services such as infrastructure, health and education services are added, service inputs have a share of over 60 per cent of inputs into value creation of businesses in the US economy (Figure 8.1).

Traditionally, economic theory held that services lag in productivity behind manufacturing and therefore inhibit economic growth – a phenomenon called 'Baumol's disease' (Baumol, 1967). This argument may apply to certain consumer services where productivity may not be the purpose of service delivery or is hard to measure (e.g., a hair stylist, a fine-dining restaurant or an opera house). However, if services are supplied to businesses, they can significantly contribute to productivity growth of manufacturing as well as that of the overall economy (e.g., Fixler and Siegel, 1999; Oulton, 2001; Wirtz and Ehret, 2009). Indeed, empirical research shows that business services used by manufacturing firms are the most important drivers of productivity growth in developed economies, followed by the use of IT (Triplett and Bosworth, 2003). Figure 8.2 illustrates that in today's advanced economies connected enterprises can hire almost any conceivable business activity and asset as a service.

Why would business services improve the productivity of a manufacturing firm? Consider the following example. A manufacturing firm runs its own canteen with 100 workers, who in the national statistics are all classified as 'manufacturing employees' and who produce 'manufacturing output' (their output is captured in the added value created by their employer, that is, the manufacturing firm). However, how good is a manufacturing firm in buying ingredients for cooking, designing and running kitchen

Creating and capturing value 131

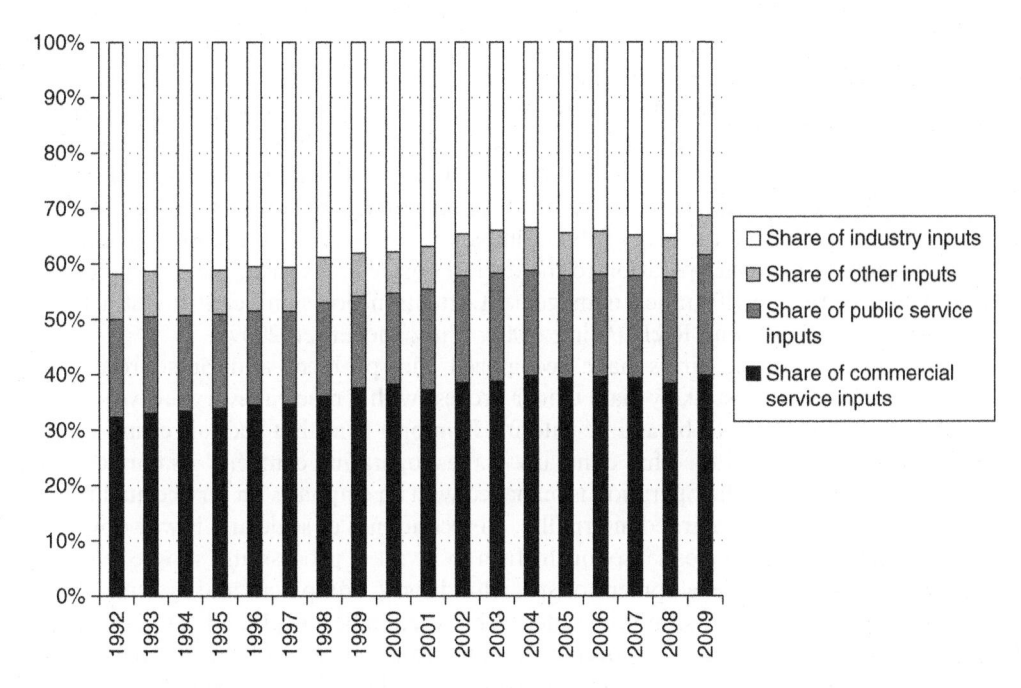

Source: OECD STAN database; accessed 30 April 2012.

Figure 8.1 The growing share of services as inputs into commercial activity in the US economy

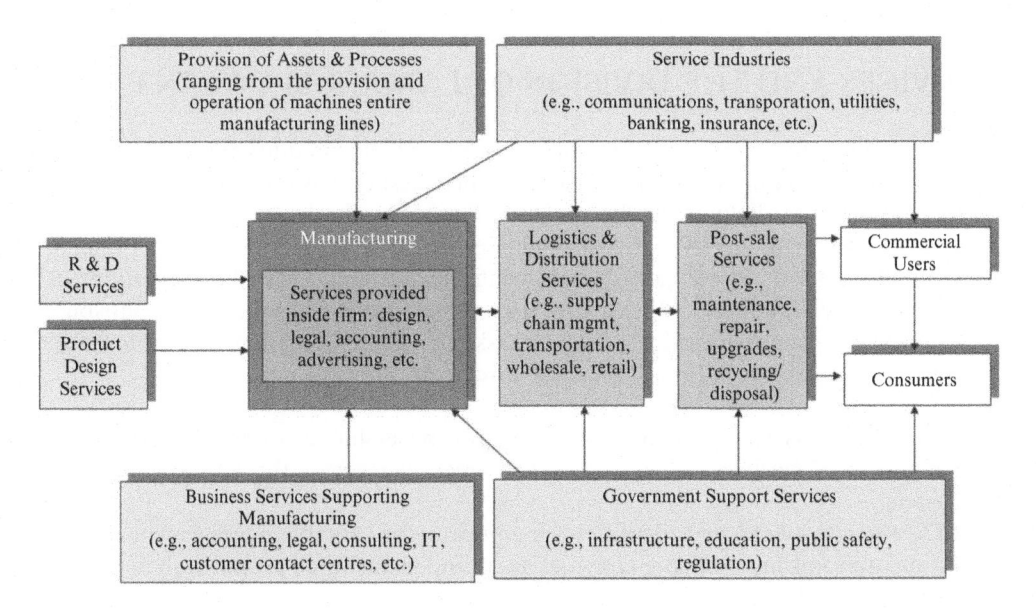

Figure 8.2 Business to business services – a growth engine for the service sector

processes, motivating chefs, and controlling quality and costs in a canteen? The general answer is that it would probably neither produce fantastic food nor be very cost effective. The reasons for this are threefold. First, the operation lacks economies of scale and is high on the learning curve. Second, the manufacturer does not have a lot of experience catering to many sites, which makes management, cost and quality control, and benchmarking difficult and expensive. Third, the firm has little incentive to improve processes or conduct R&D on that aspect of its business, mainly because of the low volume and low criticality of canteen operation to the overall business. As such, the canteen operation would neither justify much management attention nor significant investments in process improvements and R&D (Wirtz, 2000; Wirtz and Ehret, 2009).

Many manufacturing firms have recognized this problem and outsourced their canteen operations, most likely via a tender process with a renewal every few years. The winning bidder is likely to be a large catering firm or a firm that specializes in running canteens across many sites. That company makes 'operating canteens' its core competency. This means that the operation is managed with an emphasis on service quality and costs (sites can be benchmarked internally), has economies of scale and is way down the learning curve. It also makes sense for the firm to invest in process and service redesign, and in innovation and development of specialized tools, equipment and systems, as the benefits can be reaped across multiple sites. What used to be a neglected support activity within a manufacturing firm has become a management focus and core competency of an independent service provider calling for professional management and entrepreneurial responsibility. This logic applies to all non-core functions within firms and has led to more and more outsourcing, even among service organizations (e.g., Heracleous, Wirtz and Pangarkar, 2009; Heracleous and Wirtz, 2010).

What are the underlying drivers of this division of labour between companies? A non-ownership perspective provides a partial answer to this question, as discussed next.

SERVICES AND THE DIVISION OF LABOUR BETWEEN FIRMS

Why does it pay for firms to use external service providers in addition to or as an alternative to their own operations? In economics, this question has puzzled several research streams in the wider field of the theory of the firm. Why do firms exist at all within an efficient market economy? What costs and benefits affect their organizational boundaries? What events or forces call for redrawing these boundaries? These questions call for a clear understanding of what a firm does. They also explain the value contributed by external service providers for taking on tasks or value co-creation. We explore three major streams of research that shape these core questions:

1. *Property rights theory* highlights the costs of ownership as a crucial factor playing in favour of business service providers which can act as the efficient owners of assets.
2. *The resource-based view* identifies the unlocking of valuable management capacity from unpromising non-core activities as an important value proposition of business service providers.
3. Finally, the *entrepreneurial theory of the firm* conceives the use of external service

providers as an important way for a firm to navigate its organizational boundaries to its most promising business opportunities.

Property Rights Theory: Services as an Alternative to Ownership

From its early days, service research has highlighted the value of services as an alternative to owning goods for obtaining value. A firm has the alternative to use its assets and employees to produce the services it needs or to buy services from external providers (Lovelock and Gummesson, 2004). Property rights theory analyses factors and conditions for optimal own-versus-rent decisions. At its heart, this theory is concerned with the efficient size and boundaries of the firm. The boundaries of the firm are defined by the ownership titles it holds for assets like machines, inventories and intellectual property (Grossman and Hart, 1986). The important implication for service research is the development of a framework for deciding when a firm should use external service providers rather than its own people, assets and processes.

Property rights theory was developed for the analysis of economic issues arising from the shared use of assets. Assets are valued for their potential services (Barzel, 1997). For example, commuting or leisure driving are part of a car's service potential, while valuable output produced is that of a machine. Property rights contain (1) the right to use an asset (*ius usus*), for example use a machine for manufacturing; (2) to change its form and substance (*ius abusus*), for example to change parts and components of the machine; (3) to obtain income or other benefits (*ius fructus*), for example to rent the machine to a third party; and (4) to transfer all residual rights, for example to sell the machine (*ius succesionis*) (Furubotn and Pejovich, 1972).

Contracts can be used to share valuable assets and define the terms of property rights across several parties, and thereby put assets to their most valuable use. This works under the assumption that contracts accurately reflect the different valuations of the various services of an asset to the sharing parties, and that enforcing the terms of contracts is costless. In such a perfectly known world, the institution of ownership would not matter as all economic actors simply rent what they need according to their valuation, thus ensuring the highest-valued use of an asset. But according to Coase (1960) this is unlikely to hold as writing and enforcing contracts is costly, and fundamentals of valuations are exposed to uncertainty. Ownership is a social invention for organizing economic activity in situations when actors refrain from value-creating activities because of prohibitive costs of contracting.

When is it beneficial *not* to own an asset? Building on Coase (1960), property rights theory highlights factors that render owning an asset inefficient and play in favour of external service providers. In the main, two types of costs decide if using a service provider is more efficient than ownership, that is, measurement and governance costs.

First, measurement costs need to be incurred when determining the value a collaborating partner contributes in order to enforce the terms of a contract. If the output of an activity can easily be measured and enforced, service contracts tend to be the more efficient solution. If measurement costs are high, or measurement is unfeasible at all, the firm is better off by assuming ownership and managerial authority (Barzel, 1997). Thus, industries tend to favour vertical integration to explore value mechanisms in early stages of their life-cycle, while the share of externally sourced services rises once critical value drivers are well understood and performance measures are easily established and maintained. Once managers are able to define performance indicators, establish measurement methods and enforce contract terms, external sourcing of a service becomes a feasible option.

Second, allocating rights to users of assets like machines or equipment implies governance costs for specifying and enforcing contracts in the face of potential opportunism like the hold-up of a powerful supplier with the aim to reap profits from its customer. Ownership titles grant their holder residual decision rights and residual profit (Grossman and Hart, 1986). Property rights theory claims that ownership of an asset should be allocated to the business for which it is most 'specific' in the sense that the firm finds itself in the most favourable position to maximize the asset's value. Ownership is only efficient for these 'specific' assets, while the firm can move towards higher efficiency by hiring services from non-specific assets by the means of service contracts. For example, if you are in the business of inventing an 'eating experience' with revolutionary cooking processes like the legendary El Bully restaurant (Chesbrough, 2011), you have strong reasons to own specialized kitchen equipment. Ownership of kitchen equipment is much less imperative for a multinational car company.

Property rights theory provides an organizing principle by which ownership of an asset is efficient in the hands of that economic party that is in a position to maximize its value. If not in this position, firms should use external service providers to contract for the asset's services. In maturing industries, assets tend to lose their specific character over time and companies become more capable of measuring value contributions. This leads to an increased division of labour between companies, where downstream companies tend to source a growing share of services from upstream service providers which specialize in asset ownership. Industrial manufacturing is a case in point, where a growing range of assets is being managed by external service providers. For example, facility managers manage office and factory buildings, contract manufacturers undertake production, performance contractors guarantee performance levels, for example for the operation of an engine or the heating of a building.

To summarize, property rights theory holds as a guiding principle that firms should assume ownership over specific assets and crucial but hard to measure elements of the value creation process. Otherwise they should hire external business service providers.

Resource-Based View: Freeing Up Management Capacity to Focus on Growth Opportunities

The resource-based view emphasizes the aim for growth as a driving factor of the division of labour between firms, and highlights the role of management in shaping the competitive position of a firm (Wernerfelt, 1984, 1995; Prahalad and Hamel, 1990). According to the resource-based view, the firms themselves are the tools and sources for differentiation. In their pursuit of growth, firms strive to build unique capabilities in order to capture rents not available in undifferentiated markets. Penrose (1980) pioneered this approach by providing a conceptual framework for investigating the key factors that affect a firm's growth. She started from an assumption similar to property rights theory: resources are bundles of different uses and, consequently, resource value is derived from the services to which they are applied (Penrose, 1980). Firms differentiate themselves by developing unique capabilities for the use of resources. This perspective makes management (in a broader sense) the decisive force that differentiates a firm and affects its growth.

One important strand of the resource-based view investigates how companies can cultivate resources that drive their differentiation. These resources include 'those (tangible and intangible) assets which are tied semi-permanently to the firm', such as brand names, in-house knowledge of technology, employment of skilled personnel, trade contacts, machinery, efficient procedures and capital (Wernerfelt, 1984, p. 172). A key force driving the growth of the firm is based on the perception of growth (or differentiation) opportunities by the firm's management (Penrose, 1980). Management shapes the growth opportunities of a firm in two ways: (1) a firm can only target that fraction of its growth opportunities that its management capacity allows it to address – unlocking management capacity is imperative for taking on growth (Penrose, 1980); and (2) the limited capacity requires a firm to prioritize its management attention on areas with the most promising growth opportunities and to delegate remaining areas to external service providers.

The resource-based view highlights the role of business services as building blocks of the intelligent enterprise where managers focus on entrepreneurial opportunities and delegate all complementing activities to world-class external service providers (Quinn, 1992).

The Entrepreneurial Theory of the Firm: Designing Boundaries in Order to Navigate towards Opportunities

Property rights theory and the resource-based view provide snapshots of situations where firms provide value. The entrepreneurial theory of the firm goes a step further by proposing the firm as the tool of entrepreneurs for exploring and exploiting business opportunities.

Research in economics (e.g., Schumpeter, 1934; Schmookler, 1966; Kirzner, 1973; Baumol, 1993; Lewin, 1999) and strategic management and organization (e.g., Shane and Venkataraman, 2000; Alvarez and Barney, 2004; Foss et al., 2007) has highlighted how entrepreneurs shape organizations and how organizations support entrepreneurial action, thus providing a framework for explaining the dynamic forces that affect the boundaries of the firm and the rise of business services.

Broadly conceived, entrepreneurial action is concerned with the exploration and exploitation of profit opportunities arising from either un/under-served needs or un/under-used resources in an economy (Kirzner, 1997; Shane and Venkataraman, 2000). Firms show a Janus-face, the front showing individual perceptions and visions of business opportunities, and the back consisting of organizational resources, rules and routines that help it to shape and exploit profit opportunities (Lewin, 1999). Kirzner (1973) highlighted the role of the entrepreneur as an agile agent who identifies opportunities overseen by ordinary market participants and takes action to profit from them. Arbitrage is the simplest form – buying low from ignorant sellers and selling dear to ignorant buyers. Kirzner maintains that arbitrage is just a simplified version of a universe of profit opportunities that can be exploited by more complex commercial activities such as manufacturing, trade or R&D. Entrepreneurs enhance the range of business opportunities by mobilizing capital and knowledge, and by developing efficient routines and processes through the means of business organization within a firm (Mises, 1949; Klein, 1999).

While everyone has some potential for acting entrepreneurially, economic organization can provide a substantial leverage for entrepreneurial activity. For example, the evolution of the mass market for automobiles was not only driven by a visionary entrepreneur who perceived the potential for individual means of transportation, but also by the design of an organization that mobilized capabilities and resources for its exploitation. In a nutshell, entrepreneurs are the lifeblood directing firms to profitable opportunities, while firms provide entrepreneurs with capital, resources and an infrastructure that can enhance entrepreneurial opportunities and their exploitation (Lewin, 1999; Sautet, 2000; Foss et al., 2007).

This entrepreneurial perspective has decisive implications for the role of ownership and property rights in shaping economic growth and the demand for business services. From an entrepreneurial perspective, ownership is a tool for shaping and directing entrepreneurial processes like experimenting, exploring and exploiting business opportunities (Foss et al., 2007). Firms use ownership in order to direct resources to expected higher-valued uses, based on an entrepreneurial vision and a business model that contains a unique value proposition (Foss et al., 2007). Equity ownership is the instrument to reap the returns and bear the risk entailed in entrepreneurial projects and thus is used to attract resources for the deployment of entrepreneurial projects (Knight, 1921). Ownership is linked to the scope of entrepreneurial projects for a firm, and subsequently shapes its boundaries on resource markets. From an entrepreneurial perspective, resources and activities not related to the entrepreneurial focus of the firm should be sourced from external service providers.

In this entrepreneurial perspective, the opening up of business models and the rise of business services are flip-sides of the same coin. Their appetite for growth and the pressure of competition forces firms to direct their capital to the most promising business opportunities. This implies a continuous review of core activities and subsequent restructuring processes. Firms started with the outsourcing of routine operations and are now in a position to use external service providers for almost any conceivable function, operation or asset class. As a result, firms are transforming into 'intelligent enterprises' (Quinn, 1992) that can rent almost every conceivable activity or asset type as a service while focusing on areas of un/under-served needs of customers or underused potential of resources.

Summary of Economic Theories and the Value Propositions of Non-ownership

Service researchers are increasingly noting that services are a viable alternative to owning resources. Property rights theory holds that companies using services can avoid the costs of ownership when they are not specific for their business. The resource-based view maintains that business services free up valuable management capacity to focus on a firm's most promising opportunities. The entrepreneurial theory of the firm suggests that the legal boundaries of the firm should move with its exploration and exploitation of business opportunities. As a major implication of economic theories on non-ownership, business services empower managers to adjust the structure of their firms to business opportunities and organization costs. Business services are the backbone of service-based business models. Yet, the story untold by service research is how the offering of business services has become an entrepreneurial opportunity in its own right. We discuss these service-based business models next.

KEY ELEMENTS OF SERVICE-BASED BUSINESS MODELS

Shifting the Strategic Focus from Product Market Strategies to Business Model Design

By decoupling access to resources and competencies from ownership, business service providers open up a new strategic dimension: to organize a firm around business opportunities rather than products. Product market strategies build on cost advantages or differentiation of products. As products are becoming increasingly commoditized in hypercompetitive economies, business model thinking reverses the competitive strategy approach by starting from the opportunity in order to identify value propositions (Chesbrough, 2006; d'Aveni, Dagnino and Smith, 2010; see Table 8.1). Business services provide the backbone for business model design, by empowering their clients to delegate responsibilities to external providers.

From the business model perspective, business service providers offer their clients

Table 8.1 Business Model Versus Product Market Strategy

	Product Market Strategy	Business Model
Definition	Pattern of managerial actions that explains how a firm achieves and maintains competitive advantage through positioning in product markets	A structural template of how a focal firm transacts with customers, partners and vendors. It captures the pattern of the firm's boundary-spanning connections with factor and product markets
Main questions addressed	• How to segment the market? • Which customers to serve? • Which products to sell? • What position to adopt against competition? • What kind of generic strategy to apply (e.g., differentiated versus cost leadership)? • How to enter the market?	• How to connect with factor and product markets? • Which parties to bring together to exploit a business opportunity, and how to link them to the focal firm to enable transactions (i.e., what exchange mechanisms to adopt)? • What information or goods to exchange among the parties, and what resources and capabilities to deploy to enable the delivery of the value proposition? • How to control the transactions between the parties, and what incentives to adopt for each of the parties?
Unit of analysis	Firm	Focal firm and its exchange partners
Focus	Internally/externally oriented: focus on firm's core competencies and resulting competitive edge that then gets translated into a product market strategy	Externally oriented: focus on firm's exchanges with other providers in a network of capabilities that together deliver a powerful value proposition that is difficult for others to copy

Source: Adapted from Zott and Amit (2008), p. 5.

options for the design of their business (Pisano and Teece, 2007; Chesbrough, 2011). The non-ownership value of business services builds on the strategic empowerment of business clients. Business models entail three crucial elements:

1. The *value proposition* describes the unique contribution of the firm in the value creation process. This is the main focus of management attention, investments and operations. The value proposition relates to the domain where the firm's management identifies business opportunities with the potential for value generation at a profit.
2. The *value-capturing mechanism* describes the revenue stream related to the value proposition and associated costs. Besides the customer's willingness to pay, managers must pay attention to the appropriability of profits against competitors, suppliers and customers.
3. The management of the *value network* is an almost logical consequence of the focus on a value proposition. The value network consists of connections with partners and complementors needed to mobilize complementary assets and capabilities, such as sales channels, supply channels, value-added services and many more.

In the following sections we take a closer look at the role of business service providers in the business model.

Value Proposition

The value proposition describes the offering of the provider from the customer's perspective, most importantly the offered benefits and their role in the customer's value chain. Business service providers offer a unique type of value proposition, entailed in the non-ownership value of services: they empower their clients to design their organizational boundaries and structure. As discussed in the introductory sections, non-ownership entails at least three types of value proposition. By using services, clients delegate ownership of assets, processes or entire operations to a business service provider (property rights theory), so that they can focus on their core competencies (resource-based view) and navigate towards the most promising business opportunities (entrepreneurial theory of the firm). Understanding value propositions directly related to the division of labour opens valuable strategic insights for service providers. For example, the aircraft engine manufacturer Rolls-Royce builds its service strategy on investments into information systems favouring its measurement costs and capabilities to track and manage engine performance. Outsourcing providers such as IBM ensure that clients can focus on their core activities or competencies and on the most promising opportunities rather than being distracted by the management of internal IT services. Not least, companies may orchestrate collaborating firms in order to navigate towards business opportunities. For example, consumer goods companies like Procter & Gamble involve external suppliers and designers in their product development process (Huston and Sakkab, 2006).

Business services provide a specific type of value proposition – the non-ownership value of designing a business in line with conditions affecting its performance (Chesbrough, 2011).

Value Capturing

Value capturing describes how a business intends to monetize its value proposition. Business service contracts provide particularly powerful tools for value capturing, as companies can use them to allocate profits across a network. Profit is residual, uncertain income, showing potential for both up- and downsides. Business service contracts are especially useful when a service provider has established a unique position to handle the up- and downsides of a business operation – take Rolls-Royce aircraft engines 'power by the hour' contracts, where this is most apparent. Airlines delegate the profit impact related to the engine performance to contractors like Rolls-Royce. Because Rolls-Royce is exclusively compensated on the effective hours an aircraft is flying, it has a strong incentive to keep it in the air. By the same token, the airline shifts a substantial share of the downsides of engine performance to Rolls-Royce (see Box 8.1).

BOX 8.1   ROLLS-ROYCE SOURCE AIRPLANE ENGINES TAKES OVER RESPONSIBILITY BY MEANS OF PERFORMANCE CONTRACTS

Many manufacturing firms enhance their competitive edge by providing superior value to their customers in the form of service. Rolls-Royce is one example. Rolls-Royce is a successful company because it focuses on technical innovation and makes world-class aircraft engines. Rolls-Royce engines power about half of the latest wide-bodied passenger jets and a quarter of all single-aisle aircraft in the world today. A very important factor for its success has been the move from manufacturing to selling 'power by the hour' – a bundle of goods and services that keeps customers' engines running.

Imagine this: high above the Pacific, passengers doze on a long-haul flight from Tokyo to Los Angeles. Suddenly, there is a bolt of lightning. Passengers may not think much about it, but on the other side of the world, in Derby in England, engineers at Rolls-Royce get busy. Lightning strikes on jets are common and usually harmless, but this one has caused some problems in one of the engines. The aircraft will land safely and could do so even with the engine shut down. The question is whether it will need a full engine inspection in Los Angeles, which would be normal practice but cause delays and inconvenience hundreds of passengers waiting in the departure lounge.

In an instant after lightning hits an engine, a stream of data is beamed from the plane to Derby. Numbers dance across screens, graphs are drawn and engineers scratch their heads. Long before the aircraft is due to land word comes that the engine is running smoothly, no engineer on the ground will have to examine it, and the plane will be able to take off on time.

Industry experts estimate that manufacturers of jet engines can make about seven times the revenue from servicing and selling spare parts over the lifetime of an engine than they do from selling the engine. Since it is so profitable, many independent servicing firms compete with companies like Rolls-Royce and offer spare parts for as little as one-third of the price charged by the original manufacturer. This is where Rolls-Royce has used technology and service to make it more difficult for competitors to steal its clients. Instead of selling engines and then later parts and service, Rolls-Royce has created an attractive bundle, which it branded TotalCare®. Customers are charged for every hour that an engine runs. Its website advertises it as a solution ensuring 'peace of mind' for the lifetime of an engine. Rolls-Royce promises to maintain the engine and replace it if it breaks down. The operations room in Derby simultaneously monitors the performance of some 3500 engines, enabling it to predict when engines are likely to fail and let airlines schedule engine changes efficiently and reduce repairs and unhappy passengers. Today, about 80 per cent of engines shipped to its customers are covered by such contracts. Although Rolls-Royce had engine troubles on its A380 Trent Engines, they have fixed the problem quickly and bounced back from the incident with many more orders for their engines.

Sources: Economist (2009, 2011); Lovelock and Wirtz (2011, p. 18); www.rolls-royce.com (accessed March 2013).

The revenue base is a crucial element of the value proposition for the customer, directly affecting the financial value of the offering from the customer's perspective. At the same time, revenue constitutes the top line of the provider's profit calculation. Thus, defining the revenue base has a crucial impact for both the client's perception of the value proposition and the profitability of the provider.

Up-front investments to establish a service are another factor calling for a sound definition of the revenue base. Xerox's initial struggle to attract customers to its technically superior photocopying machines illustrates the challenge (Chesbrough, 2006). During the market introduction of the photocopier machine, Xerox struggled to attract demand. Potential customers perceived the up-front investment as a barrier, even though photocopying promised higher quality and lower costs. Xerox achieved the commercial breakthrough by installing the machines for free on its customers' office floors and charging based on the number of copies made. Xerox took responsibility by assuming ownership of the machine, thereby taking on investment and grant maintenance and repair. As Xerox is compensated on the quantity of copies made, it has an incentive to keep machines running. By the same token, its clients delegate substantial risks associated with the ownership of the machine to Xerox.

Another challenge for capturing value is the appropriability regime, which serves to protect the revenue base against free-riders and imitators (Pisano and Teece, 2007). This is particularly relevant for high-tech firms which mainly contribute knowledge, ideas and concepts to the value creation process. Legal instruments to protect this intellectual capital become toothless once knowledge is codified and easy to transfer. For the sake of value capturing, even extremely outsourced companies employ physical products to enforce their value claims. Take Qualcomm – now one of the most profitable technology companies. As a pioneer in digital mobile communication, Qualcomm started to design and build entire mobile communication networks, including handsets, antennae and networking technology. During the market expansion stage of 3G mobile services, Qualcomm licensed out the patents on its digital mobile standard to handset manufacturers and infrastructure builders, effectively offering technology as a service. For Qualcomm, the vast investments of its collaborating manufacturers and network companies work as leverage that locks out competitors and renders Qualcomm outrageous returns. While Qualcomm has reduced its investment in physical assets massively, it protects its share of the value pie by both intellectual property and physical products, like micro chips (Mock, 2005).

To conclude, clients use business services to transfer downside risks on profits to providers of business services. Providers aim to transform this risk into an opportunity by developing superior capabilities and technologies enabling them to deliver the service.

Value Network Design

Value networks of independent collaborating firms hold the potential to outperform vertically integrated firms by unlocking benefits of specialization. However, the value of sharing resources and responsibilities across a network comes at a cost of integrating all contributions to a consistent and compelling customer experience (Hunt and Morgan, 1994; Wuyts et al., 2004). To make this happen, at least three types of players must be

Table 8.2 Roles of companies within value networks

	Network Architect	Hybrid Contributor	Technology Provider
Scope	Vision for positioning the system, accountable for system performance	Differentiation in a selective section of the system (e.g., offering a specific app)	Technology specialist
Governance	Network architect, sets the rules	Compliance, follows the rules set by the network architect	Compliance, follows the rules set by the network architect
Technological interface	Defines interfaces between network participants and between network and end-users	Adapts for compatibility in the network	Adapts for compatibility in the network, may contribute to maintain interfaces
Appropriation regime	System performance with multiple appropriation options, through sales to end-customers	Core offerings (services or products) to end-customers	Intellectual property like patents plus protection mechanisms, like ownership of core assets or infrastructure

present in the network: (1) a *network architect*, who integrates the network for a coherent customer experience; (2) *hybrid contributors*, who strengthen the attractiveness of the network; and (3) *technology providers*, who manage the underlying infrastructure and facilities.

Within the value network, companies assume different roles depending on their capabilities and strategic position. For example, Apple assumes the role of the network architect in the iOS network and orchestrates contributions, shapes the network personality, and rules technical standards and governance mechanisms. App providers are hybrid network participants who design crucial elements of the user experience and thereby contribute core offerings to the value proposition of the network, but do not govern network architecture. Finally, SAP plays the role of a technology provider that enables and supports network collaboration and, in this case, maintains the payment infrastructure (see Table 8.1).

To the extent that competition moves from the domain of the individual firm to the domain of a value network, choosing the appropriate network becomes essential. Performance and attractiveness, as well as the position within a network, determine how strongly a company can thrive by network collaboration. For example, software programmer Rovio almost went bankrupt within the PC-gaming network, but thrived by offering one of the most popular iOS apps with its 'Angry Birds' app (see Box 8.2).

SUMMARY AND CONCLUSIONS

Contrary to common opinion, business services rather than consumer services are the main driver of the rise of the service economy. As economies develop, competition

> **BOX 8.2 ORCHESTRATING A NETWORK THE CASE OF THE APPLICATION ECONOMY**
>
> The application economy is a crucial backbone of Apple's leading position in the mobile communication industry and the rise of the smart phone as a computer platform. While users may love or loathe gadgets like Apple's iPhone, iPad or iPod, a diverse set of companies collaborate for the user experience. Many of those companies go unnoticed as users order software with the swipe of a finger. Without the payment systems licensed by German software vendor SAP, software houses and content providers would see little reason for placing their offerings on the iTunes store. Content providers further drive attractivity of the network by enriching the content and providing user experiences. In some cases, the application economy has opened up a new distribution channel for creative software houses like Angry Birds designer Rovio, which before could not bypass market-dominating software houses like gaming studio Electronic Arts (EA). Meanwhile, Apple orchestrates the application economy for ensuring a consistent and compelling user experience. Apple achieves this through communication and branding, upgrades and new applications stimulated by R&D and software development, and supporting collaborating firms (e.g., through developer software support and developer conferences). Many different elements have to come into place to make a network thrive. Some computer companies have had to learn this through painful lessons when they tried to catch up with Apple through quick fixes, as Hewlett Packard did through the misguided acquisition of the mobile computing company Palm Inc.
>
> *Sources:* Harper and Endres, 2010; http://investor.apple.com (accessed 5 March 2013).

intensifies and forces businesses to focus their management attention and investments into the areas with the most promising opportunities given the firm's capabilities and core competencies.

The pressure on businesses to increase focus opens up huge opportunities for business service providers. Business service providers offer their clients a unique value dimension – to direct management attention and investments to their most promising opportunities and draw support from world-class service providers. Service research has started to note this non-ownership value where providers take on responsibilities for assets, processes and operations, and even the results of operations on behalf of their clients.

Economic theory elucidates unique value propositions of business services. According to property rights theory, companies delegate governance and measurement costs to business service providers. The resource-based view highlights that business service providers empower their clients to focus their management attention and investments on the most promising opportunities. Not least, the entrepreneurial theory of the firm elucidates the contribution of business services for exploring and exploiting business opportunities.

Business models provide a strategic framework for service providers to transform the non-ownership value of their offerings into profits. Design of business models builds on three key areas: (1) the *value proposition* describes the domain where the service provider maintains unique competencies and assets in order to provide benefits to client firms; (2) the *value-capturing mechanism* describes the revenue base and how the provider company claims and protects its share; (3) the *value network design* describes the role the provider intends to play in the network. However, non-ownership value resides in integration of a network by different types of participants, such as network architects, hybrid contributors and technology providers.

The concept of non-ownership value holds substantial promise for research and management. In service research, non-ownership value provides a theoretical foundation for analysing and designing the division of labour between firms across a service system (Maglio and Spohrer, 2008). In industrial services, non-ownership value provides a theoretical lens for elucidating service infusion in industrial firms (Ostrom et al., 2010). Businesses find a growing range of opportunities by using e-services (Rust and Kannan, 2003) as elements of business models for sharing valuable assets like machines, cars or real estate (Economist 2013). Another research opportunity is to study the role of non-ownership value as a factor impacting the upgrading decision of business customers (Bolton, Lemon and Verhoef, 2008).

NOTE

1. This chapter draws on Ehret and Wirtz (2010) and Wirtz and Ehret (2013). It is an updated and extended version of these earlier publications.

REFERENCES

Alvarez, S.A. and Barney, J. 2004. 'Organizing Rent Generation and Appropriation toward a Theory of the Entrepreneurial Firm', *Journal of Business Venturing*, vol. 19, no. 5, pp. 621–635.
Barzel, Y. 1987. 'The Entrepreneur's Reward for Self-Policing', *Economic Inquiry*, vol. 25, no.1, p. 103.
Barzel, Y. 1997. *Economic Analysis of Property Rights*, Cambridge University Press, 2nd edn, Cambridge.
Baumol, W.J. 1967. 'Macroeconomics of Unbalanced Growth: The Anatomy of Urban Crisis', *American Economic Review*, vol. 57, no. 3, pp. 415–427.
Baumol, W.J. 1993. *Entrepreneurship, Management, and the Structure of Payoffs*, MIT Press, Cambridge and London.
Bolton, R.N., Lemon, K.N. and Verhoef, P.C. 2008. 'Expanding Business-to-Business Customer Relationships: Modeling the Customer's Upgrade Decision', *Journal of Marketing*, vol. 72, no. 1 (January), pp. 46–64.
Chesbrough, H. 2006. *Open Business Models: How to Thrive in the New Innovation Landscape*, McGraw-Hill, Boston and London.
Chesbrough, H. 2011. *Open Services Innovation*, Harvard Business School Press, Boston.
Coase, R. 1960. 'The Problem of Social Cost', *Journal of Law and Economics*, vol. 3, no.1, pp. 1–44.
D'Aveni, R.A., Dagnino, G.B. and Smith, K.G. 2010. 'The Age of Temporary Advantage', *Strategic Management Journal*, vol. 31, no. 13, pp. 1371–1385.
Dyer, J.H. and Singh, H. 1998. 'The Relational View: Cooperative Strategy and Sources of Interorganizational Competitive Advantage', *Academy of Management Review*, vol. 23, no. 4, pp. 660–679.
Economist 2009. 'Briefing Rolls-Royce. Britain's Lonely High Flyer', *The Economist*, 10 January, pp. 58–60.
Economist 2011. 'Per Ardua', *The Economist*, 5 February, p. 68.
Economist 2013. 'All Eyes on the Sharing Economy', *The Economist*, 9 March.
Ehret, M. and Wirtz, J. 2010. 'Division of Labor between Firms: Business Services, Non-ownership Value and the Rise of the Service Economy', *Service Science*, vol. 2, no. 3, pp. 136–145.
Fixler, D. and Siegel, D. 1999. 'Outsourcing and Productivity Growth in Services', *Structural Change and Economic Dynamics*, vol. 10, pp. 177–194.
Foss, K., Foss, N.J., Klein, P.G. and Klein, S.K. 2007. 'The Entrepreneurial Organization of Heterogeneous Capital', *Journal of Management Studies*, vol. 44 (November), pp. 1166–1186.
Furubotn, E.G. and Pejovich, S. 1972. 'Property Rights and Economic Theory: A Survey of Recent Literature', *Journal of Economic Literature*, vol. 10, no. 4, pp. 1137–1163.
Ghosh, M. and John, G. 1999. 'Governance Value Analysis and Marketing Strategy', *Journal of Marketing*, vol. 63, no. 4, pp. 131–145.

Grossman, S.J. and Hart, O.D. 1986. 'The Costs and Benefits of Ownership: A Theory of Vertical and Lateral Integration', *Journal of Political Economy*, vol. 94, no. 4, pp. 691–719.
Harper, D.A. and Endres, A.M. 2010. 'Capital as a Layer Cake: A Systems Approach to Capital and its Multi-level Structure', *Journal of Economic Behavior and Organization*, vol. 74, no. 1/2 (May), pp. 30–41.
Hart, O. 1995. *Firms, Contracts, and Financial Structure*, Clarendon Press, Oxford.
Heracleous, L. and Wirtz, J. 2010. 'Singapore Airlines' Balancing Act – Asia's Premier Carrier Successfully Executes a Dual Strategy: It Offers World-Class Service and is a Cost Leader', *Harvard Business Review*, vol. 88, no. 7/8, pp. 145–149.
Heracleous, L., Wirtz, J. and Pangarkar, N. 2009. *Flying High in a Competitive Industry: Secrets of the World's Leading Airline*, McGraw-Hill, Singapore.
Hunt, S.D. and Morgan, R.M. 1994. 'Relationship Marketing in the Era of Network Competition', *Marketing Management*, vol. 3, no. 1, pp. 18–28.
Huston, L. and Sakkab, N. 2006. 'Connect and Develop', *Harvard Business Review*, vol. 84, no. 3, pp. 58–66.
Kirzner, I.M. 1973. *Competition and Entrepreneurship*, The University of Chicago Press, Chicago and London.
Kirzner, I.M. 1997. 'Entrepreneurial Discovery and the Competitive Market Process: An Austrian Approach', *Journal of Economic Literature*, vol. 35 (March), pp. 60–85.
Klein, P.G. 1999. 'Entrepreneurship and Corporate Governance', *Quarterly Journal of Austrian Economics*, vol. 2, no. 2, pp. 19–42.
Knight, F. 1921. *Risk, Uncertainty and Profit*, Houghton Mifflin Company, The Riverside Press, Cambridge, Boston and New York.
Lay, G., Copani, G., Jaeger, A. and Biege, S. 2010. 'The Relevance of Service in European Manufacturing Industries', *Journal of Service Management*, vol. 21, no. 5, pp. 715–726.
Levitt, T. 1975. 'Marketing Myopia', *Harvard Business Review*, vol. 53, no. 5, pp. 26–183.
Levy, S. 2011. 'Jeff Bezos – CEO of the Internet', *Wired*, December.
Lewin, P. 1999. *Capital in Disequilibrium: The Role of Capital in a Changing World*, Routledge, London and New York.
Lovelock, C. and Gummesson, E. 2004. 'Whither Services Marketing? In Search of a New Paradigm and Fresh Perspectives', *Journal of Service Research*, vol. 7, no. 1, pp. 20–41.
Lovelock, C. and Wirtz, J. 2011. *Services Marketing: People, Technology, Strategy*, 7th edn, Prentice Hall, Upper Saddle River.
Maglio, P.P. and Spohrer, J. 2008. 'Fundamentals of Service Science', *Journal of the Academy of Marketing Science*, vol. 36, no. 1, pp. 18–20.
Mises, L.V. 1949. *Human Action: A Treatise on Economics*, 3rd rev. edn, Henry Regnery, Chicago.
Mock, D. 2005. *The Qualcomm Equation: How a Fledgling Telecom Company Forged a New Path to Big Profits and Market Dominance*, Amacom, New York.
OECD 2008. *Staying Competitive in the Global Economy: Compendium of Studies on Global Value Chain*, OECD, Paris.
Ostrom, A.L., Bitner, M.J., Brown, S.W., Burkhard, K.A., Goul, M., Smith, D.V., Demirkan, H. and Rabinovich, E. 2010. 'Moving Forward and Making a Difference: Research Priorities for the Science of Service', *Journal of Service Research*, vol. 13, no. 1 (February), pp. 4–36.
Oulton, N. 2001. 'Must the Growth Rate Decline? Baumol's Unbalanced Growth Revisited', *Oxford Economic Papers*, vol. 53, pp. 605–627.
Penrose, E. 1980. *The Theory of the Growth of the Firm*, 2nd edn, Basil Blackwell, Oxford (1st edn: 1959).
Pisano, G.P. and Teece, D.J. 2007. 'How to Capture Value from Innovation: Shaping Intellectual Property and Industry Architecture', *California Management Review*, vol. 50, no. 1, pp. 278–296.
Ploetner, O. 2008. 'The Development of Consulting in Goods-Based Companies', *Industrial Marketing Management*, vol. 37, no. 3, pp. 329–338.
Ploetner, O. 2012. *Counter Strategies in Global Markets*, Palgrave Macmillan, New York.
Prahalad, C.K. and Hamel, G. 1990. 'The Core Competence of the Corporation', *Harvard Business Review*, vol. 68, no. 3, pp. 79–91.
Quinn, J.B. 1992. *The Intelligent Enterprise: A Knowledge and Service Based Paradigm for Industry*, The Free Press, New York.
Rifkin, J. 2000. *The Age of Access: How the Shift from Ownership to Access is Transforming Capitalism*, Putnam, New York.
Rust, R.T. and Kannan, P.K. 2003. 'E-Service: A New Paradigm for Business in the Electronic Environment', *Communications of the ACM*, vol. 46, no. 6 (June), pp. 36–42.
Sautet, F. 2000. *An Entrepreneurial Theory of the Firm*, Routledge, London.
Schmookler, J. 1966. *Invention and Economic Growth*, Harvard Business School Press, Cambridge.
Schumpeter, J.A. 1934. *The Theory of Economic Development*, Harvard University Press, Cambridge.
Shane, S. and Venkataraman, S. 2000. 'The Promise of Entrepreneurship as a Field of Research', *Academy of Management Review*, vol. 25, no. 1, pp. 217–226.

Stigler, G. 1952. 'The Division of Labor is Limited by the Extent of the Market', *Journal of Political Economy*, vol. 59, pp. 185–193.
Triplett, J.E. and Bosworth, B.P. 2003. 'Productivity Measurement Issues in Service Industries: "Baumol's Disease" has been cured', *FRBNY Economic Policy Review*, September, pp. 23–33.
Wernerfelt, B. 1984. 'A Resource-Based View of the Firm', *Strategic Management Journal*, vol. 5, no. 2, pp. 171–180.
Wernerfelt, B. 1995. 'The Resource-Based View of the Firm: Ten Years After', *Strategic Management Journal*, vol. 16, no. 3, pp. 171–174.
Williamson, O.E. 1971. 'The Vertical Integration of Production: Market Failure Considerations', *A.E.R. Papers and Proceedings*, vol. 61 (May), pp. 112–123.
Williamson, O.E. 1985. *The Economic Institutions of Capitalism*, The Free Press, New York.
Wirtz, J. 2000. 'Growth of the Service Sector in Asia', *Singapore Management Review*, vol. 22, no. 2, pp. 37–55.
Wirtz, J. and Ehret, M. 2009. 'Creative Restruction – How Business Services Drive Economic Evolution', *European Business Review*, vol. 21, no. 4, pp. 380–394.
Wirtz, J. and Ehret, M. 2013. 'Service-Based Business Models: Transforming Businesses, Industries and Economies', in R.P. Fisk, R. Russell-Bennett and L.C. Harris (eds), *Serving Customers: Global Services Marketing Perspectives*, Tilde University Press, Melbourne, pp. 28–46.
Woelfl, A. 2005. 'The Service Economy in OECD Countries', in OECD (ed.), *Enhancing the Productivity of the Service Sector*, OECD, Paris, pp. 27–63.
Wuyts, S., Dutta, S. and Stremersch, S. 2004. 'Portfolios of Interfirm Agreements in Technology-Intensive Markets: Consequences for Innovation and Profitability', *Journal of Marketing*, vol. 68, no. 2, pp. 88–100.
Zeithaml, V.A., Parasuraman, A. and Berry, L.L. 1985. 'Problems and Strategies in Services Marketing', *Journal of Marketing*, vol. 49 (Spring), pp. 33–46.
Zott, C. and Amit, R. 2008. 'The Fit between Product Market Strategy and Business Model: Implications for Firm Performance', *Strategic Management Journal*, vol. 29, no. 1, pp. 1–26.

9. Measuring business activity in the UK[1]
Julian Frankish,[2] Richard Roberts, David J. Storey and Alex Coad

INTRODUCTION

Since small businesses and enterprise were 'discovered' by public policy in the UK during the 1960s, commencing in earnest with the publication of the Bolton Committee (1971) report, there has been considerable official interest in this form of economic activity. The Conservative Government from 1979 to 1997 initially saw enterprise as a strategy for job creation but later switched to a focus on growth. The Labour Government after 1997 continued this 'cross-party' support, although it placed greater weight upon the potential role of enterprise to address wider issues of 'social inclusion', e.g. HM Treasury (1999), than its predecessor government. The election of a Conservative-led coalition in 2010, faced with a challenging macro-economic environment, meant that once again small businesses and enterprise creation were seen as playing a critical role in a UK economy seeking to emerge from recession (Greene, 2002; Greene et al., 2008).

All the above policies rely, to a greater or lesser extent, on data that can be used to assess policy impact. Crucial to this is reliable information on the number (stock) of businesses and the entry and exit of firms. In practice this is provided from several sources which often provide divergent pictures of scale and trends. Hence it is vital to have a clear understanding of the strengths and limitations of the datasets that measure the scale of enterprise, and which are then used to assess which factors influence the performance of (small and particularly young) enterprises. This chapter seeks to provide that understanding and has three aims.

First, it provides a concise overview of some of the main sources of data on business and enterprise in the UK. These are official sources such as the Labour Force Survey (LFS) of self-employment, the number of firms in the business stock and the number of businesses on official registers with regard to certain taxes. It also covers non-official sources such as those from the Global Entrepreneurship Monitor (GEM) and data from the British Bankers' Association (BBA).

Second, it provides a detailed overview of a data source – bank records – that has received relatively little attention as a means of examining questions relating to new and small business survival and growth.[3] It highlights a dataset of start-ups compiled from the records of a large bank – Barclays – one of the leading providers of banking services to small businesses in the UK. It shows how those data were derived and compares and contrasts them with other official and unofficial sources. This dataset only covers firms in England and Wales.

Third, we demonstrate the value of the Barclays dataset by showing that bank records offer data that are not readily obtainable from any other source and so permit policy analysis that would otherwise not be possible. Three illustrations of how it has been used as a key input into providing a better understanding of enterprise policy issues are provided.

BUSINESS ACTIVITY: OFFICIAL DATA

Self-employment

The official estimate of self-employment is the broadest measure of engagement with business activity. It is estimated by the UK's Office for National Statistics (ONS) as part of its key Labour Force Survey. The LFS is based on the responses of approximately 60,000 households, with 20 per cent of the sample being replaced each quarter. In recent years the core LFS sample has been boosted to ensure that there are robust estimates available for local areas.

The design of the survey is based on a common framework agreed through the International Labour Organization (ILO). This means that the resulting estimate of self-employment has traditionally been the most comparable of enterprise statistics across OECD countries. For example, it constitutes the key building block for the COMPENDIA database on business ownership in developed countries over time (Van Stel, 2003).

The ILO framework is also notable for taking a very broad view of self-employment. In essence, self-employed status is self-assessed. This means the respondent is self-employed if they say that they are, subject to a small number of consistency checks. In other areas the definition of self-employment is more prescriptive[4] but the UK data on self-employment comply with the ILO approach. The inclusive nature of this approach has led to the introduction of additional questions on the LFS asking, among other things, whether a self-employed respondent is paid by an agency or is the director of a limited company. This has been done in order to arrive at an alternative measure of 'true' self-employment; one that corresponds more closely to the expectations of those using the data.

The official LFS time series on UK self-employment commenced with annual data for 1984. The series then moved to a quarterly basis from 1992. However, there were estimates prior to this date, stretching back as far as the mid-1950s, although these are not on a consistent basis with the current series. The self-employment estimates provide an important starting point for considering a number of the key aspects of enterprise in the UK. These include the division between full-time and part-time engagement, self-employment as a secondary activity, the gender balance of enterprise involvement and the geographic variation in self-employment across the UK.

However, these estimates do not address all aspects of a particular policy interest in enterprise. They show how things are and how they have changed, but do not provide reasons for these changes; this requires more tailored research of the type undertaken by, for example, Parker et al. (2012). Recently, however, the official measure of self-employment has become an explicit target of enterprise policy, rather than merely informing its construction. This serves to heighten the significance of the measure.

148   *Handbook of service business*

Business Stock

The estimates of self-employment say something about engagement with enterprise at the individual level. However, policy and research are equally interested in business activity at the level of the firm. In the UK the most encompassing estimate of the business stock is produced by the Enterprise Directorate of the Department for Business Innovation and Skills (BIS).

The estimates, issued under the title of Small and Medium-Sized Enterprise Statistics for the UK, have been produced annually in their current form since 1994.[5] They are published in the third quarter of each year and refer to the position of the business stock at the beginning of the previous year. This 18 month lag reflects the underlying sources used for the estimates. In arriving at them the BIS essentially combines a data source that is a count of firms with one that is an estimate.

The former is the Inter-Departmental Business Register, or IDBR (of which more below). This contains details of all independent UK firms that are registered with Companies House,[6] either for Value Added Tax (VAT) or the Pay as You Earn (PAYE) system of tax collection that operates for employees. The latter is the residual population of self-employed not working in one of the registered firms captured by the IDBR. The main issue for this group is estimating the number of firms operated by these self-employed. To do this the Department of Business Enterprise and Regulatory Reform draws upon tax data from HM Revenue and Customs to estimate the balance of partnerships and sole traders. The end result of this process of estimation is a figure for the number of private sector businesses in the UK – companies, partnerships and sole traders.

In addition to the headline figure, the small and medium-sized enterprise (SME) statistics also provide a range of other estimates regarding the composition of the stock. BIS provides a breakdown of the stock by the number of employees in the firm, with accompanying data on aggregate employees (and employment) and aggregate turnover (with the exception of that derived from financial services). Estimates are available for high level business sectors and, in a more limited form, for three digit SIC codes. Also, every other year the estimates are extended to include the business stocks of the nine English regions, Scotland, Wales and Northern Ireland. These data are the basis of virtually all the regional comparisons of enterprise in the UK (Van Stel and Storey, 2004).

While the SME statistics are the only official estimate of the business population there is a problem over the allocation of the self-employed population to the stock of independent and ultimate holding companies. There is a risk that too many of the self-employed are being allocated to the residual population mentioned above. This would reduce the estimate of the business stock to some degree, although the exact magnitude would depend upon other changes to the methodology. Notwithstanding this, the SME statistics remain the first point of reference when asking the question 'how many?' with respect to UK business numbers.

Business Registers: IDBR and VAT

In the discussion of the official estimates of UK business numbers we mentioned that they were partly based upon the IDBR. The IDBR is a count of the relevant business population, namely those firms registered for VAT and/or PAYE. It replaced the VAT register as the primary official source of data on business numbers. This change was made to meet EU requirements for improved enterprise statistics.

Before the IDBR the original business register was that relating to VAT. This was created in 1973 to accompany the introduction of the tax as part of the requirements for UK entry into the (future) EU. The VAT register provided data on the count of relevant businesses by industrial sector, region, employee numbers and turnover. This means that it provided a significantly wider range of data than available from the SME statistics. It also had the advantage of offering flow measures in addition to simple counts of the stock. For example, official publications showed the number and rate of registrations and de-registrations by geographic area.

The main limitation of the VAT register is the high turnover threshold for compulsory registration – £77,000 in the financial year 2011/12[7] (the highest in the EU). Even allowing for the voluntary registration of firms below the threshold, this meant that the VAT register captured well under 50 per cent of the total number of businesses in the UK at any given time. The high threshold also meant that it was not possible to treat the accompanying flow data as a close proxy for business start-ups and closures in a given period, although they have often been used in research studies because of the absence of any alternative. The issue of the threshold was particularly prominent in 1991 and 1993 when, in order to reduce the administrative burden placed on the smallest firms, it was increased substantially, by 38 per cent and 20 per cent respectively. This has resulted in the absence of a reasonably consistent official measure of the business stock going back more than 15 years.

The IDBR was created during the 1990s. However, it is only since 2008 that it has supplanted the VAT register as the primary official measure of the business population. By extending the count of the business stock to include all companies and those unincorporated firms with employees, it has markedly improved the coverage of UK business numbers. In 2007 the IDBR was approximately 30 per cent larger than the VAT register for the previous year. The data are also supplied in a way that covers all of the topics addressed by the VAT register, including the 'survival' rates of firms entering the IDBR. Having said this, the IDBR still represents less than 50 per cent of the total estimated business stock, albeit the majority of economic activity. This gap is a reminder of the relatively limited data available on the large number of non-employing self-employed individuals.

The development of first the VAT register, then the IDBR, highlights the tensions that have existed over the years in UK policy between wanting to better understand the business population (and business owners) and trying to limit the administrative burdens placed on firms from measures that could add to this understanding. The result is that research has often had to turn to non-official data sources to supplement the official view.

BUSINESS ACTIVITY: TWO NON-OFFICIAL DATA SOURCES: THE GEM AND BBA

Prior to providing an overview of non-official data, it is worthwhile asking why it is necessary to use such alternative sources. After all, non-official business and enterprise data would seem less likely to generate high quality data and be more prone to selection bias than those drawn from official sources, where businesses can ultimately be compelled to provide data.

While this assessment may be correct, there are also limitations with official data. As noted in the previous section, official data collation may be restricted by legal limitations, such as those applying to the analysis of tax data or to judgements about the acceptable compliance burden to be placed on firms and owner-managers. In such instances there is little option but to turn to alternative sources to fill in the gaps.

The two examples outlined in the remainder of this section – the GEM UK Survey and the BBA data on small businesses– are perhaps the most interesting of the non-official sources. They represent very different aspects of data collation on business activity – one a very large survey and the other a near census.

Before looking at each of these sources we should acknowledge one other major source of non-official data– the business databases built up by credit reference agencies. In the UK the two main databases are those constructed by Dun & Bradstreet and Experian. While these databases have significant value, generated from additional data supplied by or collated on the firms, they do not necessarily provide a better 'count' of the business population. In these terms both databases effectively look like modestly enhanced versions of either the IDBR or SME statistics and have often been developed with the assistance of UK banks. We have therefore decided not to examine them in detail in this chapter.

The GEM UK Survey

Begun in 1998, and led by Paul Reynolds (Davidsson, 2005), GEM was conceived as an international study of enterprise on a more ambitious scale than previously attempted. The aim was to ask a set of identical questions across a range of countries with sufficient sample sizes to be able to derive robust conclusions about similarities and differences in business and enterprise activity.

The headline data from GEM concern the rate of new firm formation and involvement with established businesses by national populations. However, the scope of the survey means that it also contains information about matters such as innovation, public policy and wider societal attitudes to enterprise. These are all topics that are not addressed by the official data sources mentioned earlier and which necessarily require a survey format.

The UK has participated in GEM from its inception. Official interest in GEM UK was very strong. Indeed, for certain public bodies, notably the Regional Development Agencies, the survey was used to set the targets for measuring the outcomes of enterprise policy objectives. The enthusiasm for GEM UK meant that it had, for many years, the largest survey base of any of the participating countries. For example, the sample base for the 2007 UK survey was more than 42,000 individuals aged 16–64, or more than 25 per cent of the entire global sample for GEM across 42 countries in that year. In more recent years this has decreased so that in 2011 10,573 individuals aged 16–80 were surveyed.[8]

The headline GEM metrics classify enterprise involvement into three categories – nascent, new and established. The first of these concerns activity that has been in place for no more than three months. The second is where the enterprise has been operating for between four and 42 months. All activity is also (self-) identified as either 'necessity' or 'opportunity' based, essentially undertaken for negative and positive reasons respectively.

While GEM UK provides a wide range of data on enterprise issues, it is undoubtedly the headline output of the survey that has attracted the greatest attention over the years. In some ways this is unfortunate since it is the interest in attitudes, expectations and policy where the survey perhaps adds most value, given that most of the participating countries will have existing measures of self-employment and the business stock. Indeed, the GEM UK data may have given policy makers a misleading impression of the country's relative enterprise position. An example of this with respect to the UK and US will be presented later.

Bank Data: BBA Survey of Small Businesses

An alternative source of non-official data on business activity is the organisations through which the vast majority of this activity passes, namely the Clearing Banks.[9] In the UK the representative body for banks is the BBA. Since 1991 it has co-ordinated the collection of data on the majority of UK businesses.

The origin of the BBA's Small Business Survey lies in the 1991–1992 recession. During this period the relationship between the banks and the representative bodies of small firms was strained, with accusations of poor treatment of customers, particularly with regard to access to finance in difficult times and the 'passing on' of interest cuts. With the encouragement of the Bank of England, the main providers of banking services to small firms agreed to supply aggregate data on the financial position of the sector, including the number of current accounts, deposit balances and overdraft/term lending, for those firms with a turnover of less than £1 million.

The survey has evolved over time with one major change and a number of more minor additions to the range of data collated. The major change was introduced in 2002 when the definition of relevant firms was tightened. Prior to this, inclusion was based on turnover through an individual current account. This was changed to the turnover through the combined current accounts of a business where it held more than one at a given bank. More recently the data items captured by the survey have expanded to include start-ups, new term loans and new term lending. The periods for data compilation have also changed, from every six months to every quarter and now every month.

Although it is a survey, drawn from the data of eight banks, the nature of the small business banking market in the UK means that it is, in practice, close to a census of businesses operating using a business current account. The participants almost certainly account for more than 95 per cent of all business current accounts in the UK. The main limitation is the same as that for any survey drawing on multiple sources, namely arriving at common definitions for variables that can then be implemented effectively. For example, the original account level definition of qualifying firms was based upon data system limitations among some of the banks. It is also limited by the presence of multi-banking, meaning that certain firms will appear in the data of more than one bank. This is less of an issue in relation to the original purpose of the data collation – examining small firm finance – but is more problematical for measuring the business population.

As noted, the main strength of the BBA survey lies in its regular coverage of new and small business finance and the fact that these data can be disaggregated at a sub-national level and on a monthly basis.

USING BARCLAYS CUSTOMER RECORDS AS A SMALL BUSINESS DATA SOURCE

Barclays Bank is one of the UK's main suppliers of banking services to small firms. It has more than 600,000 business customers with a turnover of less than £1m, representing approximately 18 per cent of all UK firms operating through a business current account (Barclays, 2009), although for historic reasons the bank's involvement in Scotland and Northern Ireland is more limited. The range of data held on these customers has become more extensive over the years, not least as a result of regulatory requirements.

The business current account forms the cornerstone of the day-to-day relationship between the bank and its customers. The transactions conducted through these accounts have the potential to shed considerable light on small business activity as well as a 'count' of the enterprise population. In addition to this transactional data the bank also holds data on a range of variables relating to both the business and its owner(s). These data will include items such as address information, the firm's legal form, its type of economic activity, product holdings, and the age and gender of the owner.

The range of data that is available is certainly important, but the fact that it is set within an integrated source that provides the ability to track business entities and their owners over time, typically on a month-to-month basis, is especially useful. As a result it is possible to track the dynamics of individual firm activity, for example growth, closure, changes in ownership, and so on. The data held by Barclays are currently available on a monthly basis beginning from March 1991 as part of a data warehouse facility (with some earlier data, albeit in alternative formats, available from legacy systems).

Since 1988 Barclays has produced quarterly estimates of starts, closures and the business stock in England and Wales. While the BBA survey now includes start-ups among its collected data, the Barclays estimates are still unique in that they produce estimates of business closures and stock at less than yearly intervals.

The start-up estimates are derived by grossing up the volume of start-ups with Barclays by an estimate of the bank's market share, derived from a market research survey syndicated and agreed by most of the UK Clearing Banks. The way the survey is conducted means that the estimates are revised for up to 12 months after the quarter to which they refer. As with the data for new registrations drawn from the IDBR, the start-ups captured in these estimates are only a proportion of the total number of new firms, as many businesses will start operating by using a personal bank account. However, as will be illustrated later, the estimates of business start-ups compiled by Barclays yield consistently higher volumes than the IDBR, reflecting the wider coverage.

From 2003 the start-up estimates have been expanded to include a number of firm/owner-manager characteristics – location, sector, gender and age. While the BBA data on start-ups cover location, the Barclays estimates remain the only source of start-up data covering all four characteristics on a quarterly basis.

Barclays also produces estimates of closures and of the business stock. In a similar way to start-ups, the closure estimates involve grossing up the volume of Barclays customers who have closed in a quarter. The estimation of closures is more difficult than for start-ups. As the data are drawn from a single bank there is always the possibility that firms may have switched, although the estimates try to keep the impact of this to a minimum.[10] More importantly, there are no surveys of market share with respect to business closure. The estimates attempt to get round this constraint by using data from the start-up survey. The idea is that it is possible to derive a stylised closure 'profile' – the proportion of firms of a certain age that will have closed in a quarter. This can be used in conjunction with the estimates of start-ups and market share from that prior period to estimate a share of business closures. Essentially, the end result is a weighted average of start-up market shares.

The combination of starts and closure estimates is then used to derive estimates of the business stock on a quarterly basis. The stock data allow a more direct comparison with other data sources such as the IDBR and the BBA data. In fact, the count of the business stock has an integral role in the method used by Barclays to derive quarterly starts and closures data. For the first few quarters after initial publication, the stock is simply derived from the net change of starts less closures. However, retrospectively when more historical IDBR data are published, the Barclays count of businesses is re-calculated to be consistent with the IDBR. Indeed, if required in the interests of consistency, the starts and closures flows will also be revised for the relevant historical period. This is, in effect, a practical adjustment of the data to take account of known issues such as the changing incidence of multi-banking or switching activity.

The strengths and weaknesses of the Barclays estimates are derived from the same factor: they are produced by one organisation. On the positive side, this means that there is a single definition of the data, an ability to add additional data without having to operate at the lowest common denominator, and an ability to make informed 'adjustments' over time. Set against this are the limitations relating to all estimates, together with issues around the accurate identification of certain measures, for example closure. The Barclays data are in some cases in effect 'practitioner estimates' but, in the absence of better alternatives, they are likely to continue to have a central position in understanding business activity in the UK.

To summarise, there are three main benefits to using bank data. First, they offer greater coverage of the business population than is available from either official business registers or company accounts. They contain firms that are too small to be on the former and a wider range of legal forms than are found among the latter. Second, bank records can provide greater scale than is possible from a survey. Finally, bank records provide greater data frequency, offering the ability to look at turnover on a monthly rather than the more common annual basis for the majority of business registers. This provides the opportunity to construct within-period data, as with the Barclays volatility measure, and to observe seasonal dynamics.

The primary limitation of bank data is that although they have breadth, both in terms of coverage of business types and absolute numbers, they lack the depth in certain areas that can be obtained from either a survey or accounts. For reasons of cost, perceived (lack of) value to the bank and legal considerations, there is only limited information on owners, their background and preferences for their business. Effective identification of closure and exit is a further issue. Although this is also the case for official business registers which accumulate dormant businesses, an additional consideration when drawing data from a single bank is the role of switching.

THE BARCLAYS NEW FIRM PANEL DATASET: DERIVING AND DEFINING THE 2004 COHORT

What remains clear is that, despite its acknowledged imperfections, Barclays is in a unique position to construct a dataset that would permit the analysis of survival and growth among new businesses, without being restricted to, for example, limited companies, requiring survey responses or lacking granularity. However, before considering some of these issues in more detail, we will begin with a simple question – how to define a start-up?

The 2004 Cohort: Defining a Start-up

When does a business start up? For Barclays, identifying a customer as a start-up requires a positive response to two questions. First, is this the first account for the customer with Barclays? Then, if so, is it the first account for the business at any bank? If the answer to these two questions is 'yes' then the customer is a start-up. For reasons described more fully below, our dataset was drawn from non-financial firms identified as start-ups which entered the Barclays customer base between March and May 2004; a total of 23,344 such firms.

Two important caveats to this relatively straightforward description of a start-up are required. One is that a minority of firms will have been trading prior to opening a business current account. We noted previously that companies and partnerships must trade through such accounts. However, this is not the case for sole traders; they are permitted to make use of personal accounts.[11] It is not clear how many start-ups using Barclays are in this position, although their activity levels will, in most cases, have been on a minimal scale. The other caveat is that a new business does not necessarily start trading immediately upon opening an account. Indeed, Barclays' records indicate that approximately 5 per cent of start-ups do not make use of their account in the subsequent 12 months. We therefore only include firms with financial activity in the month following entry to the customer base.[12]

The 2004 Cohort: Start-up Data

Having defined the set of start-up businesses of interest, what data were available on each of them at the time of entry into Barclays' customer records? In the first half of 2004 these were more limited than at present. In particular, data on the owner(s) consisted only of their gender and age. To supplement these we were able to make use of questions put to owners of start-ups as part of a previous related study. But the three questions, on three aspects of the owner(s) and their choices that the literature indicated could be related to survival and growth, were voluntary and not part of the mandatory account-opening process used by Barclays at the time. First, information on the highest level of educational qualification achieved by the owner was requested. This was intended as a proxy for general human capital (van der Sluis et al., 2005). Second, information was sought about prior business experience (Westhead et al., 2005), including previous ownership and/or ownership among immediate family members. This captured the specific human capital of business experience, supplementing the more general dimension reflected in education. Finally, owners were asked about the sources of advice and support they used prior to start-up since the literature suggested that these could both enhance the owner's human capital and also provide additional resources (Pons Rotger et al., 2012).

Some 27 per cent (6,240) of start-up customers answered these questions in full. We then supplemented this with data collected by Barclays as part of its general account-opening process. These were the age and gender of the owner(s), the legal form of the business, the type of activity in which it would be engaged and its location within the UK.

On-going Data

As well as start-up and structural characteristics, any analysis of survival and growth requires data on the size of firms over time. Credit turnover – the value of payments into a current account[13] – was used as our measure of size, being a very close approximation to sales revenue inclusive of taxes. The much greater granularity of turnover compared with using measures of employee numbers is a particular strength, and drawing such data from bank records also provides other advantages. One is the direct observation of data without the need to survey businesses.[14] Another is the greater frequency with which the data can be observed. Credit turnover is available monthly rather than being limited to, for example, 12 month periods. For our dataset, turnover was aggregated across six month periods from the date of start up.

The frequency of turnover data also permits the creation of relevant new variables. For example, theory regarding business survival and growth places considerable importance on a fall in firm turnover below a given 'reserve' level as a factor inducing exit; for example, Jovanovic (1982), Gimeno et al. (1997). In doing this, there is an emphasis not only on the absolute size of the business but also the variance of this size. With monthly observations it is possible to create a measure of turnover 'volatility' that approximates this element of theory. For our dataset this volatility variable was the standard deviation of turnover for each firm across a six month period divided by total turnover.[15]

In addition to the level, growth and variability of turnover, bank records also provide the opportunity to calculate measures of financial management. For this dataset the key variable was use of an unauthorised overdraft, an important financial product in the UK that has traditionally been used as the first source of working capital for small firms. With prior agreement from the relevant bank, it permits a bank customer to make payments even when the balance in their current account falls below zero. Provided their balance remains above a given amount, the customer only pays interest on the amount of overdraft used.[16] However, customers can, in most cases, exceed their overdraft limit.[17] If this occurs the bank usually applies both a flat charge and a considerably higher interest rate to the entire balance. The financial costs of exceeding an overdraft are therefore clear. Although firms may occasionally judge it is worth incurring these costs, persistent unauthorised use generally represents poor management on the part of the owner(s).

The dataset includes two variables relating to the use of unauthorised overdrafts. The first is a simple binary variable based on whether the business was in unauthorised overdraft at any point during a six month period. The second records the proportion of that period spent in that position. To ensure these measures are not simply reflecting more general overdraft use – that is, that excess use provides additional information – we also

Table 9.1   Start-up dataset: variable definitions

Dependent	
exit_x.y	= 1 if business closes in year x, half y, else = 0. Conditional on firm being active at the beginning of this period
Explanatory	
Firm	
leg_frm	Legal form – company, partnership or sole trader
sic_grp	Industry group – agriculture, manufacturing, construction, motor trades, wholesale, retail, hotels & catering, transport, property services, business services, education & health and other services
gor	UK region – East of England, East Midlands, London, North East, North West, South East, South West, West Midlands, Yorkshire, Wales, Scotland and Northern Ireland
own_xs	= 1 if the number of owners of the business is more than the minimum number for its legal form – company >1, partnership >2
turn_x.y	Credit turnover in year x, half y
vol_x.y	Volatility of credit turnover in year x, half y. Measured as standard deviation of monthly turnover divided by mean monthly turnover
odlim_use_x.y	= 1 if there was use of an overdraft limit in year x, half y, else = 0
odlim_pc_x.y	Mean percentage of overdraft limit used in year x, half y
odxs_x.y	= 1 if there was use of an overdraft beyond agreed limit in year x, half y, including where the limit was zero, else = 0
odxs_pc_x.y	Proportion of year x, half y spent beyond overdraft limit
Owner	
own_male_inv	= 1 if one or more males as owner(s), else = 0
educate_x	Highest level of educational attainment among owner(s) – age 16, age 18, age 21+
age	Age of the owner at start-up. The mean age where there are multiple owners
bexp_x	= 1 if owner(s) has business experience of type x, else = 0 – family, self
adv_x	= 1 if owner(s) had advice/support from source x, else = 0 – enterprise agency/ business link, accountant, solicitor, college/university, start right seminar, princes business trust, family, other

include two further variables. One shows whether a firm used its overdraft limit at all in a period and the second shows the mean proportion of the limit used over the period. A full listing of all variables is provided in Table 9.1. Of course, not all new firms survived and we now turn to the issue of the exit of businesses from the dataset.

Exit and Closure

Establishing when a business has closed is perhaps the most challenging aspect of any study of survival and growth. Even for datasets taken from near comprehensive official sources, the date at which exit occurs may lag by some time the actual closure. When using bank data, two main issues have to be resolved. The first is to distinguish between those businesses that have closed and those that have switched to another bank. This dataset uses Barclays closure reason codes and examines closures over a six year period. Of the initial sample, 396 firms (6 per cent) closed their account with Barclays but continued to trade. Our suspicion is that this may be an understatement because imperfections in the coding process meant that not all switchers were recorded, although the scale of the under-estimations is difficult to determine.[18]

The second issue is how to judge when a given business has actually closed. While the majority of Barclays customers ceasing to trade simultaneously close their accounts, a significant minority become dormant, that is, the account remains open but with no activity.[19] For the firms in our sample we used a simple rule – if the business had shown no turnover in consecutive six month periods then it was deemed to have closed in the first of these periods.[20]

It is important to note that this identification process relates to closures and is not limited to business 'failures', that is, those firms that cease to trade with some external financial liability. It is estimated that such firms represent less than 20 per cent of all business closures, including those that close because they may not have met the objectives of their owner(s) and others that may merely reflect the arrival or awareness of better opportunities (Watson, 2010).

THE CHARACTERISTICS OF NEW FIRMS STARTING IN 2004

Start-up Characteristics

Almost one-half of new firms when they began were sole traders (Table 9.2). This reinforces the point made previously regarding the wider coverage of the business population that is achieved by looking at bank records. The majority (72 per cent) had only one owner, with 16 per cent having more than the minimum requirement for their legal form.[21] More than 80 per cent of the start-ups had at least one male owner, in line with or slightly higher than official data and previous studies. The predominant age range for those involved with the start-ups was 25–44, accounting for nearly two-thirds of owners, and this was also in keeping with other data sources (Greene et al., 2008).

The first of the voluntary questions revealed that more than one-quarter of owners had a university degree or equivalent qualification, with 33 per cent having no or limited formal educational attainment. The second question highlighted a marked response bias towards those with prior business experience, either personal (72 per cent) or through family (63 per cent). These high proportions contrast with previous studies that typically suggest that around one-third of owners have prior business experience (Westhead et al., 2005). Finally, the question on pre-start advice and support reveals a large minority of owners (38 per cent) who judged that they did not require any external input. Those that did mainly sought either professional advice from accountants (36 per cent) or more informal advice from family and friends (30 per cent).

Post-start Characteristics

A primary motivation for compiling the dataset was to examine business survival over an extended period. Twenty-eight per cent of firms (1,763) remained in the Barclays

Table 9.2 *Variable characteristics and change over time: variables at start-up and end Year 6 (% of surviving firms)*

		Start-Up	Year 6
leg_frm	company	37.4	49.3
	partnership	13.3	11.2
	sole trader	49.4	39.5
sic_grp	agriculture	1.0	1.1
	manufacturing	5.2	5.6
	construction	14.9	16.7
	motor trades	3.5	3.1
	wholesale	2.2	2.2
	retail	12.0	9.4
	hotels & catering	9.2	6.0
	transport	3.8	2.9
	property services	3.5	5.1
	business services	25.9	27.5
	education & health	2.5	3.3
	other services	16.2	17.1
gor	East of England	15.7	18.0
	East Midlands	7.3	5.6
	London	22.1	20.8
	North East	3.7	3.7
	North West	6.7	6.7
	South East	12.6	12.9
	South West	9.9	10.6
	West Midlands	9.6	9.0
	Yorkshire	5.9	6.4
	Wales	6.5	6.4
own_xs	yes	15.8	22.6
own_male_inv	yes	81.5	85.0
age		38.97	40.77
educate_x	age 16	32.9	33.5
	age 18	17.0	17.2
	age 21 +	27.1	27.6
bexp_x	family	63.2	61.7
	self	72.1	75.2
adv_x	enterprise agency/business link	10.4	9.6
	accountant	36.3	39.4
	solicitor	4.9	5.1
	college/university	4.2	4.5
	Start Right seminar	0.8	0.3
	Prince's business trust	1.4	0.5
	family	30.0	27.3
	other	6.7	5.7

Measuring business activity in the UK 159

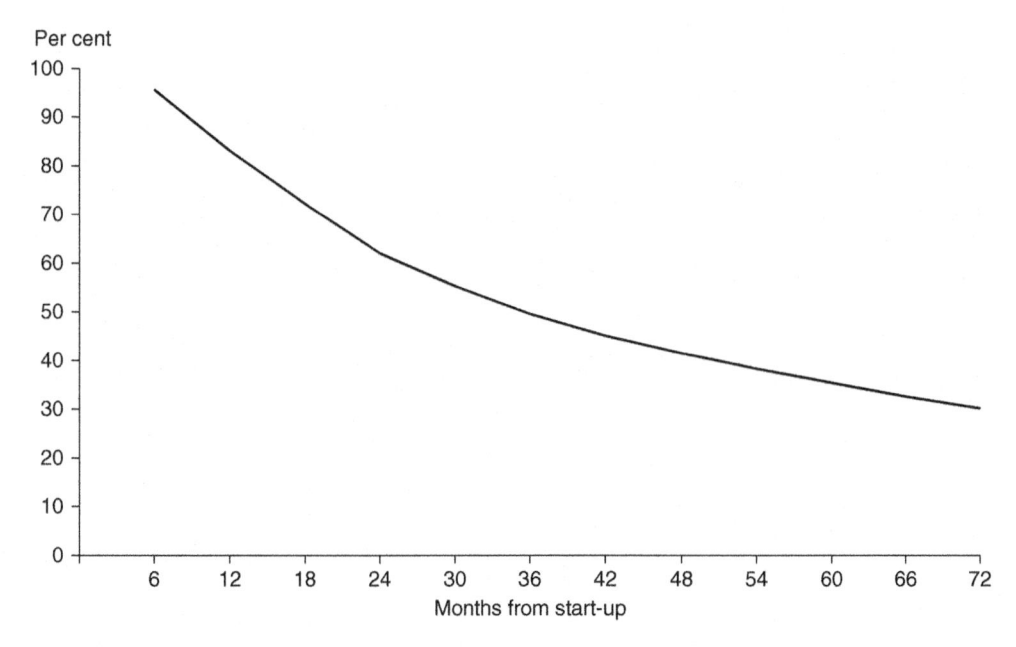

Per cent

[Figure: line graph showing survival rate declining from ~95% at 6 months to ~30% at 72 months, x-axis labeled "Months from start-up" with values 6, 12, 18, 24, 30, 36, 42, 48, 54, 60, 66, 72]

Source: ONS.

Figure 9.1 Survival rates

customer base after six years. Excluding 396 switching customers, approximately 30 per cent of businesses continued to be active after this length of time. This survival rate is not surprising given that the volume of start-ups (and closures) is around 12 per cent of the existing stock in a typical year. Figure 9.1 shows the conditional closure rate profile for non-switching customers over six month periods. The highest closure rates are between 18 and 30 months confirming the inverted u-shape familiar from previous studies (e.g. Frank, 1988), albeit that it is perhaps a little flatter, with closure rates more than five years after start-up remaining above 7 percent. Although not the focus of the dataset, the equivalent closure rate profile for switching customers is that the peak is later and the decline more pronounced. This is a reflection of the market for small firm financial services in the UK, where start-ups are not charged for their accounts for a period, usually 12–18 months, after opening.

The exit within six years of more than two-thirds of the firms in the dataset results in some marked changes in the composition of the remaining stock compared with that presented above. Table 9.2 shows the share of limited companies among surviving businesses increases significantly along with a linked rise in the proportion of multiple owner firms. At the same time the average age of owners increased by more than six years, while the share of owners under 35 years of age at start-up falls from 39 per cent to 30 per cent, implying that new firms with older founders were more likely to survive.

The other key objective of the dataset was to look at the factors shaping firm growth over the medium term. Here the data corroborate the results from previous studies in that the majority of surviving firms remains relatively small. After rising in the first

160 *Handbook of service business*

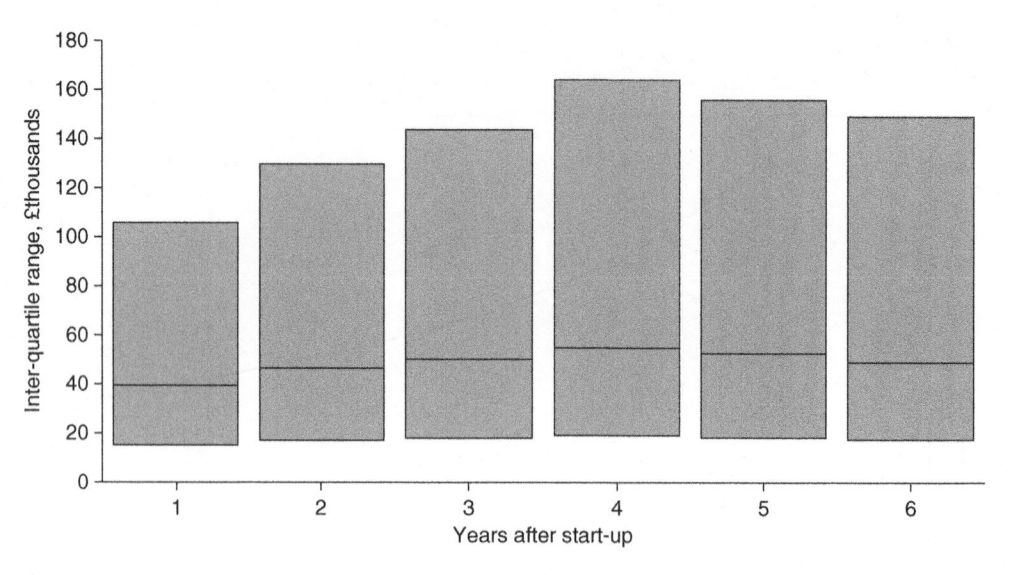

Figure 9.2 Turnover

couple of years, median turnover remains roughly stable at about £50,000 over the following four years (Figure 9.2). In each year more than one-quarter of firms had turnover of less than £20,000. Turnover of this scale indicates that the majority of new firms comprise just their owners (they have no employees). The threshold of the upper quartile of turnover increases from 2.7 to three times that of the median over the data period, standing at £150,000 in the sixth year after start-up.[22]

Year 5 of the dataset corresponded to the deepest recession in the UK economy in more than 60 years. The impact on growth among firms in the dataset was equally marked. For the first four years slightly more surviving firms grew than contracted (53–56 per cent). In year 5 a majority (58 per cent) contracted, with this proportion increasing in year 6 (to 62 per cent). The impact of the recession may also account for the flattening off in conditional closure rates noted above.

An important advantage of bank data is its greater frequency. The volatility measure is scaled to the turnover of the firm and thus allows for the modest changes in size distribution outlined above. Median volatility declines by 13 per cent in the first two years, after which it remains steady. Lower and upper quartile volatility also fall in the first two years, although the reduction is greater for the former (19 per cent) than for the latter (7 per cent). This implies a stretching out of the distribution as volatility is reduced following a 'shake-out' of firms.[23]

We also find that the measures capturing financial management point to the use of authorised overdrafts rising from less than 20 per cent of firms to more than 30 per cent over the first three years after start-up, with approximately one-third of surviving businesses in this position by years five and six. This increase reflects both a rising demand for overdrafts and a greater willingness on the part of the bank to supply the facility as a firm becomes more established. When firms make use of their overdraft limit the median proportion used also increases in the first couple of years to around 25 per cent,[24] while for the upper quartile this proportion moves to 50 per cent. Data on unauthorised overdraft use are striking. For all but the final year covered by the dataset more than 40 per cent of surviving firms have some unauthorized balances. Even in year six this proportion only falls to 38 per cent. However, the extent of such use is somewhat more limited. The median length of time spent in this position was around 5 per cent of the year, although one-quarter of such firms had excess balances for more than one day in five. The extensive, if mainly modest, use of overdrafts in this way may indicate that access to credit is more important for businesses than the cost of that credit.

USING BANK DATA TO ADDRESS KEY POLICY-RELATED QUESTIONS

The main sources of data on UK business activity were explored earlier in this chapter – stock, starts and closures from both official and non-official sources. It was argued that such data were crucial for a better understanding of policy but, as this section shows, the picture of even the most basic issues appears very different depending on which dataset is explored. To illustrate this we will examine five 'simple' questions that have been posed by the Minister responsible for Enterprise:

- How many businesses are there in the UK?
- Has the UK become more enterprising in recent years?
- How does enterprise involvement in the UK compare with other countries, notably the US?
- How likely is a new firm to continue to trade after a given period of time?
- What factors influence the growth and survival of a new enterprise?

These questions are explored in turn, drawing upon the different datasets.

How Many Businesses Are There in the UK?

Table 9.3 sets out data for all six sources on the stock of enterprise activity in the UK. For illustrative purposes, the data in the table all refer to the start of 2007 and show both the absolute value of each stock measure, with the exception of the GEM UK survey, and each measure scaled to both the population of working age and all individuals aged 16 and over.

The range of stock measures for enterprise involvement is clear, even allowing for the fact that some of the measures relate to individuals and some to businesses. The smallest measures are those derived from the two business registers. The high threshold for registration means that the VAT stock consisted of only 1.7 million firms at the beginning of 2007. The inclusion of PAYE firms in the IDBR increases this volume by more than 450,000. However, even this measure is only a modest proportion of the estimates for self-employment (3.8 million) and SMEs (4.7 million). Both of the bank-derived measures lie in between these extremes.

Scaled by the population measures, the data sources produce rates of business involvement ranging from a low of 3.5 per cent for the VAT stock to a high of 9.7 per cent for all firms. The fact that there are these variations does not necessarily matter if there is clarity about the different sources, their strengths and limitations. However, it does mean that

Table 9.3 Six data sources on the business stock, start-2007

Measure	Source	Size (thousands)	% of population: 16 and over	% of population: working age*
Self-employment[1]	ONS	3780	7.8	10.1
SMEs	BERR	4679	9.7	12.5
VAT	ONS	1699	3.5	4.5
IDBR	ONS	2131	4.5	5.8
New & established[2]	GEM		6.0	7.7[3]
Current account holding	BBA	3795	7.0	9.1
Current account holding[4]	Barclays	2927	6.8	8.8

Notes:
* Males aged 16–64, Females aged 16–59.
1. Average of Q4 2006 and Q1 2007.
2. GEM UK 2007 Survey.
3. Based on population aged 18–64.
4. England and Wales only.

there is no clear answer to the 'simple' question of how many businesses there are in the UK or the rate of individual involvement with enterprise.

Has the UK Become More Enterprising in Recent Years?

There is wide variation in estimates of the UK business/enterprise population and the same is true of measures of the inflow into this population (Table 9.3). Table 9.4 shows start-up estimates for 2007 using five of the measures previously discussed.

Again, the high thresholds for the business registers result in relatively low formation rates for these measures; at 200,000 (VAT) and 300,000 (IDBR) new registrations in each case are less than 1 per cent of the relevant populations. In contrast, bank-based measures produce numbers and rates approximately twice as high, at 400,000–500,000 and

Table 9.4 Entry rates and numbers

Measure	Source	Size (thousands)	% of population: 16 and over	% of population: working age*
VAT	ONS	206	0.4	0.5
IDBR	ONS	302	0.6	0.8
Nascent[1]	GEM			2.9[2]
Current account holding	BBA	559	1.2	1.5
Current account holding[3]	Barclays	432	1.0	1.3

Notes:
* Males aged 16–64, Females aged 16–59.
1. GEM UK 2007 Survey.
2. Based on population aged 18–64.
3. England and Wales only.

Measuring business activity in the UK 163

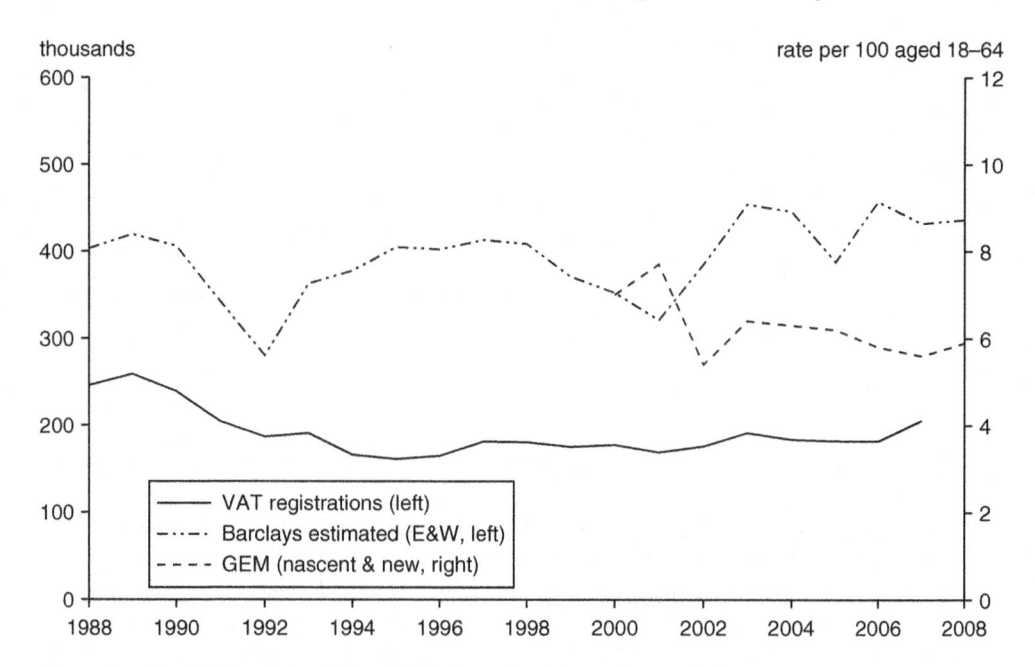

Note: Break of VAT series in 1991 and 1993 due to large change in threshold.

Source: ONS, Barclays, GEM UK.

Figure 9.3   *UK Business start-ups: a selection of measures*

between 1 per cent and 1.5 per cent. Perhaps most interestingly, the GEM UK measure is significantly higher than either of those drawn from business registers or bank records. This is in contrast to the Survey's stock measure, which lay in the middle of the range of estimates. This may indicate a disjoint between the measures, something that will be examined in the next section.

The contrasting measures of the stock and flow of enterprise activity are not just an issue of levels but also one of change. Over time the figures produced by these sources have both moved by differing amounts between years and sometimes in differing directions (Figure 9.3). This means that studies of enterprise activity over time require a clear understanding of how underlying factors can shape the resulting data.

How does Enterprise Involvement in the UK Compare with Other Countries, Notably the US?

Of the data sources considered above, two are available on consistent bases in both the UK and the US – the GEM Survey and the ILO-based measure of self-employment – although both capture different aspects of business/enterprise activity. While the self-employment estimates are more clearly a stock measure, those derived from GEM are a composite of stock and flow. GEM defines enterprise activity in three ways – nascent, new and established. Until recently the emphasis was on the combination of the first two categories. When taken together with data on self-employment, the result

164   *Handbook of service business*

Table 9.5   Measures of enterprise: US and UK

Measure	Source	Country	2001	2002	2003	2001–2003 mean
Nascent and new[1]	GEM	UK	7.8	5.4	6.4	6.5
		US	11.6	10.5	11.9	11.3
Self-employment[2]	ONS	UK	7.1	7.1	7.6	7.3
	BLS	US	6.6	6.5	6.8	6.6

Notes:
1. % population aged 18–64.
2. % of population aged 16 and over.

points to ambiguity regarding the relative scale of enterprise involvement in the two countries.

A consistent feature of GEM is that the rate of entrepreneurial activity (among those aged 18–64) has been substantially higher in the US than the UK. That is, a higher proportion of the US sample states that they are either engaged in an enterprise that has paid wages for less than three months (nascent enterprises) or for between four and 42 months (new enterprises). For the three years 2001–2003 the US GEM rate averaged 11.3 per cent, more than 50 per cent higher than the equivalent UK rate of 6.5 per cent (Table 9.5).

Data on self-employment are less straightforward to compare than the GEM results. The standard US self-employment figures exclude the owner-managers of limited companies, a group present in the UK definition. However, comparable US data are occasionally published (e.g. Hipple, 2004). For the same three year period, 2001–2003, this showed a US self-employment rate averaging 6.6 per cent among the population aged 16 and over. In comparison, the UK self-employment rate was 7.3 per cent over the same period (Table 9.5). Therefore, on this measure, the UK had a higher level of enterprise involvement, a marked difference from the picture painted by the GEM Survey.

To highlight these contrasting results is not to suggest that one or other of these sources is the definitive view of enterprise in these countries. However, it is an example of the need to interpret data carefully. In the case of the GEM Survey the differences between the UK and the US are primarily in the 'nascent' rates, while the 'new' and 'established' rates are similar. This suggests that the first category maybe capturing broader societal views of enterprise rather than actual engagement.

While this may seem a relatively trivial issue, it is important to understand that there are quite well set views about the relative enterprise position of the UK with respect to the US among those shaping policy.[25] Of course, it may still be the case that the US provides a better environment for, say, high growth firms. What considering this alternative data does is to provide a different perspective on broad enterprise involvement.

How Likely is a New Firm to Continue to Trade after a Given Period of Time?

In the UK, the official source of data on survival rates is the IDBR, with the VAT register having performed this role prior to 2008. The measure of survival looks at entry into,

Table 9.6 Survival and closure rates

Year	Survival rates* IDBR[1]	Survival rates* Barclays[2]	Closure rates† IDBR[1]	Closure rates† Barclays[2]
1	92.9	88.6	7.1	11.4
2	79.3	73.0	14.7	17.7
3	62.9	59.5	20.7	18.4
4	51.9	49.0	17.4	17.7
5	44.9	41.8	13.5	14.7

Notes:
* % of initial firms surviving after given period.
† % of firms surviving at start of given period closing by the end.
1. Firms entering the IDBR in 2002.
2. New start-up firms in March and September 2003.

and exit from, the IDBR. For example, of the firms first entering the IDBR during 2002 92.9 per cent remained after 12 months, 62.9 per cent after three years and 44.9 per cent after five years (Table 9.6). The same data can be used to obtain conditional 'closure' rates, that is, the proportion of firms leaving the IDBR in the next period given that they remain at the current time. These show that closure rates for the 2002 cohort peaked 24–36 months after entry at more than 20 per cent. The IDBR data are also available on a geographic and industry basis, permitting comparisons across the economy.

While the IDBR data are an important source of information, it will be recalled that they have limitations resulting from the way in which the register is constructed. The two most important relate to the entry threshold and to the identification of business closure. Given the high turnover threshold for compulsory VAT registration and the relatively small number of employing firms, it is likely that many of the firms entering the IDBR are not recent start-ups. In a similar way, firms leaving the IDBR may not have closed or, indeed, closed firms may not have exited.

A clearer idea of whether these factors influence survival rates can be seen when they are compared with the Barclays Bank data. Since 1992 the bank has estimated survival rates based on the opening and closure of business current accounts. Although this is a lengthy time series it is restricted by the fact that the accounts are not limited to new firms. From 2003 the original series has been supplemented by estimates based on the entry (and exit) of start-up customers. Over the period 2003–2008 the Barclays data consistently show lower survival rates than the official data – 86.6 per cent after one year, 59.5 per cent after three years and 41.8 per cent after five years (Table 9.6). As a result, early stage closure rates are higher, although the peak period for the closure rate remains between two and three years after start-up, the same as indicated by the official data.

What Factors Influence the Growth and Survival of a New Enterprise?

The Barclays New Firm Panel Data set sheds valuable light on the above question. Frankish et al. (2013) and Coad et al. (2013) examine the factors influencing both survival and growth. They show that the choice of legal form significantly influenced survival in three out of the four time periods, with firms choosing limited company status being much more likely to survive than those choosing either sole trader or partnership status; secondly, none of the 'standard' human capital variables exerted a consistent influence over survival; thirdly, that taking business advice appeared as likely to lower as to raise survival rates.

Instead it was the 'account behaviour' variables that dominated any explanation of business survival. Surviving new firms were significantly less likely to exhibit month-to-month variations in cash flows. They were also much less likely to exceed their overdraft or borrowing limits and, if they did, they were more likely to spend a longer period 'in excess'. Nevertheless, the picture that emerges is that there is a strong chance component in which new firms survive and which firms do not. Alternatively expressed, we can say that, at the level of the individual business, the role of good or bad luck is powerful.

Frankish (2013) also shows there are strong links between survival and growth. As with the survival results, the 'usual suspects' of entrepreneurial human capital – education, business experience, access to advice – play only a tiny role in explaining which firms survive and grow, which firms merely survive and which firms do not survive at all. What emerges as critical for growth is access to financial resources. Hence early survival is higher among firms that start large. This effect becomes considerably less important after two years, with periods of sales growth then playing a much bigger role. In essence the survival of a young firm is enhanced considerably when it has a period when sales rise sharply – even if that burst of growth does not continue.

CONCLUSION

This chapter has reviewed the measurement of business and enterprise activity in the UK and its link with policy. It has provided brief overviews of official and non-official data sources, highlighting some of the strengths and weaknesses associated with each source and concluding that on balance the bank-based data sources were superior on most grounds. This is a far from trivial finding since the different data sources point to striking contrasts between the scale and nature of business activity in Britain and how the UK compares internationally. Finally, we have highlighted the insights provided from analysing the outcomes from the Barclays 2004 new firms cohort.

The central lesson that emerges is the importance of understanding data sources. Although it may appear trivial, an inadequate appreciation of what a particular source is actually measuring all too easily leads to misleading conclusions regarding various aspects of enterprise activity. As we have seen, even an apparently simple question such as 'Is the US more entrepreneurial than the UK?' is not as straightforward to answer as might be expected.

What the issue of measurement also makes clear is that there is rarely a definitive view of any of the concepts associated with enterprise. For example, is the focus of interest businesses, the individuals behind them or both? Different data sources will have relative strengths in each of these areas. The end result is a need to draw effectively on a range of data to answer questions from the simple to the complex.

The effective analysis of business and enterprise rests on the quality and appropriateness of available data. The position in the UK today is vastly different from that faced by the Bolton Committee (1971) which noted, 'we were surprised to discover that no comprehensive census...had ever been carried out' (1971: 264), with the result that 'assessing the gravity of the decline of the sector must be largely a matter of subjective judgement' (1971: 87).[26] Today's problem is deciding which of the multiplicity of sources – none of which is perfect – is most appropriate for the task in hand.

NOTES

1. This chapter includes references to analyses of Barclays' customer records undertaken by the authors with the permission of the bank. All research was conducted in a manner consistent with data protection obligations. No personal details were released to individuals outside of the Barclays Group.
2. Julian Frankish and Richard Roberts write in a personal capacity and the chapter does not seek to represent the views of Barclays Bank.
3. Indeed, we are aware of only one prior dataset from such a source that has been used to consider these issues (Cressy, 1994, 1996).
4. For example, HM Revenue & Customs will apply a 'control' test to those claiming self-employed status (Storey and Greene, 2010: 466–467).
5. There are estimates prior to this point, going back as far as 1979, but not on a comparable basis.
6. And the equivalent organisations in Scotland and Northern Ireland.
7. Another limitation is the impact of VAT exemption on certain activities. This results in the exclusion of a high proportion of firms in particular industry sectors, e.g. health.
8. We are grateful to Jonathan Levie for this information. It should be noted that the major reduction was in England, which provided half of respondents in 2007 and only 25 per cent in 2011.
9. Those banks that are responsible for paper and electronic clearing services in the UK.
10. Research indicates that the incidence of switching is much lower than closure. Only about 5 per cent of established UK SMEs actually switch banks over a three year period (Fraser, 2005).
11. For sole traders with low turnover an important consideration is that personal current accounts operate on a 'free in credit' basis, whereas equivalent business accounts have explicit charges after an initial period.
12. We also include a small proportion of firms which did not show activity in their first full month, but in either May or June 2004. In these cases the start month of the firm was recorded as the month prior to activity.
13. Excluding payments from related accounts, for example deposit accounts held by the business.
14. This can also be obtained from business accounts, although it can be time consuming to access these for small firms and, in the case of the UK, a large proportion of the corporate population is not required to supply this information.
15. This scales the measure to the size of the business.
16. Although there may be a periodic charge to maintain it and the bank is able to change or withdraw it at short notice.
17. Including where they have no agreed limit.
18. Market research indicates that c. 4 per cent of the established mainstream business population in the UK switches bank each year.
19. Indeed, some of these may have switched rather than closed.
20. Some Barclays customers can show little or no activity for a number of months before seeing turnover return to non-negligible levels. This may reflect the nature of many 'micro' businesses.
21. Two for partnerships and one for limited companies.
22. Firms with turnover of at least £1 million were 3.5–4 per cent of the surviving population in years 4 to 6. This level of turnover corresponds to approximately 8–12 employees.
23. More than 35 per cent of firms in the dataset close within 24 months of start-up.
24. That is, if a firm's overdraft limit is £1,000 then its mean overdrawn balance across a year is £250.
25. Indeed, one of the authors included the data mentioned above in a presentation to representatives of the, then, Department of Trade and Industry, to be met with considerable surprise on the part of the participants.
26. However, the Committee's recommendation in that publication for an 'enterprise census' has never been taken up.

BIBLIOGRAPHY

Audretsch, D.B. and Mahmood, T. (1995). 'New Firm Survival: New Results Using a Hazard Function', *Review of Economics and Statistics*, 77(1), 97–103.
Barclays (2009). *Sustainability Report 2008*, Barclays, London.
Bolton Committee (1971). *Report of the Committee of Inquiry on Small Firms*, HMSO, London.
Coad A, Frankish, J.F., Roberts, R.G and Storey, D.J. (2013). Growth Paths and Survival Chances: An Application of Gambler's Ruin Theory, *Journal of Business Venturing*, 28(5), 615-632.
Cressy, R.C. (1994). 'Staying With It: Some Fundamental Determinants of Business Startup Longevity', CSME Working Paper 17, Warwick Business School, University of Warwick.
Cressy, R.C. (1996).'Pre-entrepreneurial Income, Cash-Flow Growth and Survival of Startup Businesses: Model and Tests on UK data', *Small Business Economics*, 8(1), 49–58.
Cressy, R.C. (2006). 'Why Do Most Firms Die Young?', *Small Business Economics*, 26, 103–116.
Davidsson, P. (2005). 'Paul. D. Reynolds: Entrepreneurship Research Innovator, Co-ordinator and Disseminator', *Small Business Economics*, 24(4), 351–358.
Department of Business Enterprise and Regulatory Reform (2008). *Enterprise: Unlocking the UK's Talent*, Department of Business Enterprise and Regulatory Reform, London, March.
Ericson, R. and Pakes, A. (1995). 'Markov-Perfect Industry Dynamics: A Framework for Empirical Work', *Review of Economic Studies*, 62, 53–82.
Frank, M.Z. (1988). 'An Intertemporal Model of Industrial Exit', *Quarterly Journal of Economics*, May, 333–344.
Frankish, J., Roberts, R., Storey, D.J. and Coad, A. (2012). 'Growth Paths and Survival Chances', SPRU Working Paper 195, University of Sussex.
Frankish, J., Roberts, R., Coad, A., Spears, T.C. and Storey, D.J. (2013). 'Do Entrepreneurs Really Learn? Evidence from Bank Data', *Industrial and Corporate Change*, 22(1), 73–106.
Fraser, S. (2005). *Finance for Small and Medium Sized Enterprises*, Bank of England, Warwick Business School, Warwick.
Gimeno, J., Folta, T.B., Cooper, A.C. and Woo, C.Y. (1997). 'Survival of the Fittest? Entrepreneurial Human Capital and the Persistence of Underperforming Firms', *Administrative Science Quarterly*, 42, 750–783.
Greene, F.J. (2002). 'An Investigation into Enterprise Support for Younger People, 1975–2000', *International Small Business Journal*, 20(3), 315–336.
Greene, F.J., Mole, K.F. and Storey, D.J. (2008). *Three Decades of Enterprise Culture*, Palgrave Macmillan, Basingstoke.
Hipple, S. (2004). 'Self-employment in the United States: An Update', *Monthly Labor Review*, July, BLS, Washington, DC.
HMTreasury (1999). *Enterprise and Social Exclusion*, HM Treasury, London.
Holmes, P., Hunt, A. and Stone, I. (2010). 'An Analysis of New Firm Survival using a Hazard Function', *Applied Economics*, 42(2), 185–195.
Jovanovic, B. (1982). 'Selection and Evolution in Industry', *Econometrica*, 50(3), 649–670.
Konings, J. and Xavier, A. (2002). 'Firm Growth and Survival in a Transition Country: Micro Evidence from Slovenia', Discussion Paper 114/2002, LICOS, Centre for Transition Economics, K.U. Leuven.
Mata, J. and Portugal, P. (1994). 'Life Duration of New Firms', *Journal of Industrial Economics*, XLII(3), 227–345.
Parker, S.C., Congregado, E. and Golpe, A.A. (2012). 'Is Entrepreneurship a Leading or Lagging Indicator of the Business Cycle? Evidence from UK Self-employment Data', *International Small Business Journal*, 30(7), 736–753.
Penrose, E.T. (1959). *The Theory of the Growth of the Firm*, Blackwell, Oxford.
Persson, H. (2004). 'The Survival and Growth of New Establishments in Sweden, 1987–1995', *Small Business Economics*, 23(5), 423–440.
Pons Rotger, G., Gørtz, M. and Storey, D.J. (2012).'Assessing the Effectiveness of Guided Preparation for New Venture Creation and Performance: Theory and Practice', *Journal of Business Venturing*, 27, 506–521.
Segarra, A. and Callejón, M. (2002). 'New Firms' Survival and Market Turbulence: New Evidence from Spain', *Review of Industrial Organization*, 20(1), 1–14.
Storey, D.J. and Greene, F.J. (2010). *Small Business and Entrepreneurship*, Pearson/FT, London.
Strotmann, H. (2007). 'Entrepreneurial Survival', *Small Business Economics*, 28(1), 87–104.
Van der Sluis, J., van Praag, M. and Vijverberg, W. (2005). 'Entrepreneurial Selection and Performance: A Meta-analysis of the Impact of Education in Developing Economies', *World Bank Economic Review*, 19(2), 225–261.
Van Stel, A.J. (2003). *Compendia 2002:2. A Harmonized Data Set of Business Ownership Rates in 23 OECD Countries*, EIM, Zoetermeer.

Van Stel, A.J. and Storey, D.J. (2004). 'Is there a Upas Tree Effect?', *Regional Studies*, 38(8), 893–910.
Watson, J. (2010). *SME Performance: Separating Myth from Reality*, Edward Elgar, Cheltenham and Northampton, MA.
Westhead, P., Ucbasaran, D., Wright, M. and Binks,M. (2005). 'Novice, Serial and Portfolio Entrepreneur Behaviour and Contributions', *Small Business Economics*, 25, 109–132.

10. The growth of information-intensive services in the US economy
Uday Apte, Uday Karmarkar and Hiranya Nath

INTRODUCTION

Most of the large economies in the world are already dominated by services, in that they contribute more than 50 percent of national income. We are now in the midst of another major evolutionary trend, that from a material or physical to an information economy. This change is most visible in developed economies but is occurring everywhere. In this chapter, we explore the confluence of these two trends by examining the double dichotomy of products versus services and information-intensive versus material-intensive (or non-information) industries, which divides the economy into four super-sectors (Karmarkar 2008; Apte et al. 2008). Figure 10.1 provides some illustrative examples of industries in the four super-sectors. Note that certain physical manufacturing and service examples (for example, computers, telecom) fall in the information sector following the definition by Porat (1977). It should also be pointed out that many industries do not really lie entirely inside one cell. For example, the health care industry breaks down just about evenly across the material and information sectors.

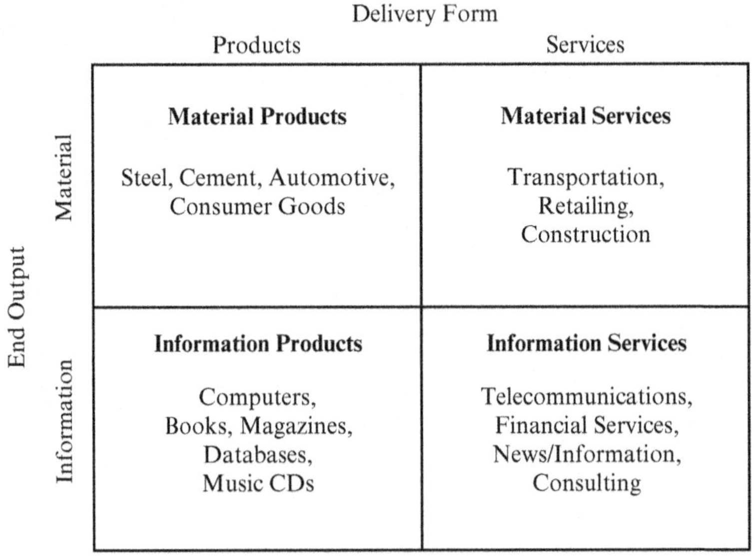

	Delivery Form	
	Products	Services
End Output: Material	**Material Products** Steel, Cement, Automotive, Consumer Goods	**Material Services** Transportation, Retailing, Construction
End Output: Information	**Information Products** Computers, Books, Magazines, Databases, Music CDs	**Information Services** Telecommunications, Financial Services, News/Information, Consulting

Source: Apte, Karmarkar and Nath (2012a), Figure 1.1, page 3.

Figure 10.1 A 2×2 decomposition of the US economy with sector examples

Our analysis of the broad changes in the US economy uses sector-level data organized by the Standard Industrial Classification (SIC) and North American Industrial Classification System (NAICS) codes. We also examine the evolution of jobs and wages in the US. This research is different from the GNP studies and gives a more detailed perspective. Wages are a major part of GNP but are not exactly the same.[1] Also, the number of jobs can and does distribute differently across the economy, since average wage rates differ substantially across sectors. Furthermore, the GNP data are aggregated at the level of SIC/NAICS codes. But in fact, companies and jobs often cut across the boundaries of the super-sectors we are examining. The data on jobs and wages thus present a different perspective at a finer level of resolution. We analyze data on the US job market (employment and wages by more than 800 occupational categories) since 1999 and extended up to 2007. One major finding is that information workers in services now account for the largest share of total US jobs, in a significant change from the historical pattern, in which non-information workers in services were the largest segment of the labor market.

It is fair to say that the US economy is now an information economy in terms of GNP, jobs and wages. What is more, the largest component of the US economy is now information-intensive services. While the economic crises in the last decade have resulted in some moderation of these trends, we can expect continued movement in the same direction in the near future.

This transformation to information and information-intensive services has a wide array of consequences. First, the economics of information-intensive sectors are different in certain specific ways that have an impact on competition. Entry barriers are low, and simple economies of scale are less pronounced. Physical location is less of a differentiator across competing firms. All of these tend to make competition more intense. On the other hand, barriers like network externalities can have the opposite effect. Furthermore, there may be opportunities for finding niche audiences. While trade in services is generally difficult due to their intangibility, trade in content-based services and even transactional services is now feasible. There are many other structural implications for economies. One systematic phenomenon is the de-integration occurring in many verticals such as music and publishing. Another is the appearance of mechanisms such as open sourcing and of direct exchange and barter, which can create a kind of demonetization effect. Finally, the structural changes extend below the sector level down to organizations and even jobs and tasks.

In the next section, we review research on the growth of information-intensive services in the US and survey the literature on related topics. Then we present an analysis of the two-way breakdown of the US economy based on GNP data (as in Figure 10.1) to document the growth of information-intensive services. In the next section, we present the trends in employment and wages, capital investments in information and communication technology (ICT) and the changing patterns of international trade in information services. The subsequent section presents a brief discussion on the drivers of the economic evolution that has been documented in the previous sections, along with some of the consequences of this evolution for industries. We conclude with a summary and a description of our ongoing research on these topics.

LITERATURE REVIEW

The research on various aspects of the information economy belongs to a wide range of disciplines.[2] In this literature review, we focus only on those studies that are relevant to the issues discussed in this chapter. Apte et al. (2008) first proposed dividing the economy into four super-sectors, displayed in Figure 10.1. This decomposition of the overall economy, applied to the US national income data, allows them to demonstrate that the share of information services has not only been rising but also has become the largest segment of the economy. By 1997, about 56 percent of total US GNP was generated in the information services super-sector. This share increased from about 36 percent in 1967 to 49 percent in 1992 and subsequently to 56 percent in 1997. They also examined the labor market by measuring the information intensity of various occupations and calculating full-time equivalent (FTE) employment and wages of information and non-information workers. According to their definition, information workers are involved in creating, processing and communicating information. However, they presented employment and wage data only for 1999. Apte et al. (2012a) updated and extended this study, analyzing the trends until 2007, when information services accounted for 55 percent of GNP. This study also showed that, by 2007, information workers in service industries were the largest segment of the labor market in terms of total employment and wages. It also analyzed the recent trends in international trade and documented how in recent times the US trade in services, particularly in information-intensive services, has been the largest component of overall services trade in which it has a trade surplus.

The analysis presented in the above studies crucially hinges on the separation of the information from the material component of the economy. The seminal contributions of Machlup (1962, 1980) and Porat (1977) described methods for measuring the information or knowledge component of the total value generated in the economy. Fritz Machlup's 1962 study was one of the first attempts to assess and to present a comprehensive statistical profile of what he called the "knowledge industry". His study provided a conceptual framework for research into quantitative as well as qualitative aspects of knowledge-based information activities. In 1977, Marc Porat undertook an extensive study of information-based activities in the US economy on behalf of the US Department of Commerce.[3] He used a conceptual framework similar to that of Machlup but in order to define and measure the information economy adopted an approach based on the national income accounting framework, which is quite distinct from the one used by Machlup. According to his definition, the information domain (sector) of the economy is "involved in transforming information from one pattern to another", whereas the material domain is "involved in the transformation of matter and energy from one pattern into another" (Porat 1977, vol. 1, p. 2).

The operational definition of information used by Porat includes "all workers, machinery, goods, and services that are employed in processing, manipulating, and transmitting information" (1977, vol. 1, p. 2). Thus, the measure of the information segment of the economy consists of two major components. First, it includes the total value added by industries producing goods and services that are directly used in processing, manipulating and transmitting information (called the primary information sector by Porat 1977). Examples of this include computers, books, music CDs, telecommunications and financial services. Second, the information economy measure also includes value-added contributions of workers and capital that are involved in informational activities but employed in industries that produce material goods and services (called the secondary information sector by Porat 1977). For example, the value-added contributions of the designers and computers in the apparel industry or the contributions of the reservation clerks, computers and phones in the hotel industry are a part of the information sector. According to Porat's (1977) calculations, the information component accounted for about 46 percent of the US GNP in 1967.

The Organisation for Economic Co-operation and Development (OECD) (1981, 1986) used Porat's methodology to measure the size of the information economy in the US along with eight other member countries. It showed that the share of the information sector in US GNP was about 43 percent in 1958 and about 49 percent in 1972 (OECD 1981). In a more recent study, Apte and Nath (2007) reported that the share of all information activities in total GNP grew from about 56 percent in 1992 to 63 percent in 1997. Apte and Nath (1999) examined this decomposition of information versus material for a number of service industries and noted the shift of the US economy toward information services in the early 1990s. Choi et al. (2009) conducted a similar study for Korea, with results that mirror those for the US economy.

There are a few studies that examined employment patterns along the information versus material (non-information) dichotomy. Osberg et al. (1989) identified three broad occupational sectors within modern advanced economies such as those of the US and Canada: the goods sector, composed of occupations that directly involve the manipulation or transformation of goods; the personal service sector, consisting of occupations that involve service to other individuals; and the information sector, involving the production or manipulation of symbolic information. The information sector is further subdivided into two types of occupations: the data processors, who are engaged in routine manipulation, storage and transfer of information within previously defined categories (for example, clerical work), and the knowledge producers, who are engaged in the establishment of original categories or analyses (for example, engineering or computer programming). They argued that the growth of these three sectors could be explained by the inherent unbalanced growth of the economy.[4] According to them, labor productivity in goods and data production grew steadily over time due to the increasing capital intensity of production and the impact of ICT advances. In contrast, labor productivity in the personal services and knowledge sectors did not tend to increase over time, as labor time was the output in personal services occupations, and the human creativity essential to knowledge production demonstrated little tendency to increase over time. These differences in productivity growth explained the relative increase in the employment of personal service and knowledge workers. The authors of this study analyzed occupational data between 1960 and 1980 in the US to demonstrate a relative shift in employment toward knowledge-based occupations.

Using decennial census data on employment by detailed occupations and industries between 1950 and 2000, Wolff (2006) extended this analysis and found that information workers (knowledge producers and data processors) increased in the US from 37 percent of the workforce in 1950 to 59 percent in 2000. His analysis further showed that the growth of information workers was not attributable to a change in tastes for information-intensive goods and services. Instead it was partly due to changes in production technologies that made it possible to substitute goods and service workers for information workers and partly due to differential rates of productivity movements among the industries of the economy, thus fitting the framework of unbalanced growth.[5] Freeman (2002) further discussed various labor market outcomes of ICT extension to economic activity.

Studies of the employment of information or knowledge workers can also be related to the recent literature that has focused on the increasing wage gap between low-skilled and high-skilled workers. Skill-biased technological innovation has been shown to be the most plausible explanation for this development. For example, Autor et al. (1998a) argued that increased use of computers (which is used as a proxy for skill-biased technological change) and skill upgrading accounted for growth in demand for and wages of skilled workers. Autor et al. (1998b), Autor et al. (2003), Autor (2007) and Acemoglu and Autor (2010) further explored the relationship between technological change and employment and earnings of skilled (information) workers. Berman et al. (1998) and Machin and Reenen (1998) examined empirical evidence for skill-biased technological change for several advanced countries, including the US, and established that skill-biased technological progress had increased the demand for skilled workers. Focusing on IT professionals, Mithas and Krishnan (2008) further documented that investment in human capital and IT intensity of firms led to substantially higher compensation for these workers. Furthermore, Nath (2011) showed that ICT investment has lowered the employment of information workers and concludes that this is the net effect of two factors working in opposite directions. He argued that the increased demand for information workers due to skill-biased technological change has been more than offset by the decrease in demand due to ICT-enabled service innovations such as automation and offshore outsourcing.

Related to this literature, but considered from the operations management perspective, Apte and Mason (1995) and more recently Mithas and Whitaker (2007) focused on global disaggregation of information-intensive services. Apte and Mason (1995) developed a classification framework to identify the services and jobs that were most amenable to service disaggregation. Building on this classification scheme, Mithas and Whitaker (2007) proposed and empirically validated a theory of service disaggregation, which argued that service-generating high-information-intensity jobs were relatively more amenable to disaggregation. They also found that high-information-intensity occupations that required higher skill levels had experienced higher employment growth but a decline in salary growth. Furthermore, occupations with a higher need for physical presence had also experienced higher employment growth and lower wage growth. However, the scope of these changes in the production and delivery of services (particularly information-intensive services) was far greater than just spatial disaggregation of the supply chain, as noted by Karmarkar (2004).[6] This line of research continued to examine so-called "service industrialization", which included the link between the global information economy and service industrialization (Karmarkar 2008, 2010). In a recent article, Apte et al. (2012b) explore potential schemes for representing complex processes in information-intensive services.

Behind the growth of the information economy are ICT advances and the investment in ICT capital. The latter is broadly divided into three categories of fixed assets (computer and peripheral equipment), software and communications equipment. Several studies have examined the impact of ICT capital on productivity and employment. For example, Stiroh (2002) investigated how ICT capital investment affected labor productivity growth in the US, while, as discussed above, Nath (2011) further examined the impact of ICT investment on the employment of information workers. However, data on various ICT capital assets as a separate category have been available only recently, and further research on these trends is warranted.

Trade in services in general, and information-intensive services in particular, has received some attention in recent years. Apte and Nath (2012) have undertaken a comprehensive account of growth and patterns of US trade in information-intensive services, the largest segment of services trade in the US. However, there are several strands to this literature on trade in services; some studies examine the determinants of international trade and investment in services (for example, Polese and Verreault 1989; Freund and Weinhold 2002; Grunfeld and Moxnes 2003; Kimura and Lee 2006; Co 2007; Mann and Civril 2008), others focus on gains from trade in services in terms of productivity and growth (for example, Mattoo et al. 2006; Hoekman and Mattoo 2008; Amity and Wei 2009), and yet other studies discuss policy issues related to services trade (for example, Bhagwati 1987; Hoekman 1996; Deardorff 2001; Hoekman et al. 2007; Deardorff and Stern 2008). Perhaps the most comprehensive review of these different strands of the literature was provided by Francois and Hoekman (2010).

STRUCTURAL CHANGES AND THE GROWTH OF INFORMATION SERVICES

As Apte et al. (2012a) showed, the share of information services in the overall economy has been rising for more than four decades and they are now the largest segment of the US economy. Despite a dearth of detailed evidence, there are strong indications that this shift has also been under way in other advanced economies.[7] Even some emerging market economies such as India are likely to experience a similar long-run structural shift in the near future.

The Broad Structural Changes

In order to document this trend for the US, we use a decomposition approach whereby the US economy is disaggregated into four super-sectors (Figure 10.1). As mentioned earlier, Porat (1977) had originally measured the size and structure of the US information economy for 1967. Following the same methodology, Apte and Nath (2007) and Apte et al. (2008, 2012b) updated those measures for 1992, 1997, 2002 and 2007, and also combined this decomposition of the economy with a decomposition along products versus services. This approach presents a new view of the broad structural changes that have taken place in the US.[8]

The decomposition essentially involves classifying the industries into the products and services categories and aggregating the material and information value-added data separately to construct a 2 × 2 matrix similar to Figure 10.1. As an illustration, we show this 2 × 2 decomposition for 2007 in Figure 10.2. Each cell in this figure includes (1) actual value in current prices and (2) percentage distribution of shares in total GNP.

Figure 10.3 presents the percentage shares of these four super-sectors in the US GNP

176   *Handbook of service business*

	Delivery Form	
	Products	Services
Material (End Output)	1,450,469 (10.2%)	4,194,413 (29.6%)
Information (End Output)	750,487 (5.3%)	7,798,062 (54.9%)

Note: Values are in millions of current USD. Percentage shares are in parentheses.

Figure 10.2   A 2 × 2 decomposition of the US economy in 2007

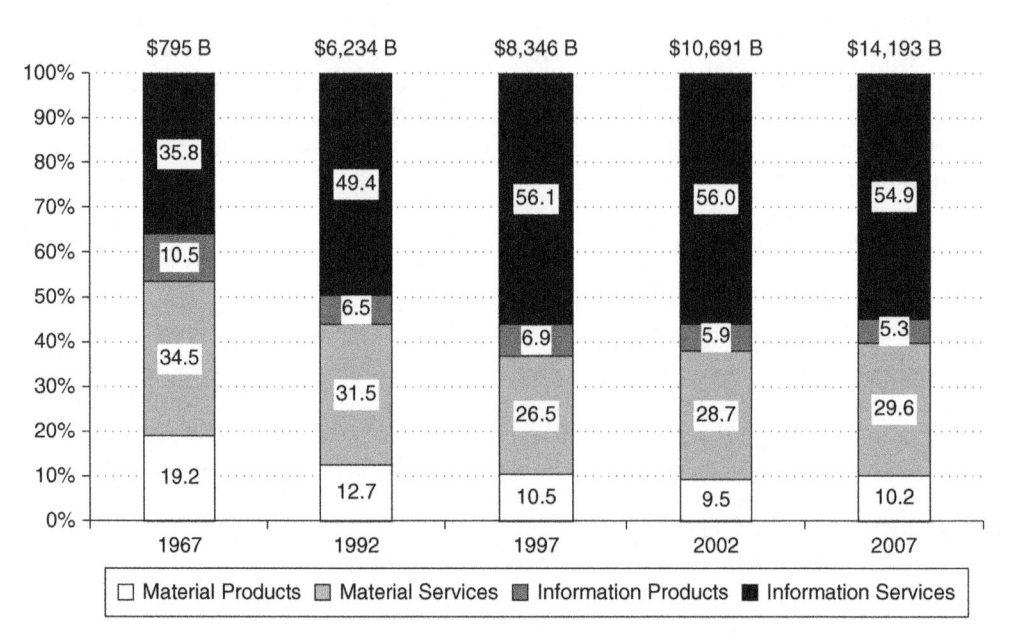

Year	GNP	Material Products	Material Services	Information Products	Information Services
1967	$795 B	19.2	34.5	10.5	35.8
1992	$6,234 B	12.7	31.5	6.5	49.4
1997	$8,346 B	10.5	26.5	6.9	56.1
2002	$10,691 B	9.5	28.7	5.9	56.0
2007	$14,193 B	10.2	29.6	5.3	54.9

Note: GNP values (in billions of current USD) are shown at the top of the bar.

Figure 10.3   Percentage shares of 4 super-sectors in the US GNP

for 1967, 1992, 1997, 2002 and 2007. The decompositions for the first three years were taken from Apte et al. (2008) and for the last two years from Apte et al. (2012b). In 1967, material services and information services had very similar shares. But over the years, the share of information services has significantly increased and by 2007 they accounted for about 55 percent of the US GNP. In contrast, information products have been the smallest sector and their share has steadily declined from more than 10 percent in 1967 to slightly more than 5 percent in 2007. When we compared the percentage shares of these four super-sectors in 2002 and 2007 with those in previous years, we made two important observations. First, the shares of information products and information services continuously declined from 1997 through 2007. Second, while the GNP share of material products declined continuously until 2002, there was a slight increase between 2002 and 2007. In contrast, the share of material services declined until 1997 and then rose steadily. Although this may look puzzling at first glance, there are some interesting facts at the disaggregate level that shed some light on factors underlying these recent trends.

Recent Trends in Information-intensive Services

Table 10.1 presents some relevant facts about the relative size, structure and growth of the information-intensive services (including the government) for 2002 and 2007. We use an ad hoc definition of information-intensive industries: service industries with an information share of more than 50 percent are considered to be information-intensive services. Accordingly, there are 22 private service industries, roughly at the two-digit level of industrial classification, and government services and agencies that accounted for about 59 percent of the GNP in both 2002 and 2007. The GNP shares are shown in columns 1 and 5 of Table 10.1. Columns 2 and 6 present the shares of information value added for each industry and, as they indicate, 11 industries are entirely information services. The largest three (in terms of GNP share) are two financial service industries – *Federal Reserves, banks, credit intermediation and related activities* and *Insurance carriers and related activities* – and *Broadcasting and telecommunications*, with GNP shares greater than 2 percent. The other relatively large information-intensive services include *Miscellaneous professional, scientific and technical services*, *Ambulatory health care services*, *Wholesale trade* and *Retail trade*.

The growth rates of value added between 2002 and 2007 are shown in column 9. These rates represent growth of quantity or volume. Note that during this period, the US economy grew by 14 percent. Among these 23 industries, 13 grew faster than the overall economy during the same period. Four of them grew by more than 40 percent and they include *Broadcasting and telecommunications, Insurance carriers and related activities, Funds, trusts and other financial vehicles* and *Computer systems design and related industries*. The percentage changes in prices for various industries during 2002–2007 (column 10) provide another perspective on the changes that took place during this five-year period. Note that some of these information-intensive services experience substantial negative price changes during this period that are primarily due to increases in productivity facilitated by ICT advances and may partially explain the decline in the share of the information economy (Table 10.1).

Government at all levels – local, state and federal – accounted for about 12 percent of GNP and, within them, information activities contributed about 74 percent in 2002,

178   *Handbook of service business*

Table 10.1   Industry share, information share, changes in real value added and prices for 22 information-intensive private service industries and government in 2002 and 2007

| Sl. No. | Industry title | 2002 ||||
		Industry share in total GNP	Information share in industry value added	Industry share in total information value added	Industry share in total material value added
		(1)	(2)	(3)	(4)
1	Motion picture and sound recording industries	0.48	100.00	0.78	0.00
2	Broadcasting and telecommunications	2.51	100.00	4.05	0.00
3	Information and data processing services	0.55	100.00	0.88	0.00
4	Federal Reserve banks, credit intermediation and related activities	4.08	100.00	6.59	0.00
5	Insurance carriers and related activities	2.32	100.00	3.76	0.00
6	Funds, trusts and other financial vehicles	0.24	100.00	0.38	0.00
7	Legal services	1.44	100.00	2.33	0.00
8	Computer systems design and related services	1.03	100.00	1.66	0.00
9	Management of companies and enterprises	1.67	100.00	2.70	0.00
10	Educational services	0.92	100.00	1.49	0.00
11	Performing arts, spectator sports, museums and related activities	0.52	100.00	0.84	0.00
12	Publishing industries (includes software)	1.14	99.61	1.83	0.01
13	Miscellaneous professional, scientific and technical services	4.26	98.51	6.78	0.17
14	Securities, commodity contracts and investments	1.44	96.73	2.25	0.12
15	Administrative and support services	2.53	87.26	3.57	0.84
16	Ambulatory health care services	3.14	76.30	3.87	1.95
17	Other transportation and support activities	0.69	69.70	0.77	0.55
18	Social assistance	0.56	66.28	0.60	0.49
19	Wholesale trade	5.75	59.81	5.56	6.06
20	Retail trade	6.84	58.76	6.50	7.39
21	Hospitals and nursing and residential care facilities	2.77	57.63	2.58	3.08
22	Rental and leasing services and lessors of intangible assets	1.38	51.44	1.15	1.76
23	Government	12.65	73.87	15.11	8.67
	Total/average	58.89	83.16	76.03	31.09

Information-intensive services in the US economy 179

	2007			2002–07	2002–07
Industry share in total GNP	Information share in industry value added	Industry share in total information value added	Industry share in total material value added	Percentage change in real value added	Percentage change in prices
(5)	(6)	(7)	(8)	(9)	(10)
0.44	100.00	0.73	0.00	8.9	11.6
2.44	100.00	4.05	0.00	40.2	−11.2
0.48	100.00	0.80	0.00	28.8	−11.2
3.40	100.00	5.65	0.00	4.7	6.0
2.77	100.00	4.59	0.00	40.7	17.2
0.32	100.00	0.53	0.00	86.7	−9.7
1.49	100.00	2.47	0.00	6.3	31.0
1.13	100.00	1.87	0.00	54.2	−8.3
1.80	100.00	2.98	0.00	−5.6	48.6
0.97	100.00	1.60	0.00	3.8	34.9
0.52	100.00	0.86	0.00	10.2	22.6
1.03	99.64	1.70	0.01	21.9	−1.9
4.60	98.37	7.51	0.19	22.7	20.6
1.42	100.36	2.36	−0.01	3.8	27.2
2.66	86.95	3.85	0.87	31.4	8.4
3.23	76.08	4.08	1.94	23.1	13.5
0.68	71.22	0.81	0.49	22.3	9.7
0.58	64.05	0.61	0.52	19.8	17.7
5.75	60.24	5.75	5.75	21.5	11.3
6.29	56.61	5.91	6.86	9.5	12.6
2.81	56.67	2.65	3.07	13.3	21.7
1.40	50.26	1.17	1.75	17.5	16.9
12.42	69.59	14.35	9.49	5.2	25.0
58.61	82.92	76.88	30.94	20.45	13.09

declining to 70 percent in 2007. Government grew more slowly than the economy during this period.

TRENDS IN EMPLOYMENT, WAGES, ICT INVESTMENT AND TRADE IN INFORMATION-INTENSIVE SERVICES

Trends in Employment and Wages

We now move to discuss the recent trends in employment and wages in the US, focusing on how these have evolved for information and non-information workers. Conceptually, information workers are those involved in creating, processing and communicating information. The others, who are primarily employed in the production of material goods and physical services, are classified as non-information workers. Since many occupations combine information tasks with material or physical tasks, each occupation is decomposed into information and non-information tasks and aggregated to FTE information and non-information workers at the industry levels. The detailed methodology for calculating FTE information and non-information workers is discussed by Apte et al. (2008) and Apte et al. (2012a). Primarily because of data limitations, our analysis focuses on the period since 1999.

Broad trends

Along with the structural shifts that we have discussed above, there have been important changes in the labor market. As Apte et al. (2008) and Apte et al. (2012a) showed, if occupations are classified into information and non-information categories and they are distributed between products and services industries, they reveal some interesting trends in the US labor market. Apte et al. (2008) and Apte et al. (2012a) used a classification scheme – based on the framework developed by Apte and Mason (1995) and used by Apte et al. (2008) – which recognizes that every occupation uses information at various degrees of intensity.

Using four different categories of workers in total employment in the US, in 1999 non-information workers in service industries were the largest segment of the US workforce, accounting for more than 45 percent of total employment (Figure 10.4). This declined to less than 45 percent in 2007. In contrast, the share of information workers steadily increased from about 41 percent in 1999 to 45 percent in 2007. The respective shares of information and non-information workers in product industries were relatively small and they have been steadily declining throughout the period, as shown in Table 10.2.

Using average wages (in current dollars) for the four categories of workers, these were highest in the product sector (Figure 10.5). Within this sector, the earnings of information workers were on average 1.8 times higher than those of non-information workers. In the services sector, the earnings of information workers were about 1.4 times higher than those of non-information workers. Overall, the average wage of information workers was one and one-half times greater than that of non-information workers. However, there are sectoral differences in the earnings of information workers. On average, they earned about 1.3 times more in the products sector than in the services sector. This, however, appeared to reflect the fact that most information workers in the products

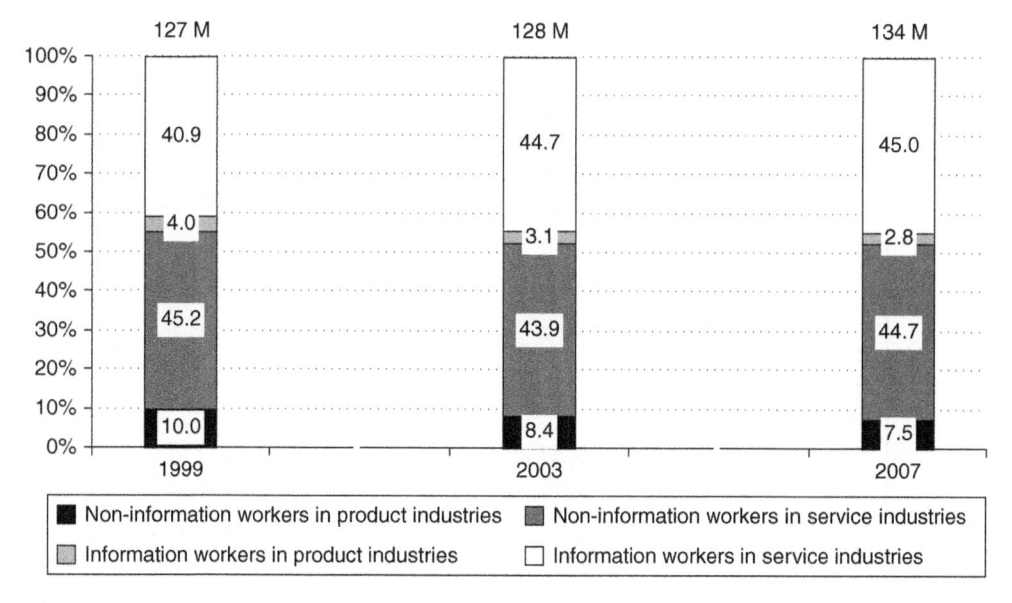

Figure 10.4 Employment shares of information and non-information workers in product and service industries

sector were engaged in high-end information jobs, while a large number of information workers in the service sector were engaged in low-end information jobs.[9] We also found that non-information workers generally earned more in product industries than in services, although the difference was not large.

Another way of looking at growth of employment and wages during this period is by plotting graphs. Figure 10.6 depicts the employment growth of information and non-information workers in the products and services sectors. Employment of both types of workers in the products sector fell steadily through this period. There was a sharp decline in 2002, immediately after the recession of 2001. The services sector exhibited a different pattern in that the employment of non-information workers declined sharply between 1999 and 2000, while that of information workers rose significantly during the same period. Between 2000 and 2001, employment of both types of workers declined slightly, only to rise steadily thereafter. The growth of services employment accelerated after 2004.

Figure 10.7 depicts the growth of average real wages for both information and non-information workers in the products and services sectors.[10] The average real wage of information workers in the products sector rose steadily until 2002, remained almost constant until 2004 and fell for two consecutive years (Figure 10.7). Between 2006 and 2007, it grew again. The average real wage of non-information workers, however, increased until 2002 and then steadily declined. The average real wage of information workers in the services sector rose sharply between 2001 and 2002, declined slightly until 2005 and then started rising. The average real wage of non-information workers in services also exhibited a similar pattern of growth over this period of time.

Table 10.2 Employment share, information intensity, overall employment growth and growth of information employment in 34 private service industries, postal services and governments (in averages over 2002–2007)

Sl. No.	Industry title	Industry share in total employment (1)	Information share in total industry employment (2)	Annual average growth rate of employment (3)	Annual average growth rate of info. employment (4)
1	Internet service providers, web search portals and data processing service	0.30	84.95	−3.49	−3.32
2	Securities, commodity contracts and other financial investments and related activities	0.61	83.49	1.55	2.11
3	Internet publishing and broadcasting	0.02	83.01	4.42	3.46
4	Insurance carriers and related activities	1.64	82.67	0.37	0.44
5	Credit intermediation and related activities	2.16	82.12	1.43	1.54
6	Funds, trusts and other financial vehicles	0.07	81.76	0.08	−1.25
7	Educational services	9.29	79.60	1.21	1.52
8	Monetary authorities – central bank	0.02	77.87	−2.85	−2.69
9	Professional, scientific and technical services	5.33	77.65	2.54	2.73
10	Other information services	0.04	77.52	3.47	3.61
11	Lessors of nonfinancial intangible assets (except copyrighted works)	0.02	76.71	0.28	0.87
12	Management of companies and enterprises	1.35	76.39	2.61	3.05
13	Clothing and clothing accessories stores	1.08	72.46	3.12	3.33
14	Postal service	0.40	71.60	−0.79	−0.82
15	Sporting goods, hobby, book and music stores	0.51	70.10	0.03	0.13
16	Publishing industries (except internet)	0.71	66.18	−1.15	−0.74
17	General merchandise stores	2.23	65.83	2.52	1.82
18	Miscellaneous store retailers	0.71	65.11	−1.95	−1.91
19	Electronics and appliance stores	0.42	64.57	1.04	1.31
20	Wholesale electronic markets and agents and brokers	0.55	63.98	4.88	6.32
21	Building material and garden equipment and supplies dealers	0.96	61.56	2.50	2.85
22	Furniture and home furnishings stores	0.44	61.49	1.54	1.91

23	Gasoline stations	0.67		-0.78	-0.41
24	Religious, grant-making, civic, professional and similar organizations	1.00	60.95	0.52	0.29
25	Broadcasting (except internet)	0.25	60.81	-0.42	-0.72
26	Telecommunications	0.80	60.70	-2.70	-2.83
27	Social assistance	1.59	59.84	3.11	2.17
28	Merchant wholesalers, durable goods	2.30	59.78	0.90	1.00
29	Non-store retailers	0.34	56.67	-0.16	-0.44
30	Rental and leasing services	0.49	55.81	-0.30	-0.71
31	Health and personal care stores	0.73	55.25	1.20	0.07
32	Museums, historical sites and similar institutions	0.09	55.01	1.72	1.22
33	Food and beverage stores	2.18	54.93	-0.18	0.10
34	Merchant wholesalers, nondurable goods	1.56	53.84	0.51	0.66
35	Federal, state and local government (OES designation)*	7.46	53.40	-1.18	-1.66
36	Real estate	1.09	53.16	1.68	3.17
	Total	49.43	51.92	0.74	0.76
			65.37		

Note: * OES, Occupational Employment Statistics.

Source: Authors' calculations based on publicly available data from the Bureau of Labor Statistics (BLS) website (www.bls.gov).

184 *Handbook of service business*

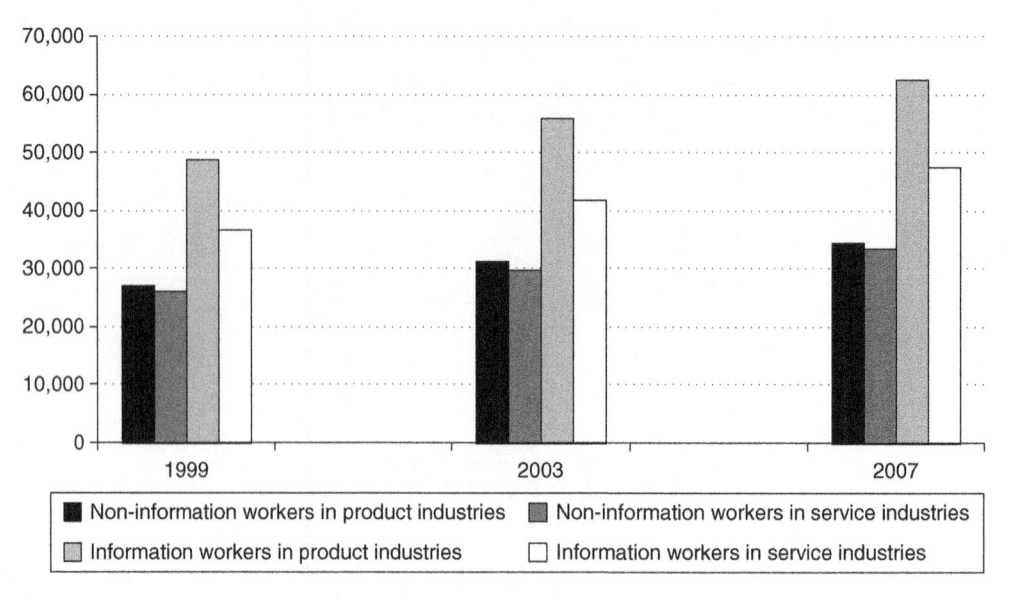

Figure 10.5 Average wages earned by information and non-information workers in product and service industries

Sectoral employment patterns
Employment in 36 service industries with information workers constituting more than 50 percent of their respective employment accounted for about half of the total employment in the US economy during the period 2002 to 2007 (Table 10.2). The numbers shown in Table 10.2 are annual averages for 2002 to 2007.[11]

Six of these 34 private industries and the government accounted for more than 2 percent of the total employment (column 1). *Educational services*, with more than 9 percent of total employment, *Professional, scientific and technical services*, with more than 5 percent, and the government, with more than 7 percent, are among the largest employers. For 15 of these service industries, information workers account for more than 70 percent of total employment. The industries that have experienced average annual employment growth of above 3 percent include *Internet publishing and broadcasting, Other information services, Clothing and clothing accessories stores, Wholesale electronic markets and agents and brokers* and *Social assistance*. In contrast, employment in *Internet service providers, web search portals and data processing services, Monetary authorities – Central Bank*, and *Telecommunications* actually declined by more than 2 percent per year. In relation to the average annual growth rate of information workers by industries (column 4), *Other information services, Management of companies and enterprises, Clothing and clothing accessories stores, Wholesale electronic markets and agents and brokers*, and *Real estate* have not only experienced high (3 percent) growth of information workers but these growth rates are also higher than the respective total employment growth rates for those industries.

Information-intensive services in the US economy 185

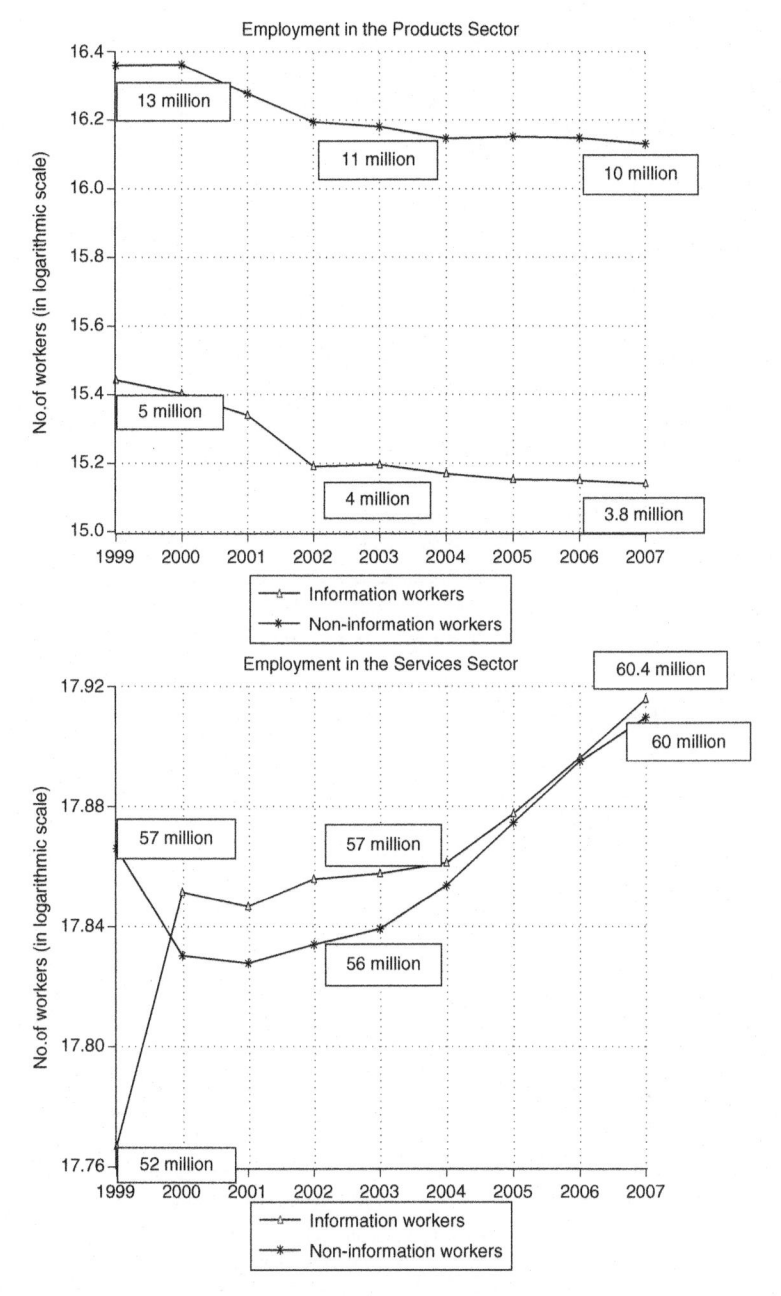

Note: Differences in logarithmic values between successive period (×100) approximate percentage change. Actual employment numbers are shown in textboxes for 1999, 2003 and 2007.

Source: Apte, Karmarkar and Nath (2012a), Figure 4.2, page 35. Modified to include actual numbers for three specific years.

Figure 10.6 Employment growth in the US: 1999–2007

186 *Handbook of service business*

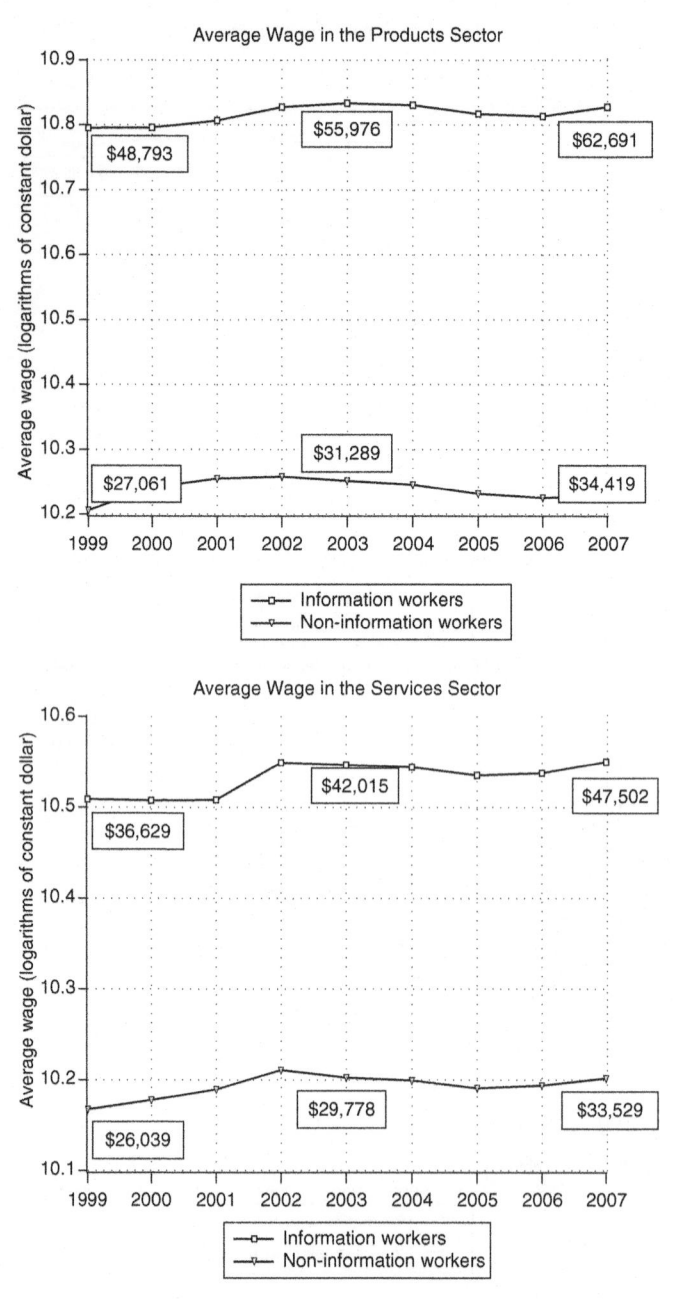

Note: Differences in logarithmic values between successive period (×100) approximate percentage change. Actual dollar values in current USD are shown for 1999, 2003 and 2007.

Source: Apte, Karmarkar and Nath (2012a), Figure 4.3, page 36. Modified to include actual USD values for three specific years.

Figure 10.7 Growth of average wages in the US economy: 1999–2007

Information-intensive services in the US economy 187

Investment in ICT Capital

For the purpose of examining the major trends in ICT capital investment we designate 12 out of 42 different types of fixed assets (listed under the heading "equipment and software" by the Bureau of Economic Analysis (BEA)) as ICT capital assets that include three types of software (prepackaged, custom and own account), eight types of hardware (mainframes, PCs, Direct Access Storage Devices, printers, terminals, tape drives, storage devices and system integrators) and communications equipment.[12] Annual data on investment in these fixed assets for the period from 1992 to 2007 are available from the *Fixed Assets Accounts Tables* at the BEA website (http://www.bea.gov/national/FA2004/index.asp). ICT capital accounted for about 34 percent of total investment in 1992 and this ratio gradually increased to about 42 percent in 2002 and then declined to about 39 percent in 2007 (Figure 10.8). The distribution of ICT investment among three major categories – computer and peripheral equipment, software and communications equipment – is shown in Figure 10.9. Note that software has not only been the largest component of ICT investment but its share has persistently increased so that in 1992 it accounted for about 39 percent of the total ICT investment, rising to about 57 percent in 2007.

The top 20 service industries according to their average shares in total private investment in equipment and software in 2002 and 2007 are notable for the fact that nine of these accounted for more than 65 percent of total investment in equipment and software

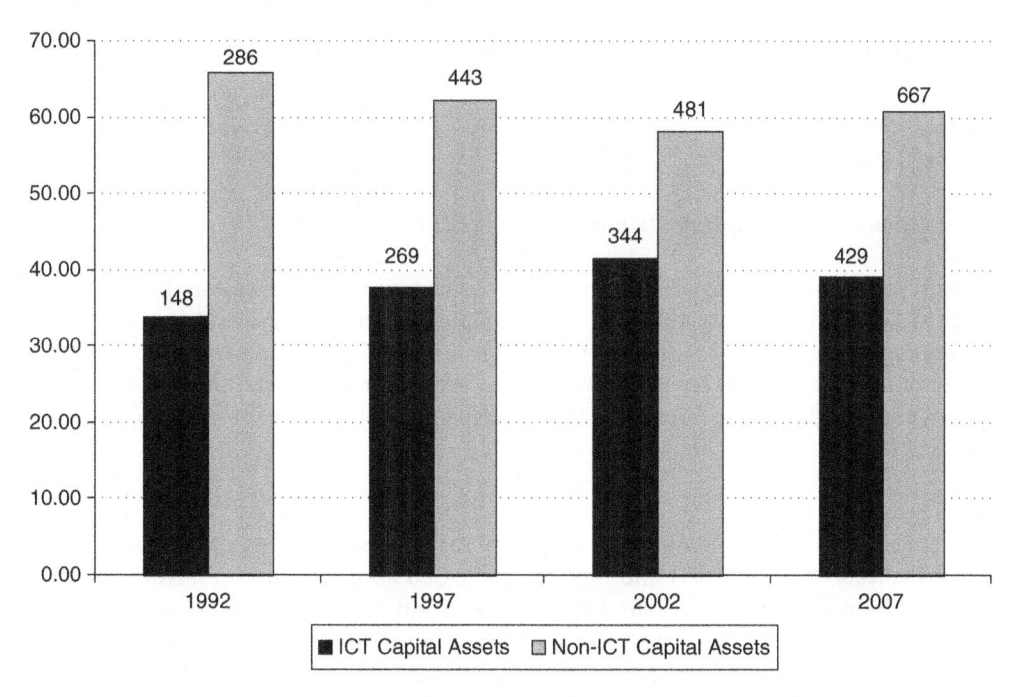

Note: Actual values (in billions of current USD) are at the top of the bar.

Figure 10.8 Investment in ICT and non-ICT capital assets

188 *Handbook of service business*

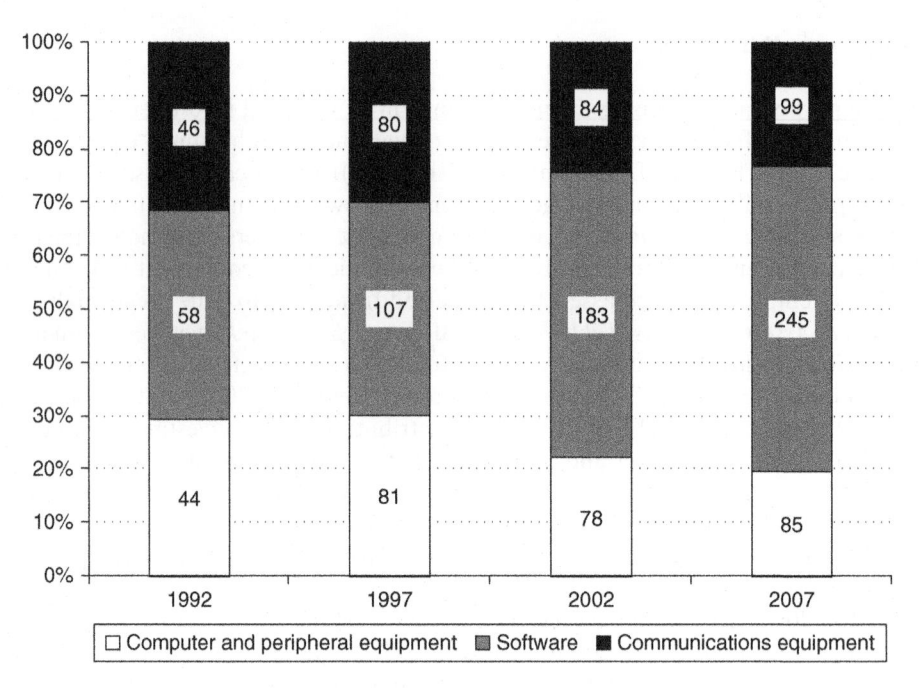

Note: Actual values (in billions of current USD) are shown inside the bar.

Figure 10.9 Investment share of three categories of ICT capital assets

(Table 10.3). Furthermore, the share of ICT investment declined for all these industries between 2002 and 2007.

International Trade in Information-intensive Services

Finally, we turn to some of the important recent trends in international trade in the United States, focusing on the trade of information-intensive services.[13] It is important to note at the outset that, although the US has long had an overall trade deficit, it has more or less consistently reported a surplus in services trade which has grown in recent years, not least due to the faster than average expansion of trade in information-intensive services since the mid-1990s.

Broad trends
As Apte and Nath (2012) and Apte et al. (2012a) showed, the dollar value of US merchandise trade (both exports and imports) increased about seven times, from less than half a trillion USD to more than 3 trillion USD, and the dollar value of services trade increased almost ten times, from barely 100 billion USD to about one trillion USD, between 1980 and 2010. Thus, the US goods trade was more than three times larger than services trade in 2010 (Figure 10.10).[14]

Figure 10.10 plots the annual dollar value of trade balances (exports–imports) for both goods and services in the United States between 1980 and 2010. Several observa-

Table 10.3 ICT investment shares for 20 top service industries by their shares in total investment in equipment and software

Sl. No.	Industry title	Industry share in total investment in equipment and software (average between 2002 and 2007)	Share of ICT capital investment 2002	Share of ICT capital investment 2007
		(1)	(2)	(3)
1	Credit intermediation and related activities	8.18	35.03	31.56
2	Broadcasting and telecommunications	6.81	93.26	92.63
3	Retail trade	5.33	34.65	32.13
4	Rental and leasing services and lessors of intangible assets	4.96	12.38	10.71
5	Miscellaneous professional, scientific and technical services	4.86	67.32	65.86
6	Utilities	4.67	12.65	12.58
7	Wholesale trade	4.50	46.92	42.24
8	Construction	4.12	13.15	10.19
9	Management of companies and enterprises	3.38	85.74	86.37
10	Hospitals	3.34	19.19	16.50
11	Insurance carriers and related activities	2.77	83.76	83.49
12	Ambulatory health care services	2.50	17.11	13.71
13	Administrative and support services	2.43	70.59	68.67
14	Information and data processing services	2.19	95.61	95.56
15	Publishing industries (including software)	2.17	90.53	90.31
16	Securities, commodity contracts and investments	1.62	89.65	89.80
17	Air transportation	1.36	13.67	14.50
18	Truck transportation	1.29	15.46	13.20
19	Educational services	1.26	49.53	47.11
20	Computer systems design and related services	1.17	92.32	91.44
	Total	68.90		

Source: Authors' calculations based on publicly available data from the Bureau of Economic Analysis (BEA) website (www.bea.gov).

tions follow. First, while the US has been a net importer of goods, it was a net exporter of services throughout the sample period. In 2010, the US ran a deficit of about 700 billion USD in merchandise trade but had a surplus of more than 150 billion USD in services trade. Second, as the merchandise trade deficit grew significantly over the years, so did the services trade surplus. There was a steady and rapid increase in the services trade surplus between 1985 and 1997, followed by a steady decline between 1997 and 2003 and then a renewed rapid rise. Apart from a slight decline between 2000 and 2001, the

190 *Handbook of service business*

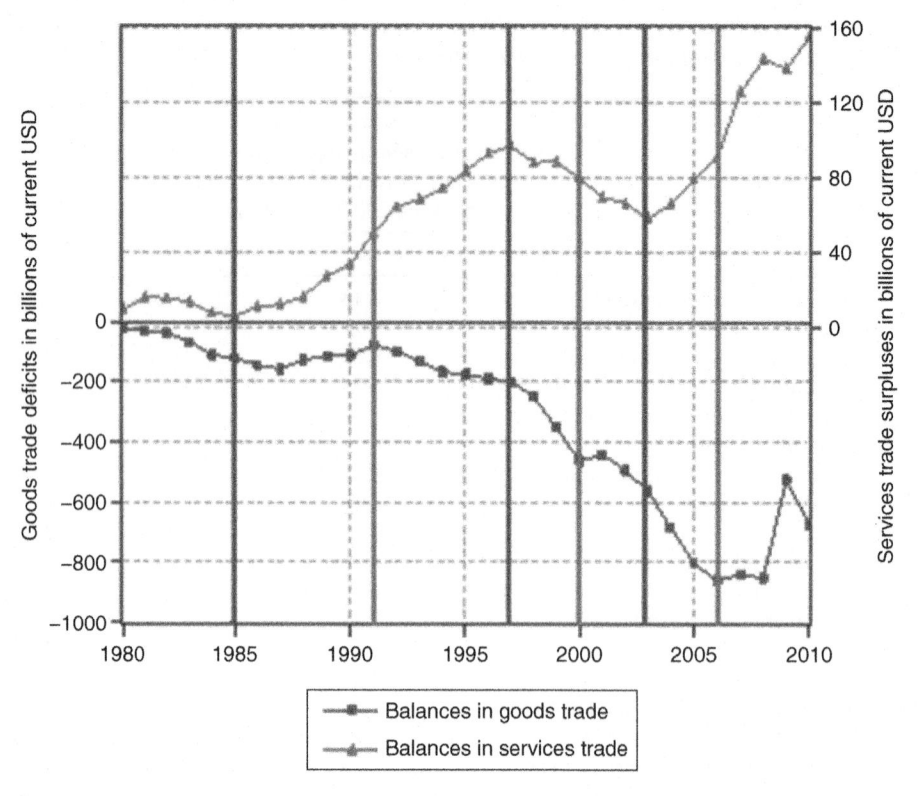

Note: Trade balances for goods and services are calculated using data from Table 1.1.5 of the BEA's National Economic Accounts. Services include both private and government services.

Source: Apte, Karmarkar and Nath (2012a), Figure 5.1, page 42.

Figure 10.10 US Trade Balances (exports − imports) in goods and services: 1980–2010

deficit in merchandise trade steadily increased between 1991 and 2006. By 2005, the merchandise trade deficit had surpassed 800 billion USD and stayed there for the next three years, before falling drastically to about 500 billion USD in 2009. Although the deficit in goods trade increased again in 2010, it did not reach the 2005 to 2008 level. Third, while the merchandise trade balance appeared to be sensitive to business cycle fluctuations, this was not the case for the balance in services trade. For example, the steady increase in the services trade surplus between 1985 and 1997 was not dented by the recession of the early 1990s. Similarly, the decline in the services trade surplus during the 2001 recessionary cycle seemed to be a continuation of the downward trend between 1997 and 2003. Furthermore, the drop in 2009 was moderate.

We further examine the patterns in services trade by looking at its share in total trade. Figure 10.11 presents the share of services in total trade and also the export and import shares of services separately. Trade in services accounted for about 17 percent of all US trade in 1980. This share increased to about 24 percent in 1992 and then

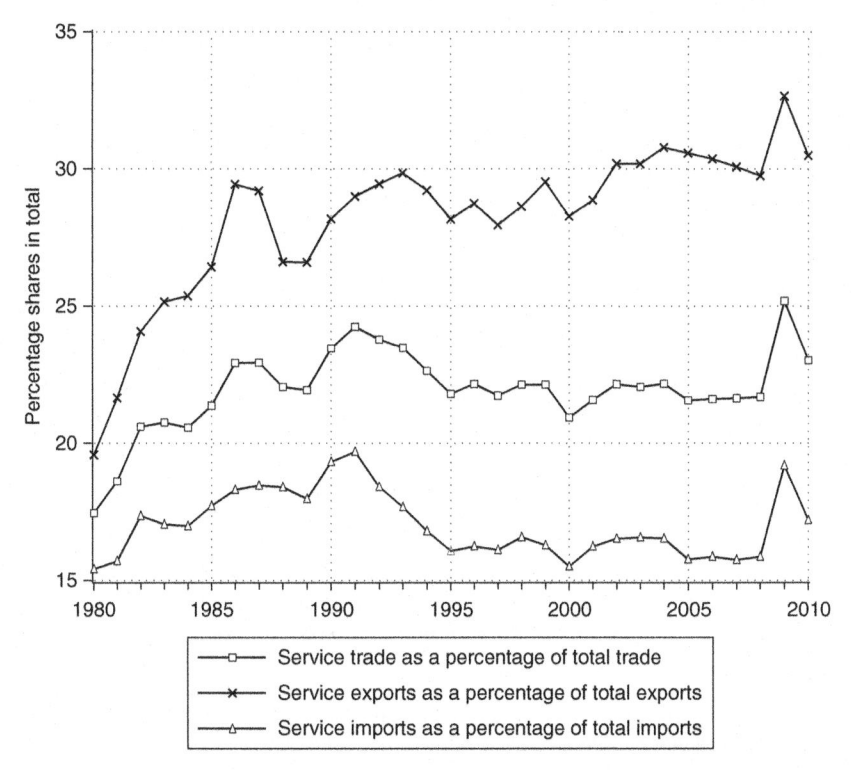

Information-intensive services in the US economy 191

Note: Shares of services trade, exports and imports are calculated using data from Table 1.1.5 of the BEA's National Economic Accounts. Services include both private and government services.

Source: Apte, Karmarkar and Nath (2012a), Figure 5.2, page 44.

Figure 10.11 Trade shares of services trade, services exports and services imports: 1980–2010

steadily decreased to about 20 percent in 2000, after which it increased slightly during 2000 to 2008, and significantly to about 25 percent in 2009. In 2010, trade share of services dropped to about 23 percent. The export and import shares of services followed similar patterns, although export shares had been much larger than import shares. The export share of services fluctuated between a minimum of 19.6 percent (in 1980) and a maximum of 32.7 percent (in 2009), while the import share fluctuated between 15.4 percent (in 1980) and 19.7 percent (in 1991).

Disaggregate trade patterns
We now present the export and import shares of five major categories of private services during 1992 to 2009, in Figures 10.12 and 10.13 respectively. The more detailed breakdown into travel, passenger fares, other transportation, royalties and license fees, and other private services clearly indicates that *Other private services* experienced the highest growth in both export share and import share (Figures 10.12 and 10.13).[15] The export share of *Other private services* (in total exports of private services) increased from about

192 *Handbook of service business*

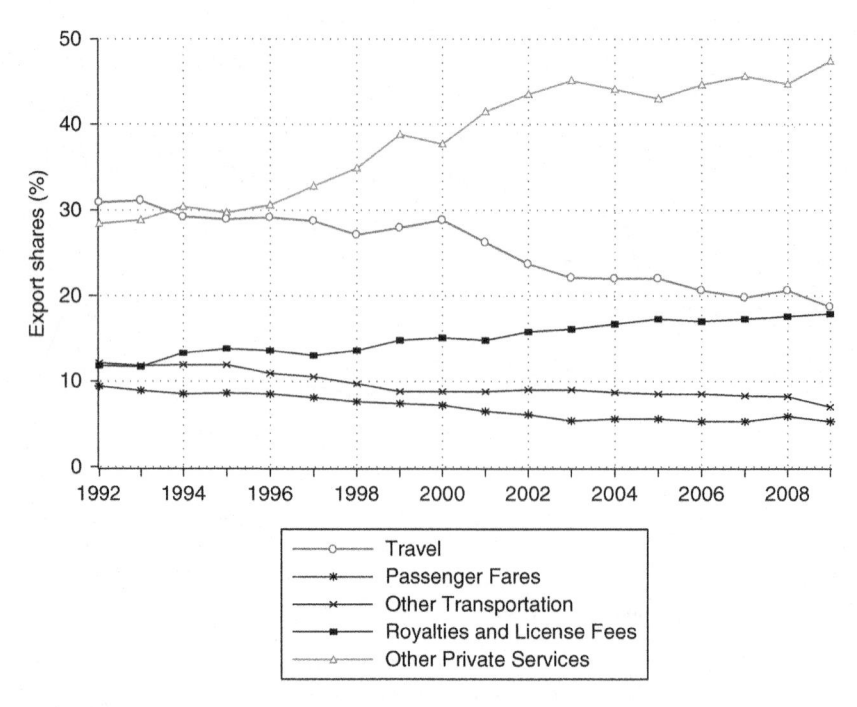

Note: Export and import shares of major categories of services are calculated using data from Table 1 (Trade in Services), Detailed Statistics for Cross-Border Trade under US International Services, BEA.

Source: Apte, Karmarkar and Nath (2012a), Figure 5.4(a), page 46.

Figure 10.12   Shares of five major categories of services in total services exports from the US

31 percent in 1992 to about 49 percent in 2009, while the import share increased from about 25 percent to about 50 percent during the same time period. Further, *Royalties and license fees* experienced growth during this period. For example, its export share increased from about 13 percent in 1992 to about 19 percent in 2009. The import share also increased from about 5 percent to about 8 percent during this period.

Note that the services included within these two broad categories are primarily information-intensive services.[16] The service category *Royalties and license fees* includes the following detailed subcategories: industrial processes; books, records and tapes; broadcasts and recordings of live events; franchise fees; trademarks; general-use computer software; and other intangibles. Thus, exports of *Royalties and license fees* are essentially the royalties and license fees received by the US for the use of the intangible items listed above in foreign countries. Similarly, imports are such payments by the US for the use of these intangible items developed and produced in foreign countries.

Furthermore, the services included in the category *Other private services* are also primarily information-intensive services.[17] This category incorporates a broad range of activities: education; financial services; insurance; telecommunications; business, professional and technical services; and a residual *others*. The category *Business, profes-*

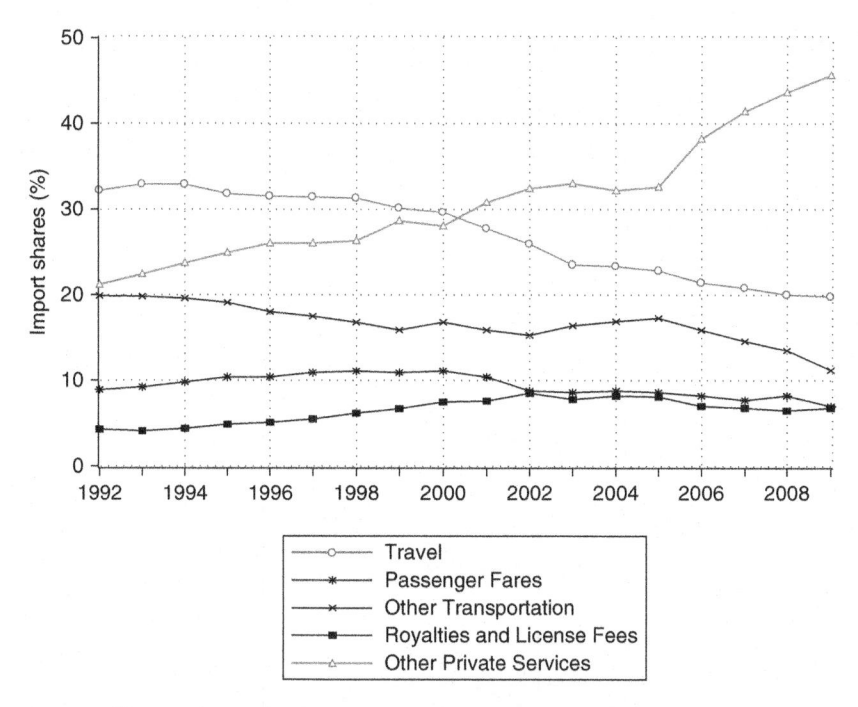

Note: Export and import shares of major categories of services are calculated using data from Table 1 (Trade in Services), Detailed Statistics for Cross-Border Trade under US International Services, BEA.

Source: Apte, Karmarkar and Nath (2012a), Figure 5.4(b), page 46.

Figure 10.13 Shares of five major categories of services in total services imports into the US

sional and technical services is further subdivided into advertising; computer and data processing; database and other information services; research, development and testing services; management, consulting and public relations services; legal services; construction, engineering, architectural and mining services; industrial engineering; installation, maintenance and repair of equipment; and other business, professional and technical services. Except for two categories – *Construction, engineering, architectural and mining services* and *Installation, maintenance and repair of equipment* – which do not entirely involve creating, processing and communicating information and require some physical activity, the list comprises services that are highly information intensive.

As discussed above, the export and import shares of these two major categories of information-intensive services, *Royalties and license fees* and *Other private services*, increased significantly between 1992 and 2009. Together they accounted for about 68 percent of total private services exports from the US in 2009. Similarly, the combined import share of *Royalties and license fees* and *Other private services* was about 58 percent of total imports of private services in 2009.

Total trade data for the detailed subcategories under *Royalties and license fees* are available only from 2006 (Table 10.4). Two major items, industrial processes and

Table 10.4 Shares of various sub-categories in total exports and imports of royalties and license fees, 2006–2009 (in percentages, unless otherwise stated)

	Exports				Imports			
	2006	2007	2008	2009	2006	2007	2008	2009
Industrial processes	45.8	43.0	42.5	39.7	70.3	66.8	63.0	65.3
Books, records and tapes	2.1	1.8	1.6	1.6	3.2	3.0	3.1	3.2
Broadcasting and recording of live events	0.6	0.7	0.6	0.7	4.3	0.8	3.9	0.9
Franchise fees	4.6	4.7	4.8	4.8	0.8	0.7	0.9	0.8
Trademarks	14.7	13.7	13.2	13.0	8.2	9.0	9.4	9.5
General-use computer software	32.0	36.0	37.2	40.1	12.6	19.2	19.2	19.8
Other intangibles	0.1	0.1	0.1	0.1	0.5	0.4	0.7	0.5
Total value of trade in royalties and license fees (millions of current USD)	70,727	84,580	93,920	89,791	23,518	24,931	25,781	25,230

Note: Data are available only for unaffiliated trade before 2006 and therefore they are not comparable with the figures since 2006.

Source: Apte, Karmarkar, and Nath (2012a), Table 5.3, page 50. Calculations based on data from Table 4 (Royalties and License Fees), Detailed Statistics for Cross-Border Trade under US International Services, BEA.

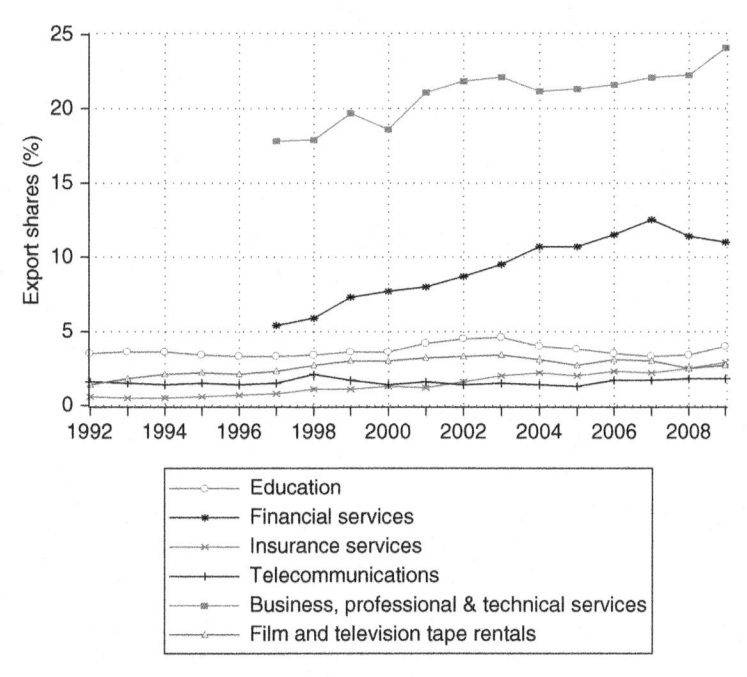

Figure 10.14 Shares of six subcategories of Other Private Services in total exports of private services from the US

general-use computer software, together accounted for about 80 percent of total exports of *Royalties and license fees* and more than 80 percent of total imports into the US. While the share of *Industrial processes* declined, that of *General-use computer software* increased during this four-year period. Overall, the total export value of this broad category was more than three times higher than its import value.

In Figures 10.14 and 10.15, we present the shares of major subcategories of services under *Other private services* in total services exports and imports respectively. Note that for *Financial services* and *Business, professional and technical services*, data were available only since 1997. Among the export categories, *Business, professional and technical services* and *Financial services* were the two largest subcategories, with about 24 percent and 12 percent of total private services exports from the US, and their share rose from about 18 percent and 5 percent respectively in 1997. While, during the recent financial crisis, the share of *Financial services* dropped from its peak in 2007, the exports of *Business, professional and technical services* continued to grow. Among the services imports, the share of *Business, professional and technical services* grew from less than 15 percent in 1997 to about 25 percent in 2009. The other service that experienced significant growth in its share, particularly since 2000, was *Insurance*. In 2000, imports of *Insurance* accounted for about 5 percent of total private services

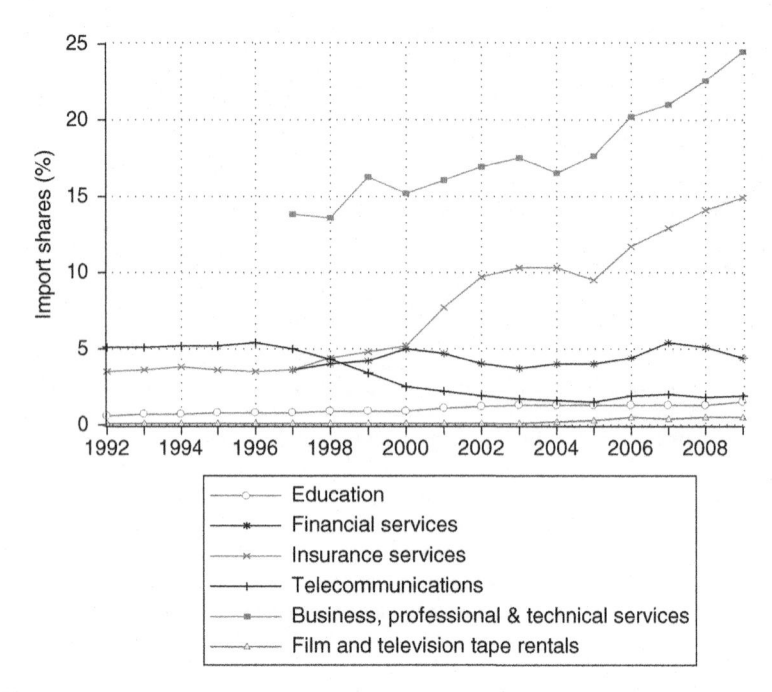

Note: Export and import shares of major categories of services are calculated using data from Table 5 (Other Private Services), Detailed Statistics for Cross-Border Trade under US International Services, BEA.

Source: Apte, Karmarkar and Nath (2012a), Figure 5.5(b), page 51.

Figure 10.15 Shares of six subcategories of Other Private Services in total imports of private services into the US

imports into the US. It grew to about 15 percent in 2009. Bermuda was the largest exporter of insurance to the US.

We now examine a few detailed subcategories within *Business, professional and technical services*. Table 10.5 presents the percentage shares of ten different subcategories in total export and import values of *Business, professional and technical services* for four years between 2006 and 2009. *Management consulting and public relations services* was the largest subcategory, accounting for about one-quarter of total exports and more than one-quarter of total imports under the broad category. This was followed by *Research, development and testing services*, with about 15 percent of exports and more than 15 percent of imports. The import share has also increased over time.

Overall, it is likely that the dramatic increases in export and import shares of financial services and insurance reflect greater global financial integration through the use of ICT. Further, the more than doubling of the import share of *Computer and information services* is associated with the growth of offshore outsourcing of these services. Finally, a significant decline in the import share of telecommunications is probably the consequence of a substantial cost reduction in providing these services due to technological advances.

In general, although the global pattern of services trade resembles that of merchandise trade (primarily due to the predominance of physical services over information-intensive

Table 10.5 *Shares of various sub-categories in total exports and imports of business, professional and technical services, 2006–2009 (in percentages, unless and otherwise stated)*

	Exports				Imports			
	2006	2007	2008	2009	2006	2007	2008	2009
Advertising	4.37	3.94	3.57	3.40	3.07	3.07	2.67	2.85
Computer and data processing services	6.64	6.94	7.34	7.35	20.82	20.34	19.11	19.83
Database and other information services	5.03	4.58	4.25	4.12	0.95	1.12	1.24	1.12
Research and development	14.83	15.06	15.12	15.63	15.03	18.51	19.72	19.21
Management consulting and public relations services	24.80	26.18	25.25	24.17	30.09	27.65	27.08	27.14
Legal services	6.08	6.17	6.36	6.22	1.98	2.18	2.41	2.07
Construction engineering, architectural and mining	6.30	5.78	6.17	5.82	2.26	2.15	2.30	2.19
Industrial engineering services	4.52	3.67	3.28	4.27	2.18	3.89	4.40	4.49
Installation, maintenance and repairing services	8.88	8.44	8.24	9.59	7.43	7.40	7.15	7.52
Other business, professional and technical services	11.40	12.39	13.70	12.80	14.53	12.37	12.80	12.26
Total value of trade in business, professional and technical services (millions of current USD)	86,390	103,765	115,229	116,629	61,698	70,413	82,537	81,995

Source: Apte, Karmarkar, and Nath (2012a), Table 5.4, page 52. Calculations based on data from Table 7 (Business, Professional, and Technical Services), Detailed Statistics for Cross-Border Trade under US International Services, BEA.

services for most countries), the pattern for information-intensive services is different. While trade in products and physical services is affected by distance, trade in information services is not. However, language and culture play an important role (Karmarkar 2004). For example, the presence of a large English-speaking populace has been the driving force behind the significant growth of India's information-intensive services trade with the US and the UK. Further, the low cost of logistics for trade and the ability to modularize information-intensive services and content production would potentially lead to higher volumes of trade in information-intensive services. This growth is likely to be higher in the largest information services markets such as the United States. One of the implications of this is that the largest geographically distributed information services market is composed of countries with large English-speaking populations (Karmarkar 2004; Apte and Karmarkar 2007).[18]

DRIVERS OF ECONOMIC EVOLUTION AND THE CONSEQUENCES FOR INDUSTRIES

A fundamental factor in the structural evolution of economies is technologically driven productivity changes. These have both absolute and relative effects. Productivity increases of course raise outputs and create economic growth. Relative changes in productivity growth across sectors can cause their relative shares to grow or decline with concomitant changes in employment and wage distributions (Baumol 1967; Karmarkar and Rhim 2012). It appears that the growth of services and the relative growth of information-intensive services can, at least in part, be traced to this mechanism.

Productivity growth is itself the result of many specific developments in processes that are often, though not necessarily always, technology driven. One can think of many of these developments under the umbrella of industrialization to emphasize the analogy with the Industrial Revolution after the mid-19th century (Karmarkar 2008). As in that case, industrialization in services has involved superior processing power, the automation of processes (capital for labor substitution), increased precision in measurement and specification of outputs, standardization of processes (as in software) and the availability of low-cost logistics (storage and transportation). These capabilities in turn lead to a set of business strategies, including outsourcing and off-shoring (Apte and Karmarkar 2007), process design and innovation, new services and the geographical deployment of processes up to a global scale.

The consequences for industries and industry sectors are substantial and often quite disruptive. In the case of information-intensive services, there are significant changes in process economics, leading to the restructuring of operations and processes in the small (that is, within firms), as well as chains and networks of production and delivery in the large (that is, at the level of industries and sectors). Unlike the physical world, industrialization in information-intensive processes has not always meant increases in scale and size. In fact, increased processing power has been associated with Moore's Law and miniaturization, and economies of scale for hardware are low (Karmarkar 2010). The appearance of infrastructure as a service (IaaS) and cloud computing means that firms can scale up information processing operations linearly. Due to process automation, variable costs can be low and fixed costs are also low. The result is increased intensity of competition and the possibility of increased variety and differentiation. Reengineering processes and service designs is not difficult, resulting in the threat of commoditization. Low-cost logistics mean that the arena for competition is much wider in terms of both markets and production. At the same time, barriers to entry still exist but they are driven more by network economics and other positive externalities on the market side, rather than scale and fixed costs of production. What is more, since the new technologies allow for easy capacity expansion, even relatively small fixed costs (including some on the customer side) can create natural monopolies and customer lock-in.

A crucial phenomenon which is characteristic of information-intensive sectors is that of "convergence" across hardware, processes and chains, in a way that is not possible in the physical world. So while steel and wood chains can never really coalesce, we indeed see the potential for the convergence of many different content-driven service sectors, which were historically considered distinct. For example, different media sectors were segregated by the physical nature of their publishing medium. Even within paper-based publishing, different frequencies of delivery led to distinct models and sub-sectors. So daily, weekly, monthly and quarterly news magazines did not compete very directly. Today those distinctions are fading, as "currency" becomes a common currency on the internet and web. Similarly, we see convergence across different media as TV, music and print publishing start to coalesce on the web. Even more broadly, we can see that heretofore distinct businesses now start to overlap and confront each other in an increasingly common arena. Apple has rapidly become a leading consumer product company, as well as a retailer of content. Intel and Samsung are migrating to software, Google and Amazon have mobile appliances, and in business-to-business markets many technology firms from IBM to HP are moving from hardware and software to business services.

The consequence of these changes is the restructuring of services industries from the atomic scale of tasks and operations up to the level of industries and across industries. Organizations are beginning to become more virtual, both in the sense of structure and geography (Karmarkar and Mangal 2007). Many sectors such as entertainment and publishing are exhibiting vertical de-integration and lateral bundling (Karmarkar 2004, 2008; Apte et al. 2012b). In many other service sectors these changes are still incipient but inevitable and inexorable.

SUMMARY AND CONCLUSIONS

We have presented a picture of the transition (one might say transformation) of the US economy toward services, information and their nexus: information-intensive services. These now dominate the economy in terms of GNP, employment and wage bill. Although this transition was somewhat slowed by the recent recessions (after 2001), we believe it will continue into the near future. Correspondingly, trade patterns also show a similar trend toward an increased share of services and information, though trade overall continues to be dominated by products. Nevertheless, the relative importance of information-intensive services and knowledge-based assets in trade will likely continue to grow. In corporate investments, the proportion of budgets going toward computers and communication is growing overall, with the share of software also showing steady growth. It is not easy to predict whether this trend will continue into the future or what the main causes will be. On the one hand, the ongoing evolution in enterprise systems may mean continuing expenditures for the replacement of software assets. On the other hand, the trend to IT outsourcing and Software as a Service may mean a net shift from capital expenditures to operating expenses. With hardware, the trend seems clearer; in-house capacity is being exchanged for hosted server capacity, "cloud computing" and IaaS, while the costs of all types of equipment continue to fall.

These trends will have implications for policy makers and managers. It is possible that the overall shift in investment patterns toward information and communication technologies, coupled with the shift to cloud computing, will mean a shift in the cost patterns of some sectors. Entry into many industry sectors will be easier as the requirements for fixed investments decrease. The normal expectation would be for a larger number of entrants in a sector but this is confounded by network effects, which seem to be leading toward concentration in certain services like online search and social communication. We would conjecture that there is likely to be more volatility in sector composition and firm valuations, which could be reflected in capital markets and stock prices.

Some implications of wage rate differentials seem clear. The information and knowledge-based sectors show higher total wages, share of the wage bill and individual wages. Correspondingly the share of jobs has also increased. One obvious consequence of the shift in the numbers of jobs combined with differential wage rates is increased income inequality, though there could be other factors at play. It could also be inferred that the trend to information intensity favors returns to education. However, there is a potential lurking hazard created by the accompanying industrialization of information-intensive services (Karmarkar 2010). A combination of automation, outsourcing and off-shoring is already threatening low-end information-intensive (white collar) occupations. Transaction-oriented sectors like financial services are visible examples, as are consumer services such as ticket sales, retail sales and counter transactions.

In addition to the changes at the scale of the economy, there are also ongoing changes at the level of tasks, jobs, organization and industries. These show the depth of the change in the economy. They also suggest an ongoing pressure to reorganize companies in order to operate more effectively in a changing environment. Survey results from the UCLA Business and Information Technologies (BIT) project (Karmarkar and Mangal 2007; Karmarkar 2008) reveal that firms are indeed becoming more virtual, with broader and flatter organizations, higher demands for information and business intelligence, and higher requirements for technology skills for all employees. Sector studies in that project show significant changes in the overall organization of industries such as entertainment and financial services.

All of these trends create a need for rethinking national and regional economic policies and company strategies. There are of course substantial opportunities for research as well. In our own ongoing research we are continuing to track the US economy, especially to understand what effect the recent recessions will have on the secular trends we have described. We will also take a more detailed look at the pattern of employment and wages down to sector and industry level. In the global BIT project (Karmarkar and Mangal 2012), the focus is shifting from country-level surveys of technology impact on business practice, to sector studies examining the restructuring of industries and shifting patterns of competition.

ACKNOWLEDGEMENTS

The authors would like to thank the editors, John R. Bryson and Peter Daniels, for providing useful comments on an earlier draft of the chapter. The authors are also grateful to Now Publishers Inc. for granting permission to reproduce some of the tables and figures from Apte, Karmarkar and Nath (2012a).

NOTES

1. One of the methods used in national income accounting is the income method that measures GNP as the sum of all incomes received by the owners of various resources used in production. Such income payments are known as factor payments, because they are paid to various factors of production such as labor, land, capital, and so on. The factor payments include employee compensation, rental income, proprietary income, corporate profits, interest income, depreciation and indirect business taxes. Wages are the largest component of employee compensation and account for a significant share in GNP. For example, wages contributed 43.5 percent of the US GNP in 2011.
2. The ways that information is conceptualized vary across disciplines. Even within a discipline, there are differences among scholars about how information is conceived and, as a result, how its larger impact in society has been analyzed. Schement and Curtis (2004) discussed various conceptualizations from across disciplines to examine the ascendance of the idea of information throughout history and to discuss its role in the emergence of the so-called information society.
3. This study uses data for 1967.
4. In his seminal contribution to the growth literature, Baumol (1967) first explained the concept of unbalanced growth. Baumol (1985) and Baumol et al. (1985) further applied this concept to explain low productivity growth and higher employment in services.
5. The classifications of information and non-information workers used by Apte et al. (2008, 2012b) are different from those proposed by Porat (1977) and Osberg et al. (1989).
6. Chase and Apte (2007) discussed the evolution of service operations from a historical perspective and identify the big ideas around which major changes have taken place.
7. Using the same methodology as Apte and Nath (2007), Choi et al. (2009) concluded that, while having a relatively larger manufacturing sector than the US, Korea was effectively an information economy by 2000.
8. The proposed classification scheme is based on three distinct criteria:

    (1) *Market transaction or delivery mode*: Products are in standard units, not differentiated by customer, priced by unit and pre-produced, while services are processed, produced and customized on demand and priced by process rather than by unit.
    (2) *Form*: Products are tangible, while services are intangible or experiential.
    (3) *Production process*: Products are produced entirely by suppliers, while services are often co-produced with the customer present.

9. By "high end" information jobs, we refer to those jobs that require high cognitive skills and innovative ideas, such as managerial jobs, scientists and designers, and by "low end" information jobs, we refer to those jobs that are routine or repetitive in nature and do not require very high cognitive skills or innovative ideas, such as travel agent or customer service representative.
10. Real wage is obtained by adjusting nominal (current dollar) wage for inflation calculated from the annual US city average of all urban consumer price indices published by the Bureau of Labor Statistics.
11. The selection of the sample period 2002–2007 for the analysis at the sectoral level is dictated by two considerations. First, the data for 1999 through 2001 are available by SIC industry codes while, for 2002 through 2007, they are available by the new NAICS codes, which are not quite comparable at the disaggregate industry level. Secondly, the years between 2002 and 2007 represent a period of business cycle expansion between two recent recessions in the US.
12. Stiroh (2002) defined IT capital to include these 12 types of fixed assets. We will call them ICT capital or ICT capital assets or ICT fixed assets interchangeably.
13. For a more detailed discussion with a special focus on the US trade in information-intensive services, see Apte and Nath (2012).
14. Note that goods trade includes a large number of information products such as computers, mobile phones and various machines and equipment used for creating, processing and communicating information.

15. The data on services trade have been highly aggregated. Disaggregated data by detailed categories of services are available only for recent years.
16. Co (2007) referred to this category as knowledge-intensive services.
17. Markusen (1989) modeled this category as a capital-intensive service.
18. For a more detailed discussion, see Apte et al. (2012a).

BIBLIOGRAPHY

Acemoglu, D. and D.H. Autor (2010), "Skills, Tasks and Technologies: Implications for Employment and Earnings", *Mimeo*, MIT.
Amity, M. and S.-J. Wei (2009), "Service Offshoring and Productivity: Evidence from the US", *The World Economy*, **32** (2), 203–220.
Apte, U.M. and U.S. Karmarkar (2007), "Business Process Outsourcing and "Off-shoring": The Globalization of Information-Intensive Services", in U.M. Apte and U.S. Karmarkar (eds), *Managing in the Information Economy: Current Research Issues*, New York, NY: Springer Science + Business Media, LLC, 59–81.
Apte, U.M. and R.O. Mason (1995), "Global Disaggregation of Information-Intensive Services", *Management Science*, **41** (7), 1250–1262.
Apte, U.M. and H.K Nath (1999), "Service Sector in Today's Information Economy", *Conference Proceedings, Service Operations Management Association*, 106–111.
Apte, U.M. and H.K Nath (2007), "Size, Structure and Growth of the US Information Economy", in U.M. Apte and U.S. Karmarkar (eds), *Managing in the Information Economy: Current Research Issues*, New York, NY: Springer Science + Business Media, LLC, 1–28.
Apte, U.M. and H.K. Nath (2012), "US Trade in Information-Intensive Services", in V. Mangal and U. Karmarkar (eds), *The UCLA Anderson Business and Information Technologies (BIT) Project: A Global Study of Business Practice 2012*, Singapore: World Scientific Publishing Co., 117–144.
Apte, U.M., U.S. Karmarkar and H.K. Nath (2008), "Information Services in the US Economy: Value, Jobs and Management Implications", *California Management Review*, **50** (3), 12–30.
Apte, U.M., U.S. Karmarkar and H.K. Nath (2012a), "The US Information Economy: Value, Employment, Industry Structure, and Trade", *Foundations and Trends in Technology, Information and Operations Management*, **6** (1), 1–179.
Apte, U.M., U.S. Karmarkar, Y.T. Leung and C. Kieliszewski (2012b), "Exploring the Representation of Complex Processes in Information Intensive Services", *International Journal of Services, Operations and Informatics*, **7** (1), 52–78.
Autor, D.H. (2007), "Structural Demand Shifts and Potential Labor Supply Responses in the New Century", paper prepared for the Federal Reserve Bank of Boston Conference, *Labor Supply in the New Century*, 19–20 June.
Autor, D.H., L.F. Katz and A.B. Krueger (1998a), "Computing Inequality: Have Computers Changed the Labor Market?", *Quarterly Journal of Economics*, **113** (4), 1169–1213.
Autor, D.H., F. Levy and R.J. Murnane (1998b), "Upstairs, Downstairs: Computers and Skills on Two Floors of a Large Bank", *Industrial and Labour Relations Review*, **55** (3), 432–447.
Autor, D.H., F. Levy and R.J. Murnane (2003), "The Skill Content of Recent Technological Change: An Empirical Exploration", *Quarterly Journal of Economics*, **118** (4), 1279–1333.
Baumol, W.J. (1967), "Macroeconomics of Unbalanced Growth: The Anatomy of Urban Crisis", *American Economic Review*, **57** (3), 415–426.
Baumol, W.J. (1985), "Productivity and the Service Sector", in R.P. Inman (ed.), *Managing the Service Economy: Prospects and Problems*, New York, NY: Cambridge University Press, 301–317.
Baumol, W.J., S.A.B. Blackman and E.N. Wolff (1985), "Unbalanced Growth Revisited: Asymptotic Stagnancy and New Evidence", *American Economic Review*, **75** (4), 806–817.
Berman, E., J. Bound and S. Machin (1998), "Implications of Skill-Biased Technological Change: International Evidence", *Quarterly Journal of Economics*, **113** (4), 1245–1279.
Bhagwati, J.N. (1987), "Trade in Services and the Multilateral Trade Negotiations", *World Bank Economic Review*, **1** (4), 549–569.
Cairncross, F. (1997), *The Death of Distance*, Boston, MA: Harvard Business School Press.
Chase, R.B. and U.M. Apte (2007), "A History of Research in Service Operations: What's the Big Idea?", *Journal of Operations Management*, **25**, 375–386.
Choi, M., H. Rhim and K. Park (2009), "New Business Models in Service and Information Economies: GDP and Case Studies in Korea", in U. Karmarkar and V. Mangal (eds), *The UCLA Anderson Business and*

Information Technologies (BIT) Project: A Global Study of Business Practice, Singapore: World Scientific Publishing, 271–298.

Clark, C. (1940), *The Conditions of Economic Progress*, London: Macmillan.

Clarke, G.R.G. and S.J. Wallsten (2006), "Has the Internet Increased Trade? Developed and Developing Country Evidence", *Economic Inquiry*, **44** (3), 465–484.

Co, C.Y. (2007), "US Exports of Knowledge-Intensive Services and Importing-Country Characteristics", *Review of International Economics*, **15** (5), 890–904.

Deardorff, A.V. (2001), "International Provision of Trade Services, Trade, and Fragmentation", *Review of International Economics*, **9** (2), 233–248.

Deardorff, A.V. and R.M. Stern (2008), "Empirical Analysis of Barriers to International Services Transactions and the Consequences of Liberalization", in A. Mattoo, R.M. Stern and G. Zanini (eds), *A Handbook of International Trade in Services*, Oxford: Oxford University Press.

Francois, J. and B. Hoekman (2010), "Services Trade and Policy", *Journal of Economic Literature*, **48** (3), 642–692.

Freeman, R.B. (2002), "The Labour Market in the New Information Economy", Working Paper 9254, Cambridge, MA: National Bureau of Economic Research.

Freund, C. and D. Weinhold (2002), "The Internet and International Trade in Services", *American Economic Review*, **92** (2), 236–240.

Freund, C. and D. Weinhold (2004), "The Effect of the Internet on International Trade", *Journal of International Economics*, **62**, 171–189.

Grunfeld, L.A. and A. Moxnes (2003), "The Intangible Globalization: Explaining the Patterns of International Trade in Services", NUPI Working Paper 657, Oslo: Norwegian Institute of International Affairs.

Hoekman, B. (1996), "Assessing the General Agreement on Trade in Services", in W. Martin and L.A. Winters (eds), *The Uruguay Round and the Developing Countries*, Cambridge: Cambridge University Press.

Hoekman, B. and A. Mattoo (2008), "Services Trade and Growth", in J.A. Marchetti and M. Roy (eds), *Opening Markets for Trade in Services: Countries and Sectors in Bilateral and WTO Negotiations*, Cambridge: Cambridge University Press.

Hoekman, B., A. Mattoo and A. Sapir (2007), "The Political Economy of Services Trade Liberalization: A Case for International Regulatory Cooperation?", *Oxford Review of Economic Policy*, **23** (3), 367–391.

Karmarkar, U.S. (2004), "Will You Survive the Service Revolution?", *Harvard Business Review*, **82** (6), 101–107.

Karmarkar, U.S. (2008), "The Global Information Economy and Service Industrialization: The UCLA BIT Project", in B. Hefley and W. Murphy (eds), *Service Science, Management and Engineering: Education for the 21st Century*, New York, NY: Springer Science + Business Media, LLC, 243–250.

Karmarkar, U.S. (2010), "The Industrialization of Information Services", in P.P. Maglio, C.A. Kieliszewski and J.C. Spohrer (eds), *Handbook of Service Science*, New York, NY: Springer Science + Business Media, LLC, 419–435.

Karmarkar, U.S. and U.M. Apte (2007), "Operations Management in the Information Economy: Products, Processes and Chains", *Journal of Operations Management*, **25**, 438–453.

Karmarkar, U.S. and V. Mangal (2007), "The UCLA Business and Information Technologies (BIT) Survey – Year 2", in U.S. Karmarkar and V. Mangal (eds), *The Business and Information Technologies (BIT) Project: A Global Survey of Business Practice*, Singapore: World Scientific Press.

Karmarkar, U.S. and V. Mangal (2012), *The UCLA Anderson Business and Information Technologies (BIT) Project: A Global Survey of Business Practice*, Singapore: World Scientific Press.

Karmarkar, U.S. and H. Rhim (2012), "Industrialization, Productivity and the Effects on Employment, Wealth, Income Distribution and Sector Size", BIT Working Paper, Los Angeles, CA: UCLA, Anderson School.

Kimura, F. and H. Lee (2006), "The Gravity Equation in International Trade in Services", *Review of World Economics*, **142** (1), 92–121.

Machin, S. and J.V. Reenen (1998), "Technology and Changes in Skill Structure: Evidence from Seven OECD Countries", *Quarterly Journal of Economics*, **113** (4), 1215–1244.

Machlup, F. (1962), *The Production and Distribution of Knowledge in the United States*, Princeton, NJ: Princeton University Press.

Machlup, F. (1980), *Knowledge: Its Creation, Distribution and Economic Significance, Volume 1: Knowledge and Knowledge Production*, Princeton, NJ: Princeton University Press.

Mann, C.L. and D. Civril (2008), "US International Trade in Other Private Services: Do Arm's Length and Intra-Company Trade Differ?", *Mimeo*, Waltham, MA: Brandeis University, Brandeis International Business School.

Markusen, J. (1989), "Trade in Producer Services and in Other Specialized Intermediate Inputs", *American Economic Review*, **79** (1), 85–95.

Mattoo, A., R. Rathindran and A. Subramanian (2006), "Measuring Services Trade Liberalization and Its Impact on Economic Growth: An Illustration", *Journal of Economic Integration*, **21** (1), 64–98.

Mithas, S. and M.S. Krishnan (2008), "Human Capital and Institutional Effects in the Compensation of Information Technology Professionals in the United States", *Management Science*, **54** (3), 415–428.

Mithas, S. and J. Whitaker (2007), "Is the World Flat or Spiky? Information Intensity, Skills, and Global Service Disaggregation", *Information Systems Research*, **18**, 237–259.

Nath, H.K. (2011), "ICT Capital and Employment of Information Workers: Are They Related?", *International Journal of Engineering Management and Economics*, **2** (2/3), 111–131.

OECD (1981), *Information Activities, Electronics and Telecommunications Technologies: Impact on Employment, Growth and Trade*, Volumes I and II, Paris: OECD.

OECD (1986), *Trends in the Information Economy*, Paris: OECD.

Osberg, L., E.N. Wolff and W.J. Baumol (1989), *The Information Economy: The Implications of Unbalanced Growth*, Halifax: The Institute for Research on Public Policy.

Polese, M. and R. Verreault (1989), "Trade in Information-Intensive Services: How and Why Regions Develop Export Advantages", *Canadian Public Policy*, **15** (4), 376–386.

Porat, M.U. (1977), *The Information Economy*, Office of Telecommunications Special Publication 77-12, Washington, DC: US Department of Commerce.

Schement, J.R. and T. Curtis (2004), *Tendencies and Tensions of the Information Age*, New Brunswick and London: Transaction Publishers.

Shy, O. (2001), *The Economics of Network Industries*, Cambridge: Cambridge University Press.

Stiroh, K.J. (2002), "Information Technology and the US Productivity Revival: What Do the Industry Data Say?", *American Economic Review*, **92** (5), 1559–1576.

Stoops, N. (2004), "Educational Attainment in the United States", http://www.census.gov/prod/2004pubs/p20-550.pdf (accessed 10 September 2012).

Wolff, E.N. (2002), "Productivity, Computerization, and Skill Change", *Federal Reserve Bank of Atlanta Economic Review*, **Third Quarter**, 63–87.

Wolff, E.N. (2006), "The Growth of Information Workers in the US Economy, 1950–2000: The Role of Technological Change, Computerization, and Structural Change", *Economic Systems Research*, **18** (3), 221–255.

11. Service and experience
Jon Sundbo

INTRODUCTION

This chapter explores the experience aspect of services, both as a concept within service management and marketing theory (e.g. Eiglier and Langeard, 1988; Normann, 1991; Grönroos, 2000; Heskett et al., 1990; Vargo and Lusch, 2006) and as a phenomenon that is being theorised in its own right. The topic is approached through Kuhn's (1970) concept of paradigms, which suggests that a scientific field, including certain research objectives, basic theoretical assumptions and methods, is accepted and used by a scientific community until a new paradigm emerges. The emergence of a new paradigm does not necessarily have anything to do with changes in the empirical world, but it is possible that it may. Primarily, paradigms are theoretical constructions that attempt to provide an understanding of a complex reality by making different, simplified, theoretical assumptions. A paradigm shift is a meta-scientific discussion about social constructivism (Burr, 1995; Hacking, 1999). This shift might reflect, or be triggered by, changes in the studied real or empirical, phenomena, but not necessarily.

Service theory is one such paradigm, while an increasingly independent experience theory may be another. Within the service marketing tradition, experience has long been used as a concept to characterise customers' assessments of a received service (see for example Berry et al., 2006; Kwortnik and Thompson, 2009; Verhoef et al., 2009; Helkkula, 2010). Other concepts such as perceived quality (Brown et al., 1991; Edvardsson et al., 1994; Grönroos, 2000) and customer satisfaction (Helkkula, 2010; Irani and Hanzaee, 2011) have also been used. However, during the last decade the notion of experience has received particular attention and has consequently become a separate topic in some of the scientific literature (particularly by Pine and Gilmore, 1999, but see O'Dell and Billing, 2005; Boswijk et al., 2007; Sundbo and Darmer, 2008; Sundbo and Sørensen, 2013). Experience has been seen as a special phenomenon in people's lives that has created a certain demand that is fulfilled by special goods, services, events or just aesthetic impressions. Thus the theoretical understanding of this part of the market and provider–customer relations now has its own paradigm (Mossberg, 2007; Hulten et al., 2009) and the concept of experience has gained its own strong identity. Customer relationships and care are highlighted in the service marketing literature but an older theoretical understanding must be added to this new paradigm – one of experience (Dewey, 1929, 1934, 1938; Valberg, 1992; Snel, 2011).

In this chapter, this development will be analysed to answer two questions in a theoretical manner. First, what is the experiential element of the service relation and, secondly, has the concept of experience gained an independent life (as advocated, for example, by Jensen, 1999 and Pine and Gilmore, 1999)? The second question can be sub-divided into two: is there a new and independent content in the empirical reality (cf. Sundbo and

Sørensen, 2013) and is there a new life in the theoretical scientific understanding, that is, has it become the core of a new paradigm (cf. Kuhn, 1970)?

The structure of the chapter is as follows. First an overview of how the concept of experience has been used in the service literature is presented. Then the two questions will be discussed from a Kuhnian perspective and for that purpose a meta-scientific model is established. This implies a deeper discussion of what the experience aspect of service marketing is and whether experience as a theoretical idea has become independent and thus the basis for a new paradigm. Finally, there is a discussion of how the possible existence of this new experience paradigm might develop in the future and how the service paradigm fits into this development.

EXPERIENCE IN SERVICES

Definition of Services and Experience

First, in order to know what we are talking about it is necessary to define services and experience. To explain why experience could be an independent phenomenon that may be the basis for a new scientific paradigm requires that it is defined in a way that differentiates it from services.

Experience is an aspect of customers' assessment of a delivered service but the experience is different from the service. We might define a service as an act that aims to solve a specific problem for a customer (whether this is of a material, personal or intellectual kind). Thus, the service is an instrumental act (to use a sociological distinction between instrumental and expressive acts) (Parsons, 1951).

Experience, on the other hand, is an expressive act, which is connected to emotions and sensing (all five senses can be used when people experience) (cf. Csikszentmihalyi, 2002; Jantzen, 2013); theory has even suggested that experience should be understood as a neurophysiological process (Jantzen and Vetner, 2007). It can be defined as 'a mental journey, which leaves an immaterial memory – a learning or a psychic sense. It might be entertaining or learning, but does not need to be so' (Sundbo et al., 2011, 14). Thus the experience is created in the mind of the receiver (e.g. a service customer). The service and the service delivery process that are emphasised in service marketing and management theory can be external stimuli, even if the stimuli are not a planned part of the service delivery.

In a service delivery the customer's assessment of whether he received a good service is primarily decided by the rational, instrumental act: did the service provided solve his problem? This is the core issue for service quality and customer satisfaction. A side effect is that the customer might have an experience in the above meaning, but he does not necessarily have to.

A linguistic problem with the word 'experience' is that in English it has two meanings. It may be something that one has learned, which is a more rational and instrumental process, or it may be a mental, expressive process, as defined above. In Germanic languages (such as German, Dutch, Danish and Swedish) there are two different words for these two processes; the first process is called 'erfahrung' in German, the second is called 'erlebnis'. In service delivery, the experience aspect to which the service marketing theory refers (Verhoef et al., 2009), the assessment of the service – the instrumental process – would in German be called 'erfahrung' and the expressive experience would be called 'erlebnis'. Thus customer satisfaction in service delivery primarily depends on the instrumental aspect ('erfahrung'). The expressive aspect ('erlebnis') that in service delivery is an extra or secondary feature can be said to be gaining increased importance for customers that leads to an independent business phenomenon and scientific paradigm: the experience aspect (as stated by Pine and Gilmore, 1999).

Thus both service marketing theorists who claim that experience is an integrated part of service delivery and those advocating the experience paradigm that claims that this phenomenon is getting an independent and different 'life' may be right. Empirically both phenomena – experience as an integrated part of service delivery and as an independent phenomenon – may exist side by side.

Experience in the Service Literature

The increasing use of the notion of experience within the service literature is based on marketing theory's development from understanding buying behaviour as a purely rational information process (as traditional economics has understood it) to emphasising that emotional elements also influence the process (e.g. Holbrook and Hirschman, 1982). One of the first inspirations for this trend was Arnould and Price's (1993) paper on service customer satisfaction. They used river sailing as a metaphor to describe the feeling and impression that customers can have and which makes them satisfied. River sailing is a 'real' experience – something that is done for its own sake, not to consume a service; river sailing is extraordinary and has some emotional and hedonistic elements which might be associated with experiences (although experiences need not be hedonistic). This feeling and impression is transmitted into service delivery. The assumption is that if a customer obtains an experience, he or she will like it and be a loyal customer.

Customer experience has become a factor in service marketing research and in prescriptive, practical, service marketing tools. The service marketing literature refers extensively to the total customer experience (e.g. Verhoef et al., 2009). This is built out of experiences in the search, purchase, consumption and after-sales phases of service deliveries. Empirical research provides support for this analysis by demonstrating that service firms that have satisfied customers outperform other firms, and customers are more satisfied the better experience they have (see for example Mittal et al., 2005; Bolton et al., 2006; Meyer and Schwager, 2007; Gentile et al., 2007). As a result, prescriptive models and methods for managing customer experiences have been developed (Grove et al., 1992; Schmitt, 1999; Pullman and Gross, 2004; Zomerdijk and Voss, 2010).

Helkkula (2010) divides the service experience literature into three groups. The first group characterises service experiences phenomenologically (e.g. Prahalad and Ramaswamy, 2004; Berry and Carbone, 2007; Meyer and Schwager, 2007). This literature is mostly case based and tries to understand how experiences are developed in customers. A second group is process based and draws on the architectural composition of service actions (e.g. Edvardsson et al., 2005; Toivonen et al., 2007). This literature is more 'engineering' in its approach. The third group is centred on an outcome-based characterisation which tries to measure the outcome (e.g. Holloway et al., 2005; Aurier and Siadou-Martin, 2007). This is more positivistic in nature.

The focus on experiences has almost replaced the interest in service quality that was popular in the 1980s and 1990s (good examples are Brown et al., 1991; Edvardsson et al., 1994; Grönroos, 2000). The service quality approach also aimed to explain how service firms retain loyal customers who are willing to pay a higher price for products. Customer loyalty is achieved through high service quality, which is understood as customers' perceived quality. The perceived quality is a combination of technical quality, which is the technical content of the service (e.g. the cleaning procedures), and functional quality, which is the way in which the service is delivered (including for example the behaviour of the service personnel) (Grönroos, 2000). Perceived quality and experience are equally seen as being the factors that make customers satisfied and loyal and make them buy the service at a high price, thus ensuring a high profit for the service provider. In theoretical discussions about service quality the notion of experience has also been used (Zeithaml et al., 1996; Sousa and Voss, 2006). This has sometimes been done by using expressive terms such as emotion and hedonistic to characterise the service consumption (cf. Caru and Cova, 2003) and sometimes instrumental terms such as the effects of the service delivery, understood as whether the customer thinks that the service solved the problem or not. Experience is in the first case used in the sense of 'erlebnis' and in the second case of 'erfahrung'.

THE EXPERIENCE ECONOMY AND SERVICE EXPERIENCE

To answer the two questions raised in the introduction about how the experience concept has entered service marketing theory as well as how it has possibly developed into a new, independent paradigm, a meta-model that can explain both is suggested here. Theoretically, one may argue that both theoretical developments (introduction of experience into service theory and the emergence of a new experience paradigm) follow a general pattern. The reason why the concept of experience has been introduced into service marketing theory and why experience has become an independent theoretical and research field may be different, but both developments follow the same general pattern.

The discussion of the two questions is started in this section by looking at how the concept of experience came into service marketing theory and how it may be becoming a new and independent paradigm. That discussion will be the basis for suggesting a meta-model for the general pattern. This will be presented in a subsequent section. This meta-model will then be used to continue the discussion about services and experience within a more general and systematic framework.

Understanding of Experience

On the one hand we can see that an experience is part of the service practice and the experience concept is in turn part of service theory. On the other hand a literature has grown up that claims, or takes for granted, that an independent experience economy exists (e.g. O'Dell and Billing, 2005; Boswijk et al., 2007; Sundbo and Darmer, 2008; Lorentzen and Hansen, 2012; Sundbo and Sørensen, 2013). An important question is whether one or the other discourse is correct. One may consider whether it is all about social or theoretical constructions; thus this question only has to do with abstract social theory separated from the empirical reality. This will be part of the following discussion.

First, we may state that an experience is an independent phenomenon, in the same way as, for example, quality. It is not just a characteristic that is inextricably connected to service deliveries. It might be that an experience normally characterises a service delivery, at least within the service literature. However, the notion of an experience could be used about a phenomenon that is not connected to service delivery; it occurs as an independent phenomenon or as a characteristic of activities other than service delivery, for example aesthetics. If we do not assume this, the argument that follows is not valid. The assumption can be supported by the fact that the notion of experience has been used in relation to the consumption of art (Dewey, 1934), nature (Dewey, 1929), education (Dewey, 1938), philosophical issues (Valberg, 1992) and learning (Snel, 2011).

Services and Experience as Independent Phenomena or General Aspects of all Business Activities?

Next, we may say that the question of how to understand experience could be reformulated as whether it is an independent phenomenon with its own life that must be understood as an *independent* activity, or whether it is only an *aspect* of other activities and must be understood within other logics (such as the service-dominant logic; Vargo and Lusch, 2006). Should experience be dealt with on its own or should it be fully integrated into other theories? In the latter case, one could assume that experience is only an aspect of service delivery; however, one could also assume that experience could be an aspect arising from the use of, for example, goods or raw materials, or from participation in leisure activities outside the formal economy. Thus experience is a general aspect of all business activities. This means that the question can be reformulated as: Is experience an *independent* 'holistic' phenomenon that, for example, is produced in particular industries and particular types of social activities or is it a general *aspect* of many business and social activities, including service delivery?

This type of problem is not new. The service phenomenon and the scientific understanding of it have had the same problem. Within some service literature and theory, services are considered a specific business sector (Gadrey, 1992; Illeris, 1996). Services include different business activities that have common characteristics which distinguish them from other activities and sectors; for example the manufacturing sector that produces goods and the agriculture and fishing sector that provides food raw materials. What distinguishes different industrial sectors is the output. The output from the manufacturing sector is goods; from the service sector it is activities, for example moving objects such as transport, cleaning and catering, or providing advice to change a client's knowledge structure. The service sector is often extended to include the public sector, which is also considered service producing. It has been discussed whether the manufacturing sector increasingly emphasises the service aspect of its activity – scientifically captured by the so-called servitisation debate (Baines et al., 2009; Wilkinson et al., 2009); in this literature, services are considered an independent phenomenon (we do not discuss marketing here).

In other literature – that is, some parts of service marketing – services are considered a particular relation between the service provider and the customer (Eiglier and Langeard, 1988; Gummesson, 1994; Grönroos, 2000). According to this tradition, this particular relation has emerged because a service must be consumed at the moment of production.

The activity in this moment is called 'prosumption' (Toffler, 1981). This means that the service-providing personnel of the service provider must be present at the moment of prosumption and must interact with the customer. This interaction is the basis for the marketing of the services actually provided for other services and the service-providing company in general. The service personnel's behaviour includes a marketing aspect. The customer's probability of buying the same service or another service from the same provider depends on the interaction between the service personnel and the customer. This particular aspect of service provision is called customer relationship marketing. Some theories (e.g. Gummesson, 1994; Foss and Stone, 2001) claim that this is not only an aspect of the provision of services, but a general marketing method that is used in all sectors including the manufacturing sector (e.g. under notions such as 'service dominant logics'; Vargo and Lusch, 2006, 2008). According to this service marketing literature, service is an aspect of general marketing.

The development of experience theory has followed a parallel track. In some academic literatures experience is considered an independent phenomenon produced in a particular industrial sector (KK Stiftelsen, 2003; Erhvervs- og byggestyrelsen, 2008; Sundbo, 2011). To be correct, one could argue that the business sector does not deliver experiences since those are the mental phenomena taking place in people's heads (cf. Csikszentmihalyi, 2002; Jantzen and Vetner, 2007; Jantzen, 2013). The experience sector provides events or elements that make experiences happen within the minds of people. Examples of such elements are aesthetics such as industrial design and architecture, music, pictures, movies, videos and TV. The experience sector includes for example advertising and PR, culture, tourism, design, architecture, museums, gastronomy, sport and media. It may seem similar to what has been called the cultural economy (Power and Scott, 2004; Andersson and Andersson, 2006); however, the cultural sector is normally defined more narrowly than the experience sector. It would not for example include a zoo. Creative industries (Caves, 2000; Hartley, 2005) have also been used as a concept to analytically grasp the phenomenon that is discussed here. However, creative industries are also often defined more narrowly than the experience sector. Creative industries would often not include, for example, amusement parks or hotels in picturesque locations.

In other parts of the literature, experience is considered to be a general aspect of marketing that is used in all industrial sectors including manufacturing and the primary sectors (e.g. Pine and Gilmore, 1999; Mossberg, 2003, 2007; Schmitt et al., 2004; Milligan and Smith, 2008; Hulten et al., 2009). If people get an experience that is related to the purchase, they will be happy and satisfied with the product. Whether they will be loyal customers and without questioning will buy another good, service or experience-event/element is not well researched in the experience marketing paradigm, but it is assumed. This experience aspect is, within this paradigm, considered one that, although coming from a sub-aspect of customer relationship marketing, gets its own life and dominates an aspect of goods and services that makes people buy the product and become loyal customers. The postulate is that experience is so much in demand and valuable to people that they buy goods, services or particular experience-events and elements more if there is an experience connected to them. They will even be prepared to pay more for the good, service or experience-event/element.

A META-MODEL FOR THE EVOLUTION OF OUR UNDERSTANDING OF FUNDAMENTAL BUSINESS ACTIVITIES

We now introduce the meta-model. A further argument for introducing a meta-model can be found in the similar discussion in the 1970s and 1980s concerning the information society, which was seen as a new economic and social structure (Kumar, 1995; Castells, 2010). Information or knowledge has always existed, but it became independent and dominant as a theoretical category for understanding not only the specific development of society at that time, but also new economic and business structures. The focus on information was due, in part, to the then emerging information technology (Sundbo, 2011). The same theoretical positions as within the service and experience paradigms also existed in the information paradigm. Some scientists focused on information as an independent phenomenon produced in a specific economic sector – the information sector (Porat, 1977; Sundbo, 2011). Others saw information as something general connected to all business activities, and even much social activity (Castells, 2010).

A General Meta-model

The current discussion about services and experiences suggests the evolution of a competing scientific paradigm. These observations lead us to put forward a general meta-model of economic structures that can help in understanding why this dynamic paradigm development seems to be repeated regularly throughout the history of fundamental business theories (those that attempt to explain a fundamental economic activity carried out within a market or a public economic framework). The principles of this meta-model stem from Kuhn's (1970) theory; however, they are applied to the special field of theories relating to fundamental business activities.

The meta-model explains the development of a new business theory paradigm as follows.

The development starts with some researcher(s) discovering a specific economic phenomenon which is claimed to be of increasing importance to economic, business or sociological functions in a society (sometimes all three of them). The phenomenon may be an activity (such as a service delivery), a general material structure (such as goods in general or IT as a general technology), a general aesthetic (such as industrial design) or a fundamental value (such as sustainable environmental handling of all activities). This could be called the discovery phase. An important step in the development of the scientific theory is the naming of the phenomenon because the name identifies the objectives of the science and highlights the evolution of a new approach or form of conceptualisation.

Next the discovered phenomenon is scientifically constituted. A paradigm develops with specific fundamental assumptions, a theoretical understanding and methods. The name of the phenomenon (e.g. 'industry', 'information', 'service', 'experience') identifies the paradigm and becomes the common identification for its adherents. Within fundamental business theories, the phenomenon gets its own business sector; that is, it is defined as a premise that the phenomenon is carried out as an independent activity in the economic production and distribution system in the society. Primarily, this business activity (such as 'manufacturing', 'information', 'service', etc.) is carried out in a specific

1. Discovery phase
A phenomenon is discovered.

It might have its own paradigm developed with specific fundamental assumptions, theoretical understanding and methods. If so, the next phases emerge:

↓

2. The constitution phase
The phenomenon is ascribed its own business sector. It becomes independent.

↓

3. The imperial phase
'Colonies'are established, that is, the theoretical claims of the phenomenon can be found in all business sectors. The phenomenon is considered an aspect of many business activities.

Figure 11.1   A model of development of business paradigms

economic sector named after the phenomenon (the 'manufacturing' sector, the 'service' sector, etc.).

Later this core business activity establishes 'colonies' in other sectors (to use a metaphor). The 'colonies' are the theoretical claims that the phenomenon can be found in all, or almost all, business sectors. Generally, it is also claimed that the phenomenon ('service', 'experience', etc.) becomes increasingly important for all business activities. To use another metaphor, this phase could be called the imperialism phase (Figure 11.1).

The constitution phase does not always emerge (e.g. a scientific paradigm does not develop; the phenomenon remains a minor factor in other scientific explanations). This depends on how the market develops. If the phenomenon in its real existence, for example service or experience, becomes emphasised in the market, this is a basis for the scientific constitution phase. This situation emerges if the phenomenon becomes of high value to the buyers or if the providers are very innovative so that new forms of the phenomenon are launched.

The imperial phase requires scientists to define the phenomenon as a fundamental one that requires its own theoretical explanation. The size and importance of the specific sector can be measured. A crucial factor for the diffusion of the paradigm is whether it can be proved that the sector is growing or is a basis for general economic growth.

A logical assumption is that the constitution phase is established first, followed later by the imperial phase. Praxis in relation to the paradigms discussed here (service and experience, with a parallel to information) suggests that the development of the 'colonial' approach (the imperial phase) comes shortly after the specific sector approach (the constitution phase).

The colonial approach seems to expand continuously until the paradigm is met by a new, competing paradigm. For example, the service marketing, or customer relations marketing approach explanation expanded into increasingly more business activities and is argued as the basis for business activities, even in manufacturing firms (Vargo and

Lusch, 2006). However, it has now been met by another paradigmatic claim, namely that experience is becoming the most important growth factor in all business sectors (Pine and Gilmore, 1999). If the latter establishes itself as a paradigm, it will try to expand in the same way, but this will in the future also be met by a new paradigm.

Consequences for Practice

The fundamental business understanding not only influences science but it also influences practice. Managers and leaders in business always look for tools or approaches that can increase turnover and profit. This leads to specialisation, a division of labour and innovation that makes products more attractive to buyers at a favourable price. Every new fundamental understanding of business can be used for this, either because the new understanding emphasises a factor that really is important for the business, or because of the 'Hawthorne' effect first noted after studies undertaken in the 1930s (Roethlisberger and Dickson, 1939) in which the sociologist discovered that productivity increased if the management showed interest in the employees by introducing new production principles. Managers thus become interested in practical versions of the new paradigm and this contributes to the progression of the paradigm. This development of practice seems to be an expression of a general human search for something better (bringing more happiness or profit) and easier (with less exertion or capital investment).

SERVICE AND EXPERIENCE RESUMED

We can now return to the service and experience discussion and use the meta-model to explain in greater depth what currently happens to the concept of experience in relation to services. Experience is connected to the imperial phase of the service paradigm. Throughout the service theory discovery and constitution phases, experience became increasingly important and when the service theory entered the imperial phase, experience became a central aspect of customer relationship marketing that is connected to service delivery. Generally speaking, it is not clearly stated whether customer experience means instrumental 'erfahrung' or expressive 'erlebnis', but one may logically argue that implicitly meant is the 'erfahrung' aspect. This at least follows from the definition of service used in this chapter as a rational act aimed at solving a problem. The experience phenomenon that is claimed to be independent in the new experience paradigm is neither defined as 'erfahrung' nor 'erlebnis' (primarily because it is expressed in English, cf. Pine and Gilmore, 1999, where this distinction cannot be expressed). However, it can be argued that, implicitly, the meaning is closest to 'erlebnis'.

The service paradigm and practice have emphasised the instrumental 'erfahrung' part of service delivery. The service operation and service design and construction literature (e.g. Johnston, 1999; Goldstein et al., 2002; Hefley and Murphy, 2008) clearly focuses on the instrumental problem-solving aspects of service deliveries. Even though this is also the case in the more 'soft' service marketing literature (e.g. Grönroos, 2000; Vargo and Lusch, 2008), it does further emphasise the customer's subjective assessment of how the service was delivered. Did the customer like the service personnel? Did the service give them a side effect in the form of an 'erlebnis' experience, eventually provoked by peripheral services (such as free champagne on a flight) as stimuli? This 'soft' service marketing tradition has developed a view on the 'erlebnis' experience. This type of experience is becoming increasingly important for customers, and therefore also for their general assessment of the delivered services and the service firm. A tradition within marketing theories deals with expressive customer experiences (e.g. Holbrook and Hirschman, 1982; Pullman and Gross, 2004; Kwortnik and Thompson, 2009). In these theories phenomena such as hedonism are discussed (Kristensen, 1984; Hopkinson and Pujari, 1999; Caru and Cova, 2003; Irani and Hanzaee, 2011). Some marketing literature refers to sensory marketing (Hulten et al., 2009) as something that appeals to the customer's senses (view, smell, taste and so forth). These theories have moved towards a new experience paradigm since these sensory marketing theories are not particularly related to services, but are more general marketing theories based on investigations on how people react to different external stimuli.

One question about this paradigmatic transformation is how much the customer's 'erlebnis' experience is connected by the customer to the service. This is assumed in the service marketing theories. This fact is the basis for the statement in these theories that good customer experience creates loyal customers who buy the service again. However, the degree to which the customer connects the expressive 'erlebnis' experience with the instrumental service has only been investigated in very few studies, if any. It might be that the customers are well aware that the expressive experience is an independent phenomenon and as long as it is very valuable to them and they get it free from the service provider, they may choose this service provider and that service. However, they may just as easily connect that expressive experience to another service and another provider, or even buy it as a pure experience from an experience firm, even if they have to pay for it. An important criterion for an event or process that is considered an experience is that it is, to some degree, extraordinary; that is, it is not a part of the daily routine (such as transporting oneself to work). This can be difficult to ensure for a service provider because it requires the regular introduction of new expressive experience elements.

We can conclude that the service paradigm has entered the imperial phase, with a core sector and 'colonial' theoretical statement of service management and marketing being general to all business sectors. Manufacturing firms are even defined as service firms in the servitisation debate (Baines et al., 2009; Wilkinson et al., 2009). Experience is part of the service sector as firms that provide pure experiences, such as theatres, TV companies, professional sports clubs and architectural and design firms, statistically are categorised as service firms. Experience is also part of the 'colonial' theories as they are part of the explanation of the service provider's relation to customers.

EXPERIENCE BECOMES INDEPENDENT

Experience as the Basis for a New Emerging Paradigm

Around 2000 the concept of experience became independent and a new paradigm started to develop. This development can be linked to a discovery phase going back in time. Experience was studied much earlier and theories have been presented (e.g. Dewey, 1929, 1934; Valberg, 1992). However, the theoretical and research movement after 2000 has introduced a new, business, perspective. The emerging paradigm in most discussions is described as 'the experience economy' – much inspired by Pine and Gilmore (1999) (see also Sundbo and Sørensen, 2013).

By 2000 a constitution phase had emerged. This started slowly with analyses and theoretical approaches that presented a new view on certain business activities. Various authors and analyses suggested different factors that, when taken together, constituted a new paradigm. The most significant works are probably those of the German sociologist Gerhard Schulze (1992), who analysed cultural preferences and behaviour in the German city of Nürnberg, and the more normative business marketing book *The Dream Society* (Jensen, 1999). Schulze introduced the term 'experience' as the name for this new understanding, certainly in German – the 'erlebnis gesellschaft', but this notion underlines the expressive aspect of the experience phenomenon. Schulze's work, as well as other analyses of what were called creative industries (e.g. Caves, 2000; Department for Culture, Media and Sport, 2001), marked a constitution phase.

Jensen's (1999) work marked the imperial phase and this developed rapidly. Later, 'experience' was taken up by the marketing literature (e.g. Mossberg, 2003, 2007; Schmitt et al., 2004). The clear paradigmatic rupture with the service paradigm – in a classic paradigmatic setting – and a further emphasis on the imperial phase was most clearly demonstrated in *The Experience Economy* (Pine and Gilmore, 1999, 2011).

Since these first path-breaking works, further analyses and theory developments within the paradigmatic framework of experience have been published (particularly Boswijk et al., 2007; Gilmore and Pine, 2007; and Sundbo and Sørensen, 2013). These works emphasise the psychological nature of experiences (Csikszentmihalyi, 2002; Jantzen and Vetner, 2007; Snel, 2011; Jantzen, 2013) and the creation of experiences (O'Dell and Billing, 2005; Harsløf and Hannah, 2008; Sundbo and Darmer, 2008), both of which emphasise the constitution phase of the new paradigm. Later studies of the management of experience production and delivery within all business sectors have also appeared (Bærenholdt and Sundbo, 2007; Horn and Jensen, 2011), which emphasise the imperial phase.

Nonetheless, the paradigm still has not been widely adopted within the scientific world. Thus there are no journals dedicated to the experience economy. There have been several tourism or cultural studies journals but these are narrower in their scope than the experience economy; for example, they do not include the general marketing aspects of experience. Marketing journals do publish articles on experience and hedonistic, sensory and similar marketing aspects but these are rarely explained within a special experience economy framework. On the other hand, we can identify analyses that select a particular experience economy sector (Department for Culture, Media and Sport, 2001; KK Stiftelsen, 2003; Erhvervs- og byggestyrelsen, 2008) and this has provided the basis for industrial policy and economic growth in countries such as Sweden, the UK, Denmark, Norway and Canada. This underlines the constitution phase of an experience paradigm.

There is a case for arguing that the experience economy paradigm has now entered the imperial phase. Colonial aspects can also be observed. The importance of expressive experiences ('erlebnis') for all customers – whether they buy raw materials, goods, services or pure experiences – is emphasised in the emerging experience marketing literature (Schmitt et al., 2004; Milligan and Smith, 2008; Smilanski, 2009). Pine and Gilmore (1999, 2011) refer to experience marketing as 'business staged as a theatre' and claim that normatively this should be a general marketing principle in all business sectors.

The basis for the emerging experience economy paradigm is an assumption that experiences are very important to people in contemporary society. The arguments for this are that people have now had their physical and even intellectual needs satisfied (Jensen, 1999; Pine and Gilmore, 1999; Sundbo, 2011). Goods are no longer interesting, and therefore no longer valuable, to customers. Even knowledge services have decreasing value as they have become more or less ubiquitous and there may no longer be a belief that intellectual activities are the basis for a better life and greater happiness. People are seeking experiences, both as entertainment, emotions and as meaning creating (providing them with a higher meaning in their life, cf. Weick's (1995) concept of meaning). The arguments may be discussed, but they are nevertheless basic assumptions for the new paradigm. As Kuhn (1970) stated, the basic assumptions of a paradigm cannot be proved, but must be taken as a given condition of the paradigm's explanation of the emphasised phenomena.

Paradigmatic Criticism

Some – at least indirect – criticism of the emerging experience economy paradigm has begun to appear. It has come from, among others, service theorists, particularly the advocates of customer relationship marketing (see Vargo and Lusch, 2008; Verhoef et al., 2009; Helkkula, 2010). In essence, the criticism is that experience marketing is not new; it has always been part of service marketing; it has always been a part of the service sector, and so forth.

However, few theorists associated with the emerging experience paradigm claim that experience is something entirely new, but that this aspect has grown in importance from being a minor to a dominant feature of the customer relation situation. The development and evolution of the arguments are exactly the same as those used in service theory to argue for services being different from manufacturing. Some scholars (e.g. Cohen and Zysman, 1987) argued in the 1980s, when a paradigmatic fight took place, that service had always been there as an aspect of manufacturing. The answer from the service paradigm was that this may be true, but the service aspect has become independent, living its own life and dominant in most business. The same criticism could also be made about the information paradigm and probably even to the discussions about the new industrial paradigm in the 19th century. The same contrast can be found even further back during the 18th century around the debates about the new industrial economy (e.g. Smith, 1966; Malthus, 1997) and the physiocrates (Quesnay; see Kuczynski, 1972), who claimed that the crucial economic activity was extracting raw material from the ground and that manufacturing had always been there (in the form of craftsmanship) and was not so important. A paradigm, according to Kuhn, cannot be objectively proved. It is a choice of explanatory approach adopted by a network of scientists.

This is normal for paradigmatic revolutions (cf. Kuhn, 1970). New paradigms take over elements of the old paradigm, but often with new explanations about old objects.

WHAT WILL HAPPEN IN THE FUTURE TO THE SERVICE AND EXPERIENCE APPROACHES?

The service and the – at least emerging – experience economy paradigms coexist. They overlap, but also have their independent worlds of understanding, whether scientific or practical. They might in the future become antagonistic paradigms. The experience paradigm may attempt to emphasise its contrast to the service paradigm, just as the service paradigm has attempted to distinguish itself from the manufacturing paradigm. It may be meaningful to see services and experience as two different phenomena and paradigms if their distinctive definitions are maintained. A service is the instrumental act of solving a practical or intellectual problem. Experience is the expressive mental journey based on emotions and sensing which leaves nothing but a memory.

Another possibility is that attempts to separate the service and the experience paradigms are softened in the same way as service researchers have done in relation to the manufacturing paradigm. Goods and services are treated as a continuum that increasingly must be understood by using one explanation (e.g. within service innovation theory; Gallouj, 1998). The same scenario may apply to the service and the experience paradigm. However, manufacturing should be included in such a continuum since experience is also connected to goods (e.g. as industrial design). Even primary sector activities should be included as part of material goods production and delivery. For example, urban food markets where gastronomic-quality food is sold directly from farmers, gardeners and fishermen to consumers have undergone a renaissance. This may suggest a triangular model (Figure 11.2).

The concrete product package can be placed anywhere within the triangle. Technology can be included in all three activities. This model provides a holistic understanding of business-to-consumer business because it emphasises the situation of the receiver of the business product (the customer, citizen or user – whatever concept one prefers). The model does not take its point of departure in the traditional definition of a firm's or an industry's identity based on raw materials or competencies (as, for example, statistical

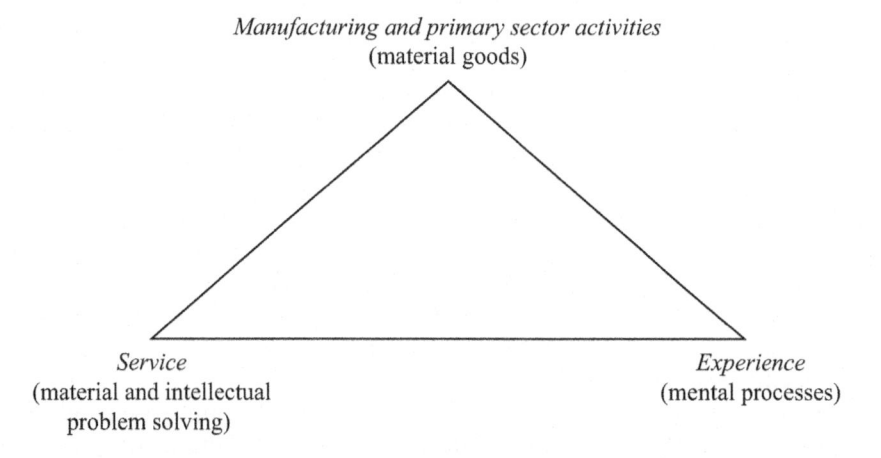

Figure 11.2   Model of goods–service–experience triangle

industry and branch categories do), but rather in what is important to the receiver, which is very often a combination of goods, services and experiences.

The consequence for practice – firms and their managers – is that their innovation activities should include all three categories and focus on the receiver's situation and what they may want. They should develop what could be called integrated innovations; that is, innovations that include goods, services and experience elements.

The model may be less adequate for business-to-business as experience is not important for business customers. That is for sure if we think of, for example, a traditional machine good or a cleaning service. However, experience is to some degree and in some situations important even to business customers. The experience element of what is traditionally considered a pure service, for example consultancy advice or catering, may be of some importance for the managers in customers' firms, as service marketing theory claims. The element of experience may also be of importance to a business customer's employees or customers.

CONCLUSION

This discussion can be concluded by answering the questions posed in the introduction. The first question was about the character of the experience element of the service relation. It has been argued that the notion of 'experience' has two meanings in the service relation: an instrumental 'erfahrung', which is the customer's assessment of whether the service delivery solved his problem, and an expressive 'erlebnis', which is a mental process in the customer's mind based on emotions and sensing. This mental process is triggered by external stimuli that could be a service delivery but could also be something else; the expressive experience is not exclusive to service deliveries.

The second question was whether experience has included a new content as an empirical reality and whether it has a new theoretical life as a new paradigm. It has been argued that experience as an expressive 'erlebnis' act or mental process has always existed. It may have achieved increased importance as a business phenomenon and in new concrete forms, but as a psychic and social empirical reality phenomenon it is fundamentally the same. However, 'experience' is getting a new theoretical life. Several authors (typically Pine and Gilmore, 1999, 2011; Jensen, 1999) argue that experience should be the basis for a new paradigm. This first aimed at explaining business development, but later it has been suggested that it could also be a paradigm for the explanation of general social, aesthetic and psychic phenomena (e.g. Schulze, 1992; Csikszentmihalyi, 2002; Andersson and Andersson, 2006; Bærenholdt and Sundbo, 2007; Jantzen and Vetner, 2007; Sundbo and Darmer, 2008; Lorentzen and Hansen, 2012; Sundbo and Sørensen, 2013).

We are currently in a situation where a paradigmatic revolution or contradiction may emerge between service approaches and experience approaches. Scientific paradigms and communities may have their own life; thus this development may be difficult to predict. However, here it has been suggested that service and experience should be seen as a continuum together with goods in a triangle model.

The implication of this suggestion is that research should emphasise goods, services and experiences as integrated elements in firms' production and delivery and in people's consumption, at least when we talk about business-to-consumer business. An important research issue is to understand what each of these elements – goods, services and experiences – means to consumers. Which needs, wishes and dreams lie behind consumers' purchase of integrated products that include goods, services and experience-provoking elements? How do consumers assess each of these elements? Another important research issue is to understand how firms develop new business models that utilise this knowledge and include all three elements in their product, marketing and innovation activities. A third research issue is to investigate what such an integrated product field means to competition on the international market, to job creation and to managerial and employee competence requirements.

This model also has consequences for managers in firms and the public sector and for industrial policy. Managers should be aware of all three elements in their development and innovation policy, in their marketing and their training policy. The current focus on business models (e.g. Osterwalder and Pigneur, 2010) makes it obvious that managers consider which of the three elements are most important in their business. This may even shift such as when experience becomes the main source of profit for a manufacturing company (as it has been for some arts industries; for example, glass factories where design and factory tourism combined with shopping centres has secured the factories' survival). Also managers of public institutions often need to analyse whether the institution mostly provides service, experience or a kind of 'production activity' (such as a museum's research). Policy makers should emphasise the combined model and support innovation within all three factors and should focus on the changed world market competition that might follow from new consumer demand patterns.

REFERENCES

Andersson, Å. and Andersson, D. (2006), *The Economics of Experiences, the Arts and Entertainment*, Cheltenham and Northampton, MA: Edward Elgar.
Arnould, E. and Price, L. (1993), River Magic: Extraordinary Experience and the Extended Service Encounter. *Journal of Consumer Research*, **20** (1), 24–45.
Aurier, P. and Siadou-Martin, B. (2007), Perceived Justice and Consumption Experience Evaluations: A Qualitative and Experiential Investigation. *International Journal of Service Industry Management*, **18** (5), 450–471.
Bærenholdt, J.O. and Sundbo, J. (eds) (2007), *Oplevelsesøkonomi. Produktion, forbrug og kultur* [Experience Economy. Production, Consumption, Culture], Copenhagen: Samfundslitteratur.
Baines, T., Lightfoot, H., Peppard, J. et al. (2009), Towards an Operations Strategy for Product-centric Servitization. *International Journal of Operations & Product Management*, **29** (5), 494–519.
Berry, L. and Carbone, L. (2007), Build Loyalty through Experience Management. *Quality Progress*, **10** (9), 26–32.
Berry, L., Wall, E. and Carbone, L. (2006), Service Clues and Customer Assessment of the Service Experience: Lessons from Marketing. *Academy of Management Perspectives*, **20** (2), 43–57.
Bolton, R.N., Lemon, K.N. and Bramlett, M.D. (2006), The Effect of Service Experiences over Time on a Supplier's Retention of Business Customers. *Management Science*, **52** (12), 1811–1823.
Boswijk, A., Thijssen, T. and Peelen, E. (2007), *The Experience Economy: A New Perspective*, Amsterdam: Pearson.
Brown, S., Gummesson, E., Edvardsson, B. and Gustavsson, B.O. (eds) (1991), *Service Quality*, Lexington, MA: Lexington Books.
Burr, V. (1995), *An Introduction to Social Constructivism*, London: Routledge.
Caru, A. and Cova, B. (2003), Revisiting Consumption Experience: A More Humble but Complete View of the Concept. *Marketing Theory*, **3** (2), 267–286.
Castells, M. (2010), *The Rise of the Network Society*, Oxford: Wiley-Blackwell.
Caves, R. (2000), *Creative Industries*, Cambridge, MA: Harvard University Press.

Cohen, S. and Zysman, J. (1987), *Manufacturing Matters*, New York: Basic Books.
Csikszentmihalyi, M. (2002), *Flow*, London: Rider.
Department for Culture, Media and Sport (2001), *Creative Industries 2001 – Mapping Document*, London: Department for Culture, Media and Sport.
Dewey, J. (1929), *Experience and Nature*, 1958 edition: New York: Dover Publications.
Dewey, J. (1934), *Art as Experience*, New York: Capricorne.
Dewey, J. (1938), *Experience and Education*, 1963 edition: New York: Collier.
Edvardsson, B., Enquist, B. and Johnston, R. (2005), Cocreating Customer Value through Hyperreality in the Prepurchase Service Experience. *Journal of Service Research*, **8** (2), 149–161.
Edvardsson, B., Thomasson, B. and Øvretveit, J. (1994), *Quality of Service*, London: McGraw-Hill.
Eiglier, P. and Langeard, E. (1988), *Servuction*, Paris: McGraw-Hill.
Erhvervs- og byggestyrelsen [Danish Enterprise and Construction Authority] (2008), *Vækst via oplevelser* [Growth via Experiences], Copenhagen: Erhvervs- og byggestyrelsen.
Foss, B. and Stone, M. (2001), *Successful Customer Relationship Marketing*, London: Kogan Page.
Gadrey, J. (1992), *L'economie des services*, Paris: La Découverte.
Gallouj, F. (1998), Innovating in Reverse: Services and the Reverse Product Cycle. *European Journal of Innovation Management*, **1** (3), 123–138.
Gentile, C., Spiller, N. and Noci, G. (2007), How to Sustain the Customer Experience: An Overview of Experience Components that Create Value with the Customer. *European Management Journal*, **25** (5), 395–410.
Gilmore, J. and Pine, J. (2007), *Authenticity: What Consumers Really Want*, Boston, MA: Harvard Business School Press.
Goldstein, S., Johnston, R., Duffy, J. and Rao, J. (2002), The Service Concept: The Missing Link in Service Design Research? *Journal of Operations Management*, **20** (2), 121–134.
Grönroos, C. (2000), *Service Management and Marketing*, Chichester: Wiley.
Grove, S.J., Fisk, R.P. and Bitner, M.J. (1992), Dramatizing the Service Experience: A Managerial Approach. In Gabbott, M. (ed.), *Contemporary Services Marketing Management*, London: Dryden Press.
Gummesson, E. (1994), Making Relationship Marketing Operational. *International Journal of Service Industry Management*, **5** (5), 5–20.
Hacking, I. (1999), *The Social Construction of What?* Cambridge, MA: Harvard University Press.
Harsløf, O. and Hannah, D. (2008), *Performance Design*, Copenhagen: Tusculum.
Hartley, J. (ed.) (2005), *Creative Industries*, Oxford: Blackwell.
Hefley, B. and Murphy, W. (eds) (2008), *Service Science, Management and Engineering*, Berlin: Springer.
Helkkula, A. (2010), Characterising the Concept of Service Experience. *Journal of Service Management*, **22** (3), 367–389.
Heskett, J., Sasser, W.E. and Hart, C. (1990), *Service Breakthroughs*, New York: The Free Press.
Holbrook, M.B. and Hirschman, E.C. (1982), The Experiential Aspects of Consumption: Consumer Fantasies, Feelings and Fun. *Journal of Consumer Research*, **9** (2), 132–142.
Holloway, B.B., Wang, S. and Parish, J.T. (2005), The Role of Cumulative Online Purchasing Experience in Service Recovery Management. *Journal of Interactive Marketing*, **19** (3), 54–67.
Hopkinson, G. and Pujari, D. (1999), A Factor Analytic Study of the Meaning in Hedonic Consumption. *European Journal of Marketing*, **33** (3–4), 273–294.
Horn, P. and Jensen, J.F. (eds) (2011), *Experience Leadership in Practice*, Roskilde: MOL Publishers.
Hulten, B., Broweus, N. and van Dijk, M. (2009), *Sensory Marketing*, Basingstoke: Palgrave.
Illeris, S. (1996), *The Service Economy*, Chichester: Wiley.
Irani, N. and Hanzaee, K. (2011), The Effect of Variety-Seeking Buying Tendency and Price Sensitivity on Utilitarian and Hedonic Value in Apparel Shopping Satisfaction. *International Journal of Marketing Studies*, **3** (3), 89–103.
Jantzen, C. (2013), Experiencing and Experiences: A Psychological Framework. In Sundbo, J. and Sørensen, F. (eds), *Handbook on the Experience Economy*, Cheltenham and Northampton, MA: Edward Elgar.
Jantzen, C. and Vetner, M. (2007), Oplevelsens psykologiske nature [The Psychological Nature of the Experience]. In Bærenholdt, J.O. and Sundbo, J. (eds), Oplevelsesøkonomi. Produktion, forbrug og kultur [Experience Economy. Production, Consumption, Culture], Copenhagen: Samfundslitteratur.
Jensen, R. (1999), *The Dream Society*, New York: McGraw Hill.
Johnston, R. (1999), Service Operations Management: Return to Roots. *International Journal of Operations & Production Management*, **19** (2), 104–124.
KK Stiftelsen (2003), *Upplevelsesindustrin 2003* [The Experience Industry 2003], Stockholm: KK Stiftelsen.
Kristensen, K. (1984), Hedonic Theory, Marketing Research, and the Analysis of Complex Goods. *International Journal of Research in Marketing*, **1** (1), 17–36.
Kuczynski, M. (1972), *Quesnay's Tableau Économique*, London: Macmillan.
Kuhn, T. (1970), *The Structure of Scientific Revolutions*, Chicago: University of Chicago Press.

Kumar, K. (1995), *From Post-Industrial to Post-Modern Society*, Oxford: Blackwell.
Kwortnik, R. and Thompson, G. (2009), Unifying Service Marketing and Operations with Service Experience Management. *Journal of Service Research*, 11 (4), 389–406.
Lorentzen, A. and Hansen, C. (eds) (2012), *The City in the Experience Economy*, London: Routledge.
Malthus, T.R. (1997), *The Unpublished Papers in the Collection of Kanto Gakuen University*, J. Pullen and T. Hughes Parry (eds), volumes I–II, Cambridge: Cambridge University Press.
Meyer, C. and Schwager, A. (2007), Understanding Customer Experience. *Harvard Business Review*, February, 117–126.
Milligan, A. and Smith, S. (2008), *See, Feel, Think, Do: The Power of Experience Marketing*, London: Cyan Books.
Mittal, V., Anderson, E., Sayrak, A. and Tadikamalla, P. (2005), Dual Emphasis and the Long-Term Financial Impact of Customer Satisfaction. *Marketing Science*, 4 (4), 544–555.
Mossberg, L. (2003), *Att skapa upplevelser* [To Create Experiences], Lund: Studentlitteratur.
Mossberg, L. (2007), A Marketing Approach to the Tourist Experience. *Scandinavian Journal of Hospitality and Tourism*, 7 (1), 59–74.
Normann, R. (1991), *Service Management*, 2nd edn, Chichester: Wiley.
O'Dell, T. and Billing, P. (eds) (2005), *Experiencescapes*, Copenhagen: Copenhagen Business School Press.
Osterwalder, A. and Pigneur, Y. (2010), *Business Model Generation*, London: Wiley.
Parsons, T. (1951), *The Social System*, New York: Free Press.
Pine, J. and Gilmore, J. (1999), *The Experience Economy*, Boston, MA: Harvard Business School Press.
Pine, J. and Gilmore, J. (2011), *The Experience Economy. Updated Edition*, Boston, MA: Harvard Business School Press.
Porat, M.U. (1977), *The Information Economy*, Washington, DC: US Department of Commerce.
Power, D. and Scott, A. (eds) (2004), *Cultural Industries and the Production of Culture*, London: Routledge.
Prahalad, C.K. and Ramaswamy, V. (2004), *The Future of Competition: Co-creating Unique Value with Customers*, Boston, MA: Harvard Business School Press.
Pullman, M.E. and Gross, M.A. (2004), Ability of Experience Design Elements to Elicit Emotions and Loyalty Behaviors. *Decision Sciences*, 35 (3), 551–578.
Roethlisberger, F.J. and Dickson, W.J. (1939), *Management and the Worker*, Boston, MA: Harvard University Press.
Schmitt, B. (1999), Experiential Marketing. *Journal of Marketing Management*, 15 (1–3), 53–67.
Schmitt, B., Rogers, D. and Vrotsos, K. (2004), *There's No Business That's Not Show Business: Marketing in an Experience Culture*, Upper Saddle River, NJ: Prentice Hall.
Schulze, G. (1992), *Die Erlebnis-Gesellschaft*, Frankfurt: Campus.
Smilanski, S. (2009), *Experiential Marketing: A Practical Guide to Interactive Brand Experiences*, London: Kogan Page.
Smith, A. (1966), *Wealth of Nations, Vol. 1–2*, London: Dent.
Snel, A. (2011), *For the Love of Experience: Changing the Experience Economy Discourse*, PhD thesis, Amsterdam: University of Amsterdam.
Sousa, R. and Voss, C. (2006), Service Quality in Multi-channel Services Employing Virtual Channels. *Journal of Service Research*, 8 (4), 356–371.
Sundbo, J. (2011), From Information Society to Experience Society. In Horn, P. and Jensen, J.F. (eds), *Experience Leadership in Practice*, Roskilde: MOL Publishers.
Sundbo, J. and Darmer, P. (eds) (2008), *Creating Experiences in the Experience Economy*, Cheltenham and Northampton, MA: Edward Elgar.
Sundbo, J. and Sørensen, F. (eds) (2013), *Handbook on the Experience Economy*, Cheltenham and Northampton, MA: Edward Elgar.
Sundbo, J., Horn, P. and Jensen, J.F. (2011), Defining Experience and Experience Leadership. In Horn, P. and Jensen, J.F. (eds), *Experience Leadership in Practice*, Roskilde: MOL Publishers.
Toffler, A. (1981), *The Third Wave*, London: Pan Books.
Toivonen, M., Tuominen, T. and Brax, S. (2007), Innovation Process Interlinked with the Process of Service Delivery – A Management Challenge in KIBS. *Economies et Sociétés*, 3 (8), 355–384.
Valberg, J. (1992), *The Puzzle of Experience*, Oxford: Clarendon Press.
Vargo, S. and Lusch, R. (eds) (2006), *The Service-Dominant Logic of Marketing*, Armonk: Sharpe.
Vargo, S. and Lusch, R. (2008), Service-Dominant Logic: Continuing the Evolution. *Journal of the Academy of Marketing Science*, 36 (1), 1–10.
Verhoef, P., Lemon, K., Parasuraman, A., Roggeveen, A., Tsiros, M. and Schlesinger, L. (2009), Customer Experience Creation: Determinants, Dynamics and Management Strategies. *Journal of Retailing*, 85 (1), 31–41.
Weick, K. (1995), *Sensemaking in Organizations*, Thousand Oaks, CA: Sage.

Wilkinson, A., Dainty, A. and Neely, A. (2009), Changing Times and Changing Timescales: The Servitization of Manufacturing. *International Journal of Operations and Product Management,* **29** (5), 425–430.

Zeithaml, V.A., Berry, L.L. and Parasuraman, A. (1996), The Behavioural Consequences of Service Quality. *Journal of Marketing,* **60** (2), 31–46.

Zomerdijk, L. and Voss, C. (2010), Service Design for Experience-Centric Services. *Journal of Service Research,* **13** (1), 67–82.

PART III

MANAGING SERVICE BUSINESSES

12. The organization of service business
Andrew Jones

INTRODUCTION: ORGANIZING SERVICE BUSINESSES IN THE TWENTY-FIRST CENTURY

This chapter begins the third major section of this *Handbook* concerned with 'managing service business'. We begin by considering the state-of-the-art knowledge about the nature of service business organization in the twenty-first century, with a particular focus on how this relates to the organization of service activity across space and between places in today's (unevenly) globalized economy. Later chapters in this section and the next will examine some of the issues that arise in more depth, such as human resource management and the nature of service business globalization, but first there is a need to provide a broader overview of how service firms in the global economy are organizing their activities. This chapter therefore focuses on organizational processes which, it is argued, have a strong (and arguably increasingly) geographical dimension to them as service industry activity integrates at a variety of scales. In that sense, the following discussion provides an important context for understanding how service firms are addressing the major challenges discussed in subsequent chapters, including those posed by economic globalization, newly emerging economies, changing information technologies and the evolving needs of clients.

The approach taken draws on a broad and interdisciplinary social science literature including work in management studies, economic geography and organizational sociology. Since the development of debates about the emergence of service-dominated economies in the latter part of the twentieth century (Bryson et al. 2004), research into the organization of service business activity across all of these disciplines has focused on the significance of service firms to advanced industrial economies (and especially knowledge-intensive business services – KIBS). During the 1970s, early studies sought to examine the numbers of service firms within regional and national economies, and assess the value of their inputs into economies overall (Beyers et al. 1985). In the 1980s, regional scientists and economic geographers began to provide insight into how service firms organized geographically with respect to serving markets (Beyers et al. 1986; Daniels 1993). Over the past 30 years a considerable body of work within economic geography has thus engaged with the locational strategies of (knowledge intensive) business services (Bryson et al. 2004). Much of the work stemmed from an interest in the vertical disintegration of manufacturing corporations and the externalization of the provision of services such as advertising and IT as part of a shift towards what have been variously classified as 'post-Fordist' production methods (Wood 1991; Faulconbridge and Jones 2011). More recently, this work has also explored the spatial organizational strategies of a range of producer service firms that provide knowledge-intensive advisory services exclusively to business. Studies have focused on sectors including accountancy (Beaverstock 1996; Daniels et al. 1988), advertising (Clarke and Bradford 1989; Grabher

2001; Faulconbridge 2006), executive search (Faulconbridge et al. 2008; Hall et al. 2009), law (Beaverstock et al. 1999; Jones 2005; Faulconbridge 2007a) and management consultancy (Jones 2003; Strom and Mattsson 2005, 2006; Glückler 2006).

However, since the 1990s it has also become apparent that the organization of service business activity is changing in complex ways, both within and beyond the level of the service firm itself (Jones 2007). Service firms have undergone significant processes of internal restructuring, and service sectors have witnessed the emergence of increasingly internationalized firms that have experimented with a number of organizational models (Jones 2003, 2007). At the same time, and especially in the twenty-first century, new information technologies and falling transport costs have led to a more radical reconfiguration of service business activity. Service sector employees in many industries have become more mobile, and internationalization has led to the need for increased worker mobility in some areas, as well as the deployment of new technologies (Beaverstock 2007; Beaverstock et al. 2010). Information technology has also allowed new organizational models with outsourcing and new, distributed, subsidiary network forms of business organization in some service sectors. Overall, therefore, in the twenty-first century many of the existing understandings of service business organization that emerged in the last decades of the previous century have become considerably more complicated and undergone some degree of transformation.

This represents the entry-point for this chapter, and the following discussion will assess the different dimensions to this contemporary complexity of service business organization in a series of stages. It draws on work that has primarily been undertaken at the level of the firm, although research into particular service sector industries also plays a part. Much of the discussion will revolve around the new geographies of service firm organization, and new processes that are used by service firms to organize their activities. This reflects the proposition that many of the transformations of service business organization reflect the underlying importance of place and locality in the production and delivery of services, and the role of globalization and cross-border business in the growth of the largest service firms.

The rest of the chapter is organized into a number of sections that address different dimensions of service firm organization in general. Examples are drawn from a range of service industries (mainly advanced business services) to develop a more detailed overview of contemporary service business organizations. This begins in the next section with a discussion of work that has examined the new and developing geographies of service business organization, paying attention to the role of places and localities in shaping how service firms operate. A key issue is the role played by cities in organizing service business, manifest in the agglomeration and localization processes at play in urban economies. The third section then examines debates about complex divisions between concentrations of different service firms and industries in different cities and regions. It also examines how an urban locational hierarchy has emerged in the case of many KIBS industries, and how that exists at both the national and international scale. It is argued that understanding the nature of urban centres and their positionality in hierarchical relations is key to understanding the contemporary geographical organization of service business. The fourth section then turns to the level of the firm, examining both the organizational form and corporate geographies of service firms. The ways in which contemporary service firms are grappling with the complexity of cross-border business activity are considered, and a range of firm-level organizational and operational responses in different service industries are shown to exist. Of particular interest is how work practices around knowledge management, information technology and employee mobility are producing complex transformations in the way service business is organized. Finally, the chapter ends by identifying a number of important future challenges that service firms face in organizing their activities in the coming decades.

EVOLVING GEOGRAPHIES OF SERVICE FIRM LOCATION

The breadth of the concept of a 'service industry' is matched by the great variety of ways in which services businesses are organized. If the broadest of service industry definitions is taken, sectors including retail, hospitality, transport and other consumer services can be included. This presents a concomitant diversity of both organizational forms and business organizational models. However, in this chapter, and in the context of this *Handbook*, the focus will be on knowledge-intensive business services, which in large measure correspond to 'producer services' provided to other firms (Daniels 1993). The organization of service business in this respect has long been shaped by issues of agglomeration, and the key role that cities as locations play for service business firms. Over recent decades, there has thus been a restructuring and reconfiguration of the geographical organization of service firms at a range of scales from the local to the global, and the significance of urban networks has evolved in new ways. We now consider what can be argued to be three of the major characteristics of service business organization in locational terms in the twenty-first century.

First is the contemporary nature of urban agglomeration and localization factors in service business activity in the global economy (O'Farrell and Hitchens 1990; Sassen 2006). At their simplest, theories of agglomeration identify the savings made in relation to the cost of key infrastructures (public transport, information communication technology) when firms co-locate. However, in relation to knowledge-intensive services, co-location benefits refer, in particular, to the need to be close to major clients. This factor distinguishes business services from many of the manufacturing corporations studied and used to develop theory in the 1960s and 1970s (see Bryson et al. 2004). Historically, KIBS have been associated with urban centres and especially within central business districts. Such forces of agglomeration and localization were well understood historically, but what was less anticipated was how new transport and information technologies (and wider 'economic globalization) have, counter intuitively, largely exacerbated this agglomeration of service firms rather than leading to dispersal. Advanced industrial economies have thus seen an increasing concentration of service business in urban centres. In the UK, for example, Wood (2006) reports that 83 per cent of UK employment in knowledge-intensive producer services is located in four cities: London (38 per cent), Manchester (16 per cent), Birmingham (13 per cent) and Leeds (16 per cent), and similar patterns are evident in other European economies (see, for example, Hermelin 2009).

The explanation lies in the strength of the reliance by much service business activity on reputation, client referral networks and geographical proximity with clients (Bryson et al. 2004; Jones 2007; Faulconbridge 2008). One of the defining features of KIBS firms is that their services take the form of advice designed to meet the very specific needs of these clients. For example, referring to research on management consultancy and law firms, Empson (2001) and Empson and Chapman (2006) contend that KIBS advice is hard to assess not only because client firms lack the in-house expertise to complete the task devolved to the chosen business service, but also because they also lack the expertise to evaluate the work of the KIBS firm employed. Moreover, it is suggested that the services provided by one KIBS firm cannot be compared with services provided by other firms because of their contingent, unique and one-off nature. As a result, meetings are the most important moments of work in business service firms (Jones 2007; Jones and Search 2009). At the beginning of KIBS projects, meetings with clients enable their needs to be understood and a trusting relationship to be developed between the client and the service firm providing advice. Occasional meetings throughout the life of a project are then used to deliver advice to clients, to maintain clear communication and to reinforce the trusting relationship between the parties. As such, face-to-face meetings overcome some of the difficulties associated with the intangibility of the work undertaken by KIBS. Clear communication and trust help mitigate a client's inability to evaluate the advice provided by professional service firms (Daniels 1993; Keeble and Nachum 2002). Consequently, locating in a city provides a major competitive advantage for producer service firms because a large number of potential clients are co-present and easily accessible when meetings are needed. Combined with the valuable travel time saved when clients are located in the same city, the ability to arrange meetings at short notice is said to be behind the way cities have become *the* sites of producer service work (Daniels 1993).

The strength of agglomeration factors in service business organization can be illustrated further using an industry example. Research by Glückler (2007) into the management consultancy industry in Germany suggests that firms benefit from reputation enhancement when located in major business cities and are more successful at attracting new clients because of the importance of reputation in their decision-making about which consultancy firm to use (see also Glückler and Armbruster 2003). Glückler (2007) also shows that presence in key business cities allows firms to benefit from business acquired by intra-city referrals – when one client in the city recommends the firm to another potential client. Work in other advanced economies reinforces this (cf. Wood 2002). In the UK, for example, Keeble and Nachum (2002) also studied locational strategies for management consultancy, finding that firms in London benefit from easy access to a large pool of clients who can easily be served thanks to the benefits of co-location – that is, the ease of meeting clients face to face. Furthermore, this study also showed that London-based firms find it easier to innovate and develop novel lines of consultancy advice because of the benefits of informal interactions with fellow consultants based in the city. This gives London firms an advantage over firms based in other smaller UK cities where there are fewer management consultancy firms (cf. Wood 2002; see also Faulconbridge and Jones 2011).

The latter issue brings us to the second and related characteristic of service business location: the role of cities as the localizing sites where service firms cluster because of industry-specific knowledge, learning and innovation. Economic geographical theorization has suggested that for KIBS firms, two city-based knowledge- and innovation-related advantages are especially important. First, the concept of 'buzz' captures the way in which social interactions lead to access to market-related knowledge. Morgan (2004) and Storper and Venables (2004) outline how such 'buzz' relies on face-to-face encounters, trusting and reciprocal relationships and the development of city-specific industrial languages and codes which together lead to knowledge spillovers. In particular it has been shown that interactions between KIBS workers in restaurants and bars (Thrift 1994), at professional associations (Faulconbridge 2007b) or at formal training and professional development events (Hall and Appleyard 2009) lead to insights being gained into new business opportunities as a result of idle chat and gossip. Grabher (2002) suggests that this buzz forms a constant background 'noise' about new business opportunities. Such noise is also said to help clients to assess the work of producer service firms. When clients of multiple service firms interact in such social, professional or educational spaces they share their experiences of working with different firms. The gossip and rumour this generates lead to the construction of a positive reputation for those firms judged by clients to provide the best services and damage the reputation of poor performers.

The third aspect to service business location we can identify concerns labour. Again, both agglomeration and localization factors are important. The presence of multiple producer service firms in a city is also associated with the development of a pool of expert labour which acts as another factor reinforcing service firm location strategies (Daniels and Bryson 2005). For example, both Grabher (2001) and Ekinsmyth (2002) study London and show that when rival business service firms in industries such as advertising or digital marketing are located in the same city, intra-city labour churn provides a competitive advantage as talented individuals can be more easily encouraged to move between firms (see Saxenian (1994) on the analogous process in engineering communities). In addition Grabher (2001) and Ekinsmyth (2002) suggest that the increasing reliance on project teams made up of individuals drawn from within but also outside of organizations' boundaries has rendered cities important strategic sites of work for some business services. In KIBS firms where each project requires a very different type of expertise, project teams often include freelancers brought in because of their particular specialism. Cities act in this sense as vortices, sucking in freelance workers who can be drawn on as and when needed to provide expertise that is crucial for a project's success, hence further making metropolitan locational strategies advantageous. At one level being in a major business city means, therefore, that the process of searching for an individual with the required expertise is simplified.

It is therefore argued that cities provide an 'ecology' of labour for many kinds of service firms (Vinodrai 2006). This ecology comprises individual workers, labour market intermediaries and multiple co-located firms from the same industry (ibid.). The ecology generates career paths for a cohort of freelance and temporary workers that are built on multiple short-term contracts, something advantageous for firms seeking to 'hire in' knowledgeable individuals as and when needed. In addition, and in a different but related way, city-based pools of freelance labour are valuable for business service firms because they can develop repeat relationships with a subset of individuals leading to the development of an institutional structure for project work (Sydow and Staber 2002). This institutional structure develops over time as a result of repeat project working and leads to norms (trust, reciprocity, etc.) being shared both by firms hiring freelancers and by freelancers themselves, something which helps smooth the project management process when individuals are 'hired in'.

Overall, therefore, cities and certain key areas within cities are crucial in service business organization. However, it is worth adding that some recent evidence from European economies suggests that the manifestation of locations and agglomeration forces within some advanced economies is not as simple as geographical concentration of service business in urban cores. Improved information and communications technology (ICT) and transportation systems suggest that business service firms in a range of industries (e.g. architects, design consultancy, digital media) can locate outside central business districts (CBDs) in the wider sphere of influence of key city regions. Where transportation connections enable easy face-to-face contact when necessary, small business service firms in Western Europe can locate in a wider city region, as shown to be the case for UK cities (London, Birmingham, Manchester) and to some extent in other European economies (France, Germany, Denmark and Sweden) (Glückler 2006; Hermelin and Rusten 2007; Hermelin 2009). Yet these decentralization trends in firm location reflect a more diffuse agglomeration in city regions, rather than a reversal or 'end of geography' (cf. O'Brien 1992) dispersal of service business activity. To a large extent, they reflect both changes in the possibilities presented to service firms of all sizes in terms of location decision by ICT, but also the greater complexity of underlying trends in the geographies of service business organization that exist beneath the scale of aggregate firm location data and theories. In order to better understand these complexities we turn in the next section to the nature of service business organization within firms themselves.

ORGANIZATIONAL DIVERSIFICATION: GLOBALIZATION, SUBSIDIARIES AND NEW MODELS OF CORPORATE FORM

Having considered the geographical dimensions to service business organization in terms of the role of cities and regions, we now need to examine the transformations and contemporary nature of service business organization at the level of the firms themselves. The fact that many service firms cluster and co-locate in cities represents only one dimension of the organization of service business, and it is important to understand how service firms in the twenty-first century are structured at the corporate level. There are a number of important trends that are common to many service industries but perhaps the most important overall trend is the context of widening and deepening economic globalization that has been a key process in recent decades. A wide body of interdisciplinary research has indicated that, in common with other industries, business service industries have become increasingly globalized (Jones 2003). In general terms, since the late 1980s firms in investment banking, management consultancy, insurance, legal services, advertising and accountancy have begun to move out of national to transnational markets and operations (Enderwick 1989; Aharoni 1993; Daniels 1993; Lewis 1999). This process has been a progressive and uneven one, varying between both different business service industries and national economies (Bryson et al. 2004; Jones 2007; Pain 2008). We can use this as a lens to understand the diversification that has taken place in organizational form amongst service firms (although of course many service businesses also continue to serve regional and national markets as well). In that respect, the major drivers behind this shift which is impacting on the organization of service business are at least threefold.

First, as transnational corporations (TNCs) have developed in all industry sectors, business activity has escaped national economies and moved into new markets at the global scale (Dicken 2011). TNCs represent the major clients (i.e. the market) for business services that have followed their market and transnationalized their activity (Bryson et al. 2004). In this respect, business service firms have been required to respond to the needs of their clients for global-scale services (Majkgard and Sharma 1998; Nachum 1999; Strom and Mattsson 2005, 2006). Second, within many business service sectors such as investment banking, the globalization of markets has also been accompanied by the development of larger transnational service firms and a concomitant greater degree of corporate globality (Jones 2003; Faulconbridge and Muzio 2009). Organic growth and acquisition of overseas firms has produced a growing number of business service firms that are themselves transnational. These service TNCs are at the forefront of the production, distribution and consumption of services in the global economy (Bryson et al. 2004). Clearly this is entwined in a complex way with the globalization of markets for these services (Roberts 1998; Warf 2001; Miozzo and Miles 2002). Third, many business service firms are embedded in economic globalization as key actors who have developed informational products whose purpose is to facilitate the globalization of markets and firms in other sectors (Roberts 2006). This driver varies between different industries but investment banking, law and management/strategy consultancy are certainly heavily involved in providing advisory services to client firms on how to transnationalize their operations and do business in markets at the global scale. Thus an important component of much professional business service advice in a range of sub-sectors is concerned with helping other firms develop, for example, effective *corporate globality* (in spheres such as operations, ICT, human resources and information management) as they transnationalize. This is essential for them to compete effectively at the global scale (cf. Jones 2005).

As the management literature has explored in some depth, the internationalization of some service businesses can be traced back many decades, and even in a nascent form prior to the Second World War (cf. Kipping 1999). However, the development of this service business globalization in recent decades has become a much more extensive and complex process as the global population of business service firms has increased, and they have established operations in an increasing number of countries (Daniels 1993; Warf 2001; Jones 2003; Bryson et al. 2004). The rest of this section thus considers the geographical factors behind the globalization of service business activity, the emergence of globalized corporate organizational structures and the way in which the key significance of social networks to service business activity has shaped transnationalization.

Geographical Factors in the Globalization of Service Business Activity

Geographers have argued that existing theories of firm globalization developed from manufacturing industries are inadequate for understanding business service globalization over the past 50 years or so (Boddewyn et al. 1986; Daniels 1993; Bagchi-Sen and Sen 1997; Jones 2003; Faulconbridge et al. 2008). Longer-standing theories of firm globalization based around the three dominant elements of firm-specific and ownership advantages, location-specific advantages and internalization advantages (known as the OLI model) (cf. Dunning and Norman 1987) are also problematic when applied to business service firms (Bagchi-Sen and Sen 1997). The key issue that renders such theories inappropriate concerns the intangibility, perishability, inseparability and heterogeneity of KIBS service industries. For many service businesses, product specialization and diversification (i.e. economies of scope) are more important as growth strategies for firms than economies of scale because of the immobility of many services (cf. Enderwick 1989). Thus the literature suggests that the bespoke knowledge-intensive nature of business service 'products' (Boddewyn et al. 1986; Wood 2002; Bryson et al. 2004) – in combination with the heavy reliance on face-to-face interaction in the work process (Jones 2003, 2007) – means that conventional models of firm globalization provide only very limited insight into the nature of (business) service firm globalization. Rather, we can identify at least two distinctive dimensions to the way in which service businesses have globalized.

First, a number of *impediments* to the globalization of business services have been identified that contrast with those identified in work on the internationalization of firms in manufacturing sectors (cf. Dicken 2011). At one level, economic geographers have shown how the regional context in which KIBS service firms operate shapes the capacities and opportunities they have for globalization. For example, O' Farrell et al. (1998) show how the ability of business service firms within different regional economies to globalize is strongly influenced by their embeddedness in a number of regionally constituted factors, including the contact networks firms can access *within* a given region that then lead to contacts amongst potential clients in overseas markets. This kind of evidence suggests that KIBS firms globalize their businesses via social contact networks that develop in regions or places where they already operate. It is not a question of starting up a new international operation from 'cold' but pursuing potential new business overseas by exploiting existing social contact networks in a current operational geography (Jones 2003; Glückler 2006; Faulconbridge 2008).

At another level, Bagchi-Sen and Sen (1997) suggest significant hurdles to globalization exist for service firms in relation to the nature of their products and regulatory issues. They argue that, in terms of products, 'the heterogeneity of the services rendered' by KIBS firms represents a barrier because products that are suited to one national economy are not necessarily suited in another (ibid.: 1153) and 'the international diversity in the rules for granting licenses to practice' (ibid.: 1153) for many service businesses means that firms often face a different set of rules with which they have to comply in each new national context. Similarly, in many countries the fact that service firms are partnerships (e.g. legal services, consultancy) is often a disincentive to invest as there is a potential for 'unlimited liability of the partners'. Other regulatory barriers also include 'restrictions induced by government or professional associations on the use of a firm's name', 'restrictions on trans-border data flow' (which inhibit much of the normal work process of consultancy), 'government regulations limiting the tradability or access to markets' and of course 'differences in professional standards' in different countries (ibid.: 1153). Bagchi-Sen and Sen (1997) thus suggest that theories of business service globalization should focus on what they categorize as 'the conditioning, motivating and controlling factors' which influence KIBS firms in overcoming these impediments (ibid.: 1153). The implication is that theories of service firm globalization need to examine in depth what has enabled some firms to successfully enter overseas markets and why they have adopted such a strategy when others have either failed or not sought to do so. Kipping (1999), for example, provides a contribution in this respect in examining how reputation and the development of host country client networks have been essential to the success of US management consultancy firms' operations in Western Europe.

Second, the globalization of many KIBS industries including advertising, marketing and accountancy and management consultancy firms has been driven by client-following activities that led firms to globalize their activities to meet the demands of clients (Bagchi-Sen and Sen 1997; Jones 2003, 2005; Faulconbridge et al. 2008; Faulconbridge and Jones 2011). This has combined with growing pressure from domestic competition, and domestic market saturation in advanced industrial economies has also led KIBS firms to seek to globalize (Bagchi-Sen and Sen 1997). However, this may be more relevant to some business service industries than others. For example, such arguments are more applicable to accountancy, which offers more standardized products (cf. Beaverstock 1996) than the bespoke informational services provided by firms in management or specialist strategy consultancies that have seen markets in developed economies such as the UK grow over recent years (Jones 2003).

The Emergence of Globalized Business Service Firms

At the level of the firm, service business is being (often radically) reorganized in order to compete in international markets. There are many industry-specific dimensions to the emergence of globalized service firms which are beyond the scope of the discussion here. However, the way in which business service firms have restructured their organizational form as they seek to globalize has a number of commonalities across industries. At the firm level, KIBS firms have generally internationalized by opening operations in new foreign markets or developing strategic collaboration, rather than via the acquisition of foreign firms (O'Farrell and Wood 1998, 1999; Wood 2002). This has produced an ongoing need to develop more effective international corporate forms. It is certainly evident that KIBS firms in several sectors (investment banking, consultancy, legal services, advertising) have experimented with developing a *transnational* or *global* model, rather than the multinational model identified amongst manufacturing and extractive firms since the 1970s (cf. Cohen et al. 1979; Dunning 1993). A key argument developed here is that differences in organizational architecture notwithstanding, KIBS firms have sought in various ways to shift towards organizational forms that represent an extension and a deepening of organizational restructuring towards 'corporate globality' (Jones 2005). During the 1990s a number of the larger firms underwent dramatic internal restructuring (Jones 2003), moving away from a geographical to a product-based divisional structure. In essence, this represents a move away from US, UK or Asian divisions of firms towards global-scale functional divisions based around communities of practice and knowledge/expertise (e.g. a global-scale retail consultancy or utilities practice community).

Thus the evolving organizational form of transnational KIBS firms is markedly different from that typical for many large manufacturing firms (Johanson and Vahlne 1990; Andersen 1993). These discussions also relate to wider debates in economic geography about the role of mobility and technology and the way in which globalization is transforming the nature of work. This literature demonstrates that, despite the globalization of management consultancy firms over the last two decades, face-to-face interaction and co-presence remain central and crucial elements of working practices and ultimately of firm success. New information technologies have played an important role in facilitating and enabling the globalization of management consultancy firms, but the nature of their work process has meant that virtual interactions, for example via videoconferencing, have not become a substitute for most forms of face-to-face interaction (Jones 2003, 2005; see also Sturdy et al. 2006). The key impact of this has been to significantly increase the levels of business travel by employees, with substantial impacts on the working practices of firms as they globalize (see Faulconbridge et al. 2009). It is to these issues that the final part of this chapter on the organization of service business will now turn. However, before doing so we need to consider a further element to the complexity of service business organizations that links service industry evolution in localities and regions to wider globalization. In the final part of this section we consider the role of transnational partnerships and subsidiaries which involve not just larger transnational service firms but also service small and medium-sized enterprises (SMEs) in the global economy.

Local Partnerships, 'Born-globals' and New Service Firm Organizational Forms

Until recent decades, the economic geography, regional studies and management literature had established an empirically founded understanding of service industry organization within regional economies that identified large firms as the primary basis for business internationalization. As discussed earlier in this chapter, service business internationalization theories were concerned with the capacity of large firms – usually located in globalizing cities – to internationalize their operations. We have also considered the development of (large firm) corporate globality (i.e. a large firm opening up offices and business in multiple countries). However, the transnationalization of service business in a range of KIBS industries (as with manufacturing or other industries) increasingly cannot be restricted to this large firm transnational model. We need to consider three important complexities to the development of internationalized service businesses that are also relevant to SMEs.

The first aspect to transnational service business organization is the role played by transnational partnerships between firms and the acquisition of subsidiaries beyond their home country market. Within KIBS industries this is much less developed than, for example, in manufacturing or electronics (Dicken 2011). However, some recent research suggests that KIBS service firms are increasingly making use of overseas partnerships to expand their businesses into international markets (Jones and Strom 2012). This takes a variety of forms but can be a good mechanism to enable KIBS industries to overcome regulatory, institutional or cultural barriers to entry in foreign markets (ibid.). Both accountancy and management consultancy firms operating in emerging markets in Asia or Latin America have, for example, sought entry by forming strategic alliances and partnerships in this way. Such an approach has the advantage of reducing the risk of starting business by utilizing local expertise to navigate through what are often complex and unfamiliar commercial and regulatory environments. Given the embodied nature and crucial significance of trust and face-to-face communication, having local employees working for a partner firm or subsidiary provides an important basis for gaining entry into markets (Jones 2007).

Second, another relatively novel dimension to contemporary service business globalization is the emergence of 'born global' SMEs in a range of KIBS industries. These firms are small start-up firms that almost immediately do business in international rather than regional or national markets (cf. Madsen and Servais 1997; Bell et al. 2001), often being set up with the specific intent to serve global markets from their inception. These kinds of SME have been typically identified in technological, pharmaceutical and software industries (Jones 2009). However, it is clear from a limited but growing body of research that born-global firms are increasingly evident in a range of professional service industries. Within advanced industrial economies, born-global KIBS SMEs are increasingly evident in a number of industries, although the specific nature of these firms and the markets in which they operate vary considerably. A good example is the globalization of architectural firms, which have a longer history of international market activity than most (cf. McNeill 2008). In architecture, a number of the larger European and North American firms have established themselves in international markets following commissions for iconic buildings planet-wide that display a distinctive firm brand (e.g. The Foster Partnership). Within this industry, however, the specific project-based nature of architecture combined with the confines of small specialist markets for specific kinds of buildings and structures has also enabled small born-global architectural practices to compete internationally. Many of these have emerged as spin-offs from the larger global architecture firms as key employees use their experience and international client contact base to set up on their own (McNeill 2008).

Third and finally, it is worth pointing to the complex interplay of business internationalization, firm-level organizational form, and information and communications technological and locational decisions in service business organization. Beyond the partnership, subsidiary and born-global manifestations of service business internationalization discussed above, it is clear, if currently under-researched in the social scientific literature, that the interaction of the four above factors is producing a complex spatial and organizational reconfiguration of service business in advanced economies (and also to some extent in emerging economies). As discussed earlier in the chapter, conventional understandings of city-regions in advanced economies have been questioned by recent empirical research identifying new spatial configurations of KIBS service firm populations that have spread from urban CBDs facilitated by transformations and improvements in ICT and transportation links. However, at the international scale, there is limited evidence of similar but equally complex new configurations of service business organization within some KIBS industries, networks of small service firms operating across a range of European countries in KIBS sectors which make use of a flexible array of semi freelance contract employees who are also highly mobile (Bryson and Rusten 2005). The organization of certain KIBS niche industry segments is again driven by client-led needs and competitiveness, but also indicates the potential for further highly complex organizational and spatial forms to develop in the future. This issue also leads us to an important final aspect of how contemporary service business organization continues to be transformed: the nature of work practices themselves.

SERVICE BUSINESS WORK IN THE GLOBAL ECONOMY

The transformation of service business organization at the urban, regional and firm levels discussed so far has also been bound up in transformation of the organization of service work at the level of individual employees. Again, there is enormous diversity dependent on the industry and job description of individuals within service firms, but it is possible to identify a number of more general trends that are common across many service businesses. These are caught up of course in the processes of industry globalization, product innovation and technological change in service business that have taken place in recent decades and which are discussed elsewhere in this book. Three trends are especially pertinent to the overview of contemporary service business organization provided by this chapter.

The first is the impact that the centrality of face-to-face interaction – already referred to above – has on the organization of business service working practices. Whilst the aggregate significance of interpersonal interaction in driving agglomeration in firm location has been discussed already, there are other important transformations occurring in the context of the ongoing globalization of business service markets and firms. At the level of the individual employee, much KIBS work involves the cultivation and maintenance of social contact networks (Glückler and Hammer 2011). This applies at both the inter- and intra-firm level and across all scales from the urban regional to global. In terms of service business working practices, professional employees are heavily caught up in interpersonal relationships that are important not just in delivering the bespoke business products to clients through face-to-face interaction, but also in the wider practices surrounding client acquisition and retention (Jones 2005; Faulconbridge 2007a). Gaining and retaining business for many KIBS is therefore caught up in a deeply social set of factors (although that is not to say that others such as price are unimportant), and the internationalization of the market for their services and many firms' operations has and continues to produce significant change to how KIBS work is organized.

The evidence for this in the literature is substantial. For example, in management consultancy both Jones (2003) and Glückler (2006) suggest that with regard to international market entry, firms rely heavily on social networks in order to globalize their operations; developing social contact networks amongst 'local elites' within new foreign markets is essential to market entry. The geography of globalizing KIBS firms is to a considerable extent shaped by the capacity of key senior managers to gain business through personal contact with key employees in client firms in different countries or regions. This is an intrinsically embodied and largely face-to-face aspect of KIBS working practices in the global economy. Jones (2003) suggests that these networks of contacts that senior employees (as 'client carriers') maintain, develop and perpetuate in the global context represent the major deciding element in whether these business service firms win or lose contracts in markets outside their home country. Meanwhile Hall et al. (2009) similarly suggest, using a study of the executive search industry, that 'iconic individuals' can be important in facilitating the globalization of an industry because of the way, through their contacts, they legitimate the activity of new arrivals in a country. Furthermore, this applies not only for large service firms but also for SMEs (Jones 2011). There is thus growing evidence that KIBS firms of all sizes in many industries are increasingly focused on developing local contact networks as they seek to cultivate clients in foreign markets.

Second, and following on, are the internal strategies that business service firms have developed in recent decades as service businesses have internationalized their operations. As KIBS firms increasingly employ workers in disparate locations around the globe, there is a significant need for internal *global socialization* which involves strategies designed to facilitate co-presence between employees of the firm located in different offices around the globe. Strategies include global training programmes, conferences and practice community meetings (see Jones (2003) and Faulconbridge (2008) for a similar example of such strategies in law firms). The maintenance of internal contact networks in business service firms is a functional necessity for these businesses in order to cross-refer business, develop global project teams, maintain organizational coherence in terms of service standards and ensure service product consistency.

Third, and caught up in all the above, are transformations in employee mobility and the use of ICT. The internationalization of KIBS markets and firms has necessitated a wider shift to 'global working' practices beyond simple conceptions of increased international business travel (Wood 2002; Jones 2003, 2005). This shift has several elements, including a substantial increase in the numbers of foreign 'expatriate' workers moving between different offices in developing global networks (Beaverstock 2006; Hall et al. 2009); the development of more extensive secondment schemes where more employees increasingly spend periods of months or years abroad; and a widening of the recruitment base of employees, where the employee profile of the company reflects rising percentages of employees from a larger number of different countries. Thus there has been a blurring and dilution of the 'home economy' component of the workforce within management consultancy firms, along with a growing prevalence of business travel and numbers of employees living in countries other than their home state (Harvey et al. 1999; Jones 2003). More recent work has also identified the significance of 'hybrid managers' and key business service professionals in acting as intercultural communicators through overseas secondments and mobility practices (Schlunze et al. 2012). The kinds of mobile working practices within KIBS firms in the global economy are thus becoming very complex with mixtures of short- and longer-term travel combined. Certain key employees within many KIBS industries have therefore become highly mobile, and for many business service professions overseas secondment and travel are increasingly the expectation and the norm in working life (Jones 2008). This of course presents some significant challenges around work–life balance for firms competing to hire and retain the kinds of highly skilled individuals who fill such roles.

CONCLUSION: RECONFIGURATION, DIVERSIFICATION AND RETRENCHMENT

This chapter has provided an overview of how service businesses are organizing their activities in an increasingly (albeit unevenly) integrated global economy in the early twenty-first century. As discussed elsewhere in this book, the concept of the service firm or industry is enormously diverse. Thus the focus here has been on what are arguably the most significant service businesses to global economic development – knowledge-intensive business services. The approach has been to move down through a range of geographical scales from the urban/regional to the level of individual KIBS employees in order to explore the organizational processes that are shaping service business activity. Many of these are inherently geographical during an era in which the globalization of service business activity is one of the most visible forces affecting contemporary service industries.

What is hopefully evident from this analysis is that, whilst significant differences exist between service firms in different KIBS industries, a range of shared common trends are evident. Foremost is that KIBS firms have undergone, and are continuing to undergo, far-reaching transformations that reflect dramatic changes in the environment in which these service industries exist compared with 30 or 40 years ago. KIBS industries have grown and diversified whilst simultaneously becoming clustered in specific places that are increasingly interlinked across a global city network. At all scales from the local to the global, the physical location of these activities has become more complex, with new ICT reconfiguring the relationship in some cases between office locations, back office or support functions, employee locations and employee mobility. Central to this has been a dilution and reconfiguration of regional service economies identified in the later decades of the last century, as service markets and firms have internationalized. However, the transformative processes discussed in this chapter are far from concluded, not least because the economic downturn that has gripped the global economy since 2007 has produced a new wave of service business reorganization. In some sectors there has been retrenchment and withdrawal from selected areas of activity, and others such as financial services are reconfiguring in the wake of new regulation or legislative intervention at the national and supranational level.

Yet in terms of the underlying trends that have been discussed, I would end by arguing that whilst, in the wake of the recent economic downturn, new patterns of service business organization are likely to develop in the coming decade, there is no evidence that internationalization is losing momentum or even reversing. Indeed in the coming decade, the challenge for understanding service business organization will be increasingly focused upon the interaction between KIBS firms and industries that have evolved in (Western) advanced industrial economies and which are encountering and seeking to enter new emerging economies in Asia, Latin America and Eastern Europe that are themselves beginning to spawn home-grown service firms. Whilst not diminishing the significance of the factors leading to service firm agglomeration and the need for interpersonal embodied work in KIBS industries, the development of emerging markets for KIBS services is likely to underpin further significant change in the ways in which existing established firms in advanced economies organize their operations.

ACKNOWLEDGEMENT

Several sections in this chapter draw on ideas developed previously with James Faulconbridge in the context of professional service firms (see Faulconbridge and Jones 2011).

REFERENCES

Aharoni, Y. (ed.) (1993) *Coalitions and Competition: The Globalization of Professional Business Services* (London: Routledge).

Andersen, O. (1993) On the internationalization process of firms: a critical analysis. *Journal of International Business Studies*, 2: 209–231.

Bagchi-Sen, S. and Sen, J. (1997) The current state of knowledge in international business in producer services. *Environment & Planning A*, 29: 1153–1174.

Beaverstock, J. (1996) Sub-contracting the accountant! Professional labour markets, migration and organizational networks in the global accounting industry. *Environment and Planning A*, 20(3): 303–326.

Beaverstock, J.V. (2004) 'Managing across borders': knowledge management and expatriation in professional legal service firms. *Journal of Economic Geography*, 4(2): 157–179.
Beaverstock, J.V. (2006) Globalization and its outcomes. *Economic Geography*, 82: 455–456.
Beaverstock, J. (2007) Transnational work: global professional labour markets in professional service accounting firms. In Bryson, J. and Daniels, P. (eds), *The Handbook of Service Industries* (Cheltenham and Northampton, MA: Edward Elgar), 409–431.
Beaverstock, J., Derudder, B., Faulconbridge, J. and Whitlox, F. (eds) (2010) *International Business Travel in the Global Economy* (Farnham: Ashgate).
Beaverstock, J.V., Smith, R. and Taylor, P.J. (1999) The long arm of the law: London's law firms in a globalising world economy. *Environment & Planning A*, 13: 1857–1876.
Bell, J., McNaughton, R. and Young, S. (2001) Born-again global firms: an extension to the 'born-global' phenomenon. *Journal of International Management*, 7: 173–189.
Beyers, W. (1988) International perspectives on producer services: the view from the United States. In Marshall, J. (ed.), *Services and Uneven Development* (Oxford: Oxford University Press), 259–262.
Beyers, W., Alvine, M. and Johnsen, E. (1985) The service sector: a growing force in the regional export base. *Economic Development Commentary*, 9(3): 3–7.
Beyers, W., Tofflemire, J.M., Stranahan, H. and Johnsen, E. (1986) *The Service Economy: Understanding Growth of Producer Services in the Central Puget Sound Region* (Seattle: Central Puget Sound Economic Development District).
Boddewyn, J., Halbrich, M. and Perry, A. (1986) Service multinationals: conceptualisation, measurement and theory. *Journal of International Business Studies*, 17(3): 4–57.
Bryson, J.R. and Rusten, G. (2005) Spatial divisions of expertise: knowledge intensive business service firms and regional development in Norway. *Service Industries Journal*, 25(8): 959–977.
Bryson, J., Daniels, P.W. and Warf, B. (2004) *Service Worlds* (London: Routledge).
Clare, K. (2007) Cool, creative and complex: exploring social networks and gender in project-based creative industries (advertising) in London. *Annual Meeting of the Association of American Geographers*, San Francisco, April.
Clarke, D.B. and Bradford, M.G. (1989) The uses of space by advertising agencies within the United Kingdom. *Geografiska Annaler B: Human Geography*, 71B: 139–151.
Cohen, R.B., Felton, F., Nkosi, M. and van Liere, J. (eds) (1979) *The Multinational Corporation: A Radical Approach. Papers by Stephen Herbert Hymer* (Cambridge: CUP).
Daniels, P.W. (1993) *Service Industries in the World Economy* (Oxford: Blackwell).
Daniels, P.W. and Bryson, J.R. (2005) Sustaining business and professional services in a second city region. *Service Industries Journal*, 25(4): 505–524.
Daniels, P.W., Leyshon, A. and Thrift, N.J. (1988) Large accountancy firms in the UK: operational adaptation and spatial development. *Service Industries Journal*, 8(3): 317–346.
Derudder, B., Beaverstock, J., Faulconbridge, J. and Witlox, F. (eds) (2010) *International Business Travel in the Global Economy* (Aldershot: Ashgate).
Dicken, P. (2011) *Global Shift: Mapping the Changing Contours of the World Economy* (London: Sage).
Dunning, J. (1993) *The Globalization of Business* (London: Routledge).
Dunning, J. and Norman, G. (1987) Theory of multinational enterprise. *Environment & Planning A*, 15: 675–692.
Ekinsmyth, C. (2002) Project organization, embeddedness and risk in magazine publishing. *Regional Studies: The Journal of the Regional Studies Association*, 36(3): 229–243.
Empson, L. (2001) Introduction: knowledge management in professional service firms. *Human Relations*, 54(7): 811.
Empson, L. and Chapman, C. (2006) Partnership versus corporation: implications of alternative governance for managerial authority and organizational priorities in professional service firms. *Research in the Sociology of Organizations*, 24: 145–176.
Enderwick, P. (1989) *Multinational Service Firms* (London: Routledge).
Faulconbridge, J.R. (2006) Stretching tacit knowledge beyond a local fix? Global spaces of learning in advertising professional service firms. *Journal of Economic Geography*, 6: 517–540.
Faulconbridge, J.R. (2007a) Relational spaces of knowledge production in transnational law firms. *Geoforum*, 38(3): 925–940.
Faulconbridge, J.R. (2007b) Exploring the role of professional associations in collective learning in London and New York's advertising and law professional service firm clusters. *Environment & Planning A*, 39: 965–984.
Faulconbridge, J.R. (2008) Negotiating cultures of work in transnational law firms. *Journal of Economic Geography*, 8(4): 497–517.
Faulconbridge, J. and Jones, A. (2011) Geographies of management consultancy firms. In Clark, T. and Kippner, M. (eds), *The Handbook of Management Consultancy* (Oxford: OUP).

Faulconbridge, J. and Muzio, D. (2009) The financialization of large law firms: situated discourses and practices of reorganization. *Journal of Economic Geography*, 9(5): 641–661.
Faulconbridge, J.R., Beaverstock, J.V., Derudder, B. and Witlox, F. (2009) Corporate ecologies of business travel in professional service firms: working towards a research agenda. *European Urban and Regional Studies*, 16(3): 295.
Faulconbridge, J.R., Hall, S. and Beaverstock, J.V. (2008) New insights into the internationalization of producer services: organizational strategies and spatial economies for global headhunting firms. *Environment and Planning A*, 40(1): 210–234.
Glückler, J. (2006) A relational assessment of international market entry in management consulting. *Journal of Economic Geography*, 6: 369–393.
Glückler, J. (2007) Geography and reputation: the city as the locus of business opportunity. *Regional Studies*, 41(7): 949–961.
Glückler, J. and Armbruster, T. (2003) Bridging uncertainty in management consulting: the mechanisms of trust and networked reputation. *Organization Studies*, 24(2): 269.
Glückler, J. and Hammer, I. (2011) A pragmatic service typology: capturing the distinctive dynamics of services in time and space. *Service Industries Journal*, 31: 941–957.
Grabher, G. (2001) Ecologies of creativity: the village, the group and the heterarchic organisation of the British advertising industry. *Environment & Planning A*, 33: 351–374.
Grabher, G. (2002) Cool projects, boring institutions: temporary collaboration in social context. *Regional Studies*, 36(3): 205–214.
Hall, S. (2007) 'Relational marketplaces' and the rise of boutiques in London's corporate finance industry. *Environment & Planning A*, 39: 1838–1854.
Hall, S. and Appleyard, L. (2009) City of London, city of learning? Placing business education within the geographies of finance. *Journal of Economic Geography*, 9(5): 597–617.
Hall, S., Beaverstock, J., Faulconbridge, J. and Hewitson, A. (2009) Exploring cultural economies of internationalization: the role of 'iconic individuals' and 'brand leaders' in the globalization of headhunting. *Global Networks*, 9(3): 399–419.
Harvey, M., Speier, C. and Novicevic, M. (1999) The impact of emerging markets on staffing the global organization: a knowledge-based view. *Journal of International Management*, 5: 157–186.
Hermelin, B. (2009) Producer service firms in globalising cities: the example of advertising firms in Stockholm. *Service Industries Journal*, 29(4): 457–471.
Hermelin, B. and Rusten, G. (2007) The organizational and territorial changes of services in a globalized world. *Geografiska Annaler: Series B, Human Geography*, 89 (Suppl.): 5–11.
Johanson, J. and Vahlne, J. (1990) The mechanism of internationalization. *International Marketing Review*, 7(4): 11–24.
Jones, A. (2003) *Management Consultancy and Banking in an Era of Globalization* (Basingstoke: Palgrave Macmillan).
Jones, A. (2005) Truly global corporations? Theorizing organizational globalization in advanced business-services. *Journal of Economic Geography*, 5: 177–200.
Jones, A. (2007) More than 'managing across borders'? The complex role of face-to-face interaction in globalizing law firms. *Journal of Economic Geography*, 7: 223–246.
Jones, A. (2008) The rise of global work. *Transactions of the Institute of British Geographers*, 33(1): 12–26.
Jones, A. (2009) Born global professionals? Examining corporate globalization in business service firms. *The Globalization and Changing Geographies of Professional Expertise ESRC seminar*, Nottingham, March.
Jones, A. (2011) Theorising international youth volunteering: training for global (corporate) work? *Transactions of the Institute of British Geographers*, 36(4): 530–544.
Jones, A. and Search, P. (2009) Proximity and power within investment relationships: the case of the UK private equity industry. *Geoforum*, 40(5): 809–819.
Jones, A. and Strom, P. (2012) BRICS and knowledge-intensive business services (kibs): a pressing theoretical and empirical agenda. *Association of American Geographers Annual Convention*, New York, February.
Keeble, D. and Nachum, L. (2002) Why do business service firms cluster? Small consultancies, clustering and decentralization in London and southern England. *Transactions of the Institute of British Geographers*, 27(1): 67–90.
Kipping, M. (1999) American management consulting companies in Western Europe, 1920 to 1990: products, reputation, and relationships. *Business History Review*, 73(2): 190–220.
Lewis, M. (ed.) (1999) *The Globalization of Financial Services* (Cheltenham and Northampton, MA: Edward Elgar).
Madsen, T. and Servais, P. (1997) The internationalization of born globals: an evolutionary process? *International Business Review*, 6(6): 561–583.
Majkgard, A. and Sharma, D. (1998) Client-following and market-seeking strategies in the internationalization of service firms. *Journal of Business-to-Business Marketing*, 4(3): 1–41.

McNeill, D. (2008) *The Global Architect: Firms, Fame and Urban Form* (New York: Routledge).
Miozzo, M. and Miles, I. (2002) *Internationalization, Technology and Services* (Cheltenham and Northampton, MA: Edward Elgar).
Morgan, K. (2004) The exaggerated death of geography: learning, proximity and territorial innovation systems. *Journal of Economic Geography*, 4(1): 3–21.
Nachum, L. (1999) *The Origins of the International Competitiveness of Firms: The Impact of Location and Ownership in Professional Service Industries* (Cheltenham and Northampton, MA: Edward Elgar).
O'Brien, K. (1991) *Global Financial Integration: The End of Geography* (New York: Council on Foreign Relations Press).
O'Brien, R. (1992) *Global Financial Integration: The End of Geography* (New York: Council on Foreign Relations).
O'Farrell, P.N. and Hitchens, D.M. (1990) Producer services and regional development: key conceptual issues of taxonomy and quality measurement. *Regional Studies*, 24(2): 163–171.
O'Farrell, P.N. and Wood, P.A. (1998) Internationalisation by business service firms: towards a new regionally based conceptual framework. *Environment and Planning A*, 30: 109–128.
O'Farrell, P.N. and Wood, P.A. (1999) Formation of strategic alliances by business service firms: towards a new client-oriented conceptual framework. *Service Industries Journal*, 19: 133–151.
O'Farrell, P.N., Wood, P.A. and Zheng, J. (1996) Internationalization of business services: an interregional analysis. *Regional Studies*, 30: 101–118.
O'Farrell, P.N., Wood, P.A. and Zhiang, J. (1998) Regional influences on foreign market development by business service companies: elements of a strategic context explanation. *Regional Studies*, 32(1): 31–48.
Pain, K. (2008) Examining 'core–periphery' relationships in a global city-region: the case of London and South East England. *Regional Studies*, 42(8): 1161–1172.
Roberts, J. (1998) *Multinational Business Service Firms: Development of Multinational Organization Structures in the UK Business Service Sector* (Ashgate: Farnham).
Roberts, J. (1999) The internationalization of business service firms: a stages approach. *Service Industries Journal*, 19(4), 68–88.
Roberts, J. (2006) Internationalization of management consultancy services: conceptual issues concerning the cross-border delivery of knowledge intensive services. In Harrington, J. and Daniels, P. (eds), *Knowledge-Based Services, Internationalization and Regional Development* (Aldershot: Ashgate), 101–124.
Sassen, S. (2006) *Cities in a World Economy* (3rd edition) (London: Sage).
Saxenian, A. (1994) *Regional Advantage: Culture and Competition in Silicon Valley and Route 128* (London: Harvard University Press).
Schlunze, R., Agola, N. and Baber, N. (2012) *Spaces of International Management: Launching New Perspectives on Management and Geography* (Basingstoke: Palgrave Macmillan).
Storper, M. and Venables, A.J. (2004) Buzz: face-to-face contact and the urban economy. *Journal of Economic Geography*, 4: 351–370.
Strom, P. and Mattsson, J. (2005) Japanese professional business services: a proposed analytical typology. *Asia Pacific Business Review*, 11(1): 49–68.
Strom, P. and Mattsson, J. (2006) Internationalization of Japanese professional business service firms. *Service Industries Journal*, 26(3): 249–265.
Sturdy, A.J., Schwarz, M. and Spencer, A. (2006) Guess who's coming to dinner? Structures and uses of liminality in strategic management consultancy. *Human Relations*, 59(7): 929–960.
Sydow, J. and Staber, U. (2002) The institutional embeddedness of project networks: the case of content production in German television. *Regional Studies*, 36(3): 215–227.
Thrift, N. (1994) On the social and cultural determinants of international financial centres: the case of the City of London. In Corbridge, S., Martin, R. and Thrift, N. (eds), *Money, Power and Space* (Oxford: Blackwell), 327–355.
Vinodrai, T. (2006) Reproducing Toronto's design ecology: career paths, intermediaries, and local labor markets. *Economic Geography*, 82(3): 237–263.
Warf, B. (2001) Global dimensions of US legal services. *Professional Geographer*, 53(3): 398–406.
Wood, P.A. (1991) Flexible accumulation and the rise of business services. *Transactions of the Institute of British Geographers*, N.S. 16: 160–172.
Wood, P.A. (2002) European consultancy growth: nature, causes and consequences. In Wood, P.A. (ed,), *Consultancy and Innovation: The Business Service Revolution in Europe* (London: Routledge), 35–71.
Wood, P.A. (2006) Urban development and knowledge-intensive business services: too many unanswered questions? *Growth and Change*, 37(3): 335–336.

13. Managing experts and creative talent
David J. Teece[1]

INTRODUCTION

It is increasingly well recognized that the (durable) competitive advantage of business firms flows from the creation, protection and deployment of difficult-to-imitate knowledge assets. Such assets include tacit and codified know-how, associated with individual employees or embedded in organizational routines. Such know-how may be protected from imitation by the instruments of intellectual property such as trade secrets, copyrights and patents, or it may be part of the organization's difficult-to-replicate culture and routines.

Innovation in service businesses has attracted increased scholarly attention in the past 20 years. Service businesses, including professional services, are quite diverse and follow a number of different innovation strategies (Tether et al., 2001). Within service businesses, initiatives to expand capabilities and accelerate growth almost always involve the generation, application and replication of new knowledge.

In recent decades, a greater emphasis on the creation and management of knowledge has served to highlight the importance of the management of expert talent (Albert and Bradley, 1997; Reich, 2002). The growing organizational and technical complexity of problems facing the business enterprise has heightened the need for both highly trained specialists and inspired generalists.

The management style required for enhancing the productivity and contribution of experts is fundamentally different from that applicable to regular line employees. Traditional authority structures often impair expert performance; a much lighter management touch is needed. In fact, the quality of modern management may well be defined by the capacity of the business enterprise to allow/enable highly talented individuals to enjoy the guided professional autonomy they seek, while simultaneously fostering collegiality and delivering intellectual stimulation and professional and financial fulfilment.

The outputs of knowledge creation include a wide range of intangible assets, such as various types of new products, intellectual property and organizational capabilities. A particular set of forward-looking capabilities called 'dynamic capabilities' (Teece et al., 1997) are tied directly to the fortunes of the enterprise. Strong dynamic capabilities can allow the organization to adapt to (or even bring about) changes in the business environment (Teece, 2009).

The thesis advanced here is that today the competitive advantage of the enterprise in most industries is rooted in the ability to motivate experts to create knowledge, help build organizational capabilities and help shape strategy. If they are combined with good intellectual property protection, control over specialized assets and a good business model, then collective efforts at knowledge creation and renewal can help the enterprise build durable competitive advantage over rival firms.

The chapter begins by laying out a conception of knowledge creation and management as a collective process. The next sections focus on experts (literati and numerati) as knowledge creators and on how they can best be managed to encourage the generation of new knowledge. The following section turns to the managers themselves, including their entrepreneurial role guiding the creation of knowledge and its application to the challenges and opportunities that the enterprise faces. A concluding section summarizes the chapter in a knowledge model of the enterprise.

KNOWLEDGE CREATION AND MANAGEMENT

A key to understanding the growth and financial success of a firm is not to study the balance sheet too closely; rather, it is to analyse how the firm creates, protects and utilizes knowledge. Knowledge, however, is a much misunderstood concept.

Early views of the role of knowledge in the enterprise were very technical. During the height of managerial capitalism in the early- to mid-twentieth century, knowledge management advocates counselled recording and quantifying as much as possible. Following World War Two, knowledge was often defined as technology, and the R&D department was seen as perhaps the only source of knowledge creation.

More recently, scholars, notably Ikujiro Nonaka, have taken a more expansive view. Nonaka (1991) sees knowledge creation as an ongoing process distributed throughout the enterprise. Service firms large and small have implicitly accepted elements of this model (Sundbo and Gallouj, 1998).

In Nonaka's conception of knowledge, new knowledge is socially created through the synthesis of the subjective tacit knowledge of multiple individuals after that knowledge has been articulated, externalized and shared. Some of the newly created knowledge will remain tacit and therefore hard for rivals to imitate.

This knowledge creation process must be guided by the company's vision for what it wants to become and the products it wants to produce. This vision of the future must be more than simply a set of financial targets. Good leaders can accelerate the growth of knowledge by creating a mission and a corporate identity that employees find attractive.

The locus of knowledge creation is considered to be a team, whose members are drawn from different functional perspectives. The team leader must guide the team to build mutual trust so that individuals contribute their tacit knowledge for synthesis into a collective output, such as a new product or service concept (Nonaka, 1994). Teams should be given autonomy to achieve their goals within budget and time constraints. Upper management must then screen the results for consistency with corporate strategy and other standards.

The management of knowledge across the enterprise must go beyond the assembly of data and discrete facts that can be stored in databases or on intranets. Facts are information but they are not necessarily knowledge. The data and documents in a database cannot generally be recombined to make new knowledge. Individual knowledge, in order to be most useful to the enterprise, must be shared.

Databases have their place. Information about the knowledge of employees (but not the knowledge itself) needs to be widely available in order to increase the likelihood that any given employee will be able to locate the necessary expertise to address a particular challenge. This level of information is vital for effective service delivery.

Sets of facts and figures and established methods for their use can also play a supporting role (Werr and Stjernberg, 2003). At Seven-Eleven Japan, for instance, front-line employees are trained to build hypotheses about what customers at their store want, and their hypotheses can be sharpened by consulting historical data from an extensive internal system that not only contains point-of-sale data and recommended product displays but also connects to company headquarters and to manufacturers (Nonaka and Toyama, 2007). The information in the database complements the knowledge of the employees. Tools to store and analyse large data sets (now called 'Big Data') have become more powerful and readily available in recent years, just as social media and networked sensors have generated oceans of information. However, numbers and their analysis will not contribute to value creation or capture unless there are creative experts able to interpret and apply the resulting insights.

Documentation of past and current projects will be more important in some settings than others. Scientists, for example, may keep the minimum required documentation and restrict their consultation of others' previous work, while lawyers will frequently ensure that every aspect of a project is documented meticulously (Robertson et al., 2003).

Different approaches to knowledge management may also reflect differences in the markets targeted. Models that emphasize the reuse of codified knowledge may be more important for firms pursuing a low-cost strategy, whereas a less-documented, expert-based approach may be appropriate where unique problems, customized solutions and higher fees are the norm (Hansen et al., 1999). Naturally, an organization's approach to managing knowledge has profound implications for organizational choices in the areas of whom to hire, how much to rely on information technology and what to pay.

KNOWLEDGE CREATORS: THE LITERATI AND NUMERATI

As this discussion of knowledge makes clear, a company adds to its stock of knowledge by guiding and motivating employees who possess skills or knowledge. Expertise is vital for the viability of all services firms, and it is central to the existence of professional services firms that need to provide fully customized solutions for their clients.

The creative, analytical and 'rainmaking' abilities of leading professionals are very valuable to an enterprise – and generally priced accordingly. Skills that can help solve complex problems, make critical decisions or resolve complex disputes will help professional service firms win business. They may also help design and develop new products and services.

This is not to say that experts and highly-credentialed professionals per se are what make a company great. To be successful, organizations must develop distinct processes, a good reputation and other intangibles. In fact, if a company becomes too dependent on one or a handful of individuals, and especially if they are remunerated extravagantly, the morale of all employees can be undermined. And hiring more experts generally cannot save a dysfunctional organization (Pfeffer, 2001).

Experts are unique bundles of skills and knowledge and are not interchangeable, with some proving far more valuable than others. Individual productivity in many fields is quite skewed. This was first observed by Alfred Lotka (1926) in a study of the authorship of articles in nineteenth-century physics journals. Lotka found that approximately 6 per cent of publishing scientists produced half of all papers. Lotka's results are reasonably robust – they have been shown to hold for many disciplines in many different time periods.

Two important categories of expert with regard to knowledge creation are the *literati* and the *numerati*. Both groups are marked by high levels of education and/or experience. A third category important to both are the integrators – professionals who help synthesize the work of others. The literati tend to have both undergraduate and graduate degrees in arts and sciences, economics, business or law. The numerati are likewise highly educated, but in engineering, mathematics or statistics; the sciences, including computer science; information systems; or accounting and finance. In some fields, such as computer science, experience can substitute for an advanced degree. In other fields, like medicine, both academic and practical (clinical) training are necessary for deep proficiency. Both groups have significant analytical skills but the literati tend to be more specialized at synthesis and the communication of ideas. The numerati excel at analysis, especially of large data sets.

Many organizations have some chance of attracting or developing top talent. Most economies are experiencing high labour mobility. In western economies, the relative decline of corporate pension plans, the weakening of strong corporate cultures and the erosion of loyalty towards employers have increased the opportunity for head hunting highly skilled employees. In fast-growing developing economies, the relative scarcity of expert talent has induced a high level of job hopping in pursuit of higher salaries or greater job satisfaction.

Some avenues for securing the services of experts may be a better fit with the firm's capabilities than others (Chambers et al., 1998). For example, hiring new graduates makes the most sense for a firm with an adequate training programme. But their positive impact on the firm may not be immediate. In a longitudinal study of large US law firms, Hitt et al. (2001) found that less experienced partners were negatively correlated with firm performance but that over time the effect became positive.

Reich (2002: 107) and many others have observed that talented people can earn more today, relative to the median wage, than could talented and ambitious people in the industrial era. Globalization has allowed the best talent and ideas to be leveraged rapidly across larger market spaces than ever before. Where the stakes are high, top talent that can make a difference is able to earn exceptional rewards.

External recruitment of top talent must be performed by experienced professionals who are able to make good assessments and are in turn made accountable for their decisions. Otherwise, attempts to compete for 'star' talent from external sources may produce a bad case of 'winner's curse'.

The same factors that have made top talent more available by decreasing allegiance to specific firms also make it harder to retain experts, so it has become all the more critical that management addresses their needs. Research shows that those with the most training, education and ability are the most likely to quit if dissatisfied (Sturman and Trevor, 2001). Incentives for motivating and retaining expert talent can be both financial and non-monetary. These are more complements than substitutes; where experts are concerned, more money will generally not make up for an unsatisfactory work environment.

Getting financial incentives right is fundamental. Suffice to note that there is ample evidence that pay for performance is associated with higher performance at both the individual (Jenkins et al., 1998) and organizational (Gerhart, 2000) levels. The resulting pay differentials are generally accepted among top talent – so long as they are truly capability/performance based.

Unfortunately, the more discretion that management has to set pay, the more energy and resources are likely to be wasted by people trying to capture a larger share of the available resources (Milgrom and Roberts, 1996). This can best be avoided by setting quantifiable performance metrics as the basis for pay, but this is not always possible (Teece, 2003).

In setting pay levels, it is important to distinguish between intrinsic talent and contextual talent. Intrinsic talent is talent which provides/commands full value on a stand-alone basis. In a professional services organization, for instance, this might represent the business that professionals can source based on their own wits and capabilities; that is, independent of the brand or platform on which they stand.

Individual contextual value can exceed intrinsic value when the individual's activities depend heavily on the other complementary assets (such as infrastructure and brand) that the organization provides. Context can be a large component of value, especially in circumstances where teams must be employed to get the job done, and when the firm's infrastructure, methods and staffing play important support roles.

A firm may not need to pay as much for an expert whose 'star' quality is so firm-specific that it would not transfer very well to other settings. An important exception is when the contextual skills and knowledge of the individual would be difficult and costly to replace if the expert departs.

Getting pay wrong can lead to a loss of competitiveness. In a professional services firm, where human capital is more important than other inputs, poorly calibrated pay can lead to the departure of key experts, possibly benefiting rivals and beginning a negative feedback process in which reputation and quality decline (Teece, 2003).

For employee retention, compensation at competitive levels may be necessary but not sufficient. Other accoutrements of the job environment that matter include the organization's culture, the quality of its management, the challenge of the work and the autonomy afforded workers. Companies that rank higher on these and similar 'quality of work life' measures outperform their peers in retention (Chambers et al., 1998: 50).

MANAGING AND ORGANIZING THE LITERATI AND NUMERATI

The presence of potentially valuable creative people and/or experts only adds value to the enterprise when they are properly managed and organized. The mode of management that is most likely to bring great results is distinctly different from that required for ordinary operations.

Two important considerations for managing creative talent and experts are (1) they respond best to a less hierarchical approach than might be used with other types of employees; and (2) their creativity is most likely to flourish in an environment of open collaboration.

Light-touch Management

With respect to the literati and numerati, strongly authoritarian management aimed at forcing people to collaborate is anathema. Management must have a light touch. Otherwise cooperative efforts will be suppressed and creativity will be compromised. Management is seldom sufficiently informed to second-guess the difficult and granular technical trade-offs and judgements of the literati and numerati with respect to solving the problem at hand.

The commonest purpose for hierarchy – to delegate tasks to 'workers' – is simply not needed for many types of creative or expert professional work. Experts tend to be substantially self-motivated and self-guided. While some degree of bureaucratic organization is inevitable in large organizations, it must support the activities of experts without trying to rigidly control them.

Accordingly, management of creative and expert talent usually needs to be decentralized or 'distributed'. Traditional notions of management that rely heavily on hierarchy and decisions driven from the centre are unlikely to work well.

Of course, strong accountability is still required from the literati and the numerati. Autonomy and accountability go hand in hand; the more easily performance can be measured, the greater the autonomy that can safely be permitted.

Self-organized cooperative activity is frequently observed in scientific and creative engineering projects. Richard Nelson (1962) studied the development of the transistor at Bell Labs and noted that creative collaboration

> requires that individuals be free to help each other as they see fit. If all allocation decisions were made by a centrally situated executive, the changing allocation of research effort called for as perceived alternatives and knowledge change would place an impossible information processing and decision making burden on top management. Clearly the research scientists must be given a great deal of freedom... (1962: 569)

Nelson also noted that 'the informality of the decision structure played a very important role in permitting speedy cooperative response to changing ideas and knowledge' (1962: 579).

Fifty years later, the same lessons – particularly the importance of decentralization and flexibility for knowledge creation – were being relearned. John Chambers, CEO of the US-based network equipment company Cisco, remarked: 'In 2001, we were like most high-tech companies – all decisions came to the top ten people in the company, and we drove things back down from there' (quoted in McGirt, 2008: 91). Cisco instituted a more decentralized and collaborative management system, with a network of councils and boards entrusted and empowered to launch new businesses, and incentives to encourage executives to work together. Chambers claimed that 'these boards and councils have been able to innovate with tremendous speed. Fifteen minutes and one week to get a [business] plan that used to take six months' (ibid.). However, over time, the structure became sclerotic and, beginning in 2009, Cisco reduced the number of councils from twelve to three, while dissolving the associated boards, in a renewed push to speed up decision making (Clark and Tibken, 2011). Similarly, Pixar studios replaced the traditional isolated studio development department with small 'incubation teams' of creative staff who work with a director to refine a film idea to the point where it can be pitched to the studio's senior filmmakers (Catmull, 2008). This process keeps the creative personnel in charge and vets the problem-solving and social dynamics of the core team should the idea move to the next stage.

The point here is a simple one: in fast-paced complex environments where there is heterogeneity in customer needs and the focus is on creative activity and/or technological innovation, it is simply impossible to achieve the necessary flexibility and responsiveness with a command-and-control organizational structure. Moreover, with a highly talented workforce, excessive centralization can shut down local initiatives.

The above admonitions are not meant to imply that top management should not guide and coordinate innovative activities. In fact, there are certain types of innovation – particularly 'systemic' innovation (Teece, 1984) – where close coordination of different groups is required because the parts are so deeply interdependent.

Managers of innovative enterprises must learn to lead without the authority that comes from a position in an organization chart or the 'C' designation in their title. This imposes new challenges for some companies and some individuals, but it is the way of the future in such contexts. The challenge is to connect individual initiatives to the overall corporate strategy without building an expensive and initiative-sapping hierarchy.

In some settings, it may even be desirable to invert the traditional hierarchy in order to create the organizational structures in which professionals can perform to their potential (Quinn et al., 1996). With an inverted hierarchy, the job of the manager is to provide support by creating incentive alignment and ensuring resource availability. The experts may even take responsibility for determining executive wages.

In purely creative environments, it is indeed the highly skilled experts that hire 'bosses' rather than the other way around. The Hollywood agency model for creative talent was an early manifestation. As explained by Albert and Bradley (1997), the stars themselves, beginning with Newman, Streisand and Poitier, broke away from the studios to create their own production company, First Artists, allowing leading actors to control their professional environment and lives. The artists put a professional manager in place, but the manager's mandate was to effectuate the artist's view of how films should be produced.

University faculties have some similar attributes. The university requires the discharge of teaching, research and service obligations by faculty, but allows faculty members considerable discretion as to whether and when tasks (other than class meetings) are performed. In some universities, most notably the major research universities on the west coast of the United States, the faculty arguably hires their Dean since the Dean generally serves at the sufferance of the faculty.

Professional services organizations in the legal, medical and other fields exhibit similar characteristics. In short, creative and highly skilled knowledge workers, be they scientists, engineers, medical doctors, professors or economists, desire high autonomy and can be self-motivated and self-directed because of their deep expertise.

Implemented properly, the distributed leadership approach is not an abdication of managerial responsibility. It is just the opposite. The executive leadership team sets strategy and goals and must retain credibility with its experts as well as being answerable to the board and to stakeholders.

While creative activities need to be organized in a distributed/decentralized manner, there are operational activities that should not be managed in this way. The accounting, finance and treasury functions are obvious examples. Intellectual property at a company with numerous valuable patents may also benefit from being managed centrally.

Managing Collaboration

Because experts are increasingly specialized, interaction among people from diverse disciplines or functional groups is almost always required to solve complex problems. In professional services firms, project-specific teams may be the dominant organizational mode for nearly all activities (Larsen, 2001).

An enterprise can hire the brightest, most creative people, but it is only through successful fostering of collaboration, sharing of information and the use of social networks inside and outside the enterprise that their creative potential will be released (Subramaniam and Youndt, 2005). The development of new business services, for example, frequently involves collaboration with suppliers and customers (Love et al., 2011). Team members external to the firm must be assessed and managed almost as carefully as the firm's own employees (Bettencourt et al., 2002).

In his study of the development of the transistor, Nelson (1962) notes that teamwork in a creative context is likely to differ from traditional collaboration aimed at a fixed goal. In the transistor project, teamwork

> meant interaction and mutual stimulation and help . . . several people outside the team also interacted in an important way . . . teamwork . . . did not mean a closely directed project . . . The project was marked by flexibility – by the ability to shift directions and by the rather rapid focusing of attention by several people on problems and phenomena unearthed by others. (1962: 578)

It does not follow that every analytical and creative activity is appropriately organized in teams. Indeed, there is a great deal about traditional teams that involves hidden and unnecessary costs. When team requirements are too heavy, decision cycles lengthen, expenses mount and the organization adopts an inward focus.

Put differently, one cannot simply assume that more is better when it comes to collaboration. Consensus and participatory leadership is not always a good thing, particularly when the issues are complex and there is considerable asymmetry in the distribution of talents on the team. The right voices need to be heard.

Unproductive collaboration can be more dangerous than missed opportunities for collaboration. Forced teaming often leads to excessive consensus building, slow decision making, and the wasting of time and money.

Constructing project teams is itself an important competence (Larsen, 2001). Teams should be kept as small and intimate as possible. And project groups that complete their task or run into 'blind alleys' should disband so that the mix of talents is ready to be reconfigured as needed to meet future demands. Assigning the wrong mix of people to a project 'because they're used to working together' is a path to failure.

In principle, the outcome from a cross-functional team can exceed the capabilities of its best individual members (Larson, 2007). However, if not managed properly, the bringing together of specialists from different parts of the organization can impede innovation (Ancona and Caldwell, 1992). Team members may be too tied to their normal functions or disciplines or be mutually mistrustful. There are numerous other ways that teams go astray. These range from unproductive conflict that leads to indecision to peer pressure that leads to a flawed consensus.

Avoiding all conflict often results in poor outcomes (Tjosvold, 1985). Emotional conflict, however, is more likely to have a negative effect than is substantive conflict over solutions to task-related problems (Pelled et al., 1999). Groups that encourage the authentic expression of minority opinions tend to make higher-quality decisions (Maier, 1970; Nemeth, 2012). Conflict is most likely to contribute to high-quality decisions when trust is high, that is, when members do not suspect anyone on the team of trying to score points at the others' expense (Dooley and Fryxell, 1999).

Dougherty (1992) suggests that the interaction and collaboration necessary for innovation in cross-functional teams are best brought out by shared learning activities, such as focus groups and user visits. Such shared activities promote group cohesiveness, which has also been shown to contribute to higher performance by R&D project teams (Keller, 1986).

Group leaders can avoid suppressing healthy disagreement (based on the issues, not on the people, involved) by not expressing their positions too early in the process (Janis, 1972). Openness should be encouraged by not dismissing any idea too quickly.

Yet it is vital to have leadership, at the team level or higher, that knows which ideas can be rejected. A key role of entrepreneurial managers, after enunciating a vision, is to permit experimentation and search, then support promising paths and close down foolish ones.

An added difficulty is that teams are often spread across organizational boundaries and/or large distances. This is increasingly true for innovation, as large and small companies have begun to tap into pools of science and engineering talent in industrializing economies. The autonomy and trust necessary for managing experts are also appropriate in the 'virtual team' context, where continuous direct leadership may not be possible due to time zone differences. To overcome the social remoteness of distance, special measures, such as a project kick-off meeting that brings everyone to a single location for a few days, must be devised to at least partially formalize the process of fostering mutual support with a shared purpose (Siebdrat et al., 2009).

While physical distance clearly forces the use of fully virtual teams, virtuality is actually a matter of degree since all teams, even those whose members are co-located, will employ some forms of computer-mediated communication. There is still much work to be done regarding the performance effect of virtuality, but one consistent finding is that virtual teams require more time to complete tasks than face-to-face teams, so they may not be suitable for the most urgent projects (Martins et al., 2004).

Whether team members are dispersed or co-located, their work must be tied to the overall strategy of the business (Wheelwright and Clark, 1992). Management of the team needs to tread the line between preventing the natural tension and creativity of innovation from descending into chaos and constraining the team by defining the goals and strategy linkage so narrowly that real innovation is impossible. Takeuchi and Nonaka (1986) call this 'subtle control', where the monitoring function leads to intervention (e.g. eliminating a team member) only when absolutely necessary.

The goals of a team can be furthered by the presence of a 'heavyweight' project manager who has both credibility within the team and power/prestige in the organization as a whole (Clark et al., 1987). The prestige is important for ensuring the team the nec-

Table 13.1 Key differences between traditional teams and virtuoso teams

Team characteristics	Traditional teams	Virtuoso teams
Membership	Members chosen based on who has available time	Members chosen based on expertise
Culture	Collective	Collective and individual
Focus	Tight project management. On time and on budget more important than content	Ideas, understanding and breakthrough thinking emphasized
Target	Conventional output	Breakthrough output
Intensity	High/medium	High
Stakes	Low/medium	High

Source: After Fischer and Boynton (2005).

essary resources and room to manoeuvre, and is also important for gaining the project manager the respect and cooperation of the literati and numerati on the team.

In some special cases, when the stakes are high or the deadlines too close, an organization may assemble a team consisting exclusively of its most able experts. The management requirements in this case are somewhat different from more traditional teams because experts are typically used to being in the leadership position themselves. It may be helpful to provide some extra initial structure to foster collaboration, such as breaking into smaller groups or even pairs that can tackle segments of the overall challenge in parallel.

The goal in such project groups, or 'virtuoso teams' (Fischer and Boynton, 2005), is not accommodation and harmony; rather, the aim is to achieve excellence by unleashing individual creativity. A higher level of (topic related) conflict is to be expected and is bounded only by the common goal and deadline. The team leader will need to be able to massage large egos without seeming patronizing. Table 13.1 summarizes some of the differences between traditional and virtuoso teams.

KNOWLEDGE LEADERS: ENTREPRENEURIAL MANAGERS

The knowledge work of the enterprise, including its creative talent and experts, must be organized and directed by managers. And in their most important, entrepreneurial role, managers are essential to the exercise of a firm's dynamic capabilities (Augier and Teece, 2009).

Entrepreneurial managers can be 'members' of the numerati or literati, but it is by no means the case that they must be. And, like the numerati/literati, entrepreneurial managers can be 'grown' internally or brought in from the outside. In the case of current employees identified as having the potential for management advancement, Martin and Schmidt (2010) recommend sharing future strategies and welcoming feedback. This shows them that they are being groomed and helps to establish a collaborative atmosphere.

Much like the founders of start-up companies, entrepreneurial managers in established firms assemble and deploy resources in pursuit of fresh opportunities.

Opportunities open constantly in the global economy; consumer needs, technological possibilities, and competitor activities are always in flux. As discussed by Teece et al. (1997), the path ahead for some emerging marketplace trajectories is easily recognized. In consumer banking, for example, the shift to digital, paperless transactions has been foreseeable for years. However, many emerging trajectories are hard to discern. For instance, the applications that will drive the mass adoption of 3D printing are not yet clear.

Entrepreneurial functions are quite different from those of the ordinary manager. The ordinary manager oversees the ongoing efficiency of established processes: that schedules are met and contracts honoured, that quality and productivity improve, and that the business model is constantly tuned. Although there are creative aspects to accomplishing these tasks, managing the operations of an ongoing business is comparatively straightforward.

Entrepreneurial managers excel at the scanning, learning, creative and interpretive activity needed to sense (and later seize) new technological and market opportunities. The requisite abilities are not uniformly distributed among individuals. The discovery (or creation) of opportunity requires specific knowledge, creativity, insight into customer decision making, and practical wisdom (Nonaka and Toyama, 2007). It involves interpreting and synthesizing information in whatever form it appears, be it a chart, a picture, a conversation at a trade show, news of a technological breakthrough, or the angst expressed by a frustrated customer. The entrepreneurial manager will use this to generate or update a conjecture about the likely evolution of technologies, customer needs and marketplace responses.

Neither the identification nor even the creation of opportunities results spontaneously in successful exploitation. Indeed, many inventions go unexploited for extended periods, and the pioneer in a market may not turn out to be its eventual leader (Teece, 1986, 2006).

Once exploitable opportunities are discerned, entrepreneurial managers must also devise a business model (preferably one that cannot readily be imitated) and a strategy for capturing a meaningful share of value that a new product or service will generate (Teece, 2010). In other words, entrepreneurial management requires both creative vision and wily pragmatism from the management team.

The pursuit of new opportunities often entails changes in the organization, which demands a certain leadership style (Bass, 1985). *Transactional* leaders are able to motivate their employees to meet expectations and accomplish set tasks that fall within ordinary capabilities. *Transformational* leaders, on the other hand, know how to inspire and challenge employees in ways that cause them to perform beyond expectations. According to Bass, transformational leadership 'is more likely to emerge in times of distress and rapid change' (ibid: 39). Subsequent research has confirmed that management teams under transformational leaders are more willing to tackle new growth opportunities (Ling et al., 2008).

Although certain individuals may stand out or become the focus of public attention, entrepreneurial management involves a team. The top management team (TMT) consists of those who report directly to the CEO. Managers at lower levels of large organizations may also play an entrepreneurial role.

The TMT tackles highly complex issues and bears responsibility for the future of the organization. Among other responsibilities, the TMT sets the strategy for the organization and oversees its implementation. When the TMT performs poorly together, the result is likely to be organizational decline (Hambrick, 1994).

A well-integrated TMT, in which members share openly and truly work together on strategic issues, has been shown to facilitate the pursuit of new concepts while not losing sight of current operations – so-called organizational ambidexterity (Lubatkin et al., 2006). Many professional services firms suffer from the opposite of ambidexterity because a single-minded emphasis on billable hours crowds out opportunities to renew the knowledge base and explore new possibilities (Jensen et al., 2010).

CONCLUSIONS

Knowledge creation in the enterprise can provide a foundation for success. The process requires employees, particularly experts, to be appropriately managed. In turn, leveraging knowledge into profitability requires managers who take an entrepreneurial approach to market opportunities.

Expert talent has become indispensable for solving problems, delivering services, developing products and making decisions in today's hypercompetitive global economy. The ideal is to hire and/or promote the best people, provide them a transparent pay-for-performance package, find managers with sufficient skill and credibility to guide their work, then provide none but the necessary guidance as they perform their tasks.

Traditional hierarchical approaches to managing the literati and numerati are unlikely to bring out their best. Narrow-band compensation systems are also unlikely to attract and retain the most-skilled experts. A growing number of organizations are finding ways to break the shackles of rigid Human Resources systems in order to create a space for experts to feel comfortable and to be productive. To do otherwise risks a downward spiral of uncompetitive knowledge generation and erosion of expertise.

The proposed management model constitutes a complete knowledge-based perspective on the enterprise. Table 13.2 contrasts this Knowledge Model (right-hand column) with the characteristics of the archetypal Industrial Model that still characterizes too many large organizations.

Table 13.2   Contrasting management models of the business enterprise

Organizational characteristics	Industrial Model	Knowledge Model (for literati and numerati)
Hierarchy	Deep	Shallow
Leadership	Centralized	Distributed
Work	Segmented	Collaborative
People viewed as	Cost	Asset
Basis of control	Authority	Influence and example
Assumptions about individuals	Opportunistic	Honourable
Financial incentives	Base salary + discretionary bonus	Metrics-based compensation; limited discretion

Knowledge alone does not create competitive advantage. The knowledge of experts must be deployed strategically by entrepreneurial managers with a deep understanding of the organization's capabilities and the market's opportunities in order to deliver long-term profitability and growth.

NOTE

1. I wish to thank Dr Greg Linden for many helpful comments and suggestions.

REFERENCES

Albert, Steven and Bradley, Keith (1997), *Managing Knowledge: Experts, Agencies and Organizations*, Cambridge: Cambridge University Press.
Ancona, Deborah Gladstein and Caldwell, David F. (1992), 'Demography and Design: Predictors of New Product Team Performance', *Organization Science*, 3(3), 321–341.
Augier, Mie and Teece, David J. (2009), 'Dynamic Capabilities and the Role of Managers in Business Strategy and Economic Performance', *Organization Science*, 20(2), 410–421.
Bass, Bernard M. (1985), 'Leadership: Good, Better, Best', *Organizational Dynamics*, 13(3), 26–40.
Bettencourt, Lance A., Ostrom, Amy L., Brown, Stephen W. and Roundtree, Robert I. (2002), 'Client Co-production in Knowledge-Intensive Business Services', *California Management Review*, 44(4), 100–128.
Catmull, Ed (2008), 'How Pixar Fosters Collective Creativity', *Harvard Business Review*, 86(9), 65–72.
Chambers, Elizabeth G., Foulon, Mark, Handfield-Jones, Helen, Hankin, Steven M. and Michaels, Edward G. III (1998), 'The War for Talent', *McKinsey Quarterly*, 1998(3), 44–57.
Clark, Don and Tibken, Shara (2011), 'Cisco to Reduce Its Bureaucracy', *WSJ.com* (6 May), http://online.wsj.com/article/SB10001424052748703859304576304890449176956.html, accessed 6 May 2011.
Clark, Kim B., Chew, W. Bruce and Fujimoto, Takahiro (1987), 'Product Development in the World Auto Industry', *Brookings Papers on Economic Activity*, 3, 729–781.
Dooley, Robert S. and Fryxell, Gerald E. (1999), 'Attaining Decision Quality and Commitment From Dissent: The Moderating Effects of Loyalty and Competence in Strategic Decision-Making Teams', *Academy of Management Journal*, 42(4), 389–402.
Dougherty, Deborah (1992), 'Interpretive Barriers to Successful Product Innovation in Large Firms', *Organization Science*, 3(2), 179–202.
Fischer, Bill and Boynton, Andy (2005), 'Virtuoso Teams', *Harvard Business Review*, 83(7), 116–123.
Gerhart, Barry (2000), 'Compensation, Strategy, and Organizational Performance', in S.L. Rynes and B. Gerhart (eds), *Compensation in Organizations*, San Francisco, CA: Jossey-Bass, 151–194.
Hambrick, Donald C. (1994), 'Top Management Groups: A Conceptual Integration and Reconsideration of the "Team" Label', in B.M. Staw and L.L. Cummings (eds), *Research in Organizational Behavior*, Greenwich, CT: JAI Press, 171–214.
Hansen, Morten T., Nohria, Nitin and Tierney, Thomas (1999), 'What's Your Strategy for Managing Knowledge?', *Harvard Business Review*, 77(2), 106–116.
Hitt, Michael A., Bierman, Leonard, Shimizu, Katsuhiko and Kochhar, Rahul (2001), 'Direct and Moderating Effects of Human Capital on Strategy and Performance in Professional Service Firms: A Resource-Based Perspective', *Academy of Management Journal*, 44(1), 13–28.
Janis, Irving L. (1972), *Victims of Groupthink: A Psychological Study of Foreign-Policy Decisions and Fiascoes*, Boston, MA: Houghton, Mifflin.
Jenkins, G. Douglas, Jr., Mitra, Atul, Gupta, Nina and Shaw, Jason D. (1998), 'Are Financial Incentives Related to Performance? A Meta-analytic Review of Empirical Research', *Journal of Applied Psychology*, 83(5), 777–787.
Jensen, Søren H., Poulfelt, Flemming and Kraus, Sascha (2010), 'Managerial Routines in Professional Service Firms: Transforming Knowledge into Competitive Advantages', *Service Industries Journal*, 30(12), 2045–2062.
Keller, Robert T. (1986), 'Predictors of the Performance of Project Groups in R&D Organizations', *Academy of Management Journal*, 29(4), 715–726.

Larsen, J.N. (2001), 'Knowledge, Human Resources and Social Practice: The Knowledge-Intensive Business Service Firm as a Distributed Knowledge System', *Service Industries Journal*, 21(1), 81–102.

Larson, James R., Jr. (2007), 'Deep Diversity and Strong Synergy: Modeling the Impact of Variability in Members' Problem-Solving Strategies on Group Problem-Solving Performance', *Small Group Research*, 38(3), 413–436.

Ling, Yan, Simsek, Zeki, Lubatkin, Michael H. and Veiga, John F. (2008), 'Transformational Leadership's Role in Promoting Corporate Entrepreneurship: Examining the CEO–TMT Interface', *Academy of Management Journal*, 51(3), 557–576.

Lotka, Alfred J. (1926), 'The Frequency Distribution of Scientific Productivity', *Journal of the Washington Academy of Sciences*, 16(12), 317–323.

Love, James H., Roper, Stephen and Bryson, John R. (2011), 'Openness, Knowledge, Innovation and Growth in UK Business Services', *Research Policy*, 40(10), 1438–1452.

Lubatkin, Michael H., Simsek, Zeki, Ling, Yan and Veiga, John F. (2006), 'Ambidexterity and Performance in Small- to Medium-Sized Firms: The Pivotal Role of Top Management Team Behavioral Integration', *Journal of Management*, 32(5), 646–672.

Maier, Norman R.F. (1970), *Problem Solving and Creativity in Individuals and Groups*, Belmont, CA: Brooks/Cole.

Martin, Jean and Schmidt, Conrad (2010), 'How to Keep Your Top Talent', *Harvard Business Review*, 88(5), 54–61.

Martins, Luis L., Gilson, Lucy L. and Maynard, M. Travis (2004), 'Virtual Teams: What Do We Know and Where Do We Go From Here?' *Journal of Management*, 30(6), 805–835.

McGirt, Ellen (2008), 'Revolution in San Jose', *Fast Company*, Issue 131, 90–93.

Milgrom, Paul and Roberts, John (1987), 'Bargaining Cost, Influence Costs, and the Organization of Economic Activity', in L. Putterman and R.S. Kroszner (eds), *The Economic Nature of the Firm*, Cambridge: Cambridge University Press, 162–174.

Milgrom, Paul and Roberts, John (1996), 'The LeChatelier Principle', *American Economic Review*, 86(1), 113–128.

Nelson, Richard (1962), 'The Link Between Science and Invention: The Case of the Transistor', in Harold M. Groves (ed.), *The Rate and Direction of Inventive Activity*, National Bureau of Economic Research, Princeton, NJ: Princeton University Press, 549–584.

Nemeth, Charlan Jeanne (2012), 'Minority Influence Theory', in P. Van Lange, A. Kruglanski and T. Higgins (eds), *Handbook of Theories of Social Psychology, Volume Two*, New York, NY: Sage, 362–378.

Nonaka, I. (1991), 'The Knowledge-Creating Company', *Harvard Business Review*, 69(6), 96–104.

Nonaka, I. (1994), 'A Dynamic Theory of Organizational Knowledge Creation', *Organization Science*, 5(1), 14–37.

Nonaka, Ikujiro and Toyama, Ryoko (2007), 'Strategy as Distributed Practical Wisdom (Phronesis)', *Industrial and Corporate Change*, 16(3), 371–394.

Pelled, Lisa Hope, Eisenhardt, Kathleen M. and Xin, Katherine R. (1999), 'Exploring the Black Box: An Analysis of Work Group Diversity, Conflict and Performance', *Administrative Science Quarterly*, 44(1), 1–28.

Pfeffer, Jeffrey (2001), 'Fighting the War for Talent is Hazardous to Your Organization's Health', *Organizational Dynamics*, 29(4), 248–259.

Quinn, James Brian, Anderson, Philip and Finkelstein, Sydney (1996), 'Managing Professional Intellect: Making the Most of the Best', *Harvard Business Review*, 74(2), 71–80.

Reich, Robert (2002), *The Future of Success: Working and Living in the New Economy*, New York, NY: Vintage Books.

Robertson, Maxine, Scarbrough, Harry and Swan, Jacky (2003), 'Knowledge Creation in Professional Service Firms: Institutional Effects', *Organization Studies*, 24(6), 831–857.

Siebdrat, Frank, Hoegl, Martin and Ernst, Holger (2009), 'How to Manage Virtual Teams', *MIT Sloan Management Review*, 50(4), 63–68.

Simsek, Zeki, Veiga, John F., Lubatkin, Michael H. and Dino, Richard N. (2005), 'Modeling the Multilevel Determinants of Top Management Team Behavioral Integration', *Academy of Management Journal*, 48(1), 69–84.

Sturman, Michael C. and Trevor, Charlie O. (2001), 'The Implications of Linking the Dynamic Performance and Turnover Literatures', *Journal of Applied Psychology*, 86(4), 684–696.

Subramaniam, Mohan and Youndt, Mark A. (2005), 'The Influence of Intellectual Capital on the Types of Innovative Capabilities', *Academy of Management Journal*, 48(3), 450–463.

Sundbo, Jon and Gallouj, Faïz (1998), 'Innovation as a Loosely Coupled System in Services', Report SI4S no. 4, The STEP Group, Studies in Technology, Innovation and Economic Policy. Oslo, Norway, http://survey.nifu.no/step/old/Projectarea/si4s/papers/topical/si4s04.pdf, accessed 7 January 2013.

Takeuchi, Hirotaka and Nonaka, Ikujiro (1986), 'The New Product Development Game', *Harvard Business Review*, 64(1), 137–146.

Teece, David J. (1984), 'Economic Analysis and Strategic Management', *California Management Review*, 26(3), 87–110.
Teece, David J. (1986), 'Profiting from Technological Innovation', *Research Policy*, 15(6), 285–305.
Teece, David J. (2003), 'Expert Talent and the Design of (Professional Services) Firms', *Industrial and Corporate Change*, 12(4), 895–916.
Teece, David J. (2006), 'Reflections on Profiting from Innovation', *Research Policy*, 35(8), 1131–1146.
Teece, David J. (2009), *Dynamic Capabilities and Strategic Management: Organizing for Innovation and Growth*, New York, NY: Oxford University Press.
Teece, David J. (2010), 'Business Models, Business Strategy and Innovation', *Long Range Planning*, 43(2–3), 172–194.
Teece, David J., Pisano, Gary and Shuen, Amy (1997), 'Dynamic Capabilities and Strategic Management', *Strategic Management Journal*, 18(7), 509–533.
Tether, Bruce S., Hipp, Christiane and Miles, Ian (2001), 'Standardisation and Particularisation in Services: Evidence from Germany', *Research Policy*, 30(7), 1115–1138.
Tjosvold, Dean (1985), 'Implications of Controversy Research for Management', *Journal of Management*, 11(3), 21–37.
Werr, Andreas and Stjernberg, Torbjörn (2003), 'Exploring Management Consulting Firms as Knowledge Systems', *Organization Studies*, 24(6), 881–908.
Wheelwright, Steven C. and Clark, Kim B. (1992), *Revolutionizing Product Development: Quantum Leaps in Speed, Efficiency, and Quality*, New York, NY: Free Press.

14. Globalization of services
Joanne Roberts

INTRODUCTION

The globalization of services is part of the wider economic phenomenon of intensifying cross-border economic activity. Since the 1990s, academics and commentators have claimed that we live in a globalized economic environment (e.g. Ohmae, 1990; Reich, 1992). From the 1990s we have seen an intensification of this globalization as its drivers continue unabated. Yet it is important to note that globalization did not suddenly appear in the 1990s; rather, its antecedents are to be found in the past 200 years, and for some activities, much earlier. For instance, shipping and finance, which facilitate trade in goods, are perhaps the most obvious examples of trade in services that has occurred for many hundreds of years. Other examples include the cross-border services provided by mercenary soldiers, as well as the international supply of educational services, whether through the movement of students or tutors. As the importance of services has grown within the domestic economy, so too has their role in the international economic environment. The past 30 years have witnessed a rapid rise in cross-border service transactions (UNCTAD, 2004), yet the overseas expansion of service businesses remains poorly appreciated (Blomstermo et al., 2006; Kundu and Merchant, 2008).

This chapter traces the globalization of services. In so doing, it will identify the drivers and barriers to globalization in the service sector as well as the links between service activity and other sectors of the economy. However, the chapter begins with a brief account of the rise of globalization in the 20th century.

GLOBALIZATION

Manifestations of globalization are very much part of everyday life in the advanced countries of the 21st century. As I sit here at my desk in England, I can contact a hire car company to arrange a vehicle for my trip to Florida, in the United States, in the spring by telephoning a call centre in Bangalore, India. Similarly, when I do my weekly grocery shopping in my local Asda supermarket, a subsidiary of the American retailer Wal-Mart, I can purchase mange tout from Kenya, avocados from Peru, lamb from New Zealand, soya sauce from Japan, Camembert cheese from France and wine from Australia. The globalization of media, including the Internet, gives anyone with a mobile telephone access to real-time information from across the world; whether it be about extraordinary tales of human endurance, weather conditions, military conflicts and associated atrocities, natural disasters or everyday incidents, we are able to share in the joys and sorrows of people located in faraway places. Moreover, through social media like Twitter and Facebook, we can now engage with others across the globe as easily as we can with our neighbours.

The idea of globalization is very much part of popular discourse in today's world. Indeed, there has been much debate about the meaning and scale of globalization (Held et al., 1999; Hirst and Thompson, 1999; Scholte, 2002; Dicken, 2011; *inter alia*). Held et al. (1999, p. 2) identify three main schools of thought on globalization. First, the hyperglobalists, including Ohmae (1990), who view globalization as a new era in which people everywhere are increasingly subject to the disciplines of the global marketplace. Secondly, the sceptics, such as Hirst and Thompson (1999), who argue that globalization is essentially an international economy increasingly segmented into three major regional blocs in which national governments remain very powerful. Thirdly, the transformationalists, like Giddens (1990, 2000), who view contemporary patterns of globalization as historically unprecedented, resulting in profound change for states and societies across the globe as they adapt to a more interconnected but highly uncertain world.

As Held et al.'s (1999, p. 2) definition underlines, globalization is not confined to the economy: 'Globalization may be thought of initially as the widening, deepening and speeding up of worldwide interconnectedness in all aspects of contemporary social life, from the cultural to the criminal, the financial to the spiritual.' Although globalization has a profound influence on all aspects of our lives, the focus of this chapter is on the globalization of services from an economic perspective. Yet it is clear that, as services have become increasingly global in scope, they have contributed to, and been influenced by, the socio-cultural impact of wider processes of globalization.

Taking an economic perspective, the Organisation for Economic Co-operation and Development (OECD, 2005, p. 11) notes that:

> The term 'globalisation' has been widely used to describe the increasing internationalisation of financial markets and of markets for goods and services. Globalisation refers above all to a dynamic and multidimensional process of economic integration whereby national resources become more and more internationally mobile while national economies become increasingly interdependent.

Economic integration has resulted in significant welfare gains as it has allowed the benefits of specialization and trade to be extended beyond national borders. This has contributed to sustained economic growth for large parts of the world. However, integration can also have negative consequences. The global financial crisis of 2008 is perhaps the most significant recent manifestation of the negative economic impact of globalization. Such is the interdependence of financial markets that the collapse of the subprime mortgage market in the USA (Jain, 2009), and the subsequent failure of the investment bank Lehman Brothers, resulted in financial losses and instability in financial markets across the globe. The repercussions of the crisis are still being felt today as many advanced countries struggle to overcome economic recession and to regain their pre-2008 rates of economic growth.

Although the term globalization only entered academic discourse in the 1990s, we can find its origins in the growth of cross-border trade and investment facilitated by transport and communication developments in the 19th century. The second half of the 20th century, and especially its last two decades, witnessed an intensification of the processes of globalization brought about by a number of converging trends. These included the growth in number, size and geographical reach of multinational corporations (MNCs). The transformative capacities of new information and communications technologies

(ICTs) have been important in facilitating the development of MNCs. However, the impact of technological developments has been more widespread and, for instance, has facilitated new forms of global economic interaction, including e-commerce.

Furthermore, the international institutions established in the aftermath of World War II, specifically the General Agreement on Tariffs and Trade (GATT)[1] (integrated into the World Trade Organization (WTO) in 1995), the International Monetary Fund (IMF) and the World Bank, which sought to promote the stability and integration of the world economy, have also contributed to globalization. Indeed, regional integration projects, foremost among these being the European Union, have intensified the process of economic integration among their member states. A further boost to integration has resulted from the demise of the former Soviet Empire, resulting in the opening up of Eastern Europe and Russia to global markets. Similarly, the adoption of an open door policy in China[2] has brought about further integration. With the collapse of Soviet socialism, the global market has experienced significant expansion, and international regulatory institutions like the WTO have witnessed a growth in membership.

Last, but not least, the processes of globalization are associated with a neoliberal economic perspective, which promotes free national and international markets with minimal state intervention (Harvey, 2005). First adopted by President Ronald Reagan in the USA and Prime Minister Margaret Thatcher in the UK from the end of the 1970s, these policies became widespread in the 1980s and 1990s. Furthermore, from the 1980s, developing countries in crisis were compelled to adopt liberalizing policies as a condition of financial assistance from the Western-dominated international institutions of the IMF and the World Bank. The location of these institutions in Washington, DC gave rise to the term 'Washington Consensus' to capture the set of policies they promoted, including trade and foreign direct investment (FDI) liberalization, privatization and deregulation, fiscal policy discipline, tax reform and public spending reforms, market-determined interest rates and competitive exchange rates, as well as the legal enforcement of property rights (Williamson, 1990, 2002).

The result of these various trends over the past 30 years has been a growing economic interdependence and integration between countries, evidenced in the increasing cross-border mobility of goods, services, capital, people and knowledge. However, the trends have been counter-balanced in some sectors by barriers to globalization. The impact of such barriers has been particularly acute in certain service sectors, where they have delayed the opening up of service activities to the global market. It is also important to note that the forces of globalization have also given rise to an anti-globalization movement, which has taken many forms and sought to resist the damaging economic, social, cultural and environmental impact of neoliberal free market policies (Klein, 2000; Kingsnorth, 2003). Before the forces driving and the barriers impeding the globalization of services are examined, it is necessary to set the context by considering the nature and scale of cross-border service transactions.

THE RISE OF CROSS-BORDER ACTIVITY IN THE SERVICE SECTOR

The internationalization of services has attracted much attention in recent years (Bryson et al., 2004; UNCTAD, 2004). This interest began in the 1980s and increased with the inclusion of services on the agenda of the Uruguay round of the GATT negotiations that began in 1986 (see, for example, Shelp, 1981; Riddle, 1986; Nusbaumer, 1987; Nicolaides, 1989; Daniels, 1993; *inter alia*). Interest further developed through the 1990s, and since the entering into force of the General Agreement on Trade in Services (GATS) in 1995, the internationalization of services has become a key issue of interest in the global economy (Aharoni and Nachum, 2000; Miozzo and Miles, 2002; Cuadrado-Roura et al., 2002; *inter alia*).

Until the 1980s, the study of services in an international context had been largely neglected. Service characteristics, such as intangibility, non-storability and the need for simultaneous production and consumption, necessitating the co-location of producer and consumer, supported the view that services were non-tradable. However, as already noted, cross-border service transactions have a long history of facilitating trade in tangible goods arising from the mobility of service producers and consumers. Nevertheless, prior to the 1980s, service activity was predominantly confined to local and national markets; yet, within these markets, they were becoming increasingly important as sources of value added and employment.

Forms of Cross-border Activity

Many classifications of the various methods of cross-border service delivery have been developed (UNCTAD, 1983; Sampson and Snape, 1985; Vandermerwe and Chadwick, 1989; Edvardsson et al., 1993; Roberts, 1998; Ball et al., 2008; *inter alia*). However, currently the most influential classification of modes of cross-border service delivery is that detailed in Article I of GATS, in which a comprehensive definition of cross-border service transactions is set out in terms of four different modes of supply: cross-border, consumption abroad, commercial presence in the consuming country and presence of natural persons (Table 14.1). These four modes of cross-border service delivery are not mutually exclusive. A firm may use several simultaneously; for example, the movement of personnel may accompany cross-border trade. The method used to deliver a service will depend on a variety of factors, including the nature of the service. For instance, the movement of the consumer often facilitates cross-border activity in the tourism sector, whereas in utility sectors, such as the provision of gas or electricity, the establishment of a presence in the overseas markets through FDI to merge with or acquire an existing supplier is the norm (Roberts, 2001). In producer services sectors, like management consultancy, cross-border activity varies according to the extent to which the service may be standardized and produced on an international basis and delivered via a range of modes (Roberts, 2006). Consultancy services that require specific knowledge of the national base are more dependent on local production, whether by local firms or subsidiaries of international firms (UNCTAD, 2002).

Each of the four modes of cross-border service delivery gives rise to specific types of barriers with a subsequent restriction on the globalization of the market for services.

Table 14.1 Modes of cross-border service supply set out in GATS

Mode	Description
1	Cross-border supply of services corresponds with the normal form of trade in goods
2	Consumption abroad, or, in the words of Article I, the supply of a service in the territory of one Member to the service consumer of another Member. Typically, this will involve the consumer travelling to the supplying country, perhaps for tourism or to attend an educational establishment
3	The supply of a service through the commercial presence of the foreign supplier in the territory of another WTO member. Examples would be the establishment of branch offices or agencies to deliver such services as banking, legal advice or communications
4	The presence of natural persons or the admission of foreign nationals to another country to provide services there

Source: Compiled from WTO (1999).

In addition to barriers to the cross-border supply of service trade in the form of tariffs and subsidies, services are also affected by regulations on the mobility of customers and service personnel as well as on flows of FDI, and particularly the right of establishment. In some service sectors there are specific regulatory barriers to internationalization. For instance, in the legal and accountancy sectors, familiarity and qualifications in national practices may be required. Even where there are generally no specific regulatory barriers to the cross-border supply, other barriers may exist. Management consultancy services, for example, are not only affected by formal restrictions on the movement of personnel and clients, which can limit access to certain markets, but also informal barriers to entry may exist in the form of access to networks (Glückler and Armbrüster, 2003).

Not all services are provided through every mechanism. Castle and Findlay (1988) claim that services consumed by households tend to be delivered by FDI or through the movement of the consumer, while producer services are supplied through the movement of the service provider, involving either a permanent base or temporary presence in the consuming country. Moreover, the internationalization of services has become more complex than merely the cross-border delivery of services. Rather, we are witnessing the internationalization of the service production process.

This internationalization of production has existed for some time in manufacturing sectors where the production process is fragmented and distributed across the globe to exploit differences in production costs and the availability of resources, including highly skilled labour (Ietto-Gillies, 2002). The division of labour thus becomes highly internationalized. This phenomenon is now developing in a range of service sectors where the final service can be delivered in a bespoke manner in a specific location yet be constructed from common intermediate service components produced in locations across the globe. Indeed, as Sundbo (1994, 2002) has observed, it is possible to deliver highly customized final services through the integration of standardized intermediate service components. By breaking down a final service into its intermediate service components, it is possible to facilitate a range of potential forms of international delivery modes. For instance, Ball et al. (2008) dissect the value chain for information-intensive services and, employing the concepts of front and back office, develop a conceptual model that offers

10 types of overseas market entry modes, all involving lower levels of involvement and resources than the establishment of a subsidiary.

Publishing services provide an excellent example of this internationalization of service production. An edited book may be commissioned from an editor in one country, produced from a collection of chapters from authors scattered across a number of countries, copyedited in another country, and typeset and printed in yet another. The production of services for consumer markets as well as for producer markets can now be the result of a highly internationalized production process. Services then become subject to the forces of global markets for labour and skills as well as other resources.

It is also important to note that the cross-border delivery of services may occur within the boundaries of the firm in the form of intra-firm trade (Roberts, 1998, 1999). For instance, manuals may be distributed, or databases accessed, throughout the global network constituting intra-firm trade; yet to facilitate the cross-border delivery of a final service to a customer requires that the knowledge embodied in the manual be extracted and applied by a consultant *in situ*. Intra-firm trade in services has expanded greatly with the practice of offshoring, which requires the organizational and technological capacity to relocate specific tasks and coordinate a spatially distributed network of activities (Levy, 2005). Offshoring has grown significantly during the past 15 years, brought about by a combination of factors. Among these, technological developments, in terms of the 'bandwidth revolution' and digital convergence, are perhaps the most obvious. Yet, as Metters and Verma (2008) note, these developments alone were not a sufficient condition for the growth of offshoring. The adoption of offshoring also required a new perspective on service processes that recognized the potential for de-coupling, supportive government policies and particular cultural conditions arising from the colonial past of Western countries, which provided common cultural understandings and languages conducive to the provision of many offshore service activities (Metters and Verma, 2008). Importantly, as Levy (2005) notes, although offshoring reduces production costs, through the creation of a global commodity market for particular skills, it also brings about a shift in the balance of market power between firms, workers and countries.

There are clearly differences in the types of cross-border delivery utilized by particular service sub-sectors. Table 14.2 attempts to capture a sense of this variety by matching a range of service sectors with common mechanisms for cross-border delivery. It should be noted, though, that even firms within the same sector might use different internationalization mechanisms, reflecting factors such as the size and experience of the business. In addition, individual firms, faced with a varied range of market and regulatory conditions in overseas locations, may also utilize a variety of foreign market entry mechanisms.

The varied nature of service transactions presents difficulties for the measurement of the scale of internationalization in the service sector and among service enterprises. For example, international transactions facilitated through the movement of personnel, trade in tangible forms, or delivered over the Internet, are not adequately reflected in the available data. In addition, international service transactions that occur within the boundaries of firms through intra-firm trade, both within the service sector and in other sectors, are not accurately tracked. Indeed, intra-firm service trade can be a cover for transfer pricing (Eden, 1998); hence, the values reported do not necessarily reflect the true value of a transaction. Measuring certain types of cross-border service transactions presents particular difficulties. Nevertheless, a brief overview of the available data on

Table 14.2 Examples of services and typical form of internationalization

Service	Typical form of internationalization
Retailing	Franchising (e.g. 7 Eleven) and FDI through mergers and acquisitions (e.g. the US retailer Walmart's takeover of the UK supermarket Asda in 1999; and the UK retailer Tesco's expansion into Eastern Europe in the mid 1990s through acquisitions and into Asia beginning in the late 1990s through majority stakeholdings)
Hotels & restaurants	In addition to ownership the main business formats in the hotel sector are: management contracts, franchising and hotel consortia (e.g. Marriott International). In the restaurant sector, franchising is popular (e.g. Subway®, McDonald's, Pizza Hut)
Telecommunications	Until the 1980s this sector was largely national; since deregulation and privatization, internationalization has developed rapidly, primarily through FDI in the form of mergers and acquisitions (e.g. the Vodafone–Mannesmann merger in 2000). The sector has also witnessed significant changes in levels of market competition brought about by the diffusion of mobile communications technologies
Transport services	The deregulation and privatization of nationally based transport sectors has initiated increased internationalization through FDI in the form of mergers and acquisitions as well as strategic alliances (e.g. in the passenger airline sector strategic alliances include oneworld and Star Alliance, established in 1999 and 1997 respectively; mergers include that between the airlines KLM and Air France in 2004)
Media & entertainment industries	Trade and FDI, in the form of mergers and acquisitions. Digital convergence has promoted significant rise in cross-border FDI in this sector as well as the entry of computer software and services suppliers into the market (e.g. Google Inc.'s acquisition of YouTube in 2007)
Education	Trade occurs through the movement of students and teachers as well as through correspondence courses and increasingly through the Internet. Franchising, joint ventures and wholly owned campuses are increasingly common in higher education (e.g. the joint venture between the University of Nottingham, UK and the Wanli Education Group, China, which established the University of Nottingham Ningbo China campus in 2004 in the city of Ningbo, China)
Utilities including water, gas & electricity	Until the 1980s, these sectors were generally government owned. Privatization and deregulation have led to the formation of internationally active firms in these sectors, often facilitated by FDI through mergers and acquisitions (e.g. the Swedish energy company Vattenfall's acquisition of a controlling stake in the Dutch energy company Noun in 2011)
Healthcare	Movement of patients, e.g. healthcare tourism, but also a growing number of cross-border mergers and acquisitions (e.g. the private British health provider Bupa's acquisition in 2007 of Health Dialog, a leading health information and disease management provider in the USA)
Business services	Establishment of a presence through greenfield sites, mergers and acquisitions, and the development of networks of national providers as well as trade through the movement of clients and staff (e.g. in 2009 the UK advertising and public relations services group WPP strengthened its presence in Africa through investments in Kenya-based Scangroup and The Jupiter Drawing Room & Partners and Smollan Group in South Africa)

international service transactions will set the context for the discussion of the factors facilitating and impeding the globalization of services that follows.

The Scale of Cross-border Service Transactions: A Brief Overview

In 2011, the value of world exports of commercial services was US$ 4149 billion, or 18.6 per cent of total trade (WTO, 2012, p. 22). These exports increased from US$ 1435 billion in 2000, representing growth of 290 per cent over the period. In the interim years commercial service trade has accounted for as much as 20 per cent of total trade, but in 2011 it fell back to the smallest share since 1990 (WTO, 2012, p. 22). Over the past 20 years, the proportion of trade accounted for by services has remained close to 20 per cent (GATT, 1994, p. 2; WTO, 1999, pp. 2–5), despite the fact that services now account for more than 60 per cent of total gross value added in most OECD countries (OECD, 2011, p. 75). The failure to take up an increasing proportion of world trade despite the growth of services in domestic markets is due to the limitations of arm's length trade as a vehicle for international service transactions.

Developed countries have a competitive advantage in the production and delivery of services due in part to their more highly developed domestic service markets. Nevertheless, the level of commercial service exports originating from the emerging economies is growing, especially from China and India. According to the World Trade Organization (WTO) (2012, p. 23) the top five exporters of commercial services in 2011 were the United States (14 per cent of the world total), the UK (7 per cent), Germany (6 per cent), China (4 per cent) and France (4 per cent). The top five importers of commercial services were the US (10 per cent of the world total), Germany (7 per cent), China (6.1 per cent), the UK (4.4 per cent) and Japan (4.3 per cent). If the European Union is treated as a single entity it becomes the top exporter of commercial services (24.8 per cent of the world total), followed by the US (18.2 per cent), China (5.7 per cent), India (4.7 per cent) and Japan (4.5 per cent). The EU also becomes the leading importer (21.1 per cent of the world trade total), followed by the US (12.9 per cent), China (7.8 per cent), Japan (5.4 per cent) and India (4.3 per cent) (WTO, 2012, p. 23).

Given the limited tradability of many services, the establishment of an overseas presence through FDI or a contractual mechanism, such as franchising and licensing, is an important method of cross-border delivery in some sub-sectors. In 2011, total global FDI flows reached $1.5 trillion (UNCTAD, 2012, p. 1); service sector FDI accounted for some $570 billion of this total (UNCTAD, 2012, p. 9). Services accounted for 40 per cent of FDI projects (cross-border mergers and acquisitions and greenfield investments) in 2011, distributed among four key areas: Electricity, gas and water; Transport, storage and communications; Finance; and Business services (UNCTAD, 2012, p. 10). Since the global financial crisis of 2008, services have seen a decline as a proportion of total FDI projects from an average of 50 per cent of all projects between 2005 and 2007 to 40 per cent in 2011 (UNCTAD, 2012, p. 10).

Some modes of cross-border delivery do not involve significant trade or investment transactions. For instance, franchising allows a service format to be internationalized through licensing and varying degrees of supervision from the franchisor. Such activity gives rise to a stream of income into the home country in the form of royalty payments, which are included in the commercial service trade transactions – accounting

Table 14.3 Top 10 global franchise rankings

Rank	Name	Country	Industry
1	SUBWAY®	USA	Sandwich & bagel franchises
2	McDonald's	USA	Fast food franchises
3	KFC	USA	Chicken franchises
4	7 Eleven	USA	Convenience store franchises
5	Burger King	USA	Fast food franchises
6	Pizza Hut	USA	Pizza franchises
7	Wyndham Hotel Group	USA	Hotel franchises
8	Ace Hardware Corporation	USA	Home improvement retail franchises
9	Dunkin' Donuts	USA	Bakery & donut franchises
10	Hertz	USA	Car rental & dealer franchises

Source: Compiled from *The Top Global Franchises Report 2012* available at: http://www.franchisedirect.com/top100globalfranchises/ (accessed 30 October 2013).

for 6.4 per cent of commercial service exports in 2011 (WTO, 2012, p. 139). Yet through the licensing of ideas the manifestation of global service brands becomes significant and widespread and their impact as signifiers of globalization may far outweigh the actual cross-border financial transactions to which they give rise.

Franchising is an important mode for the international expansion of many service businesses, particularly in the hotel and fast food restaurant sectors and portions of the retail sector.[3] This is illustrated by the list of the top 100 global franchises, which is dominated by service activity and by companies originating in the USA.[4] Table 14.3 lists the top 10 global franchises, most of which are highly familiar to individuals living in the advanced and emerging economies.

Cross-border service transactions arising from the temporary movement of individuals and Internet-based exchanges are poorly represented in current trade data. Nevertheless, elements of this activity are tracked through the trade categories 'Travel' and 'Communications services', which in total accounted for 25.6 per cent and 2.5 per cent respectively of total commercial service exports in 2011 (WTO, 2012, p. 139).

As noted earlier, the expansion of the activities of MNCs as well as the development of global value chains have increased the amount and significance of intra-firm trade flows. Services are intermediate inputs for both service and manufacturing MNCs and they are traded within the boundaries of firms. Yet, this intra-firm trade remains poorly monitored on a global basis, with data only being available for the United States and Canada. In a recent report on intra-firm trade, Lanz and Miroudot (2011) explored the available data on intra-firm trade in services and found that the share of intra-firm exports in private US services exports was 26 per cent in 2008, while the respective share for intra-firm imports was 22 per cent. They argue that a reason for these relatively low shares is that the services classified under 'Travel' and 'Passenger fares' by definition cannot be traded intra-firm. An examination of the aggregate category 'Other private services' indicates that intra-firm transactions accounted for 32 per cent of exports and 40 per cent of imports, respectively (Lanz and Miroudot, 2011, pp. 14–15). From a breakdown of service activities, Lanz and Miroudot (2011, p. 15) found that the share of intra-firm trade was highest in 'Management and consulting services' (88 per cent for exports, 86 per cent for imports), which are essential for the production and distribution networks of MNCs, and in 'Research and development and testing services' (83 per cent for exports, 73 per cent for imports), which often form the basis of an MNC's competitive strength. A similar pattern of intra-firm service trade was found for Canada (Lanz and Miroudot, 2011).

FORCES DRIVING THE GLOBALIZATION OF SERVICES

This section will examine the forces driving the globalization of services. However, it is first necessary to recognize that the factors that have led to an increase in the size of domestic service activity have also contributed to the globalization of the sector. The rise in incomes over the past 50 years, resulting from a prolonged period of economic growth following the recovery from World War II, has stimulated the demand for all sorts of services. With higher incomes, individuals can go beyond satisfying their basic needs for food and shelter and allocate increasing proportions of their resources to a growing range of consumer services, from foreign holidays and health spa subscriptions to regular dining out and elective cosmetic surgery.

Beginning with the establishment of GATT in 1947, the liberalization of trade encouraged economic growth through international specialization. However, international trade also resulted in the intensification of competition as firms competed in global rather than national markets. By the 1970s this intensification of competition, together with other economic pressures, had led to changing structures of production, with profound implications for the growth of the producer service sector. Faced with oil price rises, economic recession and increased competition from the Far East, companies in the advanced countries were compelled to develop more efficient organizational structures. This led to downsizing and a focus on core competencies (Prahalad and Hamel, 1990), with other activities, including support services, being externalized and outsourced (Perry, 1992). Production structures evolved from the 'Fordist'-style mass production of the post-war era to 'post-Fordist' or 'postmodern' structures (Amin, 1994; Linstead, 2003), including the adoption of more flexible production systems (Piore and Sabel, 1984), allowing for the supply of customized goods and services in an efficient and timely fashion. These developments were influenced by Japanese management practices, including 'just-in-time' production systems, continuous improvement and quality circles (Garrahan and Stewart, 1992).

The key sources of competitiveness for the post-Fordist or postmodern organization are knowledge and information. Hence, the ability to adapt flexibly to market opportunities depends not on vast amounts of capital equipment but on the knowledge of how to access such assets in an efficient and cost-effective manner (Roberts and Armitage, 2006). This restructuring of production since the 1970s and a shift in focus towards knowledge-based resources (Grant, 1996) have led to significant growth in the producer services sector, including, for instance, knowledge-intensive business services like marketing and advertising, training, design, management consultancy, accounting and bookkeeping, legal services, environmental services, computer services and R&D consultancy (Miles et al., 1995). Such services support clients in the development and protection of their knowledge assets.

Although the externalization of service activity from the manufacturing sector has contributed to the growth of producer services, it is also important to note that these services have also grown in their own right through, for example, the development of new activities facilitated by an expansion in the scale of production. By servicing multiple clients in various locations producer service firms develop new knowledge that can be incorporated into new service offerings. Moreover, they can exploit economies of scale and scope by supplying large national and international markets and diversifying into closely related activities.

Additionally, the globalization of economic activity in all sectors has resulted in the expansion of cross-border provision of a range of trade- and investment-related services, including finance, banking, legal, insurance, transport and distribution services. Just as the growth of multinational banks in the early 20th century followed the rise of multinational manufacturing and primary sector companies (Jones, 1993), the late 20th century saw the development of a range of international providers of producer services from advertising and accountancy to management consultants and computer service companies (Aharoni, 1993; Roberts, 1998; Nachum, 1999; Aharoni and Nachum, 2000). Hence, suppliers following their clients into overseas markets initiated the globalization of a range of business and producer services firms.

As Held et al. (1999) note, the process of globalization is affecting culture. Indeed, certain aspects of culture are being globalized through, for instance, the influence of global media and global brands. This globalization of culture is itself contributing to the growth of cross-border transactions in personal and recreational services, including tourism, education and entertainment services.

The globalization of services has also been promoted by the widespread adoption of neoliberal policies, which have resulted in the privatization of service providers that were once state owned. As nationalized sectors such as utilities, like water, gas, electricity and telecommunications, became private companies they were exposed to the full forces of the global market. Once privatized, it was not long before they became international, often through mergers and acquisitions. Hence, changes in government policies opened up certain service sectors to the forces of global competition.

Privatization was accompanied by the deregulation of certain sectors, which further promoted the opening up of these areas to the global market. For instance, the number of foreign banks present in the City of London increased significantly following the deregulation of the UK financial markets in 1986, which allowed the entry of foreign firms onto the Stock Exchange (Dicken, 2011). Consequently, privatization and deregulation facilitated the development of new global service markets. Providers once confined to national markets could gain economies of scale in rapidly forming global markets.

More recently, the global financial crisis of 2008 has led to a period of austerity among many of the developed countries. One result has been a renewed effort to cut government expenditure through the privatization and outsourcing of government services. This significantly extends the scope for the expansion of international suppliers into areas as diverse as healthcare, education, prison management, security services and social services. Healthcare and higher education sectors, for example, are already experiencing growing levels of cross-border activity (Angeli and Maarse, 2012; Li and Roberts, 2012).

The coordination of government policies has also influenced the scale of cross-border provision of services. For instance, regional integration, best illustrated by the European

Union's single market initiative of 1992, gave rise to a large and increasingly harmonized European market. As the European Union has expanded in size over the past two decades so too has its single market. However, barriers to services have persisted, and in 2006 the European Parliament and the Council adopted the Services Directive to bolster the single service market.

The coordination of government policies since the end of World War II is embodied in a number of international institutions, including the WTO, IMF and the World Bank. The activities of such institutions have promoted globalization. Of particular relevance to services is the WTO, formed as an outcome of the Uruguay GATT round of negotiations (1986–1993). Since 1947, trade in manufactured goods has been increasingly liberalized through the various rounds of GATT negotiations. By the 1980s, member countries, especially those with an advantage in service delivery like the US, were pushing for the inclusion of services in the GATT system (Egger and Rainer, 2008). A key outcome of the GATT Uruguay round of negotiation was GATS. This covers all services with two exceptions: services provided in the exercise of governmental authority; and, in the air transport sector, air traffic rights and all services directly related to the exercise of traffic rights (WTO, 2001). Since its inception, there has been a progressive liberalization of cross-border activity in services as member states sign up to the liberalization of sectors in a flexible and ongoing manner. The GATS sets out a number of rules concerning international service transactions, detailed in Table 14.1. These include the Most Favoured Nation (MFN) clause, a core principle of the original GATT agreement, which ensures that member countries may not discriminate in their treatment of other members. In addition, GATS prohibits restrictions on market access and embodies the principle of national treatment, which ensures that foreign service suppliers are treated no less favourably than domestic providers.

Although the MFN principle is a general rule, the application of the market access and national treatment principles depends on each country's commitments. Horizontal commitments apply to all sectors listed by a country. However, country- and sector-specific exceptions from the market access and national treatment principles can be specified (Egger and Rainer, 2008, p.1666). Importantly, the pace of liberalization is largely determined by each member state deciding which services to open up to global markets (WTO, 2001).

In addition to the policies and regulations of national governments and international organizations, technology has had a significant impact on the globalization of services. As Bhagwati (1984) elaborated over three decades ago, when the application of ICTs to services was in its early days, new technologies allowed services to be de-coupled from the provider and embodied in tangible forms. Moreover, with the rise of interactive real-time rich communication facilitated by the Internet and social networking technologies, the scope for the delivery of services at a distance has increased significantly. Telemedicine, for instance, enables the provision of increasingly sophisticated healthcare services at a distance (Ekeland et al., 2010).

Technology has also facilitated the increased mobility of people and artefacts. Moreover, technological innovations during the past century, like the aeroplane and the computer, have resulted in the emergence of major new service activities like the passenger airline sector and information technology consultancy. In addition, advances in communications technologies, beginning with the telephone in the mid 1870s, facilitated the mobility of information and knowledge, allowing the arm's length delivery of information-based services in a rapid and cost-effective fashion. Developments in mobile telecommunications in the late 20th century facilitated the emergence and growth of a completely new segment of the sector, which has been accompanied by new services embodied in software applications accessible and usable on mobile telecommunications devices.

One of the most significant technological developments affecting service activities has been the Internet, and especially the introduction of the user-friendly World Wide Web since the 1990s. The Internet has facilitated the cross-border provision of service activities from retailing, through e-commerce, to consumer services like entertainment and education, to producer services like information and technology support and consultancy services. Through technologies like the Internet, consumers can be provided with either standardized or personalized services, developed from standardized service components, at a distance. The ability of the Internet to unite global markets allows niche service markets that were once unprofitable to be exploited by specialist providers (Anderson, 2009). In a study of the impact of the Internet on trade in services, Freund and Weinhold (2002, p. 236) found that, after controlling for GDP and exchange-rate movements, a 10 per cent increase in Internet penetration in a foreign country is associated with around a 1.7 percentage point increase in export growth and a 1.1 percentage point increase in import growth.

The Internet is affecting the organizational structures that produce and deliver goods and services. One example related to many services is the increased scope for customers to engage in self-service activities. For instance, in the passenger transport sector, travellers can administer the ticket-purchasing process themselves through ticketing and journey planner websites, and air passengers can check in online. These activities can be undertaken from any location as long as the Internet can be accessed. Such activities not only change the nature of cross-border service transactions but also influence the organizational structure of service businesses by reducing the need for frontline staff, as face-to-face contact is replaced by face-to-screen interaction.

Not only have innovations led to new service activity but they have also enabled significant growth in the globalization of existing service activity. Consequently, tourism, which in the early 20th century was very much confined within a nation for all but the very wealthy, has now become highly globalized. Developments in transportation technology have been important in this sector, but they have also been accompanied by regulatory changes that have increased competition and reduced transport costs. For instance, in the airline sector, deregulation, beginning with the 1978 Airline Deregulation Act in the US and later reinforced by EU deregulation and various bilateral Open Skies agreements, has opened markets up to competition, which has not only seen the internationalization of formerly national carriers but also the emergence of a burgeoning low-cost airline sector.

A final factor driving the cross-border provision of services relates to rapidly changing economic, political, social, cultural, technological and environmental conditions. The uncertainty caused by these developments has stimulated an increasing demand for services such as management consultancy and environmental advisory services from private businesses and government agencies. Many of the risks and uncertainties that exist in today's world, arising, for instance, from global warming, the challenges of sustainability, international crime and terrorist threats to political stability, can only be addressed meaningfully at an international level. Hence, there is a demand for service providers that can demonstrate credibility on a global scale.

BARRIERS TO THE GLOBALIZATION OF SERVICES

As shown in the previous section, many barriers to the globalization of services have been removed over the past 30 years due to technological developments and regulatory reforms, including the liberalization of service markets. Nevertheless, barriers remain related to the features of particular services themselves and differences remaining in cultural and regulatory environments between nations as well as remaining technological limitations. In some cases, these barriers restrict the nature of the cross-border supply while in others they act as absolute barriers to foreign suppliers.

Although services are increasingly traded, many service activities are still locationally bound, either because they require close proximity between the producer and the client (e.g. personal services and highly customized services) or because they are specific to a particular location (e.g. natural tourist attractions). While it is possible to supply such services across borders through the establishment of a commercial presence, it is necessary to have the right people at the interface with consumers. The right people may be local citizens with a deep understanding of location and in possession of culturally specific knowledge. The establishment of a presence through the acquisition of a local firm ensures the presence of service providers that are familiar with the local customs and practices as well as the existing client base. In such circumstances, a shift from domestic provision to overseas provision can occur without any change in the service experience for the client.

Issues around the protection of proprietary knowledge can be a barrier to the globalization of service businesses. The level of standardization or customization becomes an important influence on the method of internationalization. Where a high level of standardization is possible (e.g. fast food restaurants), mechanisms such as franchising and licensing agreements can be used successfully to deliver the service to overseas markets. Contractual methods of cross-border supply are viable when it is possible to capture the qualities of the service in a codified manner (e.g. in a service manual). The knowledge used in the production of the service is standardized and it can therefore be codified and protected by a contract – as long as there are appropriate legal systems in place to enforce contracts. Clearly, service businesses expanding into overseas markets through contractual mechanisms will avoid markets with poorly developed legal systems. Lack of a developed system of protection of intellectual property will also restrict the entry of service providers through contractual mechanisms where competitiveness is underpinned by knowledge-based assets.

When a service is customized the knowledge utilized in its production is variable or idiosyncratic and dependent on the needs of individual clients. The knowledge used in the production of the service is therefore difficult to codify because much of it will be tacit in nature. The service production and delivery will require close interaction between the producer and client firm. Furthermore, it is difficult to capture such knowledge in a contract and therefore harder to protect. In such circumstances, the provision of a service must be internalized within the firm and therefore cross-border delivery is facilitated via the extension of the firm's boundaries into the overseas market through FDI. Consequently, if an overseas market is not open to FDI from an overseas service producer this barrier will prevent the cross-border supply.

The characteristics of services that limit trade can be overcome in many cases if internationalization is facilitated by the establishment of a presence in an overseas market through either FDI or related means. However, the level of concentration in the sector also influences the extent to which service suppliers will seek internationalization. Large numbers of small-scale suppliers serving local markets characterize some service sectors. Many personal services, like hairdressing and counselling services, are frequently provided on this basis. Other sectors, such as transportation and utilities, are heavily concentrated. Requiring high levels of capital investment, these sectors serve national or international markets. Still other service sectors, like business services, display a marked polarization between many small suppliers serving local and niche markets and a few very large suppliers that are international in scope (Roberts, 2003). Where concentration levels are high, and regulation permitting, there is more involvement of overseas suppliers in markets, particularly in the form of FDI facilitated through cross-border mergers and acquisitions. In contrast, when concentration levels are low there is less involvement from overseas suppliers and where such involvement is present it tends to be through contractual mechanisms such as franchising.

For many services the process of production and delivery is embedded in the producer–client relationship. Consequently, sensitivity to the cultural characteristics of the market is essential in promoting competitiveness. Despite the impact of globalization on national culture, the existence of a widespread global culture is questionable (Bauman, 1998). Even so, it may be possible to identify a global culture among certain groups, such as an elite or the international business community (Bird and Stevens, 2003). Nations, and indeed regions within them, have distinctive cultural traits and in some service sectors this can influence the degree of success of foreign-based suppliers, as well as influencing the method of foreign market entry they select. Culture is influenced by a variety of factors, including, for instance, the dominant belief system such as religion and political ideology, the degree of openness, the language and historical factors. The cultural characteristics of a nation will influence consumer preferences, business culture and practices, and may also give rise to regulation and barriers to trade; for example, the French film industry is protected as a measure to preserve French culture (Jäckel, 2007).

As previously noted, services have traditionally been more heavily regulated than other sectors of the economy. This is partly due to their intangible nature and the fact that consumers cannot judge the quality of many services. For example, patients must take on trust the advice given by a medical practitioner because they do not possess adequate medical knowledge to assess the quality of the advice. There are problems of adverse selection and moral hazard arising in the provision of services due to the asymmetric distribution of knowledge. It is for this reason that government agencies and professional associations regulate medical doctors; for instance, in the UK doctors are regulated by the General Medical Council and the British Medical Association. Other regulated service activities include those that are monopoly providers or in highly concentrated sectors, such as utilities, and areas that are of particular importance to the effective operation of the economy, like banking and finance, and which involve fiduciary responsibilities. Despite deregulation and privatization, these sectors continue to be more regulated than most other areas of the economy (White, 1999, p. 4). In addition, professional services (e.g. medical, legal, accounting, architecture and engineering) have been subject to extensive direct and indirect governmental regulation.

Regulation exists to protect consumers against monopoly power or against abuse arising because of the superior knowledge of the service provider vis-à-vis the customer. However, regulation may be a barrier to internationalization if it discriminates against foreign suppliers. Snape (1990, p. 7) notes that regulation can create barriers to the movement of suppliers and recipients of service through restrictions on the inflow of labour, FDI and service equipment; the outflow of residents; the ability of foreign professionals to practice; and taxation specific to foreign carriers. In addition, service trade may be limited through restrictions on the use of foreign contractors; restrictions on telecommunications flows (including Internet access); restrictions on placement of banking or insurance business abroad; foreign exchange controls; domestic content requirements in radio and television broadcasting, in the cinema or theatre; and restrictions in shipping.

Regulatory reform has been driven by a number of factors, including the development of new services and technological developments, which have changed the economics of specific service industries (Feketekuty, 1999, pp. 1–2). For instance, the development of mobile telephone communications has changed the economics of the telecommunications sector, introducing new competition into a sector that was once a natural monopoly. Moreover, mobile communications have given rise to a range of new affordable information services. In addition, the advances in the liberalization of service trade under the GATS agreement have further reduced regulation as a barrier to the globalization of services.

Nevertheless, regulatory barriers remain, and for good reason in certain sectors and circumstances. The global financial crisis of 2008 has stimulated reviews of financial sector regulation at the national level, with the potential for the imposition of barriers to the cross-border provision of aspects of such services. Clearly, in service sectors of central significance to a nation's economy and security, liberalization and opening up to global markets are likely to remain restricted. Moreover, GATS has attracted criticism for its emphasis on commercial interests to the potential detriment of legitimate public interest regulation and democratic decision making (Sinclair and Grieshaber-Otto, 2002).

Unsurprisingly, opposition to liberalization is often at its strongest when it concerns the privatization of publicly owned service providers. For instance, the globalization of utility companies has met with local resistance, particularly when it results from structural adjustment policies upon which financial support from the World Bank and the IMF is conditional. Such policies allow MNCs from developed countries to buy out or take control of previously nationally owned and operated assets, often at bargain prices. However, such market entry does not always run smoothly, as the case of the privatization of the water system in Cochabamba, Bolivia, demonstrates. In 1999, under pressure from international financial institutions, the Bolivian government leased SEMAPA, the public water supplier of Cochabamba, to the multinational consortium Aguas del Tunari (Aguas), whose major shareholder was the Californian-based MNC, Bechtel (Otto and Böhm, 2006). The agreement allowed Aguas to expropriate the independent water and irrigation systems and autonomous water services of Cochabamba. The result was that water prices tripled for some of the poorest people in the world without any improvement in service quality (Wolf et al., 2005; Otto and Böhm, 2006). The privatization of the water services was challenged by the coordinated resistance of the people of Cochabamba, supported by the wider Bolivian population. Protests culminated in several days of violence when the government sent soldiers into the city, and the violence only eased when the government agreed to return the service to local municipal ownership in April 2000 (Wolf et al., 2005).

Despite such cases, the harmonization of service sector regulation and the further liberalization of service markets through the current Doha round of WTO negotiations will increase the globalization of services. The liberalization of trade in services has the potential to enhance the gains from trade in goods, since many services play a critical role facilitating international trade in products other than themselves (e.g. transport, finance, insurance, communication and some professional services) (Deardorff, 1999). Moreover, increased global competition in the supply of producer services has the potential to improve the efficiency of client firms (Roberts, 2003).

Recent technological developments have also had a significant impact on the service sector. Yet technology still limits our ability to engage in cross-border transactions. While there have been important developments in the ability to move information across borders at high speed and in vast quantities, moving the tangible elements required in service delivery, including people, remains restricted by the technologies of transportation. The failure of cost-effective supersonic commercial passenger flight, for example, has placed a significant technological barrier on the movement of service providers. So, for instance, an air journey from London to New York is likely to take over six hours, a period of time that has not diminished since the widespread introduction of the commercial jet airplane in the 1960s. Additionally, environmental concerns over transport activity and increased fuel costs act as a barrier to service trade. While teleconferencing may reduce the need for travel, for some activities face-to-screen interaction cannot replace face-to-face engagement.

In the aftermath of the global financial crisis in 2008, flows of trade and FDI declined rapidly. Although these flows have rebounded, it is evident that the trade and FDI in service activity have been slower to recover than within other sectors. The economic fragility in many advanced economies is likely to continue to have a dampening influence on trade and investment flows until a firm recovery takes hold. In relation to the globalization of services, the current economic environment gives rise to two contradictory forces. On the one hand, the need for governments to follow austerity policies offers further opportunities for the liberalization and privatization of state-owned services. On the other hand, high levels of unemployment and economic hardship threaten a turn towards more protectionist trade and investment policies, particularly in areas viewed to be of importance to national culture and security.

CONCLUSION

The cross-border provision of services varies in its form and complexity according to the nature of the service and the technological, cultural and regulatory environment. Despite the globalization of services, many service activities and the businesses that facilitate their cross-border supply remain under-researched. Service businesses face many challenges in the 21st century and research activity can do much to assist both enterprises and policymakers to take the most appropriate action to ensure the maximum benefits from the globalization of service markets for consumers and suppliers. Changes in regulatory and technological environments will undoubtedly influence the evolution of cross-border supply strategies employed by service businesses. Research is essential both to enhance our understanding of service activity and to appreciate the challenges ahead, for there can be no doubt that the ongoing process of globalization, together with technological change, will promote the continued expansion of cross-border service transactions.

NOTES

1. An international agreement aimed at reducing the barriers to trade and seeking to maximize the benefits from the free flow of goods.
2. Deng Xiaoping set in train the transformation of China's economy when he announced the new 'open door' policy in December 1978.
3. As data presented by the European Franchise Federation reveal: http://www.eff-franchise.com/spip.php?rubrique9 (accessed 30 October 2012).
4. Top 100 Global Franchises 2012, The Report: http://www.franchisedirect.com/top100globalfranchises/ (accessed 30 October 2012).

REFERENCES

Aharoni, Y. (ed.) (1993), *Coalitions and Competition: The Globalization of Professional Business Services*, London: Routledge.
Aharoni, Yair and Nachum, Lilach (eds) (2000), *Globalization of Services: Some Implications for Theory and Practice*, London: Routledge.
Amin, A. (ed.) (1994), *Post-Fordism: A Reader*, Blackwell Publishers, Oxford.
Anderson, Chris, (2009), *The Longer Long Tail: How Endless Choice is Creating Unlimited Demand*, London: Random House Books.
Angeli, F. and Maarse, H. (2012), 'Mergers and Acquisitions in Western European Health Care: Exploring the Role of Financial Services Organizations', *Health Policy*, 105(2–3): 265–272.
Ball, D.A., Lindsay, V.J. and Rose, E.L. (2008), 'Rethinking the Paradigm of Service Internationalization: Less Resource-Intensive Market Entry Modes for Information-Intensive Soft Services', *Management International Review*, 48(4): 413–431.
Bauman, Zygmunt (1998), *Globalization: The Human Consequences*, Cambridge: Polity Press.
Bhagwati, J.N. (1984), 'Splintering and Disembodiment of Services and Developing Nations', *The World Economy*, 7(2): 133–144.
Bird, A. and Stevens, M.J. (2003), 'Toward an Emergent Global Culture and the Effects of Globalization on Obsolescing National Cultures', *Journal of International Management*, 9(4): 395–407.
Blomstermo, A., Sharma, D. and Sallis, J. (2006), 'Choice of Foreign Market Entry Mode in Service firms', *International Marketing Review*, 23(2): 211–229.
Bryson, John R., Daniels, Peter W. and Warf, Barney (2004), *Service Worlds: People, Organisations, Technologies*, London and New York: Routledge.
Castle, L. and Findlay, C. (eds) (1988), *Pacific Trade in Services*, Sydney: Allen and Unwin, pp. 1–15.
Cuadrado-Roura, J.L., Rubalcaba-Bermejo, L. and Bryson, J.R. (2002), *Trading Services in the Global Economy*, Cheltenham and Northampton, MA: Edward Elgar.
Daniels, Peter W. (1993), *Service Industries in the World Economy*, Oxford and Cambridge, MA: Blackwell.
Deardorff, A.V. (1999), *International Provision of Trade Services, Trade, and Fragmentation*, paper prepared for a World Bank Project, WTO 2000, 27 October.
Dicken, Peter (2011), *Global Shift: Mapping the Changing Contours of the World*, sixth edition, London: Sage Publications.
Eden, Lorraine (1998), *Taxing Multinationals: Transfer Pricing and Corporate Income Taxation in North America*, Toronto: University of Toronto Press.

Egger, P. and Rainer, L. (2008), 'The Determinants of GATS Commitments Coverage', *The World Economy*, 31(12): 1666–1694.
Edvardsson, B., Edvinsson, L. and Nystrom, H. (1993), 'Internationalisation in Service Companies', *Service Industries Journal*, 13(1), January: 80–97.
Ekeland, A.G., Bowes, A. and Flottorp, S. (2010), 'Effectiveness of Telemedicine: A Systematic Review of Reviews', *International Journal of Medical Informatics*, 79(11): 736–771.
Feketekuty, G. (1999), 'Regulatory Reform and Trade Liberalization in Services', *Proceedings of the World Services Forum Conference*, December.
Freund, C. and Weinhold, D. (2002), 'The Internet and International Trade in Services', *American Economic Review*, 92(2), Papers and Proceedings of the One Hundred Fourteenth Annual Meeting of the American Economic Association, pp. 236–240.
Garrahan, P. and Stewart, P. (1992), *The Nissan Enigma: Flexibility at Work in a Local Economy*, London: Mansell.
GATT (1994), *International Trade: 1994 Trends and Statistics*, Geneva: GATT Secretariat.
Giddens, Anthony (1990), *The Consequences of Modernity*, Cambridge: Polity Press.
Giddens, Anthony (2000), *Runaway World*, London: Routledge.
Glückler, J. and Armbrüster, T. (2003), 'Bridging Uncertainty in Management Consulting: The Mechanisms of Trust and Networked Reputation', *Organization Studies*, 24(2): 269–297.
Grant, R.M. (1996), 'Toward a Knowledge-Based Theory of the Firm', *Strategic Management Journal*, 17(S2): 109–122.
Harvey, D. (2005), *A Brief History of Neoliberalism*, Oxford: Oxford University Press.
Held, David, McGrew, Anthony, Goldblatt, David and Perraton, Jonathan (1999), *Global Transformations: Politics, Economics and Culture*, Cambridge: Polity Press.
Hirst, P. and Thompson, G. (1999), *Globalization in Question: The International Economy and the Possibilities of Governance*, second edition, Cambridge: Polity Press.
Ietto-Gillies, Grazia (2002), *Transnational Corporations: Fragmentation amidst Integration*, London: Routledge.
Jäckel, A. (2007), 'The Inter/Nationalism of French Film Policy', *Modern and Contemporary France*, 15(1): 21–36.
Jain, A.K. (2009), 'Regulation and Subprime Turmoil', *Critical Perspectives on International Business*, 5(1/2): 98–106.
Jones, Geoffrey (1993), *British Multinational Banking, 1830–1990*, Oxford: Clarendon Press.
Kingsnorth, P. (2003), *One No, Many Yeses: A Journey to the Heart of the Global Resistance Movement*, New York: Free Press.
Klein, N. (2000), *No Logo*, London: Flamingo.
Kundu, S.K. and Merchant, H. (2008), 'Service Multinationals: Their Past, Present, and Future', *Management International Review*, 48(4): 371–376.
Lanz, R. and Miroudot, S. (2011), 'Intra-Firm Trade: Patterns, Determinants and Policy Implications', *OECD Trade Policy Papers*, No. 114, OECD Publishing, available at: http://dx.doi.org/10.1787/5kg9p39lrwnn-en (accessed 3 December 2014).
Levy, D.L. (2005), 'Offshoring in the New Global Political Economy', *Journal of Management Studies*, 42(3): 685–693.
Li, X. and Roberts, J. (2012), 'A Stages Approach to the Internationalization of Higher Education? The Entry of UK Universities into China', *Service Industries Journal*, 32(7): 1011–1038.
Linstead, S. (ed.) (2003), *Organization Theory and Postmodern Thought*, London: Sage.
Metters, R. and Verma, R. (2008), Metters, R. and Verma (2008), 'History of Offshoring Knowledge Services', *Journal of Operations Management*, 26(2): 141–147.
Miles, I., Kastrinos, N., Flanagan, K., Bilderbeek, R., Hertog, P., Huntink, W. and Bouman, M. (1995), *Knowledge-Intensive Business Services: Users, Carriers and Sources of Innovation*, EIMS Publication No. 15, Innovation Programme, Directorate General for Telecommunications, Information Market and Exploitation of Research, Commission of the European Communities, Luxembourg.
Miozzo, Marcela and Miles, Ian (eds) (2002), *Internationalisation, Technology and Services*, Cheltenham and Northampton, MA: Edward Elgar.
Nachum, Lilach (1999), *The Origins of the International Competitiveness of Firms: The Impact of Location and Ownership in Professional Service Industries*, Cheltenham and Northampton, MA: Edward Elgar.
Nicolaides, Phedon (1989), *Liberalizing Service Trade*, London: Routledge.
Nusbaumer, J. (1987), *Services in the Global Market*, Boston, MA: Kluwer Academic Publishers.
OECD (2005), *OECD Handbook on Economic Globalisation Indicators*, Paris: OECD.
OECD (2011), *OECD Factbook 2011–2012: Economic, Environmental and Social Statistics*, OECD Publishing. doi: 10.1787/factbook-2011-en.
Ohmae, K. (1990), *The Borderless World: Power and Strategy in the Global Marketplace*, London: Harper Collins.

Otto, B. and Böhm, S. (2006), '"The People" and Resistance against International Business: The Case of the Bolivian "Water War"', *Critical Perspectives on International Business*, 2(4): 299–320.
Perry, M. (1992), 'Flexible Production, Externalisation and the Interpretation of Business Service Growth', *Service Industries Journal*, 12(1): 1–16.
Piore, M.J. and Sabel, C.F. (1984), *The Second Industrial Divide: Possibilities for Prosperity*, New York: Basic Books.
Prahalad, C.K. and Hamel, G. (1990), 'The Core Competence of the Corporation', *Harvard Business Review*, 68(3): 79–91.
Reich, R.R. (1992), *The Work of Nations: Preparing Ourselves for 21st Century Capitalism*, New York: Vintage Books.
Riddle, Dorothy (1986), *Service-Led Growth: The Role of the Service Sector in World Development*, New York: Praeger.
Roberts, J. (1998), *Multinational Business Service Firms: The Development of Multinational Organisational Structures in the UK Business Services Sector*, Aldershot: Ashgate Publishing Company.
Roberts, J. (1999), 'The Internationalisation of Business Service Firms: A Stages Approach', *Service Industries Journal*, 19(4): 68–88.
Roberts, J. (2001), 'Challenges Facing Service Enterprises in the Global Knowledge-based Economy: Lessons from the Business Services Sector', *International Journal of Services Technology Management*, 2(3/4): 402–433.
Roberts, J. (2003), 'Competition in the Business Services Sector: Implications for the Competitiveness of the European Economy', *Competition and Change: The Journal of Global Business and Political Economy*, 7(2–3): 127–146.
Roberts, J. (2006) 'Internationalisation of Management Consultancy Services: Conceptual Issues Concerning the Cross-border Delivery of Knowledge Intensive Services', in Harrington, J.W. and Daniels, P.W. (eds), *Knowledge-Based Services: Internationalisation and Regional Development*, Aldershot: Ashgate Publishing Ltd, pp. 101–124.
Roberts, J. and Armitage, J. (2006), 'From Organization to Hypermodern Organization: On the Accelerated Appearance and Disappearance of Enron', *Journal of Organizational Change Management*, 19(5): 558–577.
Sampson, G.P. and Snape, R.H. (1985), 'Identifying the Issues in Trade in Services', *World Economy*, 8(2): 171–181.
Scholte, J.A. (2002), "What is Globalization? The Definitional Issue – Again", CSGR Working Paper No. 109/02, available at: www2.warwick.ac.uk/fac/soc/csgr/research/workingpapers/ jascholte/ (accessed 9 April 2009).
Shelp, Ronald, K. (1981), *Beyond Industrialisation: Ascendancy of the Global Service Economy*, New York: Praeger Publishers.
Sinclair, Scott and Grieshaber-Otto, Jim (2002), *Facing the Facts: A Guide to the Gats Debate*, Ottawa: Canadian Centre for Policy Alternatives.
Snape, R.H. (1990), 'Principles in Trade in Services', in Messerlin, P.A. and Sauvant, K.P. (eds), *The Uruguay Round: Services in the World Economy*, New York: The World Bank and United Nations, pp. 45–67.
Sundbo, J. (1994), 'Modulization of Service Production and the Thesis of Convergence between Service and Manufacturing Organizations', *Scandinavian Journal of Management*, 10(3): 245–266.
Sundbo, J. (2002), 'The Service Economy: Standardisation or Customisation?', *Service Industries Journal*, 22(4): 93–116.
UNCTAD (1983), *Production and Trade in Services, Policies and Their Underlying Factors Bearing Upon International Service Transactions* (TD/B/941), New York: United Nations.
UNCTAD (1999), *World Investment Report: Foreign Direct Investment and the Challenge of Development*, New York and Geneva: United Nations.
UNCTAD (2002), *The Tradability of Consulting Services: And its Implications for Developing Countries*, New York and Geneva: United Nations.
UNCTAD (2004), *World Investment Report: The Shift towards Services*, New York and Geneva: United Nations.
UNCTAD (2012), *World Investment Report 2012: Towards a New Generation of Investment Policies*, New York and Geneva: United Nations.
Vandermerwe, S. and Chadwick, M. (1989), 'The Internationalization of Services', *Service Industry Journal*, 9: 79–93.
White, L.J. (1999), 'Reducing the Barriers to International Trade in Accounting Services', *Proceedings of the World Services Forum Conference*, December.
Williamson, J. (1990), 'What Washington Means by Policy Reform', in Williams, J. (ed.), *Latin American Adjustment: How Much Has Happened?*, Washington, DC: Institute for International Economics, available at: http://www.iie.com/publications/papers/paper.cfm?ResearchID=486 (accessed 7 November 2013).
Williamson, J. (2002), 'Did the Washington Consensus Fail?', Speech at the Center for Strategic and

International Studies, Washington, DC, 6 November, available at: http://www.iie.com/publications/papers/paper.cfm?ResearchID=488 (accessed 7 November 2013).

Wolf, A.T., Kramer, A., Carius, A. and Dabelko, G.D. (2005), 'Managing Water Conflict and Cooperation', in World Watch Institute (ed.), *State of the World 2005: Redefining Global Security*, Washington, DC: World Watch Institute, pp. 80–95, available at: http://tbw.geo.orst.edu/publications/abst_docs/wolf_sow_2005.pdf (accessed 18 November 2013).

WTO (1999), 'An Introduction to the GATS' WTO Secretariat, Trade in Services Division', October, available at www.wto.org/english/tratop_e/serv_e/gsintr_e.doc (accessed 12 July 2002).

WTO (2001), *GATS – Fact and Fiction*, Geneva: WTO, available at: http://www.wto.org/english/tratop_e/serv_e/gatsfacts1004_e.pdf (accessed 7 November 2013).

WTO (2012), *World Trade Report 2012: Trade and Public Policies: A Closer Look at Non-tariff Measures in the 21st Century*, Geneva: WTO, available at: http://www.wto.org/english/res_e/publications_e/wtr12_e.htm (accessed 15 October 2013).

15. Internationalisation of services: modes and the particular case of KIBS

Luis Rubalcaba and Marja Toivonen

INTRODUCTION

The services sector is a key player in the on-going process of globalization (Bryson et al., 2004; Bryson and Daniels, 2007). A better understanding of the internationalisation of service businesses is therefore a top priority (Bryson et al., 2012; Pla-Barber and Ghauri, 2012). Services influence globalisation in many ways with respect to productive factors, markets and consumptions. They affect productive factors by facilitating global access to capital and the production of globally competitive innovations, to the labour force and the use of new global skills in local markets, as well as to the assembly and control of global knowledge. A wide range of services is involved, from the recruitment and contracting of personnel to information and communication technology (ICT) or engineering services while financial services provide easier access to global finance capital (credit, saving and investment). In addition, services also exert influence on the globalisation of markets via transport, ICT and business services that steer the export and trade of goods (consultancy, marketing, fairs and exhibitions) towards new markets or towards the adaptation of goods to local needs, in which distributive trades and internet services also play a role. Services also facilitate and advance global corporate reputations by means of trademarks, franchises or Corporate Social Responsibility services. There is no globalisation without services.

The competitiveness of current economies and of the services economy can also be partly defined by the competitive capacity of services, which sometimes depend on price–costs factors but also on differentiation, quality and other non-cost factors (Visintin et al., 2010). The growth of business service offshoring, mainly in the last 20 years and in particular since 2000, has added a new dimension to the involvement of services in new economic growth (Rubalcaba and Van Welsum, 2007).

In this context, services are themselves international and service companies can be globalised in many different ways. Financial services or communication services become global because of their involvement with foreign direct investment (FDI). The internationalisation of knowledge-intensive business services (KIBS) is of particular significance in the wider context of the relationship between services and globalisation. It has been argued that this is one of the most important components within the general process of the globalisation of production, distribution and innovation. It has implications for the international division of labour and for the competitiveness of firms, regions and countries (Miozzo and Miles, 2003). In particular, KIBS play a significant role in the globalised structures of innovation and knowledge: internationally operating KIBS function as bridges between global, national and regional levels of economic activity by transmitting innovations and transferring knowledge between these levels (Howells and

Roberts, 2000). Active internationalisation is also considered to contribute to the growth of KIBS and to the development of the KIBS sector (Kox, 2001; Rubalcaba and Kox, 2007). The relationship between KIBS innovation and KIBS internationalisation has also been found to be important (Rodríguez and Nieto, 2012).

The ways in which firms operate in international markets have traditionally been divided into two main categories: FDI and exports (Sondheimer and Bargas, 1993). The FDI mechanism has been considered dominant in KIBS, as for all service industries, because of their need for close contact with their clients (Roberts, 1998). At the end of the 1980s, when more extensive research on the internationalisation of services became available, foreign presence through third parties was identified as the third general model; it enables better control over service delivery and service quality than exports, and requires fewer resources than FDI (Vandermerwe and Chadwick, 1989).

In the process of internationalisation, a long-established view was that service firms either followed their clients to foreign markets or, in the case of independent internationalisation, a cautious, gradual approach was adopted (Roberts, 1998). However, recent studies show that service providers may also internationalise very rapidly, in much the same way adopted by high-tech firms in the mid 1990s. A specific group therefore includes the so-called 'born globals': firms that are oriented towards international markets right from the start (Chetty and Campbell-Hunt, 2004).

Other recent studies have also emphasised the diversity of possibilities within the basic forms of international operations and within the basic process types. For instance, foreign presence can be achieved through franchising or other partner arrangements, but it can also be established through mergers and acquisitions. The development of ICT has essentially increased the options available and means that the inseparability of production and consumption is no longer the norm for many service transactions (Ball et al., 2008). For instance, instead of the traditional practice of sending individual experts to the client, KIBS can consult their clients via the internet on the basis of knowledge acquired beforehand (Roberts, 2001). In addition to actual international operations, preparatory and indirect forms are important from the viewpoint of the development of skills and contacts. These may include, for instance, serving foreign clients in the firm's home country (Roberts, 1998).

This chapter examines in more detail the versatile and sometimes indirect models that KIBS utilise for their international operations and the paths that they follow when going global. As regards the paths of internationalisation, we attempt to provide detailed information in particular about those paths that are based on KIBS' own (not a client's) initiative and represent rapid internationalisation. While the models and paths of internationalisation are our main focus, we also briefly discuss the drivers as well as the challenges of KIBS' internationalisation, and the benefits gained from foreign business.

The chapter is structured as follows. In the second section, we briefly discuss the relationship between services and globalisation before turning, in the third section, to some data on major players and exploring the different internationalisation modes and current trends across European service subsectors. In the fourth section we explore the literature on the direct and indirect forms of KIBS internationalisation and present a framework that illustrates their relationships. Both the third and the fourth sections focus on KIBS in particular, but some interesting observations about the internationalisation of services in general are also included. The fifth section relies on literature and a discussion to examine the paths to international markets, and these form the basis for the development of a three-part categorisation. A final section comprises conclusions and discussion.

SERVICE AND GLOBALISATION

The contribution of the service sector to the attainment of a global economy can be summarised by classifying the activities into three main groups (Rubalcaba, 2007a). First, there are services that make globalisation possible by establishing transport and communication networks that facilitate travel, international trade (IT), and the purchase and sale of goods and services in other places. Some of these are traditional services, for example commercial transport by road or by sea, and some others are modern services such as telecommunications that apply new technologies. All of them establish links between geographically distant places and constitute the basic infrastructure that makes globalisation possible. The improvement of these services in the last decades combined with a decrease in prices is leading to what some call the 'death of distance'; a reference to the recent exponential reduction of costs and time required to overcome physical distances. This has been an advantage not only for producers but also for consumers.

Second, there are business services that support the internationalisation of economic activity. A modern company will hardly succeed in a global economy if it does not use business-related services properly. Sometimes, the company will require services providing direct advice regarding its international strategy or the legal and tax aspects of its transactions abroad. Other clients will require assistance in foreign trade or in efficient recruitment and contracting of personnel.

Finally, there is a group of consumer services offered within the global economy of which the most globalised are audiovisual and cultural services such as cinema. The international dimension of leisure and sport services is also growing, as well as tourism-related services, which are consumer services par excellence. Obviously, distributive trades or e-commerce services are also included, facilitating the distribution of goods and other services to the global market.

Just as services are enabling globalisation, globalisation is also contributing to the internationalisation of services. Service globalisation can be considered as a four-stage process (see Rubalcaba, 2007a):

- Stage 1: *International exchange*. From the beginning of civilisation, societies and their economies have been interrelated, generating several social, economic and cultural exchanges. The fairs that took place in Europe between the 10th and 12th centuries are an example of the contribution that services made to the first stages of globalisation (Rubalcaba, 1994).
- Stage 2: *Internationalisation*. This is the link between countries, which is produced primarily through engagement in IT and the mobility of production factors. Companies become more international, as do their workers and capital. The goods industry is a leader in this stage, although general production and other professional services go hand in hand with this expansion.
- Stage 3: *Transnationalisation*. After the two World Wars a process of European reconstruction began and with this came a boom in transnational companies (later

to be called multinationals). Expansion by means of FDI supplemented simple trading relationships. The energy crisis of the 1970s led to the shift of manufacturing activities to low-cost locations and the emergence of new competitor countries (predominantly in Southeast Asia). In this context, the concept of *outsourcing* or *offshoring* was established and services continued to grow, generally within larger companies, offering alternatives to reducing costs or increasing quality to confront the competitive challenges of the emergence of global competition. Furthermore, many service companies accompanied their clients in this international adventure by deciding to follow clients overseas.
- Then, the second modern wave of globalisation, in which we are now immersed, begins. In stage 4, from the 1980s, *globalisation*, services ceased to perform a complementary role in the processes of change and market integration. Countries leading globalisation are those where the majority of value added and employment is in the service sector. Moreover, the business service sector, and KIBS in particular, emerge very rapidly, encouraged by an increasing externalisation of tertiary activities that were previously provided within, for example, the large trading companies. Globalisation implies a new way of understanding business and companies within an environment which tends to span the whole world. At the dawn of the 21st century, service companies started to extend their international strategies to the global scale, as well as services externalising some of their activities/functions to developing countries.

It is useful to consider examples of the relationships between globalisation and services (Table 15.1). The columns in the table reflect two directions of the relationship: the ways in which each service is internationalised (how globalisation affects services) and the ways in which each service contributes to the internationalisation of firms in other sectors (how services affect globalisation).

In the case of internationalising services, the first group includes the most important and traditional service industries, all of them subjected to the logic of globalisation. First, wholesale and retail distributive trades are evolving rapidly, with the implementation of large stores and chains of hypermarkets and debates about licences, permissions and opening hours in many countries that are no more than the expression of increasing competition and internationalisation of products and ways of working. Hotels, restaurants and tourism activities are industries that are clearly international since they seek to capitalise upon and arrange access to geographical, climatic, historical, cultural and economic differences within and between places (cities, regions, countries). Many customers, large hotel chains and some major tour operators target predominately foreign locations. Transport is also affected by globalisation, with decision making increasingly moving to the international level and always taking advantage of better connections and infrastructures. Many medium- or large-sized firms are affected by growing competition in road transport, while air transport is undergoing a strong process of alliances and concentration. The banking and insurance sector is confronted with a similar situation as mergers, takeovers and transnational operations continue to increase. Finally, telecommunications is also experiencing a process of strong globalisation in response to technological developments and the emergence of transnational communication companies.

Table 15.1 Relationships between services and globalisation: some examples

Sectors	Ways in which services internationalise	Ways in which services contribute to the internationalisation of all sectors
Traditional services		
Wholesale and retail distributive trades	Arrival of large stores International products Increasing competition	Use of sales potential offered by big stores Need for a comparable size to negotiate at the same level (basically food industry)
Hotels/restaurants and tourism	Foreign customers International chains of hotels Tourist concentration in international tour operators	The improvement of professional or business tourism contributes to facilitate globalisation Attraction of international businesses
Transport	Improvement of international connections Alliances and concentration in airline and road transport companies Increasing comparisons in services	High-quality and cheap transport network is key factor for goods and service globalisation
Banking and insurance	Mergers Foreign investment Accounts and operations in foreign currencies	Facilitators of transnational operations
Telecommunications	Fast incorporation of the latest technology Participation of foreign capital in phone companies Agreements between operators from different countries	Make a fast and fluid communication possible between agents, beyond existing differences Share of distance's death
KIBS		
Management consultancy	Influence of the large multinational consultancy firms	Adoption of internationally proved organisational systems Counselling in international strategy
Legal services	Collaboration agreements between lawyer firms Agreements with multinational firms	Advice on national legal system Counselling international support
Computer services	Use of standard software International suppliers	Preparation of firms for global connections and adequate use of information
Trade fair services	Organisation of trade fairs abroad	Observation of what is produced in other countries (feel global markets)
Market research	Networking Multinationals Procedure standardisation	Test on foreign markets Information improvement
Advertising	Repetition of the same ads in different countries Concentration Multinationals	Entry into foreign markets Creation of the same or similar tastes Improvement of entry potentials

Compared with these major service sectors (tourism, finance, etc.), business services, mainly KIBS, are also influenced by globalisation, although to the average consumer in a less visible way (Rubalcaba, 1999, 2007b). Consultancy, auditing, market research and advertising multinationals are the clearest examples. They offer international services to international clients and gain increasingly bigger market share. The expansion of businesses abroad is also a new characteristic of some specific activities, such as trade fairs and exhibitions, which have started to be organised outside national frontiers. Networking and collaboration agreements are characteristic of engineering services, legal services or market research, for example, and represent a useful way for small and medium-sized enterprises to follow internationalisation processes.

As regards the contribution of services to globalisation, traditional services offer abundant examples. For example, the food industry – or at least parts of it – takes advantage of the sales potential of large distribution chains, for example Lidl, Aldi and Tesco. Business travel by the managers of industrial firms involves using the hotel and tourism infrastructures of cities. A good network of transport and telecommunications is an indispensable condition for firms and their personnel to operate at a global level. Financial services facilitate transnational business operations. KIBS are also becoming essential for adapting the international flows of trade and investment to local needs in terms of market differentiation, legal and tax requirements, the internet and ICT environments, linguistic needs, and so on.

It is therefore clear that globalisation is linked in different ways to all service industries. In some cases, the particularities of services markets will be the determining factor while in some other cases, service globalisation will have closer similarities with the globalisation of manufacturing firms. In the next section particular attention will therefore be given to exploring the different forms of service globalisation.

MODES OF INTERNATIONAL ACTIVITIES IN SERVICES AND SOME EMPIRICAL EVIDENCE

Service globalisation takes many forms; some are similar to those observed in manufacturing while others are specific to services. What globalisation implies, in any case, is the ability to offer a similar product in any part of the world or to transfer products or resources from one place to another; a global conception of the market that involves the internationalisation of productive inputs and intra-firm and inter-firm multinational activities. Rubalcaba and Cuadrado (2002a) stressed the special characteristics of services linked to certain ways of globalisation. Special attention was drawn to the importance of *product differentiation* (which is never identical in all places) and *service customisation* (due to the interactive nature of services). Both aspects imply that a service will require adjustment to fit the requirements of a foreign market. This means that service *internationalisation* implies '*nationalisation*', that is to say, adaptation to the normative, economic, cultural and social parameters of foreign countries or markets. The interaction between consumers and products is at the heart of the tradability of services, but it is both an advantage and a disadvantage. There are many multinational companies in which subsidiaries retain a degree of independence as well as having multilevel management structures that provide service functions. The difference compared with manufacturing is that there is no vertical division of production in service industries; low-cost or low-technology parts of the production process cannot usually be relocated to developing countries. This means that global movements of services are not so much associated with cross-border transactions as with the transmission of processes, knowledge and techniques or with the exchange of residents and non-residents and the transfer of workers, managers and technology. The flows of international services, business services in particular, largely comprise people rather than goods, and the ideas and knowledge that they bring with them. This can explain how the international presence of service multinationals is strongly intertwined with local services and local firms (Miozzo et al., 2012) and how there are even many ownership international flows that are related to contractual relationships without any involvement of FDI (Pla-Barber et al., 2010).

Modes of service internationalisation have been analysed from different service-oriented perspectives (Noyelle and Dutka, 1988; Ruane, 1993). Vandermerwe and Chadwick (1989), for example, presented a classification system for service firm internationalisation that contains two dimensions: the relative involvement of goods in services and the degree of consumer–producer interaction. 'Exportable' services are those embodied in goods and which have low interaction. Investment-oriented services are those that require high interaction. In the last 20 years exports and FDI have been consolidated as the key modes of service internationalisation although other forms do exist, mainly contractual relationships between third parties. However, other ways of service internationalisation do exist such as those associated with brands, trademarks, franchising, knowledge flows, networking and standards (Rubalcaba, 2007b).

All these dimensions are interrelated. Although each may be developed on its own, service globalisation requires most of the following dimensions to be considered. For example, a consultant who travels from one country to another to share knowledge contributes to service globalisation through client transactions and the diffusion of a brand and a reputation associated with a firm. During this type of 'service trade', networking and knowledge transfer occur and the co-production of an international service results in adaptation to local conditions. This process of adaptation will lead to further opportunities to develop service trade, as the acquired experience may result in the provision of services to other foreign companies. A foreign firm dedicated to telecommunications will have to invest capital, sell and develop a brand, present a network adapted to local conditions and at the same time transmit international knowledge and try to partially replicate services that can increasingly be sold globally.

Among all existing modes, exports are surely the most representative and those for which more statistics are available. The United States is the leading country in service exports, even if between 1993 and 2007 its market share declined from 18.1 per cent to the current 13.9 per cent (Table 15.2). The UK and Germany follow, with market shares of around 6–6.5 per cent. China is now in fourth place, with 4.38 per cent of the world market in 2011, after impressive growth (more than doubling its share) since 1997. France and Japan have lost market share while Spain has remained steady at 3.3–3.7 per cent. Although ranked eighth, the growth of service exports from India has exceeded the Chinese rate: from 0.68 per cent in 1997 to 3.28 per cent in 2011. The leading countries in exports are in general the leading countries in imports too, but the ranking is different, with Germany and China holding second and third positions respectively. What is clear is that, in general, large or dynamic service exporters are also large or dynamic service

Table 15.2 Leading countries in services exports and imports

	Exports	1997	2007	2011		Imports	1997	2007	2011
	World	100	100	100		World	100	100	100
1	United States	18.13	13.88	13.93	1	United States	11.80	10.67	10.00
2	United Kingdom	7.59	8.33	6.57	2	Germany	10.02	8.10	7.31
3	Germany	5.87	6.32	6.08	3	China	2.15	4.07	5.98
4	China	1.86	3.56	4.38	4	United Kingdom	5.86	6.14	4.31
5	France	6.07	4.33	4.00	5	Japan	8.64	4.68	4.19
6	Japan	5.29	3.71	3.42	6	France	4.86	4.04	3.63
7	Spain	3.29	3.71	3.37	7	India	0.95	2.22	3.13
8	India	0.68	2.53	3.28	8	Netherlands	3.44	3.07	2.99
9	Netherlands	3.63	3.19	3.20	9	Ireland	1.18	2.97	2.89
10	Singapore	2.15	2.48	3.09	10	Italy	4.58	3.73	2.88
11	Hong Kong, China	2.81	2.47	2.91	11	Singapore	1.74	2.34	2.88
12	Ireland	0.46	2.70	2.62	12	Canada	2.91	2.57	2.53
13	Italy	5.04	3.23	2.52	13	Korea, Republic of	2.27	2.64	2.49
14	Switzerland	1.87	1.88	2.26	14	Spain	1.96	3.01	2.36
15	Korea, Republic of	2.00	2.09	2.25	15	Russian Federation	1.55	1.79	2.22
16	Belgium	n.d.	2.11	2.10	16	Belgium	n.d.	2.15	2.14
17	Sweden	1.33	1.83	1.82	17	Brazil	1.12	1.09	1.85
18	Canada	2.33	1.86	1.79	18	Australia	1.45	1.23	1.51
19	Luxembourg	n.d.	1.88	1.74	19	Denmark	1.07	1.66	1.42
20	Denmark	1.07	1.77	1.55	20	Hong Kong, China	1.99	1.34	1.41
21	Austria	1.64	1.57	1.47	21	Sweden	1.51	1.49	1.41
22	Russian Federation	1.07	1.14	1.28	22	Saudi Arabia, Kingdom	1.12	1.45	1.39
23	Australia	1.44	1.16	1.22	23	Thailand	1.33	1.19	1.29
24	Taipei, China	1.29	0.96	1.10	24	United Arab Emirates	n.d.	1.05	1.23
25	Norway	1.18	1.17	1.00	25	Switzerland	1.09	0.99	1.19

Source: WTO, December 2012 data.

286 *Handbook of service business*

importers, indicating a clear complementarity between imports and exports. India and China are growing with respect to both indicators, although China is importing more services than it exports while the reverse is true for India.

A comparison of the top performers in market share (Figure 15.1a) and annual growth (Figure 15.1b) of commercial services exports in 2011 reveals some similarities: the

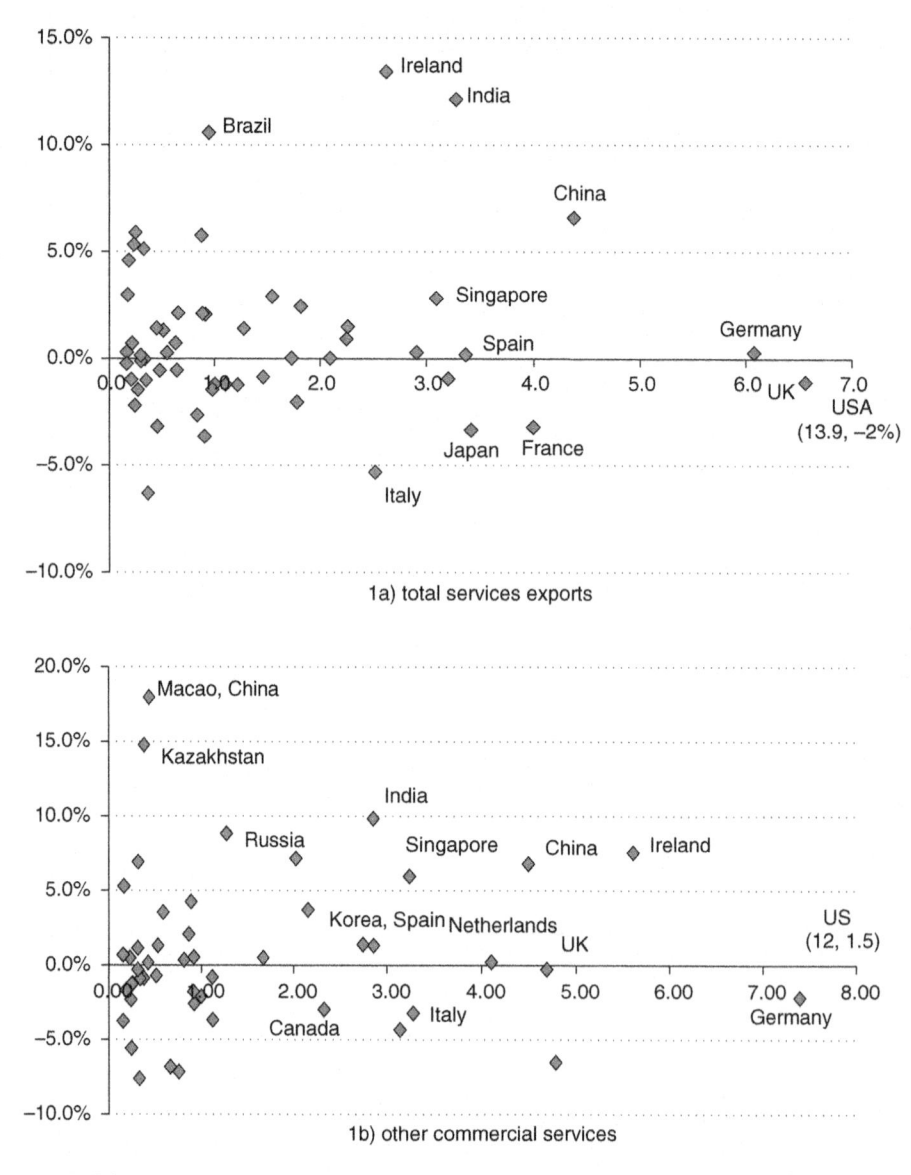

Source: WTO.

Figure 15.1 Top performers in commercial services exports: exports in other commercial services. X=Market Share (2011). Y=AGR97/11

leading position of the US, the large shares of Germany and the United Kingdom, the relatively poor recent performance of Italy and Japan, and the very large increases for India, China and Singapore.

Figure 15.1b provides an indication of the real competitive capacity of different countries based on strong service industries. The very prominent position of Ireland is notable as well as the dynamic performance of China, India, Singapore and Russia. The increase in exports in these countries is related to their increased participation in service offshoring by firms in the advanced economies during the last decade.

The key global players in services exports are identified in Figure 15.1, but this is not the full picture. Other dimensions of service internationalisation also play a role; the growth of service exports has its limits, as indicated by the fact that its share in total trade has remained more or less static at 20–25 per cent for at least three decades. On the one hand, this means that the high growth rates of service exports have been rather similar to the growth rates of trade in goods; services have fully participated in the latest globalisation wave and at similar pace to the globalisation of goods. On the other hand, service exports are held back by factors such as the limited use of ICT in many services; the heavy dependence on personalisation and even simultaneity required in many services (e.g., non-tradable public or personal services); the lack of homogeneous service products; the high degree of differentiation in services; the monopolistic market structures in many segmented service markets where asymmetric information is high; and the legal, cultural, linguistic and economic barriers that are encountered. This explains why trade is not necessarily the best way to internationalise many services.

It is therefore the case that FDI is actually the most important source of internationalisation for many services. Sometimes it is an alternative, sometimes it complements trade. At an international level, complementarities seem to be more important than a substitution effect. Those countries with high relative flows and stocks of FDI are also those countries with more exports and imports (Figure 15.2). In some countries the relative weight of IT is relatively more important than the weight of FDI (Nordic countries, for example) while for other countries the opposite is true (France, Germany and in particular Switzerland). However, an overall correlation prevails, suggesting the existence of strong complementarities between the two modes – FDI and trade in services.

The differences between exports and FDI can be further analysed by comparing different weights and trends in different service subsectors for Europe (EU27). Figure 15.3 shows the absolute numbers for the two indicators and shows how FDI stocks are important in services (FDI is more than 4 times higher than IT in 2011) while in goods it is exactly the opposite (2011 IT is about 3.5 times higher than FDI stocks). FDI stocks in services are also larger than the 2011 figure for IT in goods but FDI stocks in goods are only slightly larger than the 2011 figure for IT in services. This means that service internationalisation has a higher relative importance than goods internationalisation in absolute quantitative terms. However, with respect to relative weights in the economy, goods are more internationalised than services (Table 15.3).

In relation to different subsectors, there is a clear difference between services for which international trade is more important than FDI, such as transportation, travel/tourism and royalties, and services for which FDI is significantly more important: finance, communication, construction and housing services, information services and other services. KIBS are clearly FDI oriented.

288 *Handbook of service business*

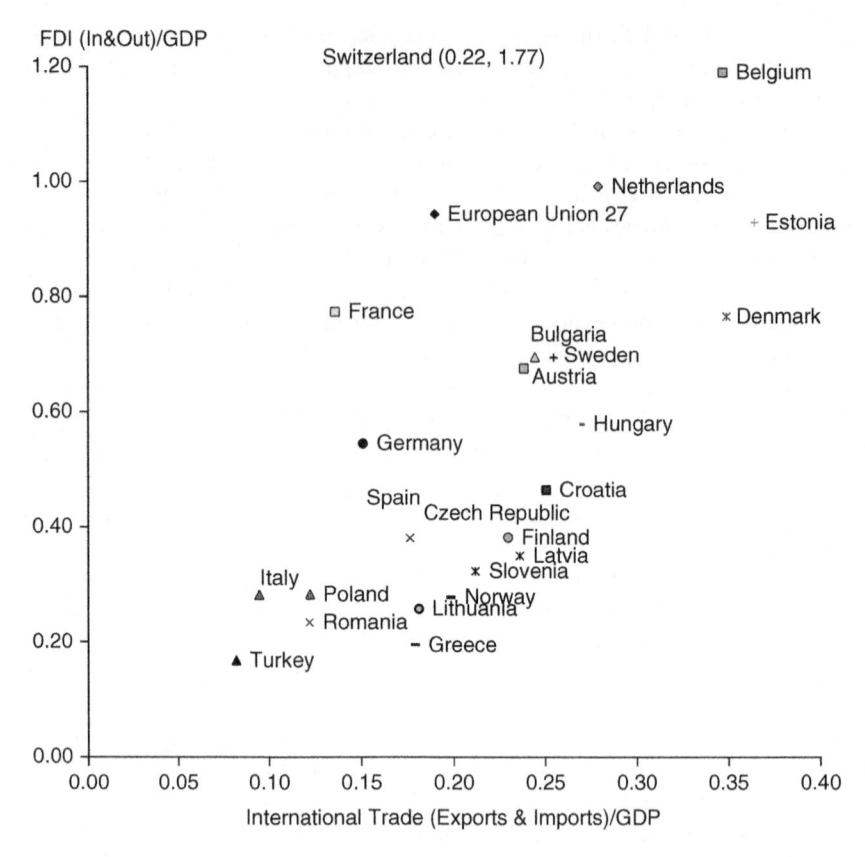

Sources: Based on Eurostat data on international trade in services and EU direct investment positions.

Figure 15.2 Relative importance (%GDP) of international trade (exports & imports) and FDI (inwards & outward stocks) in services by country, 2009

An internationalisation index, defined as a proxy variable to measure the different propensity of sectors to be international, provides an alternative way of exploring service internationalisation (Table 15.3). The index is based on the share each sector represents in total IT and total FDI compared with the share of a given sector in total value added. There are three indexes for these shares in value added (size index), exports and imports (relative trade) and inward and outward FDI (relative FDI) (see Table 15.3). The internationalisation index weights the relative trade index and the relative FDI index by the percentages of international trade and FDI in the total internationalisation for each sector.

The results show that the most internationalised sectors in the economy are the financial sector, the goods sectors (agriculture and manufacturing), the royalties/R&D world, the tourism/travel sector and other business services (many KIBS can fall under this category) (Table 15.3). However, computer and information services are not as internationalised. This could be explained by the fact that many ICT services promote the internationalisation of other sectors so that their own internationalisation could

Internationalisation of services 289

Figure 15.3 Trade (exports & imports) and FDI (inward & outward stocks). Absolute values in last available year for Europe 27

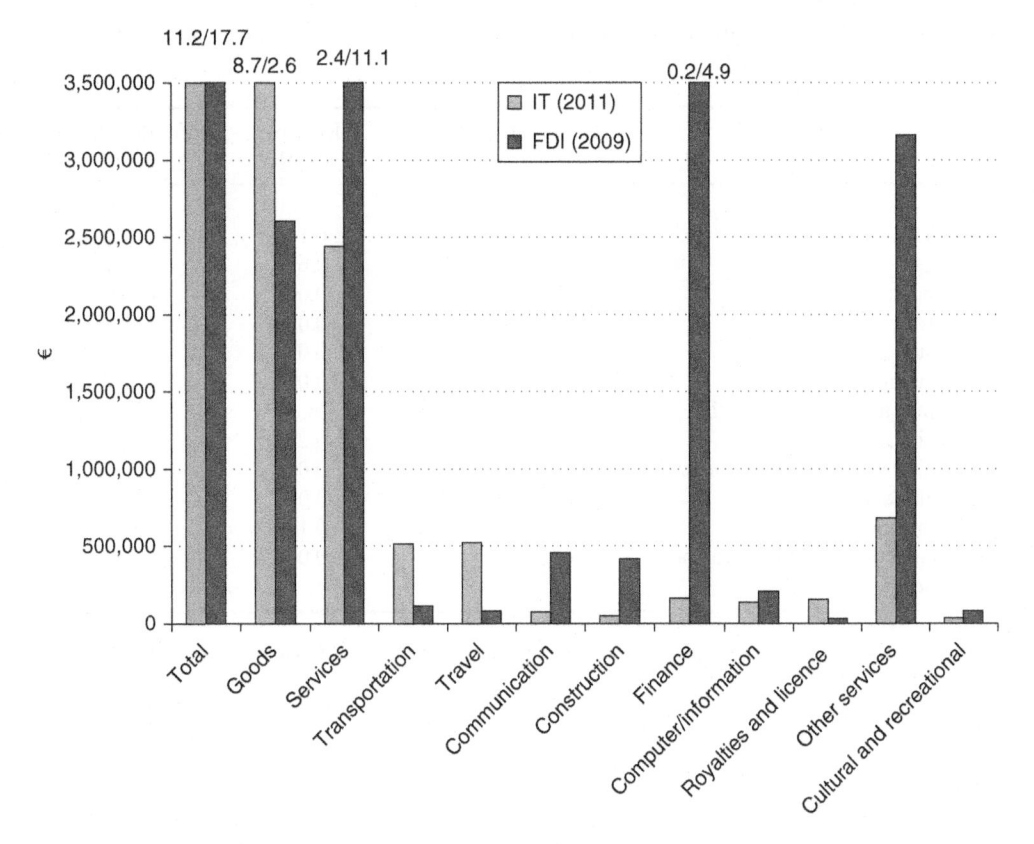

Sources: 1) Eurostat; Balance of Payments; International Trade in Services, 2) Eurostat; Balance of Payments; EU Direct Investment Positions.

be partly underestimated in the current statistics. On the other hand, it could also be explained by the limited volumes of FDI stocks compared with other service sectors and the limited volumes of trade generated by exports and imports. Both are important but have surely been overemphasised following the expansion of service ICT offshoring from 2000 onwards (see Stare and Rubalcaba, 2009).

The rates of IT/FDI in 2009 and the different trends for the period 2004–2009 are compared in Figure 15.4. There are four groups of sectors: first, sectors that are IT oriented but FDI is increasing in relative terms (travel, transport, goods); secondly, sectors that are very FDI oriented after years with very high increases in FDI (computer and cultural and recreational); thirdly, FDI oriented sectors with shifts towards more FDI at a rate higher than the average for all services (construction and other business services); and fourthly, FDI oriented services but with recent trends towards relatively higher IT investment (communication and financial services). The latter is the only group in which IT is gaining relative to FDI. In general terms, services are FDI activities with a

290   *Handbook of service business*

Table 15.3   The internationalisation of different economic sectors/subsectors in the EU27: a proxy approach

	Size index	Relative trade	Relative FDI	Internationalisation index
Total	1.000	1.000	1.000	1.000
Goods	0.172	0.756	0.147	3.401
Services	0.740	0.244	0.625	0.761
Transportation	0.045	0.048	0.006	0.880
Travel	0.030	0.055	0.004	1.598
Communications services	0.022	0.008	0.026	1.047
Construction services	0.118	0.006	0.023	0.183
Insurance and pension funding	0.009	0.006	n.d.	n.d.
Financial services	0.040	0.016	0.279	6.717
Computer and information services	0.019	0.013	0.012	0.634
Royalties and licence fees	0.007	0.015	0.002	1.880
Other business services	0.081	0.067	0.178	1.992
Personal, cultural and recreational services	0.020	0.003	0.004	0.207

Sources: Based on Eurostat data about international trade in services and EU direct investment positions.

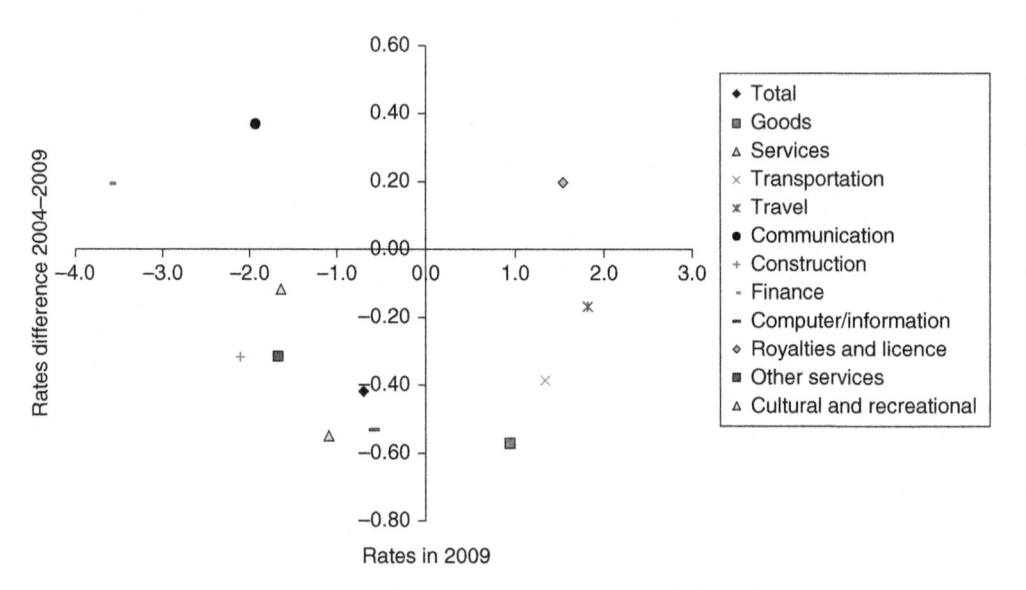

Source: Based on Eurostat data on international trade in services and EU direct investment positions.

Figure 15.4   Rates IT (exports & imports)/FDI (outward & inward stocks): levels 2009, and rates differences 2004–2009

tendency to favour FDI rather than IT, much like the total economy, but since goods (manufacturing and agriculture) are still more IT oriented sectors the total economy is somewhere between the positions for services and goods.

THE INTERNATIONALISATION OF KIBS REVISITED

We now turn to an analysis of the diverse operational forms that KIBS can utilise in international markets. Following Vandermerwe and Chadwick (1989), we divide international activities into three basic models: FDI, exports and foreign presence through third parties. All these models are applicable to KIBS. However, each of them can be implemented in several ways, and at this more detailed level some solutions are typical of KIBS whereas others are rare.

Due to their nature as an enabler of face-to-face contact, many forms of *foreign investments* are common to KIBS: this may mean a wholly owned branch or subsidiary, or a majority or minority share of a subsidiary (Blomstermo et al., 2006). An internationalising company also has two alternatives regarding entering a foreign market: it may establish a subsidiary itself (so-called 'green field internationalisation') or it can become an owner through a merger or an acquisition. The latter practice has become increasingly common in recent years and is typical of KIBS (see, for example, Roberts, 1998). The third factor that, in addition to the scope and birth of ownership, can vary in the context of FDI is the degree of centralisation: the ways of operation may be planned centrally in a head office or the subsidiaries may have quite considerable freedom in the development and conduct of their businesses (Howells and Roberts, 2000).

In the case of *exports*, services can be delivered in a material form (a letter, a report, a software disk), through the movement of persons and through telecommunications networks – primarily the internet (Roberts, 1998). For KIBS, personal interaction is again very important: travelling is the predominant model. However, due to the costs involved, it is increasingly being replaced with wired exports in cases where knowledge may be coded. Thus, KIBS companies can deliver training and consultancy material to their international clients over the internet (Javalgi et al., 2004), which has also displaced the delivery of services in material forms. There is, however, a specific type of movement – the relocation of the producer to the client – which has 'survived' this shift; the so-called 'export projects' that are an intermediate form between FDI and exports. Export projects are common in engineering consultancy in particular, and mean a temporary presence in the target market: when the assignment is completed, the office is closed and the consultants move to other markets. Thus, instead of expansion in a specific country, the aim is to use the same expertise with slight modifications in many different countries (Sharma and Johanson, 1987).

International operations through third parties may take the form of franchising, licensing or other partner arrangements. In franchising, the focal company provides other companies with the right of use of a tested service concept for a contract period against a fixed payment. The franchisees benefit from the efficiency and the cost reduction that the ready-made concept offers, and the provider benefits from the rapid growth of business that the entrepreneurship-based network enables (Alon, 2005). However, franchising has not been popular in the KIBS sector given the difficulties of protecting intellectual property rights. In expert services like KIBS, the control typical of franchising may not be a suitable way of working. Licensing is also rare in KIBS, and all in all it is much more common in manufacturing than in services. Here one company grants another company the right to represent it in a target market on specified terms, often exclusively (Grönroos, 1999). While franchising and licensing are not extensively used in KIBS, there are other partner arrangements, which vary from non-equity cooperation to contracts that include limited ownership. The cooperation may include a common brand, and it may also involve the common acquisition of clients and contacts, common subcontracting, common R&D and training, and even partially common ways of working. Deeper forms of cooperation, such as strategic alliances, are also possible (Tapscott et al., 2000).

In addition to actual international operations, there are preparatory and indirect forms of internationalisation. These may be the first steps that later lead to the selling of services in foreign markets or they may support existing business in these markets. In KIBS, there are four types of indirect activities: serving foreign clients in the firm's home country; being a member of an international chain or network; international recruitment of experts; and the offshoring of services.

Serving foreign clients in the firm's home country is often the first activity that familiarises KIBS with issues of internationalisation (Roberts, 1998). It induces them to acquire knowledge about regulations, business practices and culture in different countries. Serving the subsidiaries of international companies may also be a way to specialise (Toivonen, 2004). In addition, KIBS often play a central role in supporting domestic clients in their internationalisation process (OECD, 1999) and this forces them to acquire versatile knowledge of global markets.

The second indirect form of internationalisation is *membership of an international chain or network* (this form differs from operations through third parties because here the company is only a member, not the focal company). KIBS usually select this alternative to increase the efficiency of business. However, the common brand and the common training and R&D, typical of chains, inevitably link KIBS to the international community. In professional non-equity networks the linkage is looser, but from the viewpoint of contact formation and knowledge accumulation they are equally important.

Information technology companies, in particular, have emphasised *the international recruitment of experts* as a factor that improves the preparedness of a company seeking to enter foreign markets (Toivonen, 2002). These experts often provide useful contacts in target countries, and the multicultural personnel in a company create an innovative atmosphere (Saxenian, 2000). From the viewpoint of international contacts, *offshoring* may also be useful – even though it means an activity opposite to the company's own sales. Offshoring is a special form of outsourcing of services, its specificity being the location of service production in a foreign country. KIBS also use this model to some extent, seeking cost reductions in those functions that are outside their core business; call centres and book-keeping are typical examples (Massini and Miozzo, 2010).

A summary of direct and indirect models for the international activities of KIBS can be represented as a framework (Figure 15.5). Because providing services to foreign clients in the home country and membership of international chains and networks are closer to actual international business than international recruitment and offshoring, the former activities are separated with dashed lines and the latter with solid lines.

Internationalisation of services 293

```
                    ┌─────────────────────┐
                    │    INTERNATIONAL    │
                    │ RECRUITMENT OF EXPERTS │
                    └─────────────────────┘

    ┌──────────┐         EXPORTS              ┌──────────┐
    │ BEING A MEMBER OF AN INTER │            │ SERVING FOREIGN CLIENTS │
    │ NATIONAL CHAIN OR NETWORK  │  FOREIGN DIRECT  │ IN THE FIRM'S HOME COUNTRY │
    │          │    INVESTMENTS (FDI)         │          │
    │          │                              │          │
    │          │       FOREIGN                │          │
    │          │   PRESENCE THROUGH           │          │
    │          │    THIRD PARTIES             │          │
    └──────────┘                              └──────────┘

                    ┌─────────────────────┐
                    │ OFFSHORING OF SERVICES │
                    └─────────────────────┘
```

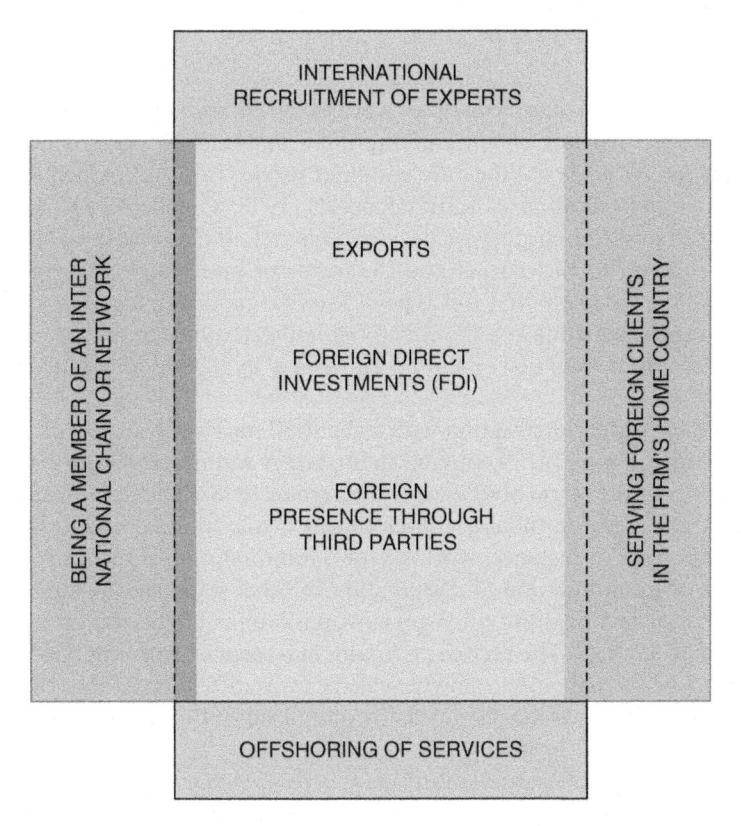

Figure 15.5 The basic models of international activities in KIBS: the actual business operations (the centre) and the preparing and supporting activities (the outer areas)

The model that is appropriate for an individual firm depends both on *the resources and skills of the firm* and *the nature of the service*. All models require contacts and material resources as well as know-how in international business (Glucker, 2004). Trusted partners and existing contacts are most critical in operations through third parties, whereas material resources are emphasised in FDI. Some models require specific skills: for instance, know-how in project management is a prerequisite for successful export projects. As regards the nature of the service, the most important question is the extent to which production and consumption are separable, locally and temporally. A categorisation into soft (inseparable) and hard (separable) services is commonly used to describe the basic difference (Erramilli and Rao, 1990; Majkgård and Sharma, 1998). The former services require that the provider must be present for the delivery, which can be achieved by using FDI, export projects or travel by the experts to the client. For the latter services, more options are available: exports (often via the internet) and operations through third parties are also possible. In KIBS, soft services are typically offerings where the elicitation and interpretation of tacit, context-specific knowledge plays a central role; hard services include information of a more generic nature.

PATHS TO INTERNATIONAL MARKETS

A traditional argument has been that services internationalise because their clients operate in foreign markets or are entering these markets. Newer studies have shown that *following the clients* is not the only path that service firms take when they go global, although it is quite common in KIBS (Roberts, 1998; Glucker, 2004; Kautonen and Hyypiä, 2009). Here, the strategy of the client restricts the alternatives open to a KIBS company in relation to the market and the forms of international operations. On the other hand, studies indicate that this type of internationalisation does not simply mean steps taken in the way defined by the client, but it also requires initiative, versatile activities and the development of new skills and know-how by the service company (Toivonen, 2004).

A benefit that internationalisation with a client offers is the prospect of a reduction of risk, as the service company has contacts in the target country and information about its markets right from the start (Majkgård and Sharma, 1998). The first big client often has an important influence on the later development of the internationalisation process of a service company. For instance, working for a multinational enterprise in one country often leads to the continuation of cooperation in other countries. A challenge emerges from the fact that local culture often remains unfamiliar to the service company when its client is an 'outsider'. The service company may neglect forming new contacts with local operators or form them too slowly, which causes difficulties if the volume of orders of the original client is seasonal or gradually diminishes (O'Farrell et al., 1998; Erramilli and Rao, 1990; Glucker, 2004).

The second form of internationalisation involves *an independent path* where a service firm advances gradually from less risky towards more demanding decisions (Johanson and Wiedersheim-Paul, 1975; Erramilli, 1991). A factor that favours this kind of cautious process is the importance of experience in international operations, and the fact that its accumulation takes time. Also, the building of credibility and trustworthiness in client and partner relationships, particularly when foreigners are in face-to-face contact, demands resources and time, and tends to favour a step-by-step approach (Majkgård and Sharma, 1998; Contractor et al., 2003).

Studies indicate that cautiousness may be visible both in the selection of markets and in the forms of operation in target markets. The former means that companies start their international activities on a small scale in those countries where the 'psychic distance' is small; that is, markets that resemble the domestic market as regards the industrial structure, business habits, culture, and so on. Neighbouring countries are often the first choice (Glucker, 2004). In the case of KIBS, the gradual process has also been argued to derive from the local orientation of these companies; that is, KIBS would not be interested in international activities in the first place. When they have formed a stable position in local markets, they reach out to a wider national market and only thereafter possibly go abroad (Roberts, 1998). Cautiousness in the forms of foreign operations is manifest in the transition from operations that require a small commitment step by step to the establishment of a subsidiary (Majkgård and Sharma, 1998).

The idea of a gradual process has been negatively critiqued in recent years; first of all, researchers have emphasised that it is not the only route for a firm pursuing independent internationalisation. For instance in KIBS export projects show that the ultimate aim of an internationalising company is not always the establishment of a subsidiary. Furthermore, European engineering offices have often started their internationalisation in distant places, such as developing countries, where the business environment is radically different from that of the home country (Sharma and Johanson, 1987). It is not rare either for KIBS to 'skip' some stages in the internationalisation process or that companies may follow the stages in a different order (O'Farrell et al., 1998).

At the end of the 1980s researchers started to pay attention to the acceleration of the pace of internationalisation by high-tech firms (Young, 1987). They found that IT firms, in particular, set up foreign business very rapidly and do not follow any kind of gradual model (Bell, 1995). Rapid internationalisation has subsequently been identified in other service sectors (Madsen and Servais, 1997). A specific group includes 'born globals' – companies whose business model includes international operations right from the start (Rennie, 1993; Rönkkö, 2001). Typically, internationalisation occurs in these companies simultaneously in many different forms, including (wired) exports, subsidiaries, strategic alliances and non-equity networks (Toivonen, 2004).

Some researchers characterise 'born globals' as companies that perceive the world as one market; country borders are not seen as a hindrance to business. These companies, which are usually small, focus on some niche activities or services where they aim to attract as their clients pioneers or first movers throughout the world. Thus, the number of clients in any one country need not be large and a typical approach linked to achieving rapid progress is making 'tests' in various markets (a so-called 'sow and reap' strategy). A high tolerance for mistakes is necessary in this context; indeed 'born globals' actually consider mistakes useful for learning. An important prerequisite for rapid internationalisation is that the entrepreneurs possess deep know-how; they have typically acquired international experience before the establishment of the company (Madsen and Servais, 1997; Chetty and Campbell-Hunt, 2004).

There are therefore three paths to internationalisation (Figure 15.6). However, these paths are not mutually exclusive and companies may combine features from more than one path. They may also take a different path in different countries or at different stages of the internationalisation process.

An example of the 'mixing' of the different paths is provided by the perception that 'born globals' may take their first international steps in countries with a short 'psychic distance', and only after that do they rapidly penetrate global markets (Madsen and Servais, 1997; Chetty and Campbell-Hunt, 2004). Correspondingly, cautious companies often speed up their internationalisation process when they have accumulated sufficient experience or the company is involved in fierce competition (Erramilli, 1991; Roberts, 2001). It is also possible to act in different ways in different countries and to engage with the internationalisation process at different speeds. In addition, the paths may be combined such that following a client and operating independently might go hand in hand: a service company may follow a client to some specific target market but select at least some independent forms of operations (Majkgård and Sharma, 1998). Thus, even though the idea of internationalisation comes from a client and serving this client is the primary task, a service company may simultaneously develop its own strategy in the target country and perhaps also in other countries (O'Farrell et al., 1998).

following a client to international markets

| KIBS company | A client/clients 'draws/draw' the service provider to international markets. Target countries and forms of operation depend to a greater or lesser extent on the needs of the client(s). ◄─────────── | client company |

independent gradual internationalisation

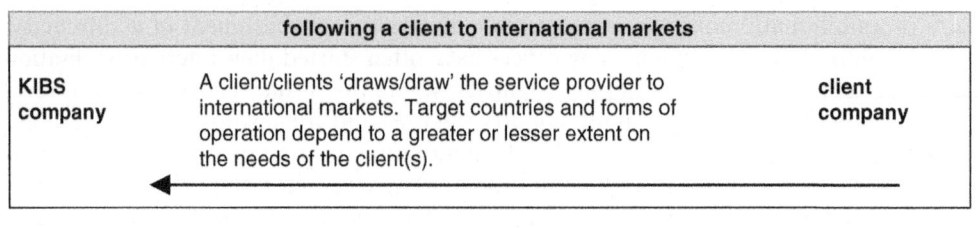

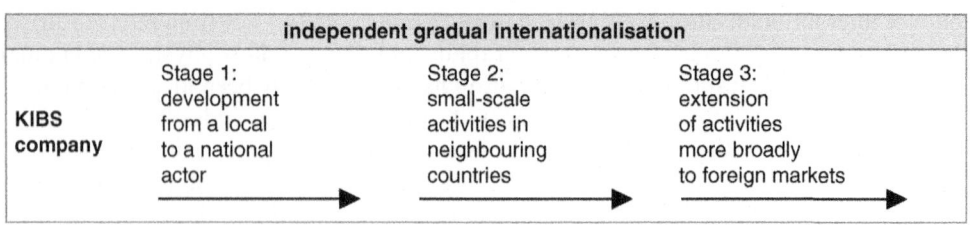

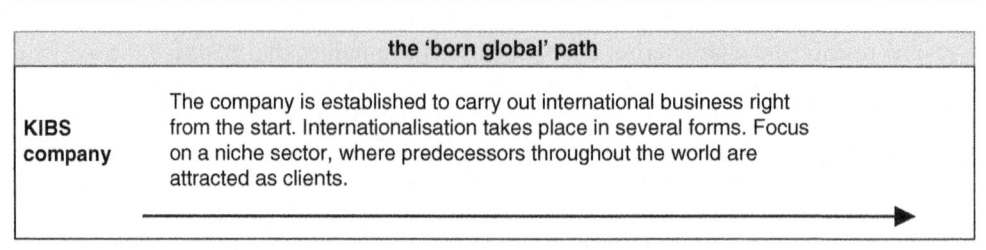

| KIBS company | Stage 1: development from a local to a national actor ──► | Stage 2: small-scale activities in neighbouring countries ──► | Stage 3: extension of activities more broadly to foreign markets ──► |

the 'born global' path

| KIBS company | The company is established to carry out international business right from the start. Internationalisation takes place in several forms. Focus on a niche sector, where predecessors throughout the world are attracted as clients. ──────────────────────► |

Figure 15.6 The Basic paths in the internationalisation of KIBS

DISCUSSION AND CONCLUSION

This purpose of this chapter is to contribute in three ways to the growing body of literature on service internationalisation.

First, the relationship between services and globalisation has been summarised and the main modes illustrated using empirical data. We have shown how FDI is the principal mode for services internationalisation in general, and for KIBS in particular. Current trends in Europe corroborate this, even if telecom and financial services have shown some signs of a relative shift to international trade and away from FDI since 2004. The growth of FDI used by KIBS as an internationalisation strategy is explained both from the limits to international trade in services (including all types of legal and natural barriers) and from the importance of FDI as a complementary rather than just an alternative mode.

Second, we have structured the topic by separating different models of international operations from the paths of going global, and by categorising both the models and paths on the basis of the latest academic research. For the models, we have utilised an existing categorisation of actual operations in foreign markets (Vandermerwe and Chadwick, 1989): FDI, exports and operations through third parties. Our own contribution is to supplement this categorisation with four preparatory or indirect models of international activities: serving foreign clients in the provider's home country, being a member of an international chain, recruiting international experts, and the offshoring of services. Each of these models has been discussed separately in earlier literature rather than being linked systematically to the framework of services internationalisation. In addition, the discussion of the paths has been dispersed and dominated by efforts to identify one typical path rather than at least the three different paths identified here, with each of them relevant in some context.

Third, an overview of the models and paths of internationalisation, *particularly in KIBS*, has been provided. This follows important earlier work written 15 years ago (Roberts, 1998), but much new knowledge has accumulated since then. Based on that literature we have identified models and paths that are typical of KIBS, as well as those that KIBS use only rarely. The factors that have made some models and paths more suitable than others have been examined.

The changes that are taking place because of advances in international trade have also been explored. On the one hand, service delivery via the internet and other wired media clearly reduces the need for experts to travel. On the other hand, FDI as the main model of internationalisation is retaining its position, and several companies undertaking wired exports also have subsidiaries. However, some care is needed when evaluating the extent to which ICT allows the breakdown of the inseparability of production and consumption in services. The core issue is the role of co-production in the service process – *a service can be co-produced also via the internet and in this case the separability increases only in relation to the location, not the time*. We are inclined to think that the extent of pre-planned elements in a service is a more decisive factor than the use of ICT as such. Exports in the form of licensing is an example of a model that enables quite large separability due to the uniform pre-planned service concept, upon which a third party builds the actual interaction with clients.

In relation to the drivers of internationalisation, our study highlights the opportunities provided by emerging markets and by new innovative service concepts. A third driver is the scalability of services which are delivered over the web or where virtual working and web tools play a central role. The main benefits achieved through internationalisation are the growth in business and the accumulation of skills and know-how; the world-wide accumulation of knowledge, credibility based on a well-known brand, and common development of products and tools.

The internationalisation of services has some policy implications. First, the importance of the topic is self-evident. The delay during the Doha Round of service liberalisation and multilateral GATS-like agreements is not justified by the economic reality. Service internationalisation is certainly very important even if some service sectors engage in only limited international activity. The role of KIBS in particular is outstanding since they are not only participating themselves in internationalisation, but also are instrumental to the process undertaken by companies belonging to any kind of economic sector. Secondly, it seems international trade and FDI are complementary, rather than substitutes, mainly at country level. This means the GATS negotiations should cover all types of services provision involved in internationalisation. It does not make much economic sense to focus just on one mode if there is also a clear relationship with other modes used by firms to internationalise their activities. Finally, the existence of different paths and explanatory factors behind service internationalisation suggests that there is scope for different models of policy development. There is no one model that can fit all. This is valid both for policies to promote service internationalisation and for policies to promote international liberalisation agreements. The specificities of different types of services are huge; service diversity matters. The differences in services are also useful for understanding the variegated nature of capitalism and the transformation of economic landscapes such as in Europe (Daniels et al., 2011). In this sense, policy makers should adapt to the specific profiles of different services landscapes.

From the managerial viewpoint, a crucial issue is how to balance the risks linked to resource investments and the possibility of controlling service quality. FDI includes the highest risks but also the greatest control. For exports and the use of partners, the risks are small but controlling service quality is more difficult. Flexibility and readiness of utilising new opportunities are the second important point: the directions selected have to be continuously evaluated. Internationalisation is not a straightforward process; it often includes trial and error. Consideration of various models in international activities and in the paths of going global is the third point. In FDI the number of subsidiaries may vary considerably: opening several offices in many countries may be successful in some cases, whereas in others it is more sensible to establish only a few subsidiaries in the main markets and use them as the 'base' for serving clients in other countries. As regards the paths to foreign markets, a cautious start followed by accelerated development has been suggested by several studies. Within the 'born global' path, it is possible that a company is internationally oriented right from the start but its business model is linked to operations in a certain region.

REFERENCES

Alon, I. (2005), *Service Franchising: A Global Perspective*, Springer, New York.
Ball, D.A., Lindsay, V.J. and Rose, E.L. (2008), 'Rethinking the paradigm of service internationalisation: Less resource-intensive market entry modes for information-intensive soft services', *Management International Review*, 48 (4), 413–431.
Bell, J. (1995), 'The internationalization of small computer software firms – A further challenge to "stage" theories', *European Journal of Marketing*, 29 (8), 60–75.
Blomstermo, A., Sharma, D.D. and Sallis, J. (2006), 'Choice of foreign market entry mode in service firms', *International Marketing Review*, 23 (2–3), 211–229.
Bryson, J.R. and Daniels, P.W. (eds) (2007), *The Handbook of Service Industries in the Global Economy*, Edward Elgar, Cheltenham and Northampton, MA.
Bryson, J.R., Daniels, P.W. and Warf, B. (2004), *Service Worlds: People Organisations and Technologies*, Routledge, London.
Bryson, J.R., Rubalcaba, L. and Ström, P. (2012), 'Services, innovation, employment and organisation: Research gaps and challenges for the next decade', *Services Industries Journal*, 32 (4), 641–655.
Chetty, S. and Campbell-Hunt, C. (2004), 'A strategic approach to internationalization: A traditional versus a "born-global" approach', *Journal of International Marketing*, 12 (1), 57–81.
Contractor, F.J., Kundu, S.K. and Hsu, C. (2003), 'A three-stage theory of international expansion: The link between multinationality and performance', *Journal of International Business Studies*, 34 (1), 5–18.
Daniels, P.W., Rubalcaba, L. Stare, M. and Bryson, J.R. (2011), 'How many Europes? Varieties of capitalism, divergence and convergence and the transformation of the European services landscape', *Journal of Economic and Social Geography*, 102 (2), 146–161.
Erramilli, K. (1991), 'The experience factor in foreign market entry behavior of service firms', *Journal of International Business Studies*, 22 (3), 479–501.
Erramilli, M.K. and Rao, C.P. (1990), 'Choice of foreign market entry modes by service firms: Role of market knowledge', *Management International Review*, 30 (2), 135–150.
Glucker, J. (2004), *A Relational Account of Business Service Internationalization and Market Entry – Theory and Some Evidence*, The University of Birmingham, School of Geography, Earth and Environmental Sciences, Working Papers on Services, Space, Society No. 15.
Grönroos, C. (1999), 'Internationalization strategies for services', *Journal of Services Marketing*, 13 (4), 290–297.

Howells, J. and Roberts, J. (2000), 'Global knowledge systems in a service economy', in Andersen, B., Howells, J., Hull, B., Miles, I. and Roberts, J. (eds), *Knowledge and Innovation in the New Service Economy*, Edward Elgar, Cheltenham and Northampton, MA, pp. 248–266.

Javalgi, R.G., Martin, C.L. and Todd, P.R. (2004), 'The export of e-services in the age of technology transformation: Challenges and implications for international service providers', *Journal of Services Marketing*, 18 (7), 560–573.

Johanson, J. and Wiedersheim-Paul, F. (1975), 'The internationalization of the firm – Four Swedish cases', *Journal of Management Studies*, 12, 305–322.

Kautonen, M. and Hyypiä, M. (2009), 'Internationalising business services and the national innovation system: The Finnish business services sector in a European comparison', *Journal of Services Technology and Management*, 11 (3), 229–246.

Kox, H. (2001), *Exposure of the Business Services Industry to International Competition*, CPB (Netherlands Bureau for Economic Policy Analysis), Report 010, The Hague.

Madsen, T.K. and Servais, P. (1997), 'The internationalization of born globals: An evolutionary process?', *International Business Review*, 6 (6), 561–583.

Majkgård, A. and Sharma, D.D. (1998), 'Client-following and market-seeking strategies in the internationalization of service firms', *Journal of Business-to-Business Marketing*, 4 (3), 1–41.

Massini, S. and Miozzo, M. (2010), *Outsourcing and Offshoring of Business Services: Challenges to Theory, Management and Geography of Innovation*, 30 July, Manchester Business School Research Paper No. 604.

Miozzo, M. and Miles, I. (2003), 'Introduction', in Miozzo, M. and Miles, I. (eds), *Internationalization, Technology and Services*, Edward Elgar, Cheltenham and Northampton, MA, pp. 1–11.

Miozzo, M., Yamin, M. and Ghauri, P. (2012), 'Strategy and structure of service multinationals and their impacts on linkages with local firms', *Service Industries Journal*, 32 (7), 1171–1192.

Noyelle, T.J. and Dutka, A. (1988), *International Trade in Business Services*, American Enterprise Institute, Ballinger, Washington, DC.

OECD (1999), *Business Services: Trends and Issues*, OECD, Paris.

O'Farrell, P.N., Wood, P.A. and Zheng, J. (1998), 'Regional influences on foreign market development by business service companies: Elements of a strategic context explanation', *Regional Studies*, 32 (1), 31–48.

Pla-Barber, J. and Ghauri, P. (2012), 'Internationalisation of service industry firms: Understanding distinctive characteristics: Introduction to the special issue', *Service Industries Journal*, 32(7), 1007–1010.

Pla-Barber, J., Sánchez, E. and Madhock, A. (2010), 'Investment and control decisions in foreign markets: Evidence from service industries', *British Journal of Management*, 21, 736–753.

Rennie, M.W. (1993), 'Born global', *McKinsey Quarterly*, 4, 45–52.

Roberts, J. (1998), *Multinational Business Service Firms: The Development of Multinational Organisational Structures in the UK Business Services Sector*, Ashgate, Aldershot.

Roberts, J. (2001), *Challenges Facing Service Enterprises*, PREST Discussion Paper: A Global Knowledge-Based Economy – Lessons from the Business Services Sector. Paper 01-03, January.

Rodríguez, A. and Nieto, M.J. (2012), 'The internationalisation of knowledge-intensive business services: The effect of collaboration and the mediating role of innovation', *Service Industries Journal*, 32(7), 1027–1076.

Rönkkö, P. (2001), 'Growth and internationalization of technology-based new companies: Case study of 8 Finnish companies', in Paija, L. (ed.), *Finnish ICT Cluster in the Digital Economy*, ETLA B 176. Taloustieto Oy, Helsinki, pp. 45–79.

Ruane, F. (1993), 'Internationalization of services: Conceptual and empirical issues', *European Economy*, 3, 109–124.

Rubalcaba, L. (1994), *Fairs and Exhibitions in European Economy*, European Commission, Brussels/Luxembourg.

Rubalcaba, L. (1999), *Business Services in European Industry: Growth, Employment and Competitiveness*, European Commission DGIII-Industry, Brussels/Luxembourg.

Rubalcaba, L. (2007a), *The New Service Economy: Challenges and Policy Implications for Europe*, Edward Elgar, Cheltenham and Northampton, MA.

Rubalcaba, L. (2007b), *Business Services in the Global Economy: New Evidence from a European Perspective*, RESER and SERU Working paper 2/2007.

Rubalcaba, L. and Cuadrado, J.R. (2002a), 'Services in the age of globalization: Explanatory interrelations and dimensions', in Cuadrado, J.R., Rubalcaba, L. and Bryson, J. (eds), *Trading Services in the Global Economy*, Edward Elgar, Cheltenham and Northampton, MA, pp. 27–57.

Rubalcaba, L. and Cuadrado, J.R. (2002b), 'A comparative approach to the internationalization of service industries', in Cuadrado, J.R., Rubalcaba, L. and Bryson, J. (eds), *Trading Services in the Global Economy*, Edward Elgar, Cheltenham and Northampton, MA, pp. 78–108.

Rubalcaba, L. and Kox, H. (eds) (2007), *Business Services in European Economic Growth*, Palgrave Macmillan, London.

Rubalcaba, L. and Van Welsum, D. (2007), 'Globalisation and global sourcing in business services', in

Rubalcaba, L. and Kox, H. (eds), *Business Services in European Economic Growth*, Palgrave Macmillan, London, pp. 213–230.
Saxenian, A. (2000), 'Networks of immigrant entrepreneurs', in Lee, C.-M., Miller, W.F., Hancock, M.G. and Rowen, H.S. (eds), *The Silicon Valley Edge – A Habitat for Innovation and Entrepreneurship*, Stanford University Press, Stanford, CA, pp. 248–268.
Sharma, D.D. and Johanson, J. (1987), 'Technical consultancy in internationalization', *International Marketing Review*, Winter, 20–29.
Sondheimer, J.A. and Bargas, S.E. (1993), 'U.S. international sales and purchases of private services', *Survey of Current Business*, 73 (9), 120–156.
Stare, M. and Rubalcaba, L. (2009), 'International outsourcing of services – What role for Central and East European countries?', *Emerging Markets Finance and Trade*, 45 (5), 31–46.
Tapscott, D., Ticoll, D. and Lowy, A. (2000), *Digital Capital: Harnessing the Power of Business Webs*, Nicholas Brealey Publishing, London.
Toivonen, M. (2002), 'Internationalization of knowledge-intensive business services in a small European country: Experiences from Finland', in Miozzo, M. and Miles, I. (eds), *Internationalization, Technology and Services*, Edward Elgar, Cheltenham and Northampton, MA, pp. 206–226.
Toivonen, M. (2004), *Expertise as Business: Long-Term Development and Future Prospects of Knowledge-Intensive Business Services (KIBS)*, Helsinki University of Technology, Department of Industrial Engineering and Management. Dissertation for the degree of Doctor of Philosophy; available at http://lib.tkk.fi/Diss/2004/isbn9512273152/, accessed 3 December 2014.
Vandermerwe, S. and Chadwick, M. (1989), 'The Internationalisation of services', *Service Industries Journal*, 9 (1), 79–93.
Visintin, S., Maroto, A., Di Meglio, G. and Rubalcaba, L. (2010), 'The role of cost related factors in the competitiveness of European services', *Global Economy Journal*, 10 (3), 4.
Young, S. (1987), 'Business strategy and the internationalization of business: Recent approaches', *Managerial and Decision Economics*, 8, 31–40.

16. In pursuit of creative compliance: innovation in professional service firms
Timothy Morris, Michael Smets and Royston Greenwood

INTRODUCTION

In this chapter, we discuss the nature of innovation in professional service firms. We argue that the distinctive characteristics of these firms affect their innovation drivers, sources, processes and outcomes. We suggest that, paradoxically, innovation in professional services is primarily driven by a pursuit of compliance, rather than differentiation. Professionals are typically required to resolve situations in which their clients aim to comply with external constraints, but seek to do so in a creative way that is least compromising for their business goals. Thus, innovation in professional service firms is primarily driven by *creative compliance*.

This driver of innovation is radically different from the differentiation ambitions that typify the disruptive or frame-breaking innovation that has dominated the technology and manufacturing literature (Bower and Christensen, 1995; Christensen and Overdorf, 2000; Tushman and Anderson, 1986; Tushman and Romanelli, 1985). As innovation is driven by articulated client needs, rather than the ambition to conceive of a break-through product that customers did not even know they wanted, in professional service firms it occurs primarily through the development or recombination of existing knowledge, often in close collaboration between professionals and with the clients whose problems are to be solved. As the innovation process unfolds, its outcomes can reverberate throughout the organization, challenging the roles of professionals, the internal organizational structure of the firm and in some cases the boundaries of the firm.

In the next section, we outline the characteristics of professional service firms and explain how they differ from other types of organization. Then, we consider the nature of innovation in these firms, arguing that it has several distinctive characteristics in this setting. We explore the inter-relationship of service and process innovations and how they link to the development and application of knowledge. We outline the symbolic and political implications of the process of innovation before considering how organizational changes may influence the development of knowledge and therefore of innovation.

THE PROFESSIONAL SERVICE FIRM

Initially, the professional service firm label was reserved for organizations in which core staff were members of a recognized profession such as law, accounting or healthcare (e.g. Brock et al., 1999). As such, their status was supported, and their operations governed, by statutory frameworks and strong socialization into professional norms (Abbott, 1988; Freidson, 1986). Recently, though, the label has been applied more broadly (e.g.

von Nordenflycht, 2010). Definitional focus has shifted from formal accreditation and professional status to the knowledge intensity of services. As a result, a variety of occupations that either lay claim to professional status or that also provide services based upon some form of expert knowledge have been included under the professional service firm umbrella. This wider group includes management consultants and a number of other business-based occupations such as marketing and IT consultants (Alvesson, 1993; Engwall and Kipping, 2002; Greenwood et al., 2006; Gross and Kieser, 2006; McKenna, 2006; Reihlen, Smets and Veit, 2010).

In this chapter, we focus on traditional professional firms comprising members of accredited professions. In particular, we focus on the two dominant business professions – accounting and law – which have attracted much scholarly interest (e.g. Brock et al., 1999; Greenwood et al., 1999; Sherer, 1995; Smets et al., 2012a; von Nordenflycht, 2010). Professionals in these firms are accountable to the organization that employs them or in which they have an ownership share, to the clients they serve and to the profession that regulates their practice. They pursue commercial goals in seeking to win client business and deliver services profitably while at the same time observing professional standards and constraints, attending to standards of work and ethics that are defined by their professional institutions. Sometimes these two goals will come into conflict, which impacts on service delivery and innovation.

Although some professional firms have recently adopted different ownership and governance models involving public listing or external shareholders (see von Nordenflycht, 2010), the characteristic form of ownership and control in professional services is and remains the 'partnership' (Empson and Chapman, 2006; Greenwood and Empson, 2003; Pinnington and Morris, 2003). In partnerships of professionals the owners of the firm, the partners, are also key producers and involved in the day-to-day running of the business to varying degrees. Hence, the partnership form of organization and governance provides a means of organizing and running the firm in accordance with the preferences of these owner-producer-managers, and, critically, of entwining the pursuit of commercial goals with professional values of meritocracy and relative equality among collaborating professionals operating as peers (Gilson and Mnookin, 1985). This organizational form is not only highly successful (Greenwood and Empson, 2003) but it is also a highly distinctive and institutionalized example of a 'hybrid' structure; that is, an organization designed to resolve the tensions of institutional complexity arising from the prescriptions of different logics (Battilana and Dorado, 2010; Greenwood et al., 2011; Pache and Santos, 2010).

In economic terms, the partnership form ensures that those holding the key assets underpinning a firm's market power, such as its expert knowledge and reputation in the client market, can secure exclusive rights to the firm's profits (Gilson and Mnookin, 1985; Sherer, 1995). These economic benefits come at the expense of relatively dispersed authority among co-owning partners. They all have a legitimate right to information about the firm's interests, a say in strategic decision making, a vote in electing new partners and involvement in the selection of partners with responsibility for managing the firm, notably the senior and managing partners. As a result, the professional partnership is said to operate on a strongly collegial basis with regard to major decisions, and to grant individual partners substantial control of their client relationships (e.g. Greenwood et al., 1990). The strategic capacity of these firms is therefore relatively weak compared with the incorporated company; hierarchy is limited and based on professional status and experience, rather than managerial authority. The professional expertise and experience of seniors and – to some extent – of the junior professionals working under their supervision are critical for the delivery of professional services.

The organizing model of the professional service firm is relatively simple (Smets et al., 2012b). It centres on transforming inputs – first and foremost the formal knowledge and practical experience of individual professionals – into outputs; that is, more or less complex, knowledge-intensive or customized solutions to client problems (e.g. Grey, 1994; Maister, 1993; McKenna, 2006; Morris and Empson, 1998). This transformation is undertaken by teams of professional staff (Hansen and Haas, 2001; Werr and Stjernberg, 2003) whose composition depends on the nature of the task. It is the responsibility of the engagement partner to match requisite types of expertise and levels of seniority to the complexity of the specific task at hand.

The resultant team then temporarily brings together professionals drawn from their 'home' practice groups, based on the nature of their professional expertise, such as tax law or commercial litigation. Collectively, these practice groups reflect the breadth of the firm's professional expertise, its extent of diversification and its portfolio of market/industry specializations (Greenwood et al., 2005). The client team, therefore, forms the core of collaboration and knowledge recombination, a role that is central to both service delivery and innovation, as we will show below.

Client teams not only bring together requisite types of expertise but also levels of seniority and, by implication, experience. Generally, professional service firms operate with a basic division of professional labour into two groups: partners (the owners) and associates, sometimes called assistants, who make up the salaried workforce (Galanter and Palay, 1991; Hansen and Haas, 2001; Malos and Campion, 2000; Sherer, 1995). Associates are qualified professionals of varying degrees of experience who work on client projects or transactions, strictly under the supervision of a partner who acts as their superior and mentor. Partners typically also 'practise', that is, they may engage in professional work for clients as well as fulfilling their supervisory obligations. However, in many cases they are primarily concerned with generating client business, maintaining client relationships and/or running the firm as owner-managers.

Team production ensures the economic viability of the firm, based on the principles of leverage and utilization, which are enshrined in this organizational form and distinctive of all professional firms (Gilson and Mnookin, 1989; Kor and Leblebici, 2005; Lowendahl, 2005; Maister, 1993; Malos and Campion, 2000; Sherer, 1995). Utilization is simply the amount of time that a professional is engaged on chargeable client activities, and serves to valorize professional expertise (Yakura, 2001). The greater a professional's utilization, *ceteris paribus*, the more profit the firm generates from that professional's efforts, assuming he or she is charged at a rate that is greater than his/her employment cost.

Associates are often engaged on several tasks simultaneously in order to maximize their utilization. To execute those tasks, associates typically also borrow, or 'leverage', the knowledge and experience of senior professionals. Leveraging senior knowledge through the labour power of juniors drives the division of labour in a professional firm and underpins its profitability. Economically, by allocating junior professionals to carry out tasks that are commensurate with their experience and charging the client at a rate that is higher than the employment cost of those professionals, the firm can generate profits, which then accrue to the partners. Therefore, leverage is measured by the ratio of partners to other professional staff. Leverage therefore requires task delegation.

INNOVATION IN PROFESSIONAL SERVICE FIRMS

The innovation literature conventionally distinguishes between product and process innovation (e.g. Adner and Levinthal, 2001; Davenport, 1992; Utterback and Abernathy, 1975). In the context of professional service firms, product or service innovation refers to the outputs of their core professional staff; that is, the customized solutions to complex client problems. Process innovation, by contrast, refers to changes in the organizing model and the team production process, outlined above, by which inputs are transformed into outputs. Despite this analytical distinction, product and process innovation are often closely interlinked in professional service firms. Pressures and procedures inherent in the organizing model motivate and facilitate product innovation that, in turn, can feed back into the organizing model.

Client-driven Product Innovation

The extent to which the distinctive organizing model of professional service delivery and product innovation are intertwined becomes apparent in the client interaction. The complex nature of their problems means that clients are frequently involved in the co-production of the solution alongside the professionals they have hired (Fosstenløkken, Løwendahl and Revang, 2003; Mills et al., 1983). They help define the problem, gather necessary information and review various options before a solution is delivered. As a result of this co-production, some degree of customization occurs even where professionals are using templates or standardized methods that they have used elsewhere. Thus, service innovation is embedded in the production process of professional services as solutions are customized in client interaction.

In practice, the extent and nature of customization varies across transactions, professional practice areas and firms, but always in close correlation with the organizing model the firm deploys in terms of team production, utilization and leverage. In any firm there appear to be limits to the extent that expert knowledge can be codified into formal methodologies, solutions, templates or algorithms (Morris, 2001). Yet, some firms seek to more extensively codify their knowledge in formal methods, techniques and templates which can be easily applied by junior staff (Hansen and Haas, 2001; Hansen et al., 1999; Maister, 1993). These firms typically seek out work which is commensurate with the application of codified knowledge. Their teams display a higher leverage ratio, as they rely relatively heavily on more junior staff leveraging seniors' existing knowledge. The application of existing knowledge by less experienced professionals in highly leveraged teams usually leaves little scope for innovation. Other firms trade on the personal or tacit knowledge of their staff and limit the codification of their knowledge. Inevitably, this limits their leverage and they will, therefore, seek out more complex assignments and rely to a greater extent on more senior staff that can deploy their experience and expertise. In law, for example, tax practices that specialize in complex tax transactions will have a relatively low leverage ratio while teams conducting more standardized tasks, such as a due diligence on a large-scale corporate transaction, will be more highly leveraged. At the extreme, the legal aspects of insurance claims, which some law firms handle, can be mass processed by large teams of paralegals and junior lawyers following codified procedures.

The relative complexity of client problems and the degree of customization of their solutions impact innovation in professional service firms in several ways. Service innovation is an inherent characteristic of much professional work, insofar as each transaction requires at least some degree of customization. At the very least, existing templates are adapted or somewhat different solutions are applied to suit the particular requirements and preferences of each client. More likely, clients present professionals with a problem in which existing knowledge has to be deployed somewhat creatively (Heusinkveld and Benders, 2005, 2012). Thus, by presenting challenging problems or by being demanding, clients prompt the development of new knowledge. However, in the process of co-producing a solution to their problem, they also contribute to the required new knowledge and, to the extent that knowledge is a strategic resource of the firm, update and extend its knowledge base (Fosstenløkken et al., 2003; Lowendahl, 2005; Løwendahl et al., 2001). In short, clients aid innovation by prompting innovation through the problems they pose, and by enabling innovation through the insights they share in the co-production process.

Inevitably, innovation is a front-line activity. It occurs in the process of delivering client solutions and derives from professionals of differing seniority levels confronting everyday work problems together in their client teams. Given their simultaneous and dispersed operation, 'improvisations bubble[d] up in multiple locations as and when transaction teams encounter[ed] specific problems' (Smets et al., 2012a: 889). Therefore, rather than being developed centrally and then distributed through to the point of production, professional service innovation 'bubbles up' from below. While the principle of leverage means it is likely that the most senior professionals, who are dealing with relatively complex problems, lead innovation, other professionals may still introduce minor forms of innovation as they resolve work problems or as they are involved in the innovation process. This front-line innovation stands in contrast to the model of innovation in many large-scale service industries and manufacturing where it is the domain of specialized units such as R&D Departments (e.g. Burgelman, 2002; Gupta et al., 2006; Tushman and O'Reilly, 1996). Indeed, the professional service firm typically does not operate with a specialist R&D unit or invest systematically in large-scale service innovation.

An interesting implication of this process, however, is that innovation is usually incremental. Because it is prompted by a particular problem and situated on the front-line, innovation sits at what might be seen as the 'Development' end of R&D, rather than entailing what might be seen as fundamental research. In other words, innovations are customized to particular problems. These innovations are likely to entail re-combinations of knowledge used elsewhere in the firm but rarely provide a substantial knowledge advance. One exception to this in law is the 'poison pill' defence, which was developed by the law firm Wachtell, Lipton and Katz and was, at the time, seen as a radical innovation (Starbuck, 1992). This defence was a legal innovation by which corporations could defend themselves against the threat of a hostile take-over bid.

Innovation for the professional service firm is not an option but an imperative, especially for those that charge premiums for their expertise. Professional service firms must continuously update their services, as it is near impossible to protect against imitation and commodification. For one thing, it is difficult to claim intellectual property rights on firm-based knowledge because in law and accounting that knowledge is derived from non-proprietary knowledge, such as legal statutes, professional standards and regulations. Also, new knowledge, once incorporated in client solutions, becomes public and is thus easily appropriable by others. For instance, innovations by a law firm become known to other parties as they are tested in court or distributed to other transaction parties where they can be extensively examined by competitors as a test of viability. Smets et al. (2012a) illustrate this process in their account of how innovations by legal advisors on syndicated finance transactions were rapidly and inevitably revealed to as many as 20 other law firms in the process of syndication; indeed, the unobtrusive nature of innovation facilitates its dissemination as it permits other firms to take on change without provoking resistance. Minor variations on any form of legal innovation will typically take place on subsequent transactions and put any novel idea beyond the scope of proprietary protection. Knowledge therefore becomes a commodity relatively quickly and although the reputation of professional service firms is built upon their organizational expertise and upon that of their partners, this commoditization process means that there is a strong incentive for firms to continually innovate – that is, to develop new knowledge – because failure to do so would undermine their ability to charge premium prices for their services. Suddaby and Greenwood (2001) argue, as well, that as expert knowledge commodifies over time, professional firms seek to 'colonize' new territories, provoking jurisdictional disputes with other professional communities such as consulting firms (see also Heusinkveld and Benders, 2005).

In sum, although professionals would not necessarily recognize it as such or use the term to describe their activities, service innovation is built into everyday professional work. It occurs in response to client demands, continuously and incrementally. As a consequence of these characteristics, innovation in professional service firms is beset by the paradox that it is inherent to the activities of core professionals, but not driven by a strategic pursuit of differentiation and competitive advantage, as it is in other sectors (Christensen and Bower, 1996; Michel et al., 2008; Miller et al., 2008). Instead, innovation occurs because of what we term 'creative compliance'. It is triggered by a suspected or anticipated misalignment between organizational practice and regulatory frameworks, which may result from one of two sources. First, innovation may be triggered by an external technological or regulatory change, which may require adjustments to existing operating procedure (Suddaby and Greenwood, 2001). Clients aiming to maintain these procedures would seek creative forms of compliance that least compromise current operations. Second, when a client considers a move that tests the boundaries of existing rules, regulations or statutes, similar creativity may be needed to ensure compliance. For example, seeking to treat an important item on the balance sheet in a novel way that portrays the company more favourably, the client turns to its professional advisors for a solution and their assurance that this treatment is legal and legitimate. In response, professionals develop a solution which permits the client to act as creatively as it wishes, but remains compliant with existing rules and regulations. The professional service firm therefore offers reputational support in its judgement. Thus, the novelty of the poison pill defence was developed in response to a client's need for protection from a hostile take-over. This emphasis on 'creative compliance' has been a particular characteristic of Anglo-American professional firms operating in common law systems in which the scope to create novel interpretations or precedents within existing legal frameworks has been highly institutionalized.

An implication of the organizational dynamics of innovation in professional service firms relates to March's (1991) well-known distinction between exploitation and exploration activities. Exploitation builds on existing knowledge and aims to better meet the needs of existing customers by incrementally improving on existing competences. Exploration, conversely, describes the search for new ideas and ways of working that depart from existing routines and drive entry into new product-market domains (Benner and Tushman, 2003). Typically, organizations are said to find it difficult to achieve both simultaneously because of the different competences and routines that underpin them (Benner and Tushman, 2003; Jansen et al., 2009; O'Reilly and Tushman, 2004; Siggelkow and Levinthal, 2003). Based on the characteristics of service innovation outlined above, this distinction is less pronounced in professional service firms. While high-leverage client teams, focused on maximizing utilization and re-using existing knowledge, are more geared towards exploitation, and low-leverage teams with a relatively high proportion of experienced professionals are better equipped to explore new uses of their knowledge, the activities by which exploration and exploitation occur are remarkably similar in the continuous, front-line innovation process described above.

In effect, the objective of maximizing utilization means that staff are allocated to solving problems which are in the domain of exploitation: this entails the use of the firm's current competences and knowledge and leads to incremental changes to these as customization proceeds. However, firms also undertake exploration through relationships with new partners that pose unprecedented and divergent demands (Beckman, Haunschild and Phillips, 2004). In this sense, professional firms use not only their internal resources but also their relationships with external parties, typically customers, to hone and extend their expertise (Fosstenløkken et al., 2003). Innovation occurs incrementally and continuously, as professionals often pursue exploitation *and* exploration in their client engagements, using knowledge in both well- and little-understood ways (Taylor and Greve, 2006).

Internally Driven Process Innovation

Service innovations that are well received in the client market and that promise to meet demand well beyond the client for whom they have been developed can trigger process innovations in which the organizing model of the firm or the team production process is modified. Intriguingly, the driver for these modifications of the organizing model rests in the model itself; specifically, in the way that partners are elected and professional staff seek to enhance their chances of promotion to partner (Anand et al., 2007). Because partners share the profits of the firm, existing partners are keen to ensure that newly elected peers do not dilute existing profit levels. Therefore, aspiring partners and those recently given that status strive to demonstrate their ability to expand the firm's client base. One way of demonstrating such profit potential is the creation of a new practice area; that is, a permanent, structural unit of professionals with a common knowledge base (Anand et al., 2007). In short, new knowledge that was originally developed by client teams drawn from multiple practice areas may support the foundation of a new practice area in which this knowledge is permanently pooled and targeted to the growth of a new market. In this way, service innovation can feed process innovation. Translating a service innovation into a modified organizing model with an extended portfolio of practice areas, though, is not straightforward. It relies on the constructive interplay of four ingredients: socialized agency, expertise, turf and organizational support.

The critical ingredient to start the process is what Anand et al. (2007) term *socialized agency*. Agency implies that there is a 'champion', an individual or group that is taking the initiative to develop a new practice area. The qualification that this agency is 'socialized', however, is important because it signals how agency is moderated through professional socialization, including expectations that candidates for partnership position themselves in this way, but that they respect norms of collegiality in the process. Hence, socialization constrains and facilitates agency.

Expertise is likely to involve a refinement and generalization of knowledge that has been developed on client engagements. Professionals typically emphasize that technical knowledge is not sufficient on its own to be valuable to clients, but has to be put into context, so that sense can be made of it in terms of a sector or geography. Such sector-specific knowledge may pertain to the way the energy industry works and how it accounts for its profits and losses and debt funding, or the development of knowledge about privatization such as that triggered by statutory changes in the UK during the 1980s and Private Finance Initiatives during the 1990s.

Turf refers to the creation of a viable organizational space in which the champion is understood to have especial expertise, and that work is channelled towards him or her from other parts of the firm and from clients because of their recognition of that expertise. Without this legitimacy, rival groups might divert the flow of work and clients not see the expert as the right point of contact.

Organizational support is also necessary because innovation requires resourcing and endorsement. The resources include marketing, financial support in the form of a budget before the practice can generate enough funding to justify its existence commercially, allocation of people from other established practices, and support against other potential political threats.

Anand et al. (2007) show that all four elements are necessary to build a new practice and that the absence of one cannot be compensated for by amplifying any of the others. Combining these elements skilfully is also critical to successful innovation. They must be combined via a two-stage process in which the first champion combines his or her agency with one of the other elements in the *emergence step*. In the second stage, the *embedding step*, the two other elements have to be successfully introduced. For example, this embedding step could involve establishing organizational 'turf', usually by building on existing client relationships, thereby showing others in the firm that the nascent practice has internal and external legitimacy. It could also involve obtaining sufficient support from the rest of the organization, in the form of people, budget and recommendations to clients that bring in a flow of work, for the practice to flourish.

New practice creation through which innovation can be established in the professional service firm is inherently political and therefore challenging to accomplish successfully. It requires the garnering and deployment of valued resources, including legitimization by powerful groups not only within the firm but also from clients and others that refer work to the firm (Glückler and Armbrüster, 2003) and relies on innovators' ability to overcome internal resistance (Heusinkveld and Benders, 2005).

CAPACITY FOR SERVICE AND PROCESS INNOVATION

As recent studies of the organizing model of law and other professional firms show, it is not only service innovations that can entail process innovation. Equally, changes to the organizing model that did not follow from service innovation affect professional firms' capacity for service innovation (Smets et al., 2012b).

This dynamic has been found to start from the very heart of the firm, namely the partnership format itself and the 'up-or-out' model by which junior professionals have been typically promoted to this rank (Galanter and Palay, 1991; Gilson and Mnookin, 1989; Malhotra et al., 2010). The simple division of labour between partners and associates is becoming more complex. Numerous reasons are offered for this change, including changes to the labour market for lawyers, the career preferences of young lawyers themselves and the opportunities offered by new technologies to outsource work more easily. Thus, firms have adapted their career structures by introducing non-partner permanent positions that do not assume the up-or-out principle (Malhotra et al., 2010), such as Counsel or Professional Support Lawyer, whose function is to provide technical support and research to fee earners. Concomitantly, firms have also gone out of their way to attract young professionals by offering an implicit contract that implies that, in return for high commitment and effort (but without the assumption of a long-term career), associates will acquire general human capital through experience of working on transactions that will stretch their professional skills. These adaptations to the traditional organizing model and employment structure of the professional service firm have implications for process and service innovation.

These implications flow from changes to the partner's role and leverage that can take place when these new roles are introduced. Given that the role of the partner is to win business and to engage in the core production process, the availability of more senior professionals occupying Counsel roles means that the leverage model changes, becoming more pronounced towards the top of the pyramid, and the dynamic of the professional team changes as well. For one thing, the partner has the incentive and time to concentrate more on winning business that is of a quality commensurate with the added experience of the professional team and with the implicit employment contract that the firm will provide opportunities for learning for its young professionals. Further, Counsel and senior associates can become more central to dealing with the complex transactions in which innovation is required.

In addition, the expansion of a cadre of Professional Support Lawyers is equivalent to the creation of an R&D capability, as these lawyers specialize in organizing and updating the firm's stock of knowledge and developing solutions to offer to their clients. Together, these innovations provide the professional service firm with the incentive and the capability to move up the value chain by seeking and discharging transactions that are more complex and thereby avoiding the trap of commoditizing knowledge (Smets et al., 2012b).

Other recent developments in the organizational structure of law and accounting firms similarly have implications for process and service innovation. In particular, the largest of these firms have started to engage in the process innovation of outsourcing and offshoring of some of their activities. In the legal sector this trend has been accompanied by the creation of new organizational forms that specialize in high-volume, lower-value work such as document review, drafting and legal research, called LPOs or Legal Process Outsourcers. Sako (2009) argues that the main motivator for the offshoring of legal work – whether through wholly owned units or by contracting with independent organizations – is wage arbitrage. Many of these LPOs are situated in relatively low-cost locations, including Indonesia and India, where there is also a substantial supply of relatively well-educated labour. The implication is that the integrated model of the firm, undertaking all areas of professional work in-house, will decompose. Professional service firms have begun applying techniques and methods used in other service and manufacturing sectors, such as process analysis and work standardization, in order to modularize tasks and define which can be outsourced and those which cannot be easily decomposed and should remain in-house. Sako (2009) notes that there are limits to offshoring relating to the nature of professional work itself, in particular the need for close client interaction and the difficulty of breaking down some parts of legal work. Another limit is that professions' jurisdictional controls mean overseas workers cannot offer advice in other national jurisdictions. These limitations mean that offshoring is unlikely to develop as quickly among traditional professions as they have in other areas such as IT.

Nonetheless, as offshoring of routine tasks proceeds, firms have an incentive to concentrate on higher-value work in which they can justify premium costs and in which they emphasize customization even more. The trend towards the decomposition of the value chain is likely to have consequences for the traditional leverage model and career structure. If lower-value work is outsourced, then the demand for junior professionals or paralegals, who have traditionally undertaken the more routine team tasks as part of the process of gaining experience, will fall and firms may, instead, seek to hire mid-career professionals who have already obtained the skills and experience to undertake more complex professional work as well as more specialist professional support staff who can support service innovation. If firms employ proportionately more senior staff and outsource routine and low-value work, there is an incentive for them to concentrate on generating higher-value innovative services for which they can charge higher prices; employing expensive senior professionals and then asking them to do routine work will be a route to lower profitability.

CONCLUSIONS

In this chapter, we have argued that the distinctive features of professional service firms influence the nature and occurrence of process and service innovation. We have argued that service innovation is a characteristic of much of the everyday work of these firms because they seek to provide customized solutions to client problems. Professional knowledge underpins the inputs and outputs of their activities and innovation involves developments in or re-combinations of individual and firm-based expert knowledge. Innovations in the form of customized solutions occur on the front-line of work rather than from specialized units such as R&D departments, are usually incremental rather than radical, and are frequent rather than periodic. Innovation, moreover, occurs as a result of client demands for 'creative compliance'; that is, creative solutions customized to the circumstances of the client, and that enable the client to accomplish a goal for which it needs assurance that it is operating legitimately and competitively. For professionals, who gain their credentials from membership of a professional body, the challenge of creative compliance is to generate solutions that provide a commercial solution for the client without falling foul of the profession's standards. The hybrid organizational form of the professional service firm has emerged to resolve this particular tension between sets of competing yet complementary prescriptions, and is associated with particular patterns of innovation because professional firms compete on their capacity for creative compliance; that is, the capacity to generate knowledge-based innovations rather than having proprietary knowledge, because once such knowledge enters the public domain it is easily and quickly imitable. The dynamic of knowledge commodification that follows from this means that, in order to avoid the economic consequences, professional service firms have to continue to develop novel solutions.

The process of creating novel solutions is linked to the structural development of the firm in the form of new practices. New practices embed and form a home for the relatively permanent development of areas of professional knowledge. Studies of new practice creation show that they are not simply the product of chance or the result of client pressure but entail the skilful combination of a set of elements through two stages of emergence and then embedding the practice. This, we have suggested, is a politically demanding task as it requires legitimation internally and externally and because it requires resources and effort by the agents who lead this process.

The organizing model of the professional service firm is based on the assumption that senior professionals, partners who own a share in the firm, lend their expert knowledge to more junior staff who can execute professional tasks on their behalf. Economically, this permits the leveraging of seniors' knowledge and reputation to secure profits. The incentive in the professional firm is therefore to maximize the amount of time professionals are engaged on client work and to distribute tasks according to each professional's experience and expertise. Thus, much of the innovation comes from more senior professionals who are allocated complex problems to resolve. One consequence is that the conventional dichotomy between exploitation, which involves incremental innovation using existing competences, and exploration, which involves more radical innovation using novel competences, is not clear-cut in professional service firms where professional teams are handling both incremental and more radical challenges as part of their everyday tasks.

We have also suggested that innovation processes and outcomes are altering as the organizing model of the professional service firms adapts. In effect, changes to leverage leave their imprint on the capacity for, and incentive to find, innovative work. Leverage changes come about through innovations to the career structure in these firms. These career innovations permit permanent, non-partner positions to exist in which knowledgeable professionals can undertake complex work, freeing up partners to seek out appropriate client transactions on which these seniors can work profitably. Another change involves the creation of specialist support positions that are not directly fee earning but that are, in effect, akin to the creation of an R&D function. We have also proposed that the trend towards the deconstruction of the integrated professional service firm via outsourcing provides an incentive for these organizations to concentrate on complex solutions in which innovation is at a premium. In this sense, we conclude that organizational changes to the professional service firm may make them appear more like other, non-professional service and mass production organizations but that the distinctive features associated with the production of complex, customized solutions mean that the characteristics of process and service innovation will persist.

Future research should build on the propositions about the distinctive innovation processes and outcomes in the pursuit of creative compliance and within the organizing model which we have discussed in this chapter. First, longitudinal studies of process and product innovations in professional service firms would offer valuable insights into the interactions of product and process innovations, the mechanisms through which they occur, and the different pathways through which they can interlink. Secondly, we have discussed in this chapter two core business professions, accounting and law, but have suggested that our propositions can be extended to other formal professions. However, we note that the type of knowledge and the jurisdictional strategies of different professions affect the ways that professional firms are organized and managed (Malhotra and Morris, 2009). Thus, it is arguable that creative compliance may mean different things to engineers or architects, for instance, than to lawyers and, accordingly, innovation processes and outcomes may differ. This opens up opportunities for fruitful cross-profession research. Third, research might be broadened to investigate the nature of innovation in occupations which are not formally professionalized but possess some of the characteristics of the professional service firm, such as advertising (von Nordenflycht, 2010), and where the balance of creativity and compliance differs. For instance, it is plausible to argue that advertising is not so concerned with the compliance challenge but primarily preoccupied with creativity. Research in arenas such as advertising could fruitfully take into account the variety of ownership models that exist to explore if and how this affects innovation. One possibility, for example, is that external shareholders drive firms to adopt different processes of innovation by instituting management models which change the role of the partner and the organization of core professional work. Finally, a fourth line of enquiry could fruitfully move beyond the level of the firm to consider institutional innovations that are driven by, or affect, the pursuit of creative compliance by regulated professionals (Smets et al., 2012a). How such innovations emerge, consolidate and percolate to the institutional level to subsequently affect professionals' capacity and motivation to innovate in future requires further empirical and theoretical elaboration.

BIBLIOGRAPHY

Abbott, A.D. (1988), *The system of professions: An essay on the division of expert labor*, Chicago, IL: University of Chicago Press.

Abbott, A.D. (1991), 'The future of the professions: Occupation and expertise in the age of organization', in P.S. Tolbert and S.B. Bacharach (eds), *Research in the sociology of organizations: Organizations and professions*, Greenwich, CT: JAI Press, pp. 17–42.

Abrahamson, E. (1996), 'Management fashion', *Academy of Management Review*, **21**, 254–285.

Adner, R. and D. Levinthal (2001), 'Demand heterogeneity and technology evolution: Implications for product and process innovation', *Management Science*, **47** (5), 611–628.

Alvesson, M. (1993), 'Organizations as rhetoric: Knowledge-intensive firms and the struggle with ambiguity', *Journal of Management Studies*, **30** (6), 997–1015.

Anand, N., H. Gardner and T. Morris (2007), 'Knowledge based innovation: Emergence and embedding of new practice areas in management consulting firms', *Academy of Management Journal*, **50** (2), 406–428.

Battilana, J. and S. Dorado (2010), 'Building sustainable hybrid organizations: The case of commercial microfinance organizations', *Academy of Management Journal*, **53** (6), 1419–1440.

Beckman, C., P. Haunschild and D. Phillips (2004), 'Friends or strangers? Firm-specific uncertainty, market uncertainty, and network partner selection', *Organization Science*, **15** (3), 259–275.

Benner, M.J. and M.L. Tushman (2003), 'Exploitation, exploration, and process management: The productivity dilemma revisited', *Academy of Management Review*, **28** (2), 238–256.

Bower, J.L. and C.M. Christensen (1995), 'Disruptive technologies: Catching the wave', *Harvard Business Review*, **73** (1), 43–53.

Brock, D., M.J. Powell and C.R. Hinings (eds) (1999), *Restructuring the professional organization: Accounting, health care and law*, London: Routledge.

Burgelman, R.A. (2002), 'Strategy as vector and inertia as coevolutionary lock-in', *Administrative Science Quarterly*, **47**, 325–357.

Christensen, C.M. and J.L. Bower (1996), 'Customer power, strategic investment, and the failure of leading firms', *Strategic Management Journal*, **17** (3), 197–218.

Christensen, C.M. and M. Overdorf (2000), 'Meeting the challenge of disruptive change', *Harvard Business Review*, **78** (2), 66–76.

Davenport, T.H. (1992), *Process innovation: Reengineering work through information technology*, Cambridge, MA: Harvard Business Press.

Empson, L. and C. Chapman (2006), 'Partnership versus corporation: Implications of alternative forms of governance in professional service firms', in R. Greenwood and R. Suddaby (eds), *Professional service firms*, Oxford: JAI Press, pp. 139–170.

Engwall, L. and M. Kipping (2002), 'Introduction: Management consulting as a knowledge industry', in M. Kipping and L. Engwall (eds), *Management consulting: Emergence and dynamics of a knowledge industry*, Oxford: Oxford University Press, pp. 1–16.

Fosstenløkken, S.M., B.R. Løwendahl and O. Revang (2003), 'Knowledge development through client interaction: A comparative study', *Organization Studies*, **24** (6), 859–879.

Freidson, E. (1986), *Professional powers: A study of the institutionalization of formal knowledge*, Chicago, IL: University of Chicago Press.

Galanter, M. and T.M. Palay (1991), *Tournament of lawyers: The transformation of the big law firm*, Chicago, IL: University of Chicago Press.

Gilson, R.J. and R.H. Mnookin (1985), 'Sharing among the human capitalists: An economic inquiry into the corporate law firm and how partners split profits', *Stanford Law Review*, **37** (2), 313–392.

Gilson, R.J. and R.H. Mnookin (1989), 'Coming of age in a corporate law firm: The economics of associate career patterns', *Stanford Law Review*, **41** (3), 567–595.

Glückler, J. and T. Armbrüster (2003), 'Bridging uncertainty in management consulting: The mechanisms of trust and networked reputation', *Organization Studies*, **24** (2), 269–297.

Greenwood, R. and L. Empson (2003), 'The professional partnership: Relic or exemplary form of governance?', *Organization Studies*, **24** (6), 909–933.

Greenwood, R. and R. Suddaby (2006), 'Institutional entrepreneurship in mature fields: The Big Five accounting firms', *Academy of Management Journal*, **49** (1), 27–48.

Greenwood, R., C.R. Hinings and J. Brown (1990), '"P2-form" strategic management: Corporate practices in professional partnerships', *Academy of Management Journal*, **33** (4), 725–756.

Greenwood, R., T. Rose, C.R. Hinings and D.J. Cooper (1999), 'The global management of professional services: The example of accounting', in S.R. Clegg, E. Ibarra-Colado and L. BuenoRodriguez (eds), *Global management: Universal theories and local realities*, London: Sage Publications, pp. 265–296.

Greenwood, R., R. Suddaby and C.R. Hinings (2002), 'Theorizing change: The role of professional associations in the transformation of institutionalized fields', *Academy of Management Journal*, **45** (1), 58–80.

Greenwood, R., S.X. Li, R. Prakash and D.L. Deephouse (2005), 'Reputation, diversification, and organizational explanations of performance in professional service firms', *Organization Science*, **16** (6), 661–673.

Greenwood, R., R. Suddaby and M. McDougald (2006), 'Introduction', in R. Greenwood, R. Suddaby and M. McDougald (eds), *Professional service firms*, Oxford: JAI Press, pp. 1–16.

Greenwood, R., R. Raynard, F. Kodeih, E.R. Micelotta and M. Lounsbury (2011), 'Institutional complexity and organizational responses', *Academy of Management Annals*, **5** (1), 317–371.

Grey, C. (1994), 'Career as a project of the self and labour process discipline', *Sociology*, **28** (2), 479–497.

Gross, C. and A. Kieser (2006), 'Are consultants moving towards professionalization?', in R. Greenwood and R. Suddaby (eds), *Professional service firms*, Oxford: JAI Press, pp. 69–100.

Gupta, A.K., K.G. Smith and C.E. Shalley (2006), 'The interplay between exploration and exploitation', *Academy of Management Journal*, **49** (6), 693–706.

Hall, R.H. (1967), 'Some organizational considerations in the professional–organizational relationship', *Administrative Science Quarterly*, **12** (3), 461–478.

Hansen, M.T. and M.R. Haas (2001), 'Competing for attention in knowledge markets: Electronic document dissemination in a management consulting company', *Administrative Science Quarterly*, **46** (1), 1–28.

Hansen, M.T., N. Nohria and T. Tierney (1999), 'What's your strategy for managing knowledge?', *Harvard Business Review*, **77** (2), 106–116.

Heusinkveld, S. and J. Benders (2005), 'Contested commodification: Consultancies and their struggle with new concept development', *Human Relations*, **58** (3), 283–310.

Heusinkveld, S. and J. Benders (2012), 'On sedimentation in management fashion: An institutional perspective', *Journal of Organizational Change Management*, **25** (1), 121–142.

Hitt, M.A., L. Bierman, K. Shimizu and R. Kochhar (2001), 'Direct and moderating effects of human capital on strategy and performance in professional service firms: A resource-based perspective', *Academy of Management Journal*, **44** (1), 13–28.

Jansen, J.J., M.P. Tempelaar, F.A.J. van den Bosch and H.W. Volberda (2009), 'Structural differentiation and ambidexterity: The mediating role of integration mechanisms', *Organization Science*, **20** (4), 797–811.

Kor, Y.Y. and H. Leblebici (2005), 'How do interdependencies among human-capital deployment, development, and diversification strategies affect firms' financial performance?' *Strategic Management Journal*, **26** (10), 967–985.

Lowendahl, B. (2005), *Strategic management of professional service firms* (3rd edn), Copenhagen: Copenhagen Business School Press.

Løwendahl, B.R., Ø. Revang and S.M. Fosstenløkken (2001), 'Knowledge and value creation in professional service firms: A framework for analysis', *Human Relations*, **54** (7), 911–931.

Maister, D.H. (1993), *Managing the professional service firm*, London: Simon & Schuster.

Malhotra, N. and T. Morris (2009), 'Heterogeneity in professional service firms', *Journal of Management Studies*, **46** (6), 895–922.

Malhotra, N., T. Morris and M. Smets (2010), 'New career models in UK professional service firms: From up-or-out to up-and-going-nowhere?', *International Journal of Human Resource Management*, **21** (9), 1396–1413.

Malos, S.B. and M.A. Campion (2000), 'Human resource strategy and career mobility in professional service firms: A test of an options-based model', *Academy of Management Journal*, **43** (4), 749–760.

March, J.G. (1991), 'Exploration and exploitation in organizational learning', *Organization Science*, **2** (1), 71–87.

McKenna, C.D. (2006), *The world's newest profession: Management consulting in the twentieth century*, Cambridge: Cambridge University Press.

Michel, S., S. Brown and A. Gallan (2008), 'Service-logic innovations: How to innovate customers, not products', *California Management Review*, **50** (3), 49–65.

Miller, R., X. Olleros and L. Molinié (2008), 'Innovation games: A new approach to the competitive challenge', *Long Range Planning*, **41** (4), 378–394.

Mills, P.K., J.L. Hall, J.K. Leidecker and N. Margulies (1983), 'Flexiform: A model for professional service organizations', *Academy of Management Review*, **8** (1), 118–131.

Montagna, P.D. (1968), 'Professionalization and bureaucratization in large professional organizations', *American Journal of Sociology*, **74** (2), 138–145.

Morris, T. (2001), 'Asserting property rights: Knowledge codification in the professional service firm', *Human Relations*, **54** (7), 819–838.

Morris, T. and L. Empson (1998), 'Organisation and expertise: An exploration of knowledge bases and the management of accounting and consulting firms', *Accounting, Organizations and Society*, **23** (5–6), 609–624.

O'Reilly, C.A. and M.L. Tushman (2004), 'The ambidextrous organization', *Harvard Business Review*, **82** (4), 74–81.

Pache, A. and F. Santos (2010), 'When worlds collide: The internal dynamics of organizational responses to conflicting institutional demands', *Academy of Management Review*, **35** (3), 455–476.

Pinnington, A. and T. Morris (2003), 'Archetype change in professional organizations: Survey evidence from large law firms', *British Journal of Management*, **14** (1), 85–99.

Reihlen, M., M. Smets and A. Veit (2010), 'Management consultancies as institutional agents: Strategies for creating and sustaining institutional capital', *Schmalenbach Business Review*, **62** (July), 317–339.

Sako, M. (2009), 'Global strategies in the legal services marketplace: Institutional impacts on value chain dynamics', Society for the Advancement of Socio-economics Annual Conference, 16–18 July, Paris, France.

Scott, W.R. (1965), 'Reactions to supervision in a heteronomous professional organization', *Administrative Science Quarterly*, **10** (1), 65–81.
Sherer, P.D. (1995), 'Leveraging human assets in law firms: Human capital structures and organizational capabilities', *Industrial & Labor Relations Review*, **48** (4), 671–691.
Siggelkow, N. and D. Levinthal (2003), 'Temporarily divide to conquer: Centralized, decentralized, and reintegrated organizational approach to exploration and adaptation', *Organization Science*, **14** (6), 650–669.
Smets, M., T. Morris and R. Greenwood (2012a), 'From practice to field: A multi-level model of practice-driven institutional change', *Academy of Management Journal*, **55** (4), 877–904.
Smets, M., T. Morris and N. Malhotra (2012b), 'Changing career models and capacity for innovation in professional services', in M. Reihlen and A. Werr (eds), *Handbook of research on entrepreneurship in professional services*, Cheltenham and Northampton, MA: Edward Elgar, pp. 127–147.
Sorensen, J.E. and T.L. Sorensen (1974), 'The conflict of professionals in bureaucratic organizations', *Administrative Science Quarterly*, **19** (1), 98–106.
Starbuck, W.H. (1992), 'Learning by knowledge-intensive firms', *Journal of Management Studies*, **29** (6), 713–740.
Suddaby, R. and R. Greenwood (2001), 'Colonizing knowledge: Codification as a dynamic of jurisdictional expanion in professional service firms', *Human Relations*, **54** (7), 933–953.
Taylor, A. and H.R. Greve (2006), 'Superman or the fantastic four? Knowledge combination and experience in innovative teams', *Academy of Management Journal*, **49** (4), 723–740.
Toren, N. (1976), 'Bureaucracy and professionalism: A reconsideration of Weber's thesis', *Academy of Management Review*, **1** (3), 36–46.
Tushman, M.L. and P. Anderson (1986), 'Technological discontinuities and organizational environments', *Administrative Science Quarterly*, **31** (3), 439–465.
Tushman, M.L. and E. Romanelli (1985), 'Organizational evolution: A metamorphosis model of convergence and reorientation', in L.L. Cummings and B.M. Straw (eds), *Research in organizational behavior*, Greenwich, CT: JAI Press, pp. 171–222.
Tushman, M.L. and C.A. O'Reilly (1996), 'Ambidextrous organizations: Managing evolutionary and revolutionary change', *California Management Review*, **38** (4), 8–30.
Utterback, J.M. and W.J. Abernathy (1975), 'A dynamic model of process and product innovation', *Omega*, **3** (6), 639–656.
von Nordenflycht, A. (2010), 'What is a professional service firm? Toward a theory and taxonomy of knowledge-intensive firms', *Academy of Management Review*, **35** (1), 155–174.
Werr, A. and T. Stjernberg (2003), 'Exploring management consulting firms as knowledge systems', *Organization Studies*, **24** (6), 881–908.
Yakura, E. (2001), 'Billables: The valorization of time in consulting', *American Behavioral Scientist*, **44** (7), 1076–1095.

17. Business and professional service firms and the management and control of talent and reputations: retaining expert employees and client relationship management
John R. Bryson

INTRODUCTION

Economies are founded upon the simple premise that a given quantity of raw material is required to produce a product and that the cost of the product will be greater than the combined cost of all raw materials (Bryson, 2008b). This is the basis of the capitalist economic system in which production systems are expected to generate profit or surplus value. Profit is invested in innovation and in improvements to existing production processes and some profit is distributed to the owners of the production process or to investors or shareholders. The distribution of profits compensates the owners or investors for the risks associated with investment. This simple mechanism of profit generation, investment and redistribution to owners or investors differentiates the current economic system from other ways of organizing the production of goods and services. Profit is essential for the continued functioning of production and importantly for providing investment to enable innovation to occur.

There are three important features of the capitalist production system that are founded upon the requirement for profit generation. The first and perhaps the most important is change. The economy and all the production units that comprise an economy are in a continual process of change or transformation. Change is fundamental to the continuing and on-going evolution of the capitalist economic system. The concept of evolution is extremely important in this context as new ways of creating value develop and as existing ways of organizing production are altered or destroyed (Nelson and Winter, 1982; Aldrich, 2004: 2; Boshma and Frenken, 2006, 2011; Boschma and Martin, 2007, 2010; Bryson and Ronayne, 2014). New organizations emerge as coalitions of people form to mobilize resources in pursuit of opportunities (Taylor, 1999, 2006). Change rather than stability tends to be the norm as the activities of existing firms are challenged by new competitors and by new ways of organizing and controlling production processes. Second, the speed of change is escalating as new technology, enhanced competition and on-going globalization create new business opportunities and, at the same time, undermine existing business models and working practices (Bryson and Rusten, 2006, 2011). Third, over time the production and sale of products and services has a tendency to become increasingly more complex. There are now more ways in which a product or service can be produced. A firm, for example, can engage in the direct production of a product or it can control a brand whilst all production processes are subcontracted to specialist manufacturing companies (Bryson and Rusten, 2008, 2011). In the former case, production is controlled directly by the company whilst in the latter case the firm is transformed into a virtual manufacturer or a co-ordinator or orchestrator of a production process comprising a sequence of tasks undertaken by other firms. These two extreme organizational forms are supported by different control systems that are underpinned by different forms of contractual relationship. Direct manufacturers employ production workers directly and must manage and enforce employment contracts as well as engage in negotiations with trade unions. Virtual manufacturers control their production system by legal contract with their subcontractors, through enforceable contract-specified penalties and by the formation of a supply chain that has within it competition. The implication is that these two forms of organizing a production system require different types of management expertise and different forms of control system.

Over time, these three features of the capitalist economic system have enhanced the quantities of expertise or knowledge that are required to produce a product as well as a service (Bryson et al., 2012). Some commentators have even argued that we are living in a new era of 'informational capitalism' and that 'what is specific to the informational mode of development is the action of knowledge upon knowledge itself as the main source of productivity' (Castells, 1997: 17). This to Castells is associated with a 'new economy' and 'new society', and a new form of flexible capitalism that through the power of networks is able to reach most parts of the globe. This account of capitalism has been criticized for the over-emphasis that it places on information and also for the ways in which it discounts the importance of information in other 'modes of development' (Bryson et al., 2004).

It is essential to distinguish between information, knowledge and expertise. Information by itself may have no commercial value as it must experience a process of transformation to convert it into a form of knowledge or understanding. Castells's 'informational capitalism' (1996) argument overstates the importance of information and fails to appreciate that information and knowledge must be transformed into a product that has commercial value. This is to highlight the distinction between knowledge and expertise. Expertise is a higher order of knowing and it is one which an individual or coalition of individuals (Taylor, 2006) can exploit to develop a commercial advantage. This is to distinguish between knowledge and expert knowledge. The production of a product requires greater quantities of expertise and increasingly the boundaries between physical products and services are becoming blurred (Daniels and Bryson, 2002) and products contain greater quantities of expert knowledge.

These three features of the capitalist production system have produced three transformations in production systems. First, the increasing complexity of production systems, enhanced competition and globalization have forced many companies to enhance the expert knowledges that are combined within their production systems. Firms are highly complex socio-technical systems formed by ever-shifting coalitions of people (Taylor, 2006; Love et al., 2011; Bryson and Ronayne, 2014), technologies and organizational systems. To survive, such systems must contain adaptive capacity and must be open to new ideas and ways of organizing production. There are many ways in which a company innovates and learns, but one consequence of the rise of expertise-ridden production systems is the growth in business and professional services (BPS) or those companies that make intermediate expert inputs into the business practices of client companies. BPS firms have a number of characteristics that imply that they are difficult firms to control and manage (Beyers and Lindahl, 1996; Beyers and Nelson, 1999, 2000; Bryson et al., 2012). Second, manufacturing has altered and has been transformed, but academic understanding of manufacturing systems has not kept pace with these transformations (Bryson et al., 2008, 2013). Simple price-based competition has been replaced or supplemented with other forms of competition that are founded on the exploitation of new forms of expertise supported by new control systems (Bryson et al., 2008). Third, traditionally it was assumed that the majority of service functions were produced and consumed locally. This is no longer the case and services can be traded and exported over distances. The implication is that services are now exposed to globalization through an on-going interplay between new forms of technology and new ways of incorporating technology into service business models (Bryson, 2007, 2008a).

The focus of this chapter is on BPS firms and issues surrounding the retention of expert employees and client relationship management. This is an important topic, but it is one that must be placed within a wider context. Firms are constantly searching for new ways to manage and control labour inputs into production systems. Traditionally it was assumed that expert labour was relatively immune from this process, but recently it has become apparent that only specific forms of expert labour appear to be protected from processes of labour or technologically enabled organizational restructuring (Fraser, 2002; Levy and Murnane, 2004; Bryson, 2007). According to Levy and Murnane (2004) those activities that involve sophisticated pattern recognition will be protected from attempts to send service functions to lower-cost labour locations or from attempts to substitute employees with computer programs. A similar point was made by Fraser (2002: 85) when she argued that white-collar workers can be

> sorted into three basic categories: those whose jobs have been 'reengineered' by technology . . .; those who are increasingly being replaced by technology (as when nearly 180,000 bank tellers were replaced by ATMs between 1983 and 1993); and those whose work lives appear – at least for now – to be resistant to such changes, typically because of the high levels of skill, experience, or creativity their jobs require.

This chapter explores some of the measures that firms have developed to try to manage this latter category of resistant workers. The chapter draws upon interviews undertaken with BPS professionals in the UK over the last ten years.

BUSINESS AND PROFESSIONAL SERVICE FIRMS

Since the 1970s one of the most important transformations in the structure of national economies has been the shift towards various forms of service employment (Bryson et al., 1993). In the economically developed world, the majority of all jobs, more than 80 per cent, involves some form of service work. Much of this work is directly related to final consumption, such as, for example, service activities related to retailing, tourism and hospitality management. In some interpretations, this shift towards service work has been considered to challenge the primacy of manufacturing as a source of innovation and economic growth (see Bryson, 1997: 93, 2008a; Bryson et al., 2004). Against this, however, the fastest growth in services in many national economies has been in 'business and professional services' such as management consultancy, computer services and technical and financial services. To the extent that these activities are inextricably linked to, if not dependent on, manufacturing reflects not the decline of manufacturing, but the growing complexity of production functions and organizations (Bryson, 1997: 93).

BPS firms have a double significance: they create wealth in their own right but they can also enhance wealth creation in their client companies (Greenfield, 1966). This means that the activities of BPS firms contribute to two types of gross value added (GVA). First, GVA produced directly by their own activities and, second, indirect GVA that is produced by client firms that can be attributed to the activities of BPS firms. It is very difficult to measure the impact BPS firms have on the competitiveness and profitability of client companies, but in some cases enhancement to GVA occurs (Bryson et al., 1999a, b). One difficulty is identifying a direct simple linear relationship between the activities of a BPS firm in a client firm and impact. Time complicates the assessment and measurement of such impacts as a BPS project might produce an impact over a long time period (Bryson et al., 1999a, b).

BPS firms exhibit a number of special characteristics that have direct implications for their management and control. In this context it is important to remember that BPS functions can be internalized within client firms or can be outsourced to specialist service providers. The shift towards employment in BPS occupations represents a shift towards expertise-based production systems in which elements of a production system rely on inputs that are difficult to control and manage. What has occurred is a shift in which control or power in the employment relationship is transferred from employers to 'expert' employees. The competitiveness of many BPS firms is founded upon three inter-related factors: expertise that is embodied in fee-earning staff, the individual reputations of fee-earning staff (Wright Mills, 1953; Greenwood et al., 2005) and the contact or relationship networks of fee-earning staff. By itself expertise is not enough; highly successful BPS professionals are identified or perhaps defined by their ability to commercialize their expertise (Bryson et al., 2008). This process of commercialization requires considerable further research. It involves two related activities. First, there is the development of new services and the ability to position them amongst clients as having commercial value and, second, there is the ability to sell a range of intangible and sometimes tangible services to clients. The latter is facilitated through relationship building and the existence of an established reputation. The quality and geographic range of an individual BPS fee earner's reputation is critical for their own personal competitiveness.

The development of business models based upon embodied expertise or embrained expertise and embodied reputations transfers power from employers to employees. But, this does not apply to all employees; only to these with established reputations amongst clients. Indicative of this transfer of power to employees are the difficulties experienced by BPS firms in managing business service professionals. The most important asset 'owned' or managed by a business service firm is its staff and their embodied expertise: this walking and highly mobile resourse leaves a BPS firm each evening and there can be no certainty that this asset will return next morning. It is very difficult for a firm to retain successful fee-earning employees. According to one advertising agency based in Birmingham, UK:

> if somebody's targeted, I had an example just over a year ago, an up and coming chap that we employed ... another two years and he would probably have been very good indeed. He

was approached by another agency and offered I think it was something like a 4 or 5 thousand pound increase. Now an increase of £5,000 is a lot of money, and he turned it down, and they came back with £10,000, and he turned it down, they came back with £14,000 and a 2 litre executive car, and then he had to go. And of course he had to move – he didn't have a choice from a financial point of view. So if somebody targets a person then nine times out of ten they will get them, and they were banking on the fact that they'd keep him for the two years before he actually, you know, got to this really good level, then he would be worth paying this sort of money. That's a really extreme example, but, you know, it is money that attracts most of the time, and position. (Interview, Birmingham Advertising Agency)

Fee-earning BPS professional employees are often extremely well connected and this means that they are known within their local BPS community. It is thus a comparatively simple exercise for rival firms to identify potential recruitment targets. It is worth noting that an individual's reputation is important for employees in some manufacturing sectors. Reputational capital matters across the economy as one way of identifying and recruiting talented individuals.

Retaining well-connected fee-earning professionals is of central concern for the majority of highly successful BPS firms. The issue is the ways in which BPS firms can dissipate the risks associated with embodied expertise and reputations. The key question that must be addressed by BPS firms is: what strategies can be developed to reduce the risks associated with staff loss and related subsequent client loss? A good example comes from an advertising agency in London in which the managing partner noted that:

demand fluctuates from the accounts of clients. The relationship waxes and wanes over time, to being very intense, to being we haven't spoken to each other for three or four months. Everything is in that relationship, because there isn't the constant contact between consultant, service provider and client all the time. It really survives on the strength of one or two personal relationships between people, which have sustained that obligation to use each other and to serve each other well over a long period of time, and the way that you develop business like that, although there is a large part which goes through a formal selection, I would say 80 to 85 per cent of our business actually comes, we call it walking in through the back door, which is exactly what it is. There is usually some formal process of selection at the end, but it's almost a foregone conclusion. The account and the business relationship has been won or lost, on those one or two contacts. (Interview, London Advertising Agency)

The difficulty is that professional employees can be headhunted by rival companies, they can retire or they can leave to establish their own firms. This means that BPS firms are constantly exposed to risks that are difficult to control and that revolve around the most important element within their business models – expert labour and personal relationships between fee-earning professionals and client employees. The same advertising company provided the example of a rival firm that had merged with another company and in this instance:

They brought them together and they have given them the name of one of the companies . . . and people, lots of people have moved, but even the new organization theoretically is as good as the old one, if not bigger and better in some ways. The [client] accounts are walking and they're following and walking because the personal relationships have been destroyed . . . But once the personal link has gone, the corporate link is a very weak one, there is no brand loyalty *per se*, only personal loyalty to individuals, and when that goes, that account becomes vulnerable. (Interview, London Advertising Agency)

It is extremely difficult, perhaps impossible, to capture the value of reputations and personal contacts that are developed by individuals and translate them into a corporate reputation. One implication is that the reputation of a BPS firm consists of an amalgam of the reputations of individual fee-earning members of staff. The corporate brand may have value related to size, geographic coverage and quality control systems that are in place, but ultimately the ability to win contracts is founded upon personal relationships and individual reputations.

It is important to remember that there are many different types of BPS firm. The majority are small and owner-managed or are small partnerships. In these smaller firms the main client contacts tend to be owned by the owner-managers or senior partners. Thus, a manager in a Birmingham, UK advertising agency noted that:

> the senior person in the agency . . . is the personality, he is the main contact, along with his wife, who joined the company about 10 years ago, and they are the two client contact personalities. If one of those left, for whatever reason, and went to another agency, we would probably lose at least one, if not two, of our three major accounts. But luckily with them being owner/managers, or whatever, it is unlikely. We do lose smaller accounts, that are handled by other people in the agency when they move, but luckily our stable base is mainly with the owner/managers.

In this case, the stability of the firm's client base is ensured by the 'ownership' of the key client contacts by the owner-managers. This raises a whole series of research questions related to the client retention strategies deployed by much larger firms. The larger firms may obtain most of their contracts from transnational firms that require transnational service provision. Client control, however, mirrors the smaller firms as partners are responsible for fronting projects with major clients and for maintaining the client relationship.

MANAGEMENT CONTROL AND BUSINESS AND PROFESSIONAL SERVICES

BPS firms are engaged in two types of competition. First, they compete for clients and this involves client retention and also acquiring new clients. Second, they compete for talented fee-earning as well as support staff. Much of the literature on BPS firms has focussed on the client marketplace (Bryson et al., 2004), but Maister (1997: 189) has argued that: 'In the next decade and beyond, the ability to attract, develop, retain, and deploy staff will be the single biggest determinant of a professional service firm's competitive success.' Attraction involves persuading the best graduates to work in a particular BPS sector and for a particular firm. At the moment, it is possible to argue that human resource systems in the majority of BPS firms are underdeveloped (Bryson and Daniels, 2008). The majority of BPS firms are small and medium-sized enterprises (SMEs) and many do not operate staff appraisal and development systems. A relatively recent development is for BPS firms to substitute higher-paid workers with partly trained lower-paid paraprofessionals. Some BPS sectors including accountancy and law have discovered that significant elements of projects can be undertaken by semi-trained employees and especially employees who may never obtain a full professional qualification. The management control issue in BPS firms resolves around two issues: first, the retention of fee-earning employees and, second, client account management. These are explored in turn.

Retaining Business and Professional Service Employees: Golden Handcuffs and Contractual Relationships

Strategies designed to retain fee-earning staff can be constructed around positive and negative incentives. Positive strategies include salary levels and the provision of a range of benefits that might include leisure facilities. Negative strategies are part of the contractual agreement that is negotiated between the firm and the employee. These strategies can include a specified period of gardening leave during which a former member of staff is unable to work and restrictive covenants against soliciting, canvassing, dealing with or accepting instructions from clients with whom the former employee has dealt with during their employment. Another strategy revolves around a system of staggered bonus payments or financial incentives; this is the golden handcuffs strategy. Golden handcuffs are a system of employee financial incentives that are designed to discourage an employee from seeking alternative employment. Examples of golden handcuffs include employee stock options that can only be acquired over a number of years and contractual obligations to repay bonuses if an employee resigns. In such a system an employee is awarded a bonus at the end of a financial year that is paid in stages throughout the following year, and resignation means that an employee loses the bonus payments that have been awarded but not paid. One advertising company based in the West Midlands (UK) had a major retention problem as most employees left after three years. The company reviewed its bonus system and

> The first thing we tried was paying it in six monthly instalments . . . So we did that one, monthly, and then we said well that's not working because if they get a decent bonus, and you split it down into six payments, it doesn't look like much, so what we will do is defer it for six months, so you get your assessment then, and then in six months' time, if you're still with the agency, you get a chunk of money that is well worth having. So if they decide to leave at any time within that period, they think about it quite seriously, because they know at the end of, you know, only three months away, there's a thousand pounds, two thousand pounds, so it's well worth it for them to think about it. (Interview, Birmingham Advertising Agency)

This strategy is deployed by firms that are dependent on embodied expertise and reputations. Golden handcuffs are only partially effective; there is nothing to stop a rival firm from agreeing to compensate a potential recruit for loss of bonus entitlements.

Some BPS firms recruit unskilled individuals and provide them with partial training in an attempt to produce an internally branded individual. The intention is to try to ensure that another firm would have great difficulty in recruiting an individual with a carefully crafted but rather limited skill set. It may also be possible to create a firm that has a different culture and one that is designed to try to ensure that fee-earning employees would be reluctant to leave. Thus, an advertising agency argued that

> We are very successful at retention, partly because we're growing, and partly because we're culturally very different to other organisations in our sector; we are extremely informal, but on the other hand we're able to maintain that informality because in terms of the general calibre of the people we've got, they're all very good. Our last five or six recruits have all been Oxbridge

graduates and we rarely take somebody without a 2:1 in a serious subject and a good track record behind them in terms of achievement with someone. We are very picky about who we recruit, and when we get them you know, it's a bit of a honey trap in this place, we pay well, it's a very relaxed atmosphere and even when they do go out, they inevitably do go out and look at their competitors, they just come back shaking their heads thinking 'crikey, they are idiots, not only are they idiots, but they live in glass boxes'. So we actually have a strategy of retention, by trying to make ourselves as alien from the popular culture of our sector as possible, so that it means that you've got to be a very brave person to undergo the sort of culture and change to move elsewhere. (Interview, London Advertising Agency)

Further research is urgently required to explore management and control issues within BPS firms. This is a difficult issue, but it is one that is being further complicated by technological developments designed to improve business networking.

The development of new technologies designed to enable social networking and relationship building is only intensifying some of the difficulties associated with the management and control of expert labour. In the UK, large BPS firms are increasingly taking former fee-earning employees to court; for example, the recruitment agency Hays Specialist Recruitment has brought a number of claims against former employees and competitor agencies to try to protect its business interests. In May 2007, Mark Ions, a former middle-ranked consultant employed by Hays, established his own special recruitment agency, Exclusive Human Resources, three weeks before resigning from Hays Specialist Recruitment. Ions commenced working for Hays on 12 January 2001 as a recruitment consultant in the field of human resources, and from September 2006 he specialized in placing training and similar personnel for a broad range of professional, public sector and commercial clients.

Hays took Ions to court over the business contacts that he obtained whilst working at Hay (Tyler, 2008: B5). Clause 18 of the contract of employment between Ions and Hays stated that:

You must not, during the course of your employment or at any time thereafter, make use of, or disclose or divulge to any person, firm or company, any trade secrets, business methods or information which you know, or ought reasonably to have known to be of a confidential nature concerning the businesses, finances, dealings, transactions, client database or other affairs of the Company or the Group or of any person having dealings with the Company which may have come to your knowledge during the course of your employment unless it is necessary for the proper execution of your duties hereunder, and you shall use your best endeavours to prevent the publication or disclosure of any such information. (Hays versus Ions, 2008: 1)

Furthermore, Clause 20 of the employment contract contained covenants against soliciting, canvassing, dealing with or accepting instructions from clients or applicants with whom Ions dealt or had contact during his employment, subject to certain restrictions. These covenants were binding during the period of employment and for a period of six months after Ions left the firm.

This was a complex case that revolved around the use of social networking sites and the ownership of an employee's work contacts. Hays encouraged its employees to use LinkedIn, an on-line knowledge network for professionals that had, at the time, over 23 million users. Hays encouraged its employees to use this network by inviting their contacts to establish on-line communities. Ions thus invited his Hays' contacts to join his on-line network, but as soon as they had joined his contacts were no longer confidential as they could be seen and contacted by anyone in his personal network. The Court Judgment concluded that:

> First, unlike the parties in *Black v Sumitomo Corporation* . . ., Hays and Mr Ions are not 'strangers'. The potential claim arises out of Mr Ions' employment with Hays. Secondly, and more importantly, Mr Ions accepts that he uploaded the addresses of business contacts to LinkedIn while he was still employed by Hays. He does not suggest that he had any of the business contacts except as a result of his employment with Hays and, given that he had been a full-time employee of Hays for over 6 ½ years, it is very unlikely that he had many, if any, independent contacts. The disclosure is therefore limited to contacts which he obtained in his capacity as an employee of Hays. (Hays versus Ions, 2008)

There is an interesting tension here between the benefits that come from participating in social networking sites and maintaining confidentiality. Sites like LinkedIn offer a means for maintaining regular contact with potential clients, but the danger is that use of these sites, in this context, exposes BPS firms to potentially uncontrollable information theft, and for these firms information is their most valuable resource. This was a landmark court case as the Hays versus Ion case 'is the first time that a [pre-action] disclosure order has been obtained from the court before legal proceedings have been issued to obtain information held on a networking site' (Tyler, 2008: B3). Ions was ordered to produce copies of all his deleted 'business contacts' from his LinkedIn account, copies of all e-mails sent to or received by his LinkedIn account from the Hays computer network and he has been asked to disclose copies of all documents including invoices that showed any use by him of the LinkedIn contacts. This case was one of the first attempts by an employer to challenge ownership of an employee's personal relationships that were established during the course of his day-to-day business activities. A key issue concerns transfer of ownership or control of key personal contacts that have commercial advantage for a firm. A BPS professional relies on contacts that have been established over a number of years. Thus, a BPS professional's contact network will contain members of his or her extended family, family friends, school and university friends, friends acquired whilst working for current and previous employers, and friends and acquaintances that are acquired during the course of an individual's everyday routines. There would appear to be boundary issues in terms of ownership of these business contacts that may be difficult to regulate.

Client Account Management: Client Intimacy and Zipper- or Velcro-Type Relationships

There is a substantial literature on client account management (Maister, 1997; Fincham, 1999) and there are many private sector courses available that attempt to provide training in client management. Client management revolves around three issues: personality, successful projects that create repeat business, and the development of a strategic relationship or partnership with some clients. In is difficult completely to remove the risks associated with losing staff and losing clients. In many cases, it is perhaps impossible for a BPS firm to regulate or control diadic relationships between employees and clients that have been built on trust and interaction over a period of time. Client relationship management is a complex process that is influenced by the nature of the service being delivered, personalities and previous experience. Many small and medium-sized BPS firms do not have a formal system of client relationship management as they do not have the time, resources or the requirements for such a system. Instead they rely on their ability to provide a quality product and service and their social networking skills. Different tasks undertaken by BPS firms may be undertaken by individuals with different personality types. Thus, one firm of solicitors based in Shropshire (UK) classifies its employees as finders, minders or grinders. Grinders never meet clients but perform back office functions that do not require direct access to clients (Daniels and Bryson, 2006: 152). A manager of a real estate company made a similar distinction when he highlighted the differences that exist between BPS employees with technical skills and those that have technical skills combined with commercial skills:

> technical skills, that is first awareness of the market – so it is personality, being able to communicate, ideas, lateral thinking, business acumen. So the ability to see opportunity, go and find it, go and acquire it and at the end of the day we are a business, we are about making money for our shareholders as well as providing a cracking service to our clients and we need those skills in our surveyors. They're not there just to do a technical job, you can only do a technical job once you've got the clients and the jobs to do it on and it is our job to get in high-quality work, process it, give the best advice, look for opportunities and maximize our position in the market and make money. That requires people with a skill set which is above just being able to do the technical work, that is where we run into problems. (Interview, Real Estate Company, Birmingham)

Technical skills enable a service to be delivered but social skills are an essential part of obtaining and retaining clients. Fee-earning professionals must have the ability to commercialize their expertise and to create products that have value. Evidently, people with technical skills are readily available, but people with technical, social and commercial skills are in short supply. This shortage of talented individuals perhaps lies behind Maister's (1997) statement regarding the relationship between staffing issues and the competitive success of professional service firms.

Social skills by themselves are not sufficient as they must be supported by the delivery of an effective product. In this context, the managing director of an advertising firm noted that client retention was

> not simply a process of contact, because I can think of many account relationships that I have had for many years, where I see the people incredibly infrequently. Personal friendship or a degree of personal like or dislike can help, but it does not matter when it comes down to the ultimate thing. Erm, given that quite often you have, if you are an outside supplier doing service supply, your direct first line of contact in the company is almost like your sponsor. They're the user but they are the sponsor, they are the one that has taken responsibility for bringing your expert in as an advisor to the company. So factors such as credibility, reliability, even if they like you as an individual, they're not going to maintain that relationship if you become a liability in terms of embarrassing them, because you are sponsored by them. So doing an extremely competent job is your best security in the account ... So, you've got to be smart, capable, presentable, good at your job, confident, you can be fielded in by your sponsor wherever, you are not going to let them down, you are going to be his [sic] pride and joy. (Interview, London Advertising Company)

So a client relationship is based upon credibility, personality and the ability to deliver a product that does not embarrass the 'sponsor'. The concept of a BPS 'sponsor' supports earlier work undertaken on organizational 'brokers' (Bryson, 1997: 97–98).

Brokers occupy 'a structural position that links pairs of otherwise unconnected actors' (Fernandez and Gould, 1994: 1455). Brokers or sponsors occupy a special position in the relationship network that links organizations together. Sponsors will tend to favour particular BPS practitioners and the BPS relationship tends to be transferred between firms as brokers are employed by other client companies (Rusten et al., 2005).

Some companies highlight the importance of developing and maintaining 'client intimacy', or in other words ensuring that the firm acquires information regarding the background of clients and, most importantly, the likes, dislikes and interests of sponsors. This information can then be used to add a personal touch to the service relationship and also to ensure that such information is captured by the firm as well as the individual. A central element of a client intimacy strategy revolves around the development of systems that try to ensure that each client appears to be treated as if it were a firm's only client (Lesky, 2008: 43). This type of service relationship follows the emphasis placed by Maister (1997) on the difference between the quality of a service and the quality of work undertaken by a BPS firm. BPS firms sell experiences and the quality of the experience is the essential component for ensuring that clients continue to utilize the services of a BPS provider. In this context, the quality of the experience is determined by combining the quality of service and the quality of work. In most cases, the quality of work provided will be determined by the professional expectations and standards of the professionals involved in delivering a work package. Most professionals would be able to provide similar qualities of work. The quality of service provided to a client, however, is one of the most important ways in which a BPS firm can differentiate itself from other firms. Systems must be developed to ensure that the highest quality of service is provided to all clients combined with the highest quality of work. This type of service experience should enhance client retention. This sort of management strategy has been implemented by a research provider located in the US, and this firm argues that 'We believe that the firm that delivers flexible, high-touch, high-tech, high value-added expert research services has a competitive advantage' (Lesky, 2008: 46). There is an important caveat, however, in that all clients have the option to select an alternative provider. In this case, cost may outweigh the benefits of an established client service provider relationship; clients will market test key service inputs to either encourage or force the current provider to reduce fee levels or to replace the current provider with a lower-cost alternative. The selection decision may not be driven completely by cost but by accessibility (geographic distance between client and service supplier), resources, social relationships, third-party referrals and reputation.

Some providers of technical services try to develop 'zipper-type relationships' with clients (AMEC, 2005: 11), sometimes known as 'velcro relationships'. In this situation a service provider will try to develop relationships with a broad range of contacts within a client company. This strategy is an attempt to break away from dependency upon a client relationship that has been established by one individual. This means that the service supplier tries to engage with the client's wider organization by developing relationships at every level of the client's management structure. In this context, a single client sponsor or BPS broker is considered to represent a risk that can be mediated by developing zipper-type relationships with client firms. The zipper metaphor describes the process by which the relationship with a firm continues to exist when a member of staff resigns – all that happens is that the service supplier slides the relationship zipper down to the next contact in the firm.

DISCUSSION AND CONCLUSIONS

The BPS sector is a complex heterogeneous sector that is dominated by SMEs, and which contains extremely creative and innovative firms and also firms that sell standardized services or recipe knowledge (Bryson et al., 2004). The processes that create wealth continue to become ever more complex. This complexity reflects the development of complex socio-technical systems that combine people, organizations and technologies in predictable and also not so predictable ways. The importance of reputational capital for BPS professionals combined with employee and client retention issues must be considered as intriguing issues for further detailed research. In this context, research should focus on exploring BPS functions that have been transformed into industrialized products by the application of new technology. This development has enabled some BPS functions to be delivered by individuals located in low-cost locations and also to break the relationship between an individual's reputation and control over client relationships. Industrialized services are delivered by firms with established reputations rather than by individuals with established reputations; the individual's reputation is replaced by technological processes and systems that are controlled and regulated by a firm. It is these technological developments combined with organizational and process innovations that have led to what Bryson has termed the '*second global shift*', or the development of a *new international division of 'service' labour* (Bryson, 2007). During the first global shift, branch plants in developing or less developed countries were associated with the assembly of products designed by and for the developed world. During the second global shift, service facilities process data as well as engage directly in ICT-mediated service interactions with consumers.

The relationship between a fee-earning professional employee and a firm or professional partnership is regulated by employment contracts. Such contracts include periods of gardening leave that are implemented as soon as the employee resigns. Gardening leave is an attempt to prevent employees transferring client relationships to other firms. The contractual relationships that have been developed to control both employee and client relationships require further academic study. This is to highlight the importance of developing a research dialogue between economic geography and economic sociology and contract law. Such a research agenda would lead to considerable advances in understanding the on-going regulation of BPS employment relationships in different sectors and countries.

This chapter has focussed on BPS functions that have not yet been industrialized. These functions are based around sophisticated forms of pattern recognition, face-to-face contact and the dynamics of interpersonal relationships. Such embodied BPS functions pose a particular challenge for academic research. It is difficult to design research methodologies that capture the complexity of the client/BPS supplier relationship. Academics have only begun to sketch out some of the elements that lie behind these relationships, but much more research is required to address the following research questions:

- How does a BPS professional establish a successful and in some cases an iconic reputation?
- How are BPS firm brands constructed and managed?
- What factors contribute to the competitiveness of BPS firms?
- What makes some BPS firms more successful than others?

- How is technology being incorporated into the business models of BPS firms?
- Is it possible to develop BPS organizational systems that will reduce the risks associated with employee and client retention?
- In what ways are BPS functions integrated into the business models of client companies? In other words, the growth of BPS firms does not represent the decline of manufacturing, but just a reordering of production processes.
- How effective are employment contracts in ensuring that employers control or regulate the activities of BPS professionals?

The enhanced importance of production systems that are dependent upon expert knowledge implies that social scientists must try to understand the ways in which different types of BPS function and are incorporated into production processes.

REFERENCES

Aldrich, H.E. (2004), *Organizations Evolving*, Sage, London.
AMEC (2005), 'Smarter partnering: developing a relationship with customers', *In Touch*, Issue 2, AMEC.
Beyers, W.B. and Lindahl, D.P. (1996), 'Lone eagles and high fliers in rural producer services', *Rural Development Perspectives*, 12, 2–10.
Beyers, W.B. and Nelson, P.B. (1999), 'Service industries and employment growth in the nonmetropolitan South: a geographical perspective', *Southern Rural Sociology*, 15, 139–169.
Beyers, W.B. and Nelson, P.B. (2000), 'Contemporary development forces in the nonmetropolitan west: new insights from rapidly growing communities', *Journal of Rural Studies*, 16, 459–474.
Blackstone, J.H., Gardiner, L.R. and Gardiner, S.C. (1997), 'A framework for the systemic control of organizations', *International Journal of Production Research*, 35: 3, 597–609.
Boschma, R.A. and Frenken, K. (2006), 'Why is economic geography not an evolutionary science? Towards an evolutionary economic geography', *Journal of Economic Geography*, 6, 273–302.
Boschma, R.A. and Frenken, K. (2011), 'The emerging empirics of evolutionary economic geography', *Journal of Economic Geography*, 11: 2, 295–307.
Boschma, R.A. and Martin, R. (2007), 'Constructing an evolutionary economic geography', *Journal of Economic Geography*, 7: 5, 537–548.
Boschma, R.A. and Martin, R. (2010), *The Handbook of Evolutionary Economic Geography*, Edward Elgar, Cheltenham and Northampton, MA.
Bryson, J.R. (1997), 'Business service firms, service space and the management of change', *Entrepreneurship and Regional Development*, 9, 93–111.
Bryson, J.R. (2007), 'A "second" global shift? The offshoring or global sourcing of corporate services and the rise of distanciated emotional labour', *Geografiska Annaler*, 89B (S1), 31–43.
Bryson, J.R. (2008a), 'Service economies, spatial divisions of expertise and the second global shift services', in P.W. Daniels, M. Bradshaw, D. Shaw and J. Sidaway (eds), *Human Geography: Issues for the 21st Century*, Prentice Hall, London: third edition, 359–378.
Bryson, J.R. (2008b), 'Value chains or commodity chains as production projects and tasks: towards a simple theory of production', in Dieter Spath and Walter Ganz (eds), *Die Zukunft der Dienstleistungswirtschaft – Trends und Chancen heute erkennen*, Carl Hanser Verlag, Munich, 265–287.
Bryson, J.R. and Daniels, P.W. (2008), 'Skills, expertise and innovation in the developing knowledge economy: the case of business and professional services', *International Journal of Services Technology and Management*, 9: 3/4, 249–267.
Bryson, J.R. and Ronayne, M. (2014), 'Manufacturing carpets and technical textiles: routines, resources, capabilities, adaptation, innovation and the evolution of the British textile industry', *Cambridge Journal of Regions, Economy and Society*, 7: 471–488.
Bryson, J.R. and Rusten, G. (2006), 'Spatial divisions of expertise and transnational "service" firms: aerospace and management consultancy', in J.W. Harrington and P.W. Daniels (eds), *Knowledge-Based Services, Internationalisation and Regional Development*, Ashgate, Aldershot, 79–100.
Bryson, J.R. and Rusten, G. (2008), 'Transnational corporations and spatial divisions of "service" expertise as a competitive strategy: the example of 3M and Boeing', *Service Industries Journal*, 28: 3, 307–323.

Bryson, J.R. and Rusten, G. (2011), *Design Economies and the Changing World Economy: Innovation, Production and Competitiveness*, Routledge Studies in Human Geography, London.
Bryson, J.R., Clark, J. and Mulhall, R. (2013), *The Competitiveness and Evolving Geography of British Manufacturing: Where is Manufacturing Tied Locally and How Might This Change?*, Department of Business Innovation and Skills, London.
Bryson, J.R., Daniels, P.W. and Ingram, D.R. (1999a), 'Evaluating the impact of business link on the performance and profitability of SMEs in the United Kingdom', *Policy Studies*, 20: 2, 95–105.
Bryson, J.R., Daniels, P.W. and Ingram, D.R. (1999b), 'Methodological problems and economic geography: the case of business services', *Service Industries Journal*, 19: 4, 1–17.
Bryson, J.R., Daniels, P.W. and Warf, B. (2004), *Service Worlds: People, Organizations, Technologies*, Routledge, London.
Bryson, J.R., Rubalcaba, L. and Strom, P. (2012), 'Services, innovation, employment and organisation: research gaps and challenges', *Service Industries Journal*, 32: 3–4, 641–657.
Bryson, J.R., Taylor, M. and Daniels, P.W. (2008), 'Commercializing "creative" expertise: business and professional services and regional economic development in the West Midlands, UK', *Politics and Policy*, 36: 2, 306–328.
Bryson, J.R., Wood, P. and Keeble, D. (1993), 'Business networks, small firm flexibility and regional development in UK business services', *Entrepreneurship and Regional Development*, 5: 3, 265–277.
Castells, M. (1996), *The Information Age: Economy, Society and Culture, Vol 1: The Rise of the Network Society*, Blackwell, Oxford.
Castells, M. (1997), *The Rise of the Network Society, The Information Age: Economy, Society and Culture Vol. I*, Blackwell, Oxford.
Daniels, P.W. and Bryson, J.R. (2002), 'Manufacturing services and servicing manufacturing: changing forms of production in advanced capitalist economies', *Urban Studies*, 39: 5–6, 977–991.
Daniels, P.W. and Bryson, J.R. (2006), *The Skill Needs of Business and Professional Services in Objective 2 Areas of the West Midlands: Final Report*. A report for the Learning and Skills Council. University of Birmingham Services and Enterprise Research Unit, Birmingham.
Fernandez, R. and Gould, R. (1994), 'A dilemma of state power: brokerage and influence in the national health policy domain', *American Journal of Sociology*, 99: 6, 1455–1491.
Fincham, R. (1999), 'The consultant–client relationship: critical perspectives on the management of organizational change', *Journal of Management Studies*, 36: 3, 335–351.
Fraser, J.A. (2002), *White-Collar Sweat-Shop: The Deterioration of Work and Its Rewards in Corporate America*, W.W. Norton & Company, New York.
Greenfield, H.I. (1966), *Manpower and the Growth and Producer Services*, Columbia University Press, New York.
Greenwood, R., Li, S.X., Prakash, R. and Deephouse, D.L. (2005), 'Reputation, diversification, and organizational explanations of performance in professional service firms', *Organization Science*, 16: 6, 661–673.
Hays versus Ions (2008), *Hays Specialist Recruitment (Holdings) Ltd (Claimants) versus Mark Ions 2. Exclusive Human Resources Ltd (Defendants)*, Royal Courts of Justice, Strand, London, EWHC 745 (Ch), [2008] IRLR 904.
Lesky, C. (2008), 'From a business and science search firm: five insights into managing an information service', *Business Information Review*, 25: 1, 40–47.
Levy, F. and Murnane, R.J. (2004), *The New Division of Labour: How Computers are Creating the Next Job Market*, Princeton University Press, Princeton, NJ.
Love, J., Roper, S. and Bryson, J.R. (2011), 'Openness, knowledge, innovation and growth in UK business services', *Research Policy*, 40: 10, December, 1438–1452.
Maister, D.H. (1997), *Managing the Professional Service Firm*, Free Press, New York.
Nelson, R. and Winter, S. (1982), *An Evolutionary Theory of Economic Change*, Harvard University Press, Boston, MA.
Rusten, G., Bryson, J.R. and Gammelsæter, H. (2005), 'Dislocated versus local business service expertise and knowledge and the acquisition of external management consultancy expertise by small and medium-sized enterprises in Norway', *Geoforum*, 36: 4, 525–539.
Taylor, M. (1999), 'The small firm as a temporary coalition', *Entrepreneurship and Regional Development*, 11, 1–19.
Taylor, M. (2006), 'The firm: coalitions, communities and collective agency', in M. Taylor and Päivi Oinas (eds), *Understanding the Firm: Spatial and Organizational Dimensions*, Oxford University Press, Oxford, 87–116.
Tyler, R. (2008), 'Court orders ex-employee to hand over social network site contacts', *The Daily Telegraph: Business:* B5, 16 June.
Wright Mills, C. (1953), *White Collar Work: The American Middle Classes*, Oxford University Press, New York.

PART IV

UNDERSTANDING SERVICE BUSINESS

PART IV

UNDERSTANDING SERVICE BUSINESS

18. How has logistics come to exert such a key role in the performance of economies, society and policy making in the 21st century?
Andrew Potter and Robert Mason

INTRODUCTION

Logistics has become one of the most progressive and important service industries in the 21st century's global economy. At a basic operational level logistics directly provides the physical glue of commerce, ensuring goods and services have time and place utility. Beyond this, however, the logistics industry has risen in strategic importance and can be seen as a critical component of competitive advantage in many current enterprises, providing crucial links in the chain of supply. In modern internationally competitive supply chains, the logistics 'cog' must be consistently robust if supply chain management (SCM) practices are to be pursued with minimal buffering inventory levels. Moreover, in the new digital age logistics is also about exploiting value from the wider industrial network, identifying and realising network and multichannel synergies so that the very best service value is provided for customers of logistics services.

In short, logistics has moved from a simplistic 'trucks and sheds' service concept to become a sophisticated management science in support of the realisation of an organisation's vision.

In addition, for many logistics has moved beyond this again to become more of an art form. Underpinning many logistics operations are people, who need to be engaged fully (and safely) if service and efficiency levels are to be consistently met. Further, judgement is often required to make trade off decisions to contingently suit the circumstances faced. Finally, logisticians must be mindful of the external effects of their activities. Increasingly logistics solutions must be developed which show a full consideration of the wider societal and environmental agenda – in short, logistics providers must act sustainably.

Logistics can be defined as follows:

> Logistics is the art of safely meeting and where possible exceeding the supply chain customers' short and long term demands for effectively managing supply chain material forward and return flows, storage and associated processes, whilst simultaneously generating efficiencies and value horizontally from the domain of the industrial network and respecting the ongoing sustainability agenda.

This chapter examines why logistics has become more strategically critical to how we all live today and the way organisations, private and public, operate. Specifically, it sets out why and how logistics has evolved into its current state. Logistics has become one of the most readily outsourced services and the chapter also explores why this has occurred and in turn how this relatively new industry has developed. Logistics continues to be a highly dynamic service and the chapter concludes by outlining some thoughts on the kinds of challenges logistics faces and proposes some possible directional signposts that may characterise how the future of logistics will pan out in the immediate decades to come.

THE EVOLUTION OF LOGISTICS – THE LAST 50 YEARS

The origins of logistics emerged from military applications, and came to mean the physical movement of troops and their supplies to the battlefield, encompassing both the planning and execution of these movements. Citing Simpson and Weiner (1989), Lummus et al. (2001) use a quote from 1898 to highlight the historical recognition of logistics: 'Strategy is the art of handling troops in the theatre of war; tactics that of handling them in the field of battle . . . The French have a third process, which they call logistics, the art of moving and quartering troops.'

However, the existence of the same principles (if not the same terminology) can be found as far back as the Roman conquests (Goldsworthy, 2003). In the art of war, logistics was seen by many Generals as being a critical component of fighting advantage. If there were too many provisions, the mobility of the soldiers would be compromised; too few, and they would be missing essential equipment. Over time, this concept has migrated into the business environment.

The Total Cost Concept

Bowersox (2007: 338) noted that, in the early 1950s, the typical manager of transport 'was expected to continuously lower the cost per hundredweight to move products and materials'. The total cost concept was first proposed by Lewis and Culliton (1956) in a paper that explored the role of air freight in physical distribution. Their analysis reshaped the argument, moving it from optimising costs associated with individual logistics activities, such as transport, to minimising the total costs of the entire delivery process. Attention was therefore shifted from a functional focus to an emphasis on minimising delivery costs across the whole firm. As noted by Bowersox (2007), a breakaway group from the American Marketing Association, including himself, was formed after discussing the total cost concept. This ultimately led to the creation of the National Council of Physical Distribution Management (NCPDM) in December 1963.

Incorporating the Customer

Beyond the narrower focus of internal processes of the firm, the need to understand that the goal was to deliver products to the end consumer began to drive an extension of the total cost concept to include external as well as internal costs. The management of a channel through which products were delivered to the end consumer and potentially containing many entities became the common view of what logistics comprised. Bowersox (2007: 338) added that this change was given great support and credence following a lecture in 1965 to the NCPDM from the strategic thinker and well respected management academic Peter Drucker. He argued that physical distribution was at the frontier of modern management and the whole process of business, and stated that many opportunities for considerably improved performance remained untapped.

The Systems Concept

In the 1970s this evolved further as businesses were forced to react to the considerable economic turbulence they faced at the time. The control of costs became even more paramount. This led to logisticians developing delivery systems which were dynamic and able to adapt to different circumstances (Mandrodt and Davis, 1992). The same authors cited companies such as Quaker Oats and Whirlpool which incorporated flexible capabilities in their physical distribution systems, combining a number of different organisations cooperating towards a common goal.

Information for Inventory

New technology development supported this expanded vision of what now began to be known as logistics and facilitated the development of further refinement and innovation. The idea of developing capability around information management that ensured that accurate and up to date stock accounts were maintained allowed for lower levels of inventory in many cases. In 1985 the NCPDM in the United States replaced the term 'physical distribution' with 'logistics' (Bowersox, 2007) and thus rebranded itself as the Council of Logistics Management (CLM).

The Customer Service Concept

Throughout the evolving vision of logistics, the importance of incorporating the customer into logistics solutions became increasingly critical. The retention of customers was viewed as vital to better optimising a firm's ongoing profitability potential. Through the 1980s and 1990s the importance of customer value, rather than a narrower focus on cost minimisation, began to develop. The classic trade off of cost versus service was increasingly focussed upon. Mandrodt and Davis (1992) argued that logistics organisations, rather than being limited in service provision to what the company could do, evolved to understanding and providing from the basis of what the customer wanted. Thus, many supply chains were turned on their heads and required a new customer service philosophy to be developed. This was termed 'service response logistics'.

The Collaborative Enterprise

The emergence of SCM from the late 1980s and through the 1990s to today further extended this thinking about optimising holistically the total system performance for the benefit of the end consumer. Supply chain integration was emphasised as critical to this endeavour (Stevens, 1990) and the concepts of supply chain collaboration and alignment emerged and were developed (Bowersox, 2007). Logistics practice was recognised by many authors as being a critical element of the supply chain and therefore the virtue of closer collaboration between the shipper and the logistics service provider was advanced (Ellram and Cooper, 1990; Skjøtt-Larsen, 2000). Interestingly though, it is still the case today that logistics service providers are invariably not conceived as conventional collaborators (Mortensen and Lemoine, 2008) in the supply chain. This is primarily due to the fact that warehouse and transport services are invariably considered to be commodities where costs should be minimised (Potter and Lalwani, 2005; Naim et al., 2006).

The Networked Era

In the last decade the evolution of logistics has continued to show great dynamism. Traditional 'bricks and mortar' firms were being reinvented and new non-asset based entities have emerged, each with the goals of leveraging opportunities from the wider industrial network, not just the supply chain network (Mason et al., 2007). These non-asset firms look to coordinate the activities of asset owning logistics operators, offering increased flexibility and a single point of contact for their customers in managing door to door logistics flows. Globalisation has continued apace, extending the importance of logistics for managing longer and more complex material movements. The criticality of logistics was also elevated as the wider concept of SCM became more widespread and more sophisticated. Logistics practice was seen to be integral to the fulfilment of the goal of integrated SCM, providing vital cogs in the chain of supply. Reflecting these changes, the CLM in the United States changed its name again and in 2005 became officially known as the Council of Supply Chain Management Professionals (CSCMP). Logistics was positioned as a supportive process in the broader field of SCM, a development which will be further explored later in this chapter.

In summary, the logistics concept has been highly dynamic and has evolved considerably as demands upon it have changed and as capabilities have grown in terms of mindset, organisational structures, and organisational cultures supported and catalysed by considerable developments in information and communications technology (ICT).

STRATEGIC SIGNIFICANCE OF LOGISTICS PROVISION TODAY

The evolution of logistics has been matched by the changing and increasingly tough demands placed upon logistics providers. In recent decades logistics has become one of the most popular supply chain support services to be outsourced. As has been shown, the providers of logistics services represent vital intermeshing cogs in the chain of supply and can be seen as important links between suppliers and their customers. In the sense that they are not only responsible for the physical transportation of products through the supply chain (the material flow), but much of the related data management (the information flow) and associated finances (the cash flow), they can play an important role in supply chains by supporting, even facilitating, the fulfilment of SCM strategies (Skjøtt-Larsen, 2000; Mason and Lalwani, 2006; Naim et al., 2006).

Today, in many sectors such as automotive, electronics and retail, the importance of goods arriving consistently on time to the right place (time and place utility) is invariably paramount. If delivery is inconsistent then this either results in sell outs, or the resultant increase in uncertainty leads to a decision to stock higher levels of inventory as a buffer. A higher level of inventory can harm competitiveness, as it eats up capital and can result in higher damage, obsolescence and theft costs; the antithesis of the SCM approach. So in the modern context, logistics provision has become strategically, as well as operationally, important in supporting strategies to build and sustain competitive advantages based on process excellence – the SCM ideal.

In more recent years the pressures on logistics service providers to optimise values from their services even further are leading to yet another redesign of the logistics process. In the emerging 'networked era', outlined earlier, innovative ways to cooperate and collaborate horizontally (Cruijssen et al., 2007; Mason et al., 2007; Hingley et al., 2011) with partners, even competitors (as well as vertically with supply chain partners), are forcing the need to reconceptualise the domain and landscape of logistics and consequently how modern logistics management should be defined. If synergies can be found and exploited in linking, for instance, return flows of cargo movement with empty return vehicles, or better optimising warehouse capacities by linking users with different seasonal patterns of trade, there can be considerable commercial as well as environmental benefit. In a world where data about cargo flows can be more easily accessed, the opportunities to convert this into useful information and support holistic enhanced logistics solutions for all is a significant prize worth pursuing. The third party logistics provider (3PL) can act as an important independent facilitator in this process within many sectors and across sector boundaries around the world.

THE WIDER SIGNIFICANCE OF LOGISTICS

Logistics also delivers wider benefits beyond the supply chains in which it is found. For any country, logistics supports economic activity, enabling raw materials to be delivered and finished goods to access markets. The World Bank produces a Logistics Performance Index (Arvis et al., 2010) that compares logistics performance between many nations of the world. It is perhaps unsurprising that there is a strong relationship between logistics performance and economic performance. Equally, logistics costs are often reflective of infrastructure and management performance in a country. Thus, many of the more developed countries have a lower value for total logistics costs as a percentage of GDP (Figure 18.1).

The logistics industry also has impacts on the population as a whole. For example, the sector is a significant employer in its own right. In the UK, for example, 4.7 per cent of the workforce is employed by the logistics sector (Freight Transport Association (FTA), 2011). However, there are also negative impacts; although the industry has taken many positive steps, logistics is a significant contributor to carbon emissions, while also contributing to noise, visual pollution and congestion.

As a result policy makers have a significant interest in the sector. On the one hand, they wish to encourage and support the efficient domestic and international movement of products, while on the other there is also a desire to introduce regulations to control their behaviour. This can manifest itself at many different levels, from local government regulations on night time deliveries to national government policies on the taxation of fuel, or regional governments' laws relating to issues such as the regulation of freight vehicle driver hours across the EU.

338 *Handbook of service business*

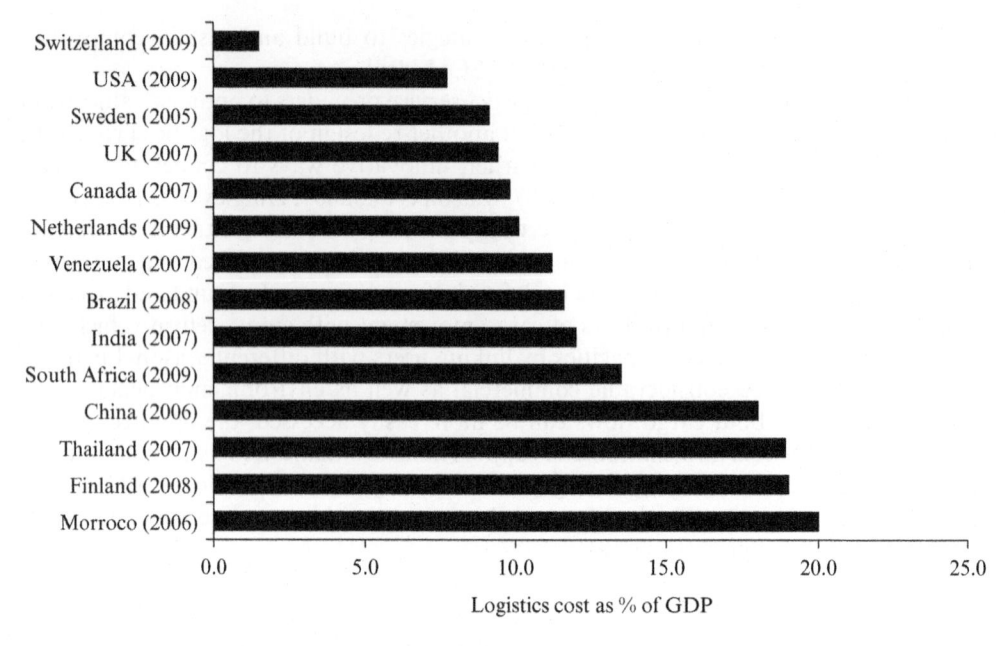

Source: Adapted from Ittmann and King (2011).

Figure 18.1   Logistics costs as a percentage of GDP

THE LOGISTICS SERVICE PROVIDERS

As has been noted above, the trend in logistics provision in many marketplaces over recent decades has been to pursue an outsourcing strategy, and this has nurtured a growing logistics industry practising in many industrial sectors. There are many potential benefits for the shipper in pursuing an outsourcing strategy. These include (Ellram, 1991; Embleton and Wright, 1998; Griffiths, 2001; Simchi-Levi et al., 2003):

- cost reductions;
- capital reductions;
- availability to focus upon production capacity and competence;
- releasing internal resources, both personnel and equipment;
- sharing risks with partners;
- quicker time to market;
- better strategic flexibility, and so on.

In addition, it permits the firm to better concentrate on its core business (Sink and Langley, 1997) and can also support this by freeing additional capital to invest.

In summary, the move to logistics outsourcing allows the shipper to transfer financial risk, improve service quality and productivity, and reduce costs through routinisation of transactions (Ellram, 1991) and size economies (Simchi-Levi et al., 2003). It can also positively affect the balance sheet, as logistics costs move from fixed to variable costs.

Compared with in house provision, the increase in flexibility can be crucial in modern markets. Logistics service providers can help smooth out fluctuating peaks by combining workloads from a range of customers or industries (Tomkins and Smith, 1998), or help manage workload troughs by restricting the exposure of the customer to under-utilised assets (Rushton et al., 2006). Service capability can also be improved as the logistics service provider may be able to create multi-user distribution centres located closer to customers, making feasible more frequent deliveries with tighter lead times (Tomkins and Smith, 1998).

The attractiveness of these benefits has led to the development in most parts of the world of a growing and dynamic new industry sector, the contract logistics industry, principally since the 1980s.

Types of Logistics Service Providers

Conventionally, if outsourced, a transactional market based approach was how logistics provision was managed. Consequently, the main focus has been on achieving the lowest possible cost. However, there are problems with an approach that uses cost as the principal value criterion. This has been particularly noticeable in supply chains, where more advanced and integrated supply chain strategies have been established. As Skjøtt-Larsen (2000) confirms, value requirements have evolved so that while competitive cost containment is still actively sought, it is not the sole, nor arguably always the dominant, value criterion. Thus, since the 1990s until today, a more network based model for logistics management has begun to emerge. Consequently, the definition of third party logistics has evolved, and can be defined as (Murphy and Poist, 2000: 121): 'a relationship between a shipper and third party which compared with basic services, has more customised offerings, encompasses a broader number of service functions and is characterised by a longer term, mutually beneficial relationship'.

Consequently, there is a need to categorise different types of 3PL, depending upon both the services they provide and the assets that they own. Berglund et al. (1999) develop a useful typology (Figure 18.2). Asset based logistics providers were typical of early players that were seen from the late 1970s and early 1980s. Owning assets such as trucks, containers and warehouses, they expanded their core business to offer wider logistics services.

Network based logistics providers emerged from the 1990s. Invariably originating as express parcel or courier services, these companies developed a global capability so that door to door shipments could be delivered with greater speed and reliability than traditional means. The ability to track deliveries and provide a proof of delivery further differentiated their capabilities and supported their aim to add value for the customer.

Information based logistics provision developed in the late 1990s, often using the phrase 'fourth party logistics' (4PL) (Gattorna, 1998). These providers moved away from owning assets (trucks, warehouses, etc.) and instead offered consultancy or coordination and information management services. They also became lead logistics providers, taking on accountability for a logistics contract while not undertaking any of the physical activities themselves. Instead, they in turn outsourced operations to subcontracted logistics players. This concept is returned to in more detail later.

340 *Handbook of service business*

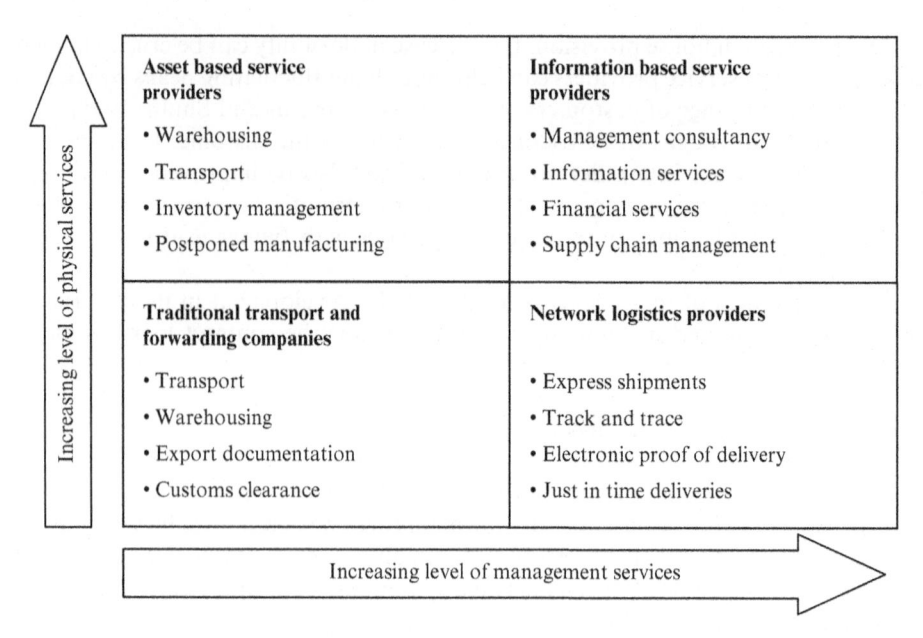

	Asset based service providers	Information based service providers
	• Warehousing • Transport • Inventory management • Postponed manufacturing	• Management consultancy • Information services • Financial services • Supply chain management
	Traditional transport and forwarding companies	Network logistics providers
	• Transport • Warehousing • Export documentation • Customs clearance	• Express shipments • Track and trace • Electronic proof of delivery • Just in time deliveries

↑ Increasing level of physical services

→ Increasing level of management services

Source: Adapted from Berglund et al. (1999).

Figure 18.2 Typology of logistics services

ALIGNING LOGISTICS PROVISION WITH THE SUPPLY CHAIN

While the Berglund et al. (1999) typology is useful for categorising logistics service providers, it does not address the issue of identifying the most appropriate type of provider for a given type of logistics service, or the nature of the relationships needed to support this. Bask (2001) suggests that a 'one size fits all' approach to third party logistics needs to be replaced with 'clearly packaged' different types of providers, with distinctly segmented service types and aligned relationship strategies. This business model highlights the importance of 'separating, classifying and prioritising processes that have the greatest impact of supply chain performance' (Bask et al., 2010: 159), so that logistics provision is aligned with the contingent supply chain strategy.

Bask (2001) explores the correlation between the complexity of the 3PL service provision and the type of relationship required to support it (Figure 18.3). The three 3PL service types are:

- *Routine*, often involving a single mode of transport, without any other additional services. The procurement of the transport is based on volume provision and is selected primarily on price. A close relationship between provider and user is not needed. Frequently, these services are provided by traditional transport companies.
- *Standard*, where some degree of customisation may be provided, such as the provision of specialist vehicle types. The service provider will have to have closer

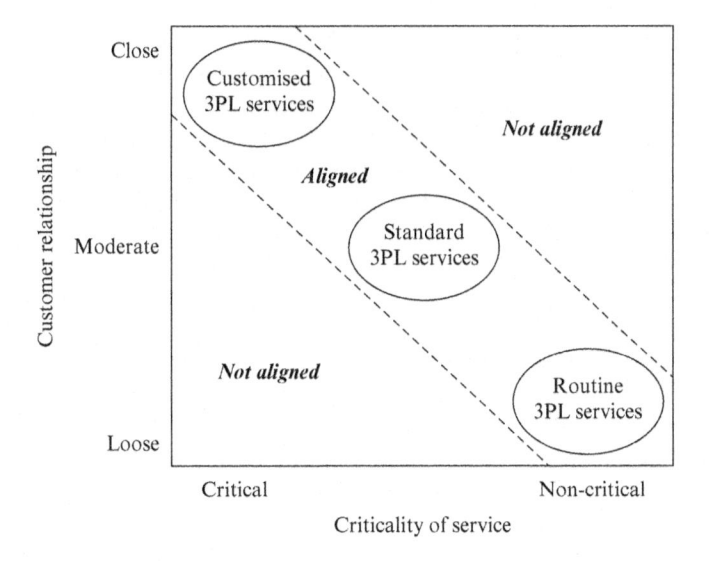

Figure 18.3 Framework for aligning logistics services

coordination and cooperation with the service user. Further, some cooperation with a complementary carrier will be required for services they cannot provide. These services are often well suited to asset based and network logistics providers.
- *Customised*, where the logistics provider supplies additional services such as warehouse provision, inventory control and ordering, product tracking and value adding activities. Partnering arrangements are established to ensure effective coordination. Services in this category may be better suited to information based providers.

What is clear from this framework is that there is a need to align the product being delivered, the nature of the logistics service required, and the type of logistics provider. The importance of such an alignment has been recognised in manufacturing environments (e.g. Fisher, 1997), but less so in logistics. Further, with the wider trends discussed earlier in the chapter, there is an increasing focus on customisation. The fourth party logistics concept is now returned to, as an enabler of this customisation.

FOURTH PARTY LOGISTICS

The term 'fourth party logistics' was first coined by the consulting firm Accenture in the late 1990s, defining it as 'an integrator that assembles the resources, capabilities and technology of its own organisation and other organisations to design, build and run comprehensive supply chain solutions' (Yao, 2011: 121). Drawn from Accenture's definition, Marino (2002: 23) reinterprets a 4PL as a consulting firm which integrates and manages 'a company's logistics resources and providers, including third party logistics providers and transportation companies'. Although Marino's definition stresses the role of monitoring 3PLs, the importance of information management is neglected. By contrast, Bade and Mueller (1999) view the concept as an evolution in supply chain outsourcing. A key aspect of many of these early definitions is the independence of the 4PL from the organisations providing the logistics services.

However, this is one aspect which has generated a more heated debate. For many, the ownership of physical logistics assets, such as trucks and warehouses, is the main divide between 4PLs and 3PLs. But the emergence of 3½PL has blurred this distinction. It is argued that the idea of having a 3PL taking on the role of a 4PL is inherently contradictory (Love, 2004; Warrilow, 2007) because the firm will optimise its own assets before looking to alternative providers. Opponents, on the other hand, believe that more flexibility is needed in defining what a 4PL is (Tierney, 2004; Biederman, 2005). As companies demanded more end to end solutions, so many 3PL providers, such as UPS, have expanded their business scope to offer consultancy services (van Hoek and Chong, 2001).

While the debate on the ownership structure of 4PLs continues, a number of key components for 4PLs can be identified (Table 18.1).

Table 18.2 summarises the main advantages and disadvantages of using a 4PL provider. Able to command the entire supply chain, the 4PL has the necessary input into all value adding activities to apply supply chain principles (Hingley et al., 2011). The 4PL reallocates the client's management time and effort to concentrate on their core activities, rationalising the management of 3PLs through a single interface (Gattorna, 1998) and normalising the performance metrics and reporting formats (Dutton, 2009). The client–4PL relationship draws upon the realisation that optimisation is only possible

Table 18.1 Key components of the 4PL concept

Factor	3PL	4PL	Key references
Asset basis	Asset based	Non-asset based (except perhaps IT systems)	Love (2004) Warrilow (2007) Win (2008) Hill (2011)
Accountability	Part (in conjunction with internal resources and/or other 3PLs)	Singular accountability	Bade and Mueller (1999) Krakovics et al. (2008) Win (2008)
Role	Logistics	Logistics and supply chain integration	Bade and Mueller (1999) Marino (2002) Schwartz (2003) Win (2008) Hill (2011)
Performance/ success measurement	Cost	Value creation within the client supply chain	Bade and Mueller (1999) Warrilow (2007) Krakovics et al. (2008) Win (2008)

Source: Adapted from Win (2008).

Table 18.2 The advantages and disadvantages of 4PLs

Advantages	Disadvantages
• Holistic management of the supply chain facilitating the application of wider supply chain principles • Frees up capital and management time for more efficient use elsewhere • A partnership and non-asset base ensures the pursuit of shared goals • Transfers the risk associated with asset ownership • Expertise and neutrality ensure solutions deliver incremental value maximisation • Reduced costs through economies of scale and scope • Provision of agility and flexibility at a reduced level of risk • Access to innovations and new technologies to monitor and enhance supply chain performance	• Increases the complexity of supply chain relationships • Risk of poorer service through damaging existing 3PL relations • May restrict the creative input of the 3PL service provider and subsequent supply chain innovation • Subject to issues of trust in relation to the sharing of sensitive information • Costs to switch 3PL providers may be high • High investment requirement which, with the lack of assets, is difficult to justify in the light of financial risk • The 4PL may find it difficult to position itself correctly • Market changes may render 4PL services obsolete • Potential restrictions to specific industries

through the pursuit of shared goals in which each party's commitment is dependent upon the level of integration. The relationship exemplifies those characteristics of a successful partnership, focussing on a long term relationship where targets for value creation hold the 4PL solely accountable for its performance.

The 4PL model enables the client to retain the benefits of asset transfer and operating cost reduction associated with 3PLs, including the reduction of risks due to demand fluctuations and payback issues (Mason et al., 2007). This freed up capital can be invested elsewhere within the business. The lack of asset ownership within the 4PL enables management decisions to be made with neutrality and objectivity, so that the system as a whole is optimised even if some activities are suboptimal (Gattorna, 1998).

Not only with regard to assets can the client be flexible. Having access to the 'best of breed' resource providers and the knowledge of multiple supply chain strategies from their client base, the 4PL is able to utilise economies of scale and scope to achieve greater agility and flexibility in response to customer needs (Gattorna, 1998; Hastings, 2001). Exposure to best practice encourages supply chain innovation and access to the latest logistics software and technology (Freibairn, 2003; Dutton, 2009).

Turning to the disadvantages, adding an extra operational layer brings additional complexity to the networks of relationships that exist between shippers, 3PLs and receivers of the products (Mason and Lalwani, 2007). The resultant lack of contact between the shipper and the 3PL may damage the relationship between these two parties, reducing the effectiveness of logistics operations. Additional services, once performed by the 3PL, may be placed back under the remit of the shipper and uncertainty over contract lengths and efficiency targets may stifle innovation by the 3PL.

The partnership required may suffer the general challenges of building the required level of mutual trust and understanding. Working with multiple clients, while necessary in the interests of perceived independence, may promote a more reserved attitude towards information sharing by the shipper (Gattorna, 1998). The necessary sharing of information, vital for the 4PL to base its decisions, seems to be at risk, particularly with oligopolistic industries where competition is fierce and information secrecy is rated highly (Hingley et al., 2011).

There are also financial burdens as a result of the 4PL's need for flexibility. Existing 3PL relationships may exist and, depending upon the contractual arrangements, can represent a substantial financial barrier due to early termination (Cabdoi, 2003). There may also be significant investments in ICT and its integration with existing systems. Such 'up front' costs can pose a financial risk unless there is a commitment to long term contractual agreements.

Overall, the above factors may dissuade the introduction of 4PLs or restrict the scope of their services. This then diminishes their advantage for holistic management, making their value proposition less clear, or even obsolete, to the client. There may also be industry-specific constraints that restrict applicability. For example, in the chemical industry 4PLs have only a limited role to play (Bertschi, 2011). Safety is of paramount importance in this industry, and 3PLs need technical and product specific expertise. Therefore, the advantages of 4PLs are significantly diminished, making 3PLs more attractive.

With the development of 4PL solutions, there are examples of the successful application of the concept. One such company to adopt a 4PL solution was Corus (now part of Tata Steel). Historically, its logistics operations in the UK were organised at a local level, with no coordination between the 16 different business units. Transport Development Group (TDG plc) were contracted to provide a 4PL solution that encompassed all activities within the UK. The outcome from this was a 40 per cent reduction in empty running, 99 per cent delivery on time performance and a 6 per cent reduction in transport costs (Freight Best Practice, 2010; TDG, 2012).

CONCLUSIONS: THE FUTURE OF LOGISTICS SERVICES

The chapter has attempted to show how the logistics service industry has constantly changed since its inception and how it has grown in maturity to become a more widely respected aspect of modern commerce. So, what of the future for the logistics service industry?

One fact that is certain is that the ongoing transformation which has occurred over the past 30 years or so will continue. As in the past, this will be fuelled by both external and internal drivers. Externally, the landscape in which logistics service providers operate will continue to evolve. Clarke (2012), in his review of the external circumstances facing his business, Tesco, which is arguably as much a logistics business as it is a retail concern, picked out four changes that he felt were most prominent: climate change, increasing commodity prices, rapid urbanisation and a growing middle class wanting to lead healthier lives. Surveying his marketplace he proposed this perspective:

> right now the tectonic plates are shifting. Customers are navigating a highly volatile landscape. On the one hand, new technology means that they have more choice, more power, and more

control than ever before. But at the same time, particularly in the developed economies, they are under enormous financial pressure. The result is that people are asking tough questions [of service providers]. (Clarke, 2012)

An outcome of this is that the channels of distribution within which logistics providers operate are also evolving. The traditional retail channels are changing as a result of technological developments, leading to a substantial increase in home deliveries. The logistics implications are the evolution from moving truckloads of products to hundreds of retail outlets to delivering parcels to many millions of homes. Achieving this cost effectively represents a major challenge, and has seen the emergence of hybrid routes such as 'click and collect', where a customer's order is delivered to the local retail outlet, and drop boxes, where deliveries are made to lockers at a location convenient to the customer.

Beyond this the industry is also changing in terms of the internal business models that are being pursued. Logistics invariably represents a significant cost to most industries so the focus on chasing efficiency gains remains a constant priority of shippers. This in turn puts great stress on margin preservation for providers, who are often operating with wafer thin margins (FTA, 2012). This is leading to providers either tailoring their offers to win as specialists or to pursue growth through acquisition and mergers to sweat economies of scale. There is also an increasing emphasis on driving operational quality so that customers can depend more confidently on logistics providers, so that wherever they may operate in the world they enjoy a consistent standard of operations and service.

What is clear is that the logistics business model is being repeatedly tested. The needs of providers of logistics services to be profitable, innovative and corporately responsible have to be managed against an agenda which has seen competition intensify, factor prices inflate, and supply chains become more vulnerable as inventory levels are reduced and dependence on extended global supply becomes more common. Solutions must be characterised by a high quality, cost competitive and personalised service which not only matches but anticipates customer requirements and respects the wider sustainability agenda.

Finally, although this chapter has aimed to demonstrate that the logistics industry has a significant contribution to bear on business, on supply chains, on regional and national economies, and on the population at large, it could also be argued that the value of this contribution is not always recognised. For instance, at the business level, while the importance of logistics has been recognised, it has yet to result in a widespread representation at board level (Wilding et al., 2010). One prominent exception was the recent accession of Tim Cook to the helm at Apple. Cook was previously famous for being the 'logistics king' of Apple before being backed by Steve Jobs to replace him. Perhaps this high profile move will signify an accelerated acceptance that logistics professionals can play pivotal roles in the leadership of firms.

This links into a wider issue of the often negative public perception of logistics (Jackson, 2011). Warehouse developments are seen as unsightly while trucks are portrayed as large and noisy. Initiatives such as the Love Logistics campaign by the FTA in the UK are attempting to address this, highlighting what life would be like without logistics. It is important that the image of the industry continues to be turned around, if only because this has a direct bearing on recruitment of high quality and well-motivated people with the appropriate skills. The foundation for successful logistics is an engaged and motivated workforce – hopefully, the industry will continue to build and will further develop so it will be able to attract the quality and quantity of personnel that it needs to sustain itself as a high quality and respected service sector in the future.

Based on the above, the future research agenda for logistics includes:

- Examining fully the implications of collaboration, both horizontally and vertically. There are limited case studies thus far and therefore a wider examination of positive and negative implications is needed. Given that there are often concerns about the legality of cooperation between operators, this issue also needs examination in the context of logistics.
- Detailed evaluation of the different distribution channels now used in retail logistics, and particularly whether certain models are more appropriate for certain situations. In the future, it may be that in certain circumstances (e.g. for rural communities) the conventional retail channel is replaced by emergent models.
- With the emergence of 4PLs, will there be an increase in multimodal transport operations? Rail operators are often criticised for not wishing to provide a door to door service where road transport is required and so 4PLs, with their broader view of networks, could be used to organise and coordinate such flows.
- The full extent of the impact of technology on logistics is not yet fully known. Will future technologies see the replacement of logistics services, through the growth of electronic distribution and 3D printing? Could technology enable dynamic scheduling, whereby customers can choose their exact delivery time, using price to encourage better schedules for the operator?
- Given the increasing importance of logistics as supply chains become global, how can logistics become more influential in the boardroom of businesses? Research is also needed to highlight the importance of logistics to local, regional, national and supra-national economies, so that the public and politicians fully appreciate the value that it can provide.

REFERENCES

Arvis, J.-F., M.A. Mustra, L. Ojala, B. Shepherd and D. Saslavsky (2010), *Connecting to Compete 2010: Trade Logistics in the Global Economy*, Washington, DC: The World Bank.

Bade, D.J. and J.K. Mueller (1999), 'New for the millennium: 4PL', *Transportation and Distribution*, **40** (2), 78–80.

Bask, A.H. (2001), 'Relationships among TPL providers and members of supply chains – a strategic perspective', *Journal of Business and Industrial Marketing*, **16** (6), 470–486.

Bask, A.H., M. Tinnilä and M. Rajahonka (2010), 'Matching service strategies, business models and modular business processes', *Business Process Management Journal*, **16** (1), 153–180.

Berglund, M., P. van Laarhoven, G. Sharman and S. Wandel (1999), 'Third-party logistics: Is there a future?', *International Journal of Logistics Management*, **10** (1), 59–70.

Bertschi, H. (2011), 'The 3PL', *ICIS Chemical Business*, 11 April, 30–31.

Biederman, D. (2005), 'Growth business', *Journal of Commerce*, **6** (23), 28–31.

Bowersox, D.J. (2007), 'SCM: The past is prologue', reprinted in Mangan, J., C. Lalwani and T. Butcher (eds) (2008), *Global Logistics and Supply Chain Management*, Chichester: John Wiley & Sons Ltd, pp. 335–343.

Cabdoi, C. (2003), 'Fourth party logistics market: A European perspective', www.frost.com/sublib/display-market-insight-top.do?id=8341069, accessed 19 September 2013.

Clarke, P. (2012), 'Winning customers in a world of change', speech to the World Retail Congress, London, 19 September, http://www.4-traders.com/TESCO-PLC-4000540/news/Tesco-PLC-Winning-customers-in-a-world-of-change-%96-Philip-Clarke-speech-15205366/, accessed 30 September 2012.

Cruijssen, F., W. Dullaert and H. Fleuren (2007), 'Horizontal cooperation in transport and logistics: A literature review', *Transportation Journal*, **46** (3), 22–39.
Dutton, G. (2009), 'The rise of the 4PL', *World Trade*, January, 20–23.
Ellram, L.M. (1991), 'A managerial guide for the development and implementation of purchasing partnerships', *International Journal of Purchasing and Materials Management*, **27** (2), 2–8.
Ellram, L.M. and M.C. Cooper (1990), 'Supply chain management, partnerships and the shipper–third party relationship', *International Journal of Logistics Management*, **1** (2), 1–10.
Embleton, P. and P. Wright (1998), 'A practical guide to successful outsourcing', *Empowerment in Organisations*, **6** (1), 94–106.
Fisher, M.L. (1997), 'What is the right supply chain for your product?', *Harvard Business Review*, **75**, 105–117.
Freibairn, J. (2003), 'Why we went 4PL with Kuehne & Nagel-Nortel', *Motor Transport*, 23 October, 8.
Freight Best Practice (2010), 'The benefits of central supply chain management: Corus and TDG', www.freight-bestpractice.org.uk/the-benefits-of-central-supply-chain-mangement-corus-and-tdg, accessed 19 September 2013.
FTA (2011), 'Logistics facts', http://www.lovelogistics.co.uk/logistics_facts/, accessed 21 September 2013.
FTA (2012), *The Logistics Report 2012*, Tunbridge Wells: Freight Transport Association.
Gattorna, J. (1998), 'Fourth party logistics: En route to breakthrough performance in the supply chain', in J. Gattorna (ed.), *Strategic Supply Chain Alignment*, Aldershot: Gower Publishing, pp. 425–441.
Goldsworthy, A. (2003), *In the Name of Rome: The Men Who Won the Roman Empire*, London: Orion Publishing Co.
Griffiths, D. (2001), 'The theory and practice of outsourcing', available at http://www.inkoopportal.com/inkoopportal/download/common/theory_and_parctice_of_outsourcing.pdf, accessed 28 September 2013.
Hastings, P. (2001), 'Party games', *Logistics Europe*, July, 38–42.
Hill, B. (2011), 'Appeal of 4PL: Proliferation and differentiation as a 5PL', *Logistics and Transport Focus*, **13** (8), 42–45.
Hingley, M., A. Lindgreen, D. Grant and C. Kane (2011), 'Using fourth party logistics management to improve collaboration amongst grocery retailers', *Supply Chain Management: An International Journal*, **16** (5), 316–327.
Ittmann, H.W. and D.J. King (2011), 'Introduction', in D.J. King (ed.), *7th Annual State of Logistics Survey for South Africa*, Pretoria: Council for Scientific and Industrial Research.
Jackson, M. (2011), 'Six years to make logistics popular', *Logistics Manager*, 1 October, www.logisticsmanager.com/Articles/17023/Six+years+to+make+logistics+popular.html, accessed 19 September 2013.
Krakovics, F., J. Eugenio Leal, P. Mendes and R. Lorenzo Santos (2008), 'Defining and calibrating performance indicators of a 4PL in the chemical industry in Brazil', *International Journal of Production Economics*, **115** (2), 502–514.
Lewis, H.T. and J.W. Culliton (1956), *The Role of Air Freight in Physical Distribution*, Boston: Division of Research, Graduate School of Business Administration, Harvard University.
Love, J. (2004), '3PL/4PL – where next?', *Logistics and Transport Focus*, **6** (3), 18–21.
Lummus, R.R., D.W. Krumwiede and R.J. Vokurka (2001), 'The relationship of logistics to supply chain management: developing a common industry definition', *Industrial Management and Data Systems*, **101** (8), 426–432.
Mandrodt, K.B. and F.W. Davis (1992), 'The evolution towards service response logistics', *International Journal of Physical Distribution and Logistics Management*, **22** (9), 3–8.
Marino, G. (2002), 'The ABCs of 4PLs', *Industrial Management*, **44** (5), 23.
Mason, R. and C. Lalwani (2006), 'Transport integration tools for supply chain management', *International Journal of Logistics: Research and Applications*, **9** (1), 57–74.
Mason, R. and C. Lalwani (2007), 'Fourth party logistics: What, why, how', in K.S. Pawar, C.S. Lalwani and M. Muffatto (eds), *Proceedings of the 12th International Symposium on Logistics*, Budapest, pp. 565–572.
Mason, R., C. Lalwani and R. Boughton (2007), 'Combining vertical and horizontal collaboration for transport optimisation', *Supply Chain Management: An International Journal*, **12** (3), 187–199.
Mortensen, O. and O. Lemoine (2008), 'Integration between manufacturers and third party logistics providers?', *International Journal of Operations and Production Management*, **28** (4), 331–359.
Murphy, P.R. and R.F. Poist (2000), 'Third-party logistics: Some user versus provider perspectives', *Journal of Business Logistics*, 21(1), 121–133.
Naim, M.M., A.T. Potter, R.J. Mason and N. Bateman (2006), 'The role of transport flexibility in logistics provision', *International Journal of Logistics Management*, **17** (3), 297–311.
Potter, A. and C. Lalwani (2005), 'Supply chain dynamics and transport management: A review', in J. Dinwoodie (ed.), *Proceedings of the 10th Logistics Research Network Conference*, Plymouth, pp. 353–358.
Rushton, A., P. Croucher and P. Baker (2006), *The Handbook of Logistics and Distribution Management*, 3rd Edition, London: Kogan Page.
Schwartz, E. (2003), 'The logistics handoff', *InfoWorld*, **25** (44), 53–58.

Simchi-Levi, D., P. Kamisnsky and E. Simchi-Levi (2003), *Designing and Managing the Supply Chain: Concepts, Strategies and Case Studies*, 2nd Edition, New York: Irwin McGraw-Hill.
Simpson, J.A. and E.S.C. Weiner (1989), *The Oxford English Dictionary 3*, Oxford: Clarendon Press.
Sink, H.L. and C.J. Langley (1997), 'A managerial framework for the acquisition of third party logistics services', *Journal of Business Logistics*, **18** (2), 163–189.
Skjøtt-Larsen, T. (2000), 'Third-party logistics – from an interorganisational point of view', *International Journal of Physical Distribution and Logistics Management*, **30** (2), 112–123.
Stevens, G.C. (1990), 'Successful supply chain management', *Management Decision*, **28** (8), 25–30.
TDG (2012), 'Reaping the benefits of fourth party logistics for Corus', www.tdg.eu.com/corus.asp, accessed 12 February 2013.
Tierney, S. (2004), 'Now there are real 4PL possibilities', *Supply Chain Europe*, **13** (5), 16–18.
Tomkins, J.A. and J.D. Smith (eds) (1998), *The Warehouse Management Handbook*, 2nd Edition, Raleigh, NC: Tomkins Press.
Van Hoek, R. and I. Chong (2001), 'Epilogue: UPS Logistics – practical approaches to the e-supply chain', *International Journal of Physical Distribution and Logistics Management*, **31** (6), 463–468.
Warrilow, D. (2007), 'Become your own 4PL', *Logistics and Transport Focus*, **9** (2), 37–40.
Wilding, R., A. Waller, S. Rossi, C. Geldard, S. Mayhew, R. Cigolini and C. Metcalfe (2010), 'Supply chain strategy in the board room', Cranfield University Working Paper, dspace.lib.cranfield.ac.uk/handle/1826/5272, accessed 27 September 2013.
Win, A. (2008), 'The value a 4PL provider can contribute to an organisation', *International Journal of Physical Distribution and Logistics Management*, **38** (9), 674–684.
Yao, J. (2011), 'Decision optimization analysis on supply chain resource integration in fourth party logistics', *Journal of Manufacturing Systems*, **29** (4), 121–129.

19. Creative systems: a new integrated approach to understanding the complexity of cultural and creative industries in Eastern and Western countries

Lauren Andres and Caroline Chapain

INTRODUCTION

The cultural and creative industries (CCIs) are now recognised as a major economic force in post-Fordist societies. First identified by Western countries as a motor of economic recovery in the 1980s (Bianchini and Parkinson, 1993) and then of economic growth in the 1990s (Department for Media, Culture and Sport, 1998, 2000; Cunningham, 2002), they have since increasingly been the focus of local and national policies in both developed and developing countries (UNDP/UNCTAD, 2010).

The transfer of concepts (e.g. creative industries or creative city) and policy (e.g. cluster) from the West to the East is perhaps problematic, as there is still an important debate on the way these industries develop economically and spatially in different places in the West (Chapain and Lee, 2009; Musterd and Murie, 2009; Leriche and Daviet, 2010; Chapain et al., 2013). On the one hand, this debate is fuelled by the inclusion in CCIs of economic activities with very different value chains and degrees of public funding in the areas of heritage (traditional cultural expressions and cultural sites), arts (performing arts and visual arts), media (publishing and printed media and audio-visual businesses) and functional creation (design, new media and creative services, creative research and development, digital and other related creative services) (UNDP/UNCTAD, 2010). On the other hand, while these industries have been explored from a variety of disciplines, such as economics (see for example Ginsburgh and Throsby, 2006; Potts et al., 2008; Towse, 2011), cultural and media studies (see for example Deuze, 2007; Bennett and Frow, 2008; Flew and Cunningham, 2010), economic geography (see for example Scott 2000, 2006; Cooke and Lazaretti, 2008; Musterd and Murie, 2009) and planning (see for example Bianchini and Parkinson, 1993; Miles and Paddison, 2005; Evans, 2009; Legnér and Ponzini, 2009), nevertheless, there has been an absence of cross-disciplinary discussions and recognition of the need to develop a more integrative approach to understanding the development of the cultural and creative industries.

Using the notion of creative systems and building on the principles of complexity theory, this chapter brings together economics, economic geography, cultural studies and planning approaches to offer a better understanding of the way CCIs develop in places with diverse social, economic and political characteristics. This implies accounting for the multi-scale network governance within which the economic and social relations of creative individuals and firms occur. As such, as noted by Healey (2006, p. 526), 'places in this relational conception emerge as nodes in one or more networks and as institutional sites, with particular material geographies'. Within this complex system material resources and values are accumulated and dispersed through relational dynamics (as in a firm's value-added chain).

Some Asian countries such as Singapore, Hong Kong and China have been rapid followers of Western countries in identifying CCIs as engines of urban development and economic growth, and developing policies intended to enhance their development (Kong et al., 2006). However, in recent years there has been an increasing concern with regard to the direct transfer of policies to support the creative industries from Western to Eastern and South Eastern Asian countries (see for example Kong and O'Connor, 2009). Indeed, the adoption of the 'creative industries' concept in some countries relies more on a willingness to be part of a global phenomenon rather than a concrete understanding of its dynamics (Kong et al., 2006). Some authors argue that there are some contradictions in the freedom inherent to the creative process and the dirigiste societies of some of these countries, such as China (Ooi, 2006; Keane, 2009). Others highlight the fact that most Asian countries are displaying more dynamic economic restructuring than Western countries, rendering the transfer of Western experiences there less effective (Mok, 2009). This chapter compares examples of creative systems in Western and Asian countries, discussing their degrees of convergence and/or divergence and the related policy implications.

The chapter is structured as follows. The next section summarises the definitional debate surrounding CCIs in Western countries and discusses how this debate has been translated in Asia. We then introduce the notion of 'creative system', which lies at the heart of our analysis. The next section compares and contrasts various creative system examples in Europe, North America and Asia. Finally we conclude by discussing the research and policy implications of our findings.

THE CREATIVE INDUSTRIES: SOME DEFINITIONAL BACKGROUND

The emergence of the concept of 'creative industries' in policy discourse in the late 1990s and the current lack of coherence around both its conceptual and operational definitions (Flew and Cunningham, 2010) need to be understood within the debate created by the genesis of the 'culture industry' concept in Germany in the 1930s and 1940s. At the time, the 'culture industry' marked a shift away from seeing 'the arts as a form of critique of the rest of life [providing] an utopian vision of how a better life might be possible' (Hesmondhalgh, 2007, p. 16) to seeing the arts as a commodity resulting from mass industrial productions such as in TV, radio, film, music, publishing... The philosophical and political tensions created by the commodification of the arts and culture have sustained a continuous academic debate since then (O'Connor, 2010). Hartley (2005) argues that the term 'creative industries' reflects a change in the way cultural products are now produced and reproduced through the impact of new technologies: 'the idea of the creative industries seeks to describe the conceptual and practical convergence of the creative arts (individual talent) with cultural industries (mass scale) in the context of new media technologies (ICTs) within a new knowledge economy for the use of newly interactive citizen-consumers' (ibid., p. 21). Other authors are more critical and see this change in terminology as a way for policy makers to take advantage of the growth in the software and new media industries to shift their policies from supporting access to culture as a merit good to supporting artists in their creative processes for economic purposes (Garnham, 2005). Nevertheless, the term has risen in popularity in the last 15 years and different definitions with some degree of overlap are now recognised worldwide (Hartley, 2005; Flew and Cunningham, 2010).

As one of the first countries to use the term in national policy documents, the UK definition has played a key role in many academic and policy definitional reflections. Thus, the Department for Culture Media and Sport (1998, p.4) defined the creative industries as

> those that are based on individual creativity, skill and talent. They also have the potential to create wealth and jobs through developing and exploiting intellectual property. The creative industries include: Advertising, Architecture, Arts and antique markets, Computer and video games, Crafts, Design, Designer Fashion, Film and video, Music, Performing arts, Publishing, Software and Television and Radio.

In contrast, the Eurostat (European Statistical Office) and United Nations' definitions include Heritage activities such as archaeological sites, museums, libraries . . . (KEA, 2006; UNDP/UNCTAD, 2010).[1] The exclusion of heritage activities tends to reflect a more market-driven approach in some countries (Hartley, 2005). In addition, many European countries have retained the term 'cultural industries' or use the term 'cultural and creative industries' instead of 'creative industries'; this emphasises the dual role of culture as both a commodity and a merit good in their policies. While some Asian countries such as Singapore, Hong Kong, South Korea and China have been rapid adopters of the creative industries discourse, they also utilise various terminologies covering different industries (Kong et al., 2006). For example, Singapore very quickly adopted the UK terminology of creative industries; however, the Singaporean definition distinguishes between arts and culture (including heritage activities), media and design (ibid.). In contrast, while a 2003 baseline study (Centre for Cultural Policy Research, 2003) showed Hong Kong directly adopting the UK 'creative industries' concept and definition, a more recent publication from the Hong Kong Government (Government of the Hong Kong Special Administrative Region, 2012) uses the term 'cultural and creative industries', including heritage activities within them. This reflects the important distinction made in China between cultural institutions and commercial cultural enterprises (Kong et al., 2006; Keane, 2009).

THE CREATIVE SYSTEM: A RELATIONAL PERSPECTIVE OF CCI DEVELOPMENT IN CITIES AND REGIONS

Due to their tendency to concentrate in space, especially in cities (Nielsén and Power, 2010), and their increasing use in urban policies in the last 30 years, research with a focus on understanding the ways CCIs develop in cities and regions has emerged in economic geography (see for example Scott 2000, 2006; Cooke and Lazaretti, 2008; Leriche and Daviet, 2010) and planning (see for example Bianchini and Parkinson, 1993; Miles and Paddison, 2005; Legnér and Ponzini, 2009). This complements existing debates in economics (Ginsburgh and Throsby, 2006) and in cultural and media studies (Bennett and

Frow, 2008). Nevertheless, there are few studies that cross disciplinary boundaries. To achieve a more integrated understanding of the way these industries develop in cities and regions, we therefore propose to develop and apply the concept of 'creative system'.

The notion of creative system, as understood in this chapter, sits within the theory of aggregate complexity (Manson, 2001), complex economics (Martin and Sunley, 2007) and relational planning complexity (Healey, 2006). Differently from algorithmic and deterministic complexity, aggregate complexity pays attention to holism and synergy resulting from the interaction of system components (Manson, 2001). It also examines the different components/agents forming the system within a complex environment undergoing multiple dynamics. Doing so, as noted by Portugali (2006), allows positioning structural changes and evolutions as key components of an analysis highly anchored in social theories (particularly the work of Lefebvre (1995), Giddens (1984), Harvey (1996) and Castells (1989, 1996)) and looking at the nature of the social production of space, including not only social components but also economic, political and cultural factors. This puts space in the core of the creative system, acknowledging that it is the product of a complex range of socio-spatial relations between various agents, such as individuals, activities or places.

In economics, the use of complexity theory has resulted in what Beinhocker (2006) names 'complexity economics'. This relies on the assumption that the economy, as an open and non-linear system, is structured around a set of *adaptive agents* which interact within various networks. This understanding of complex adaptive systems has been transferred into economic geography; see for example Martin and Sunley (2007), who note that a 'defining feature of a complex system is that it is composed of interacting subsystems and hierarchical levels'. They argue that, for example, 'the national economy can be divided into smaller territorial subsystems such as regions, cities and localities' (Martin and Sunley, 2007, p. 585). The specificities of such systems rely on the existence of a set of *subsystems* (e.g. a region or a city) which exchange energy and matter (e.g. flows of goods, services, knowledge, capital, money and people) with their environment (other regions and cities). The combination of spatial scales is thought crucial in this hierarchical architecture.

This hierarchical understanding is shared when applied to urban planning. As such, relational complexity allows for capturing 'the dynamics and tensions of relations with very different driving forces and scalar relations as these coexist in particular places and flow through shared channels' (Healey, 2006, p. 536). It also looks at the range of relations forming a territory and the complex intersections and disjunctions that can be found among and within different episodes of governance (see for example Healey, 1997, 2006; Allen, 1999, 2003; Graham and Healey, 1999). Social relations composing territories are shaped by different driving forces operating at many different spatial scales and timescales. Here, places are 'nodes in one or more networks' (Healey, 2006, p. 526). They include for example neighbourhoods, development areas, cities or regions. These places form a complex urban system (Byrne, 2003, p. 174) which is 'nested in and intersecting with regional, national, blocks (e.g. European Union) and global systems and which in turn are nested in and intersecting with individuals, households and neighbourhoods'.

Bringing together complexity economics and planning relational complexity, we understand the creative system as structured around the interplay of agents (individuals, firms, institutions) within and across three subsystems (economic, cultural and planning)

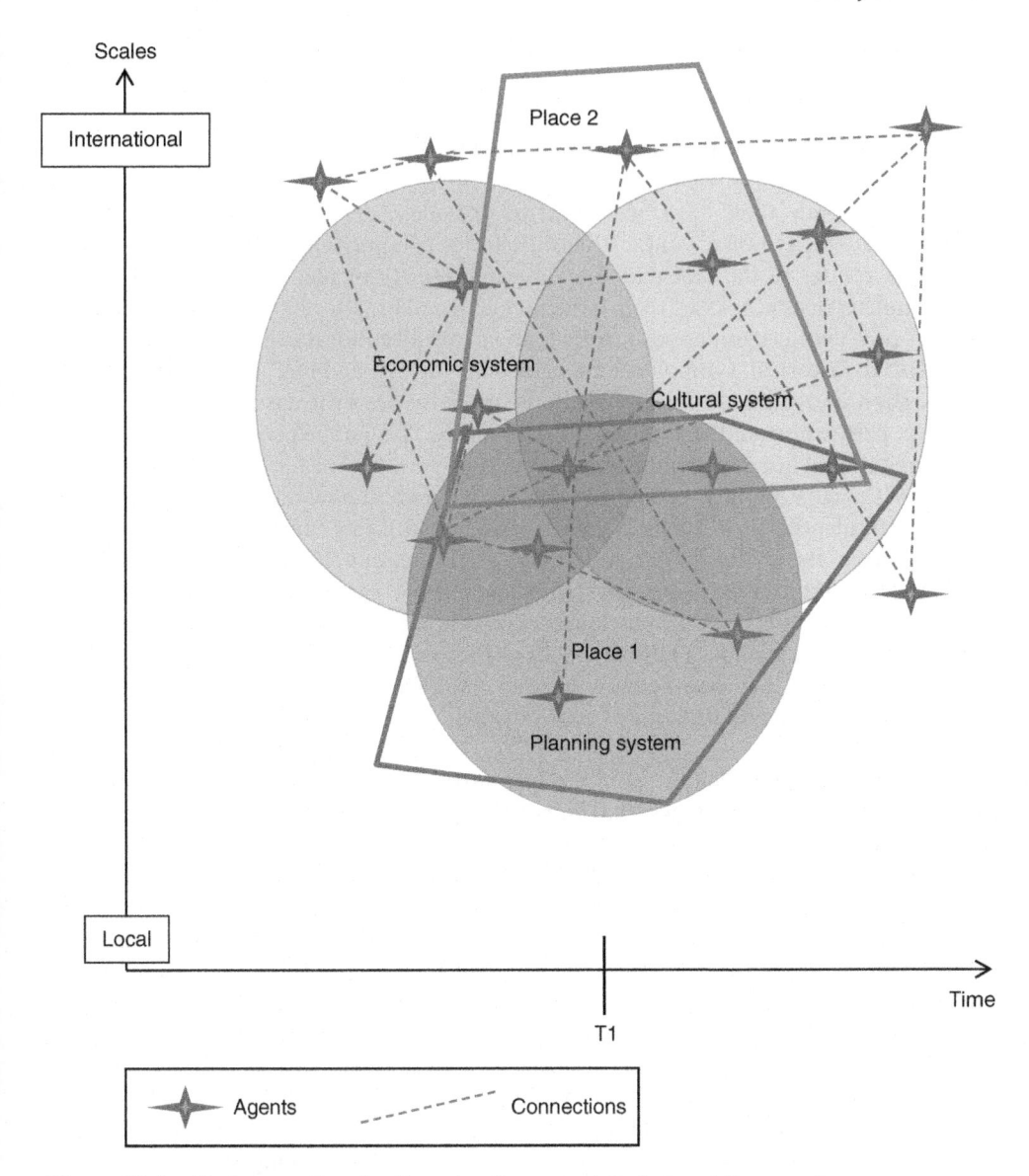

Figure 19.1 Understanding the Creative System: A Relational Perspective

in order to break the tendency of looking at the creative industries from a sole disciplinary approach – that is within a single subsystem – which in essence is extremely restrictive and partial. In addition, the creative system is multi-dimensional in that the various geographical (local, regional, national, international) and temporal (past, present, future) scales within which these agents interplay are recognised (see Figure 19.1). In terms of local and regional policies, this implies bringing together the evolving (over time) and multi-dimensional (sectoral and geographical) natures of *place*. The creative system therefore comprises the intersection of the economic and cultural subsystems as well as the planning and cultural subsystems and how these intersections are translated into space and policy.

At the Intersection of the Economic and Cultural Subsystems

Various economic tools have been used to provide a deeper understanding of CCIs (Throsby, 2008). However, one of the difficulties of applying economic tools to these industries resides in the fact that the value for society of many cultural and creative goods and services surpasses their monetary value (Throsby, 2001). This characteristic of CCIs products as merit goods resides in their overall cultural value (i.e. aesthetic, spiritual, social, historical, symbolic and authenticity values) and has justified strong public intervention and funding (ibid.). This has especially been the case for sectors such as heritage, performing and visual arts, TV and radio. This public/private duality is crucial in understanding how these industries function.

One of the most popular economic tools to understand how these industries function has been to identify their distinctive value chain (ibid.). For example, UNESCO (2009, p.19) describes the 'cultural cycle' in five stages: (1) creation, (2) production, (3) dissemination, (4) exhibition/transmission/reception and (5) consumption/participation. An earlier version of this cycle also included two additional stages: (6) archiving/preserving and (7) education/training (Throsby, 2008). These stages do not always follow a hierarchical sequence; they may actually overlap, interact and/or conflict (Hesmondhalgh, 2007) and public intervention may occur during any of them, with a particular focus on stages 6 and 7.

Based on this value chain, some models have been developed that associate some industries with its earlier stages, that is, creating artistic cultural content (such as literature, music, performing arts, visual arts, heritage) and others with later stages, that is, reproduction and dissemination (such as film, TV, radio, publishing). Yet other models have focused on the type of outputs that these industries produced (services, experiences, content or originals) (O'Connor, 2010). Nevertheless, these classification exercises tend to be arduous due to the high degree of multi-sectoral and vertical integration of these industries (Hesmondhalgh, 2007). Indeed, cultural and creative production tends to operate on a project basis, bringing together workers with diverse and specialised skills at every stage in the value chain for a specific amount of time (Caves, 2000; Hesmondhalgh, 2007). For example, the production of a film will require combining expertise in film production, post-production and distribution as well as expertise from other industries such as music, TV, advertising, publishing, digital media, performing arts. . . In addition, some cultural and creative workers have skills which can be used across CCIs sectors (i.e. actors, musicians, sound engineers. . .).

In addition to this strong multi-sectoral and vertical integration, some cultural and creative industries such as advertising, design and digital media are highly connected to other economic sectors, either directly through business-to-business interactions or indirectly through the role that their related cultural and creative occupations play in the rest of the economy. For example, Higgs et al. (2008) show that 54 per cent of all creative occupations were embedded outside the CCIs in the UK. It is also important to note that cultural and creative work takes place within for-profit and not-for-profit organisations as well as within the third community and public sectors (Markusen, 2010).

Many economic geographers have attempted to represent the way cultural and creative economic dynamics are shaping space (Chapain and Comunian, 2010). The cultural and creative industries are influenced by external and agglomeration economies and tend to cluster spatially (see ibid. for a review). As for other industries, the physical proximity inherent to spatial clustering facilitates exchanges of tacit knowledge and information and, given the specific needs of these industries, access to the specialised and diverse labour markets and supply chains necessary to all cultural and creative production (Lorenzen and Frederiksen, 2008). It is interesting to note that while some clusters are focused on mono- or pluricultural and creative production activities, some are oriented towards both cultural production and consumption (i.e. access to market), whereas others focus strictly on cultural consumption (Evans, 2009). In addition, while some clusters span across cities and regions, others tend to be highly localised within a neighbourhood (Legnér and Ponzini, 2009). Not all cultural and creative clusters are solely the products of private sector dynamics. Indeed, some clusters are supported indirectly by public interventions (general provision of cultural and creative training programmes, infrastructures, funding. . .) or direct cluster initiatives with dedicated policy and governance arrangements (Cinti, 2008). This public support can take the form of cultural, economic and planning policies and be based on a mix of economic, cultural and social rationales (Smith and Warfield, 2008; Evans, 2009). Finally, beyond clustering, some authors have highlighted the importance of international networks (Neff, 2005; Roling, 2010) and global value chains (Jansson and Power, 2010) in the way these industries function across space. Such networks have been increasingly important as the gathering pace of globalisation has resulted in a growing internationalisation of cultural and creative markets and distribution networks. This should not be left aside when exploring creative systems at the local and regional levels.

At the Intersection of the Planning and Cultural Subsystems

The intersections of the planning and cultural subsystems can be characterised by public and private strategies and initiatives targeting the development or renewal of specific scenes of the city (buildings, neighbourhoods. . .) or the city as a whole (see for example Zukin, 1995; Storper, 1997; Brown et al., 2000; Pratt, 2000, 2002; Scott, 2004). This part of the creative system sits within the role given to culture in the 1970s as a factor of social cohesion and integration (i.e. socio-cultural policy) (Bianchini, 1999), in the 1980s as a catalyst for urban and economic development as well as regeneration, city competitiveness and branding (Bianchini, 1999; Kong, 2000; Garcia, 2004), and more recently as a factor to attract workers and businesses (Florida, 2002). This evolution marked a growing recognition of the benefits of cultural industries for economic development (Pratt, 1997) and the importance of having cultural workers for both cultural and non-cultural economies (Markusen and King, 2003; Markusen and Schrock, 2006; Markusen and Gadwa, 2010). Initially, the focus was put on the acknowledgement – transferred into policy (see Scott, 2000; UNCTAD, 2004; Wiesand and Söndermann, 2005; KEA, 2006, 2009; UNESCO, 2006; Council of the European Union, 2007; UNDP/UNCTAD, 2008) – that arts and cultural physical investments as well as artists and other bohemians help revitalise specific neighbourhoods or districts (e.g. Bianchini et al., 1988; Landry et al., 1996; Lloyd and Clark, 2001; Lloyd, 2002, 2005). The narrative developed by

Florida (2002, 2005) expanded this physical approach by arguing that creativity is the driver of contemporary urban economies (across all sectors) and that creative people are the key asset in this new urban economy. This 'creative class' is mobile, discerning in its choice of where to live and work, and values cities that are characterised by their diversity and tolerance, low entry barriers, a stimulating cultural life and their authenticity (all 'soft' location factors). Despite numerous critics (see for examples Peck, 2005; Scott, 2006; Bontje et al., 2011), such tenets have been rapidly taken on board by local decision makers as complementary to the city marketing and branding policies supporting entrepreneurial and city competitiveness strategies (see for example the Rotterdam Spatial Development Strategy 2030 or Toronto 2008 Creative City Planning Framework).

The intersection of the cultural and planning subsystems thus results in a range of strategies and policies using CCIs and their spatial representations in order to position a city within a competitive network of other cities as well as fostering its economic and urban development. Such city-level (and above) strategies are formalised by local authorities, which then identify specific areas of intervention for the creative sector, the most popular of these being the creation of creative quarters or districts. Such localised initiatives have again attracted significant research interest, especially the processes of cluster creation and transformation (see for example Kong, 2009; Andres and Grésillon, 2013) and their impacts on urban transformation and gentrification (Shaw, 2005) as well as their governance arrangements and power relationships (Pruijt, 2003; Andres, 2011a) or the relationships between place, space and networking sustaining these processes (Drake, 2003).

An important distinction has been made between organic and planned clustering processes leading to the creation of a creative quarter. The first type of creative quarter (bottom-up approach) results from spontaneous and grassroots initiatives led by creative actors. Inherited from the experiences of Soho (New York – see Zukin, 1988), this artistic and creative colonisation leads to a redevelopment trajectory progressively included within a planned development process. The location and nature of those quarters are central as they are former industrial districts offering cheap or free buildings or studios in which creators create, perform and live. Most of these districts have evolved through a path of gentrification, leading to artists' displacement and a change in the nature of the creative sector (i.e. in the Berlin-Kreuzberg district – see Shaw, 2005). All these quarters have been widely included in broader strategies and policies of cultural and creative development. The second type of creative quarter is developed through a top-down approach. Driven by policy and public incentives, they are located again in formal industrial areas and are characterised by the critical mass of concentrated creative and cultural production and consumption activity. The creation and development of such clusters is typical of the way culture and arts have evolved within regeneration strategies that now include policies no longer solely dedicated to the construction of installations or key cultural events, relating to 'spectacular consumption', but to more localised policies aiming at the promotion of spaces, neighbourhoods and areas used for cultural and creative productions (Mommaas, 2004; Evans, 2009).

Having sketched an outline of the components of the creative system, we now turn to an analysis of their emergence in four cities (Montreal, Marseille, Singapore and Shanghai) by detailing the state of their cultural, economic and planning subsystems and their intersections.

EXAMPLES OF WESTERN AND ASIAN CREATIVE SYSTEMS

Originating in the UK, the recognised importance of CCIs for local development has quickly spread through North America, Europe and Asia. In this section, we discuss how CCIs have developed in key cities across these three continents by analysing the evolution of their creative systems from the 1990s onwards.

Montreal's Creative System

At the heart of the most important bilingual province of Canada, Montreal is the third Canadian metropolitan region, with around 3.7 million inhabitants; half of the population resides on the island of Montreal, constituting for the most part the City of Montreal. A strong manufacturing centre in the 1950s, Montreal has since then undergone an important restructuring of its economy towards services (Polese, 2009). CCIs have played an important role in its economic restructuring in the last 30 years, supporting its development as a centre of both cultural production and consumption. As a consequence, Montreal is ranked second behind Toronto for its concentration of CCIs (Coish, 2004). A variety of local, regional, provincial and federal cultural, economic and planning public and private actors have been active in supporting this strategy, usually in a concerted way. This is partly due to the close-knit nature of the Quebecois society, to the high concentration of CCIs on the island of Montreal, and to the combination of provincial and federal regional cultural and economic policies targeted at the island of Montreal as a region. While this strategy focused mostly on the cultural and economic subsystems in the 1990s, the role of the planning subsystem has increased in the 2000s with the development of territorially anchored cultural and creative quarters.

The Montreal creative system is strongly influenced by the presence of strong francophone, anglophone and allophone cultures. This cultural diversity is considered as an asset and the cultural and economic subsystems have been strongly intertwined in Montreal in the last 30 years. In 1986, a report from the Federal Ministerial Committee on Development of the Montreal Region recommended a focus on the local economy and on seven economic sectors, including design and the cultural industries, in order to develop Montreal as an international city (Katiya, 2011). As such, the report considered the cultural subsystem as a motor of economic growth, notably in supporting tourism. The cultural subsystem in Montreal and Quebec has tended to cover a variety of activities, including heritage, performing arts, circus, festivals, visual arts, publishing, film, TV, radio, craft and design, as well as recognising the strong CCIs' links with the education and public sectors (Leclerc, 2000; Ministère de la Culture et des Communications, 2000; Vitrine Culturelle de Montreal, 2003; Government of Québec, 2005). While recognising CCIs' economic impacts, local cultural policies put in place in the 1990s also focused on providing these industries with financial support and fostering their dissemination, cooperation, promotion and commercialization, as well as developing related education programmes (Leclerc, 2000; Ministère de la Culture et des Communications, 2000). Influenced by Florida's creative class discourse, the contribution of the cultural industries to the development of the city was reiterated at the 2002 Summit of Montreal – a large forum, organised by the City Council, bringing together key public actors with civil society to discuss the future of the city (Katiya, 2011), including the cultural and economic contributions of CCIs to local development. In response to this increasing public interest, artistic and cultural organisations in Montreal came together to create *Culture Montreal*, a non-profit organisation which aims to put 'culture at the heart of Montreal development', to support citizens' participation in cultural activities, to contribute to policy debate on the role of culture in local development and also to position the city as an international cultural metropolis. This strong synergy between the cultural and economic subsystems has supported a strong tourism industry (Tourisme Montreal, 2011). The years of the 2000s, however, marked a stronger territorialisation and branding of the cultural industries and a greater interaction of the cultural, economic and planning subsystems. While supported by the public sector, this territorialisation has emerged from needs expressed by CCIs private actors.

Indeed, since the late 1990s, Montreal has developed various parts of its territory into dedicated quarters of either creative production or consumption (Tremblay et al., 2004; Leslie and Rantisi, 2006; Tremblay and Cecili, 2009; Drevo, 2010; Rantisi and Leslie, 2010; Rantisi, 2012), bringing its cultural, economic and planning subsystems together. This was the case of one Southern part of Montreal labelled 'the City of multimedia' in 1998. This label built on the organic conversion of old manufacturing buildings into spaces for information technology firms and initiated the construction of new buildings to reinforce this concentration (Rose, 2007). In 2003, another initiative brought together public and private actors to redevelop and brand as 'le quartier des spectacles' (the performing arts district) a central area of 1 km² by regrouping 80 performing arts and exhibition venues and hosting most of the international festivals organised in the city² (Drevo, 2010). More recently, two Montreal universities have also launched the idea of developing another central part of Montreal as an innovation quarter by bringing together public and private actors in both CCIs and other economic sectors to foster innovation in the city.³ Parallel to the branding of different parts of its territory into different creative quarters, the city has also pushed forward its overall image as a centre of creative production by highlighting the presence of four strong local to provincial clusters of creative production in design, fashion, film and TV, and multimedia⁴ and other creative activities;⁵ some of them supported by metropolitan economic cluster strategies. In addition, recognising its strength in design, Montreal was officially designated a UNESCO City of Design in 2006.⁶

In summary, Montreal has successfully brought together its cultural, economic and planning subsystems to develop a creative system that positions the city as an international centre for both creative production and consumption. This has been rendered possible by the high level of interaction and the synergies between various private and public actors evolving across the different subsystems.

Marseille's Creative System

Marseille (France) is a former industrial and port city of 860,000 inhabitants. Further to its severe de-industrialisation in the 1970s the city was weakened by a sustained economic and urban crisis, leading to a demographic shrinkage, a rise of unemployment and an overall degradation and impoverishment, particularly of the central and northern districts of the city. Not only did Marseille lose many of the assets required to compete in attracting new businesses and firms, but it also lost its geographical and political status in the wider metropolitan area (Motte, 2003). For more than 20 years the city was unable to cope with overwhelming economic and social pressures, while surrounding smaller cities like Aix en Provence continued to grow. Rebranding and regenerating the city became a priority from the mid-1990s.

The creative system in Marseille is thus highly anchored in this very long trajectory of decline and subsequent progressive recovery. It also sits within an interplay of different governance levels, including the local authority, the EPAEM (the state-led organisation in charge of the 480 hectare Euromediterannee regeneration project), the département, the region and recently Marseille Provence 2013 (the organisation in charge of the 2013 European Capital of Culture Scheme). Finally, it also involves dealing with a range of multi-level challenges such as everyday deprivation issues at neighbourhood level, the regeneration of the city economy, and the rebranding and the competitive repositioning of Marseille within France and the Mediterranean area.

The urban planning and cultural subsystems are central in Marseille. Their intersection, through the built environment, has been framed and supported by the municipality since the 1970s, prior to any economic and large-scale urban strategies for recovery. Initially composed of arts and heritage facilities, it then progressively included a wide range of more bottom-led initiatives (all supported by the municipality in a fuzzy strategy of local cultural regeneration) in the 1980s and 1990s (Andres, 2011a, b), fostered by multiple derelict spaces available for temporary or more sustainable projects, as well as more recently, from the 2000s, major cultural events. Four pillars form the intersection of the cultural and planning subsystems: (1) they rely on traditional art and heritage facilities supported by the municipality; (2) they include formal organic projects that acquired flagship roles in the cultural landscape (e.g. La Friche, La Cité des Arts de la Rue); (3) they sit within an active socio-cultural strategy including a wide range of local projects and initiatives (supported by the municipality) aiming to connect communities and to foster their integration; and (4) in the last five years, they have comprised the ECOC 2013 (i.e. a range of initiatives and institutions) which in addition to fostering Marseille's rebranding strategy is highly anchored in its regeneration programme.

The intersection of the cultural and planning subsystems thus relies on the industrial morphology of the city, which created a range of derelict and underused factories and neighbourhood spaces. These spaces became the locations of temporary or long-term artistic and cultural uses. The planning subsystem is thus directly connected to two main public-led policies: the Euroméditerranée programme, which considers culture and creative industries as a way to diversify the economy and to change Marseille's image, and the urban and socio-cultural policy, which uses cultural and creative activities as a way to foster community engagement and local urban renaissance.

In contrast, the creative and cultural economic subsystem has a minor role in Marseille and mainly focuses on audio-visual, cinematographic and multimedia activities. The sector of the creative industries has only been identified as a sector of investment by the ETAP Public Amenagement Euromediterranee (EPAEM) on the one hand, and the Department and Region on the other hand, since the 2000s. Cinema and television, and to a lesser extent animation, video games and publishing, represent the sectors where an emergent local value chain of activities has been identified. Spatial clustering is however limited at the city level. Since most of these industries are at the interface between regional, national and international networks and markets, there is no need for them to be clustered in one district.

To summarise, Marseille's creative system is mainly situated at the intersection of the cultural and planning subsystems and is highly anchored in the industrial and deprivation trajectory of the city. Overall, it is characterised by a lack of coordination between these two subsystems and the economic subsystem, leading to a chaotic and uncoordinated perception of the creative spaces in the city. Typically there is no cultural and creative quarter or cluster in Marseille; only a wide, diverse range of creative facilities and activities more or less connected to each other, outside of any spatial rationale. The multi-level involvement of a range of different actors has fostered this lack of articulation between the different components of the system and has resulted in a creative system that would deserve more strategic streams, including some spatial directives.

Singapore's Creative System

Part of various empires (including the British Empire) in the past, Singapore is today a city-state and island country with a population of 5.1 million (Singapore Department of Statistics, 2012). Since its independence in 1963, Singapore has displayed a very rapid economic development, first by attracting multinational manufacturing companies to a very favourable business environment (especially low labour costs), and then by investing in local firms and specialising in finance and services when attracting multinational companies became more difficult due to growing competition from neighbouring countries during the 1980s (Beng Huat, 2009). Since the 1990s, and in line with international trends, Singapore has placed increasing emphasis on developing its knowledge industries (international finance, biosciences and pharmaceuticals, and high technologies) and its creative industries (arts and culture – including heritage activities, media and design as defined above) while also implementing a strong branding strategy to support a flourishing tourism industry (Ooi, 2002a).

It is important to note that, while benefiting from being a parliamentary republic, Singapore is characterised by a 'paternalistic' government exercising wide and deep powers over the city-state's development, including a high degree of censorship (Tan, 2007). As for many aspects of Singapore, the creative system is strongly controlled by the government and oriented internationally. There are two main phases in the evolution of Singapore's creative system from the 1990s. First, the 1990s were characterised by interactions between the cultural, economic and planning subsystems around the development of Singapore as an international centre of cultural consumption. The second phase, from around 2000, is notable for a shift towards the development of Singapore as an international centre for both creative consumption and production. However, the development of Singapore as an international centre of creative production is hindered both by the paternalistic approach of the Singaporean government and also by too much focus on the planning and cultural subsystems.

Today, the cultural and planning subsystems play an important role in the Singapore creative system. Indeed, in the 1990s, following the European trend of using culture as a motor of urban development and regeneration (Kong et al., 2006), the government decided to establish 'Singapore as "a global city for the arts" and to use culture as an anchor for building national identity' (Yun, 2008, p. 325). This objective sought to counteract a potential decline in international tourism due to an increased homogenisation in the city's architectural landscape and the need to develop its cultural offering (Beng Huat, 2009). The government therefore prepared a state plan with proposals to enhance the interest of Singaporeans in culture through educational programmes, to remedy the lack of cultural infrastructure, to increase cultural funding, and to develop and promote artistic talent (Yun, 2008). In addition, more attention was dedicated to urban conservation. These objectives were supported by adjustments to cultural public governance that involved the creation of a plethora of new cultural institutions (Beng Huat, 2009). As a result, new mega-cultural infrastructures were built, such as the new Singapore Art Museum, the Asian Civilisation Museum and the Esplanade – a concert and theatre hall (Chang, 2000). A small programme to subsidise spaces for artists and art organisations (ibid.) was also implemented. Singapore also developed, in parallel, a strong branding strategy, marketing it as the 'New Asia'[7] with the objective of 'bringing Singapore to the World and bringing the World to Singapore' (ibid.; Ooi, 2002a). This strategy helped to bring various international cultural companies (i.e. Cirque du Soleil) and stars (i.e. Michael Jackson) to perform in the city. As such, the Singaporean creative system in the 1990s was strongly dominated by the cultural and planning subsystems, with the aims of nurturing the wider local economic system and developing the city as an international centre of cultural consumption. This strategy changed from 2000 onwards, with some increasing interactions between the cultural and economic subsystems to support Singaporean creative productions.

This also marked a realignment of the strategic development of Singapore through fostering creativity and cultural production and the forging of greater links between its cultural and economic subsystems (Leo and Lee, 2004; Ooi, 2006). In doing so, Singapore followed the example of Australia and the UK in its efforts to develop as a creative hub, a strategy that was supported by three key initiatives (Entrepreneurs' Resources Centre, 2002; Lee, 2006; Monocle, 2011): (1) a limited rebranding of Singapore as 'a highly innovative and multi-talented Global City for Arts and Culture'; (2) its development as 'a hub of multimedia design capabilities', notably by injecting more funding into design education and supporting the use of design as a strategic tool by businesses; and (3) the building of a media district bringing together the media sector with the arts and heritage scene and info-communication technology in a 'work, live, play and learn' environment that supports experimentation and multidisciplinary cross-pollination (Entrepreneurs' Resources Centre, 2002). In addition, the government put more effort into supporting the development of local demand for creative outputs (Ooi, 2006). Despite the clear step towards developing Singaporean creativity, innovation and creative production, some authors argue that the censorship and strong control that the Singaporean government exercises on artistic and media outputs may actually restrict the degree of local creativity (Leo and Lee, 2004). In addition, many local creative productions still have difficulties finding a market within the Singaporean population, as much of the policy focus has been on developing mega cultural infrastructure for the international market and not enough on supporting local demand (Chang and Lee, 2003).

Shanghai's Creative System

Shanghai is the largest city in China, with a population of over 23 million inhabitants. The city has undergone a very rapid transformation and economic restructuring during the last 20 years as it has been opened up to international market forces (Daniels et al., 2012). Whereas this shift has led to the rise of the financial and commercial sectors, it has also fostered the decline of the traditional manufacturing sector. Industries mostly located in the cramped inner city were either closed down or relocated to the suburbs (Shanghai Economic Commission and Shanghai Communist Party History Research Office, 2002) to provide a cleaner and more attractive environment for foreign investment and tertiary industries (Zhong, 2012).

The rapid economic and urban development of Shanghai under the auspices of gaining and fostering its profile as a global city has been coupled with the fast development of the creative industries since 2004 (Zheng, 2010). The interest in the creative sector has incorporated a number of different features: (1) it relates to the art market and has led to the trendy transformation of former industrial buildings; (2) it targets fast-growing creative industries, particularly within the areas of entertainment and lifestyle; and (3) it also includes high-profile flagship projects such as the annual International Creative Industry Week that aims to 'forge a brand' for Shanghai's creative industries (Shanghai Creative Industry Center, 2006a; Zheng, 2011) while raising the international image of the city.

The Shanghai creative system is therefore a combination of global creative discourse and indigenous creative industry development (see Keane, 2007; Zhong, 2009). As such, the economic, planning and cultural subsystems are very much intertwined. As noted by Zheng (2010), three main public actors are driving the entrepreneurial development strategy for the creative sector in addition to the economic and urban growth of the city and its districts: the street offices, the district governments and the municipal government. With each possessing different competences, they have a crucial role in promoting places and guiding urban development. These public actors are closely twinned with private stakeholders (developers, investors, businesses and cultural/creative actors).

Shanghai economic and cultural subsystems emerged with the organic artistic clustering that took place in the late 1990s in informally rented out former industrial warehouses. This phenomenon relied on a shift in the national perception of art and culture but it did not involve a policy or strategy of socio-cultural development. As noted by Gu (2012, p.197), 'the state divested itself of much direct responsibility for art and cultural employment, and many artists were forced to make a living on the (barely emergent) market. This encouraged the commercialisation of art and culture because the barriers between the artists and the market had finally been lifted'. Within two years, these artists had transformed these warehouses into model 'art clusters' (ibid.) while accessing the international art market. However, market mechanisms also led to their displacement, the alliance of local government and property developers being the main urban transformative force in Shanghai, as in urban China generally (Wu and Yeh, 2007). In the mid-2000s, these two subsystems shifted to a very entrepreneurial strategy (Zheng, 2011) based on the municipal perception of the CCIs' potential for wealth generation and city branding. In 2005, the Shanghai Economic Commission and Shanghai Statistical Bureau targeted five creative sectors (Zhong, 2012, p.172): (1) research, development and design; (2) architectural and related design; (3) cultural activities, creation and media;

(4) consultancy and planning; and (5) fashion, leisure and lifestyle services. The municipality also labelled 75 creative industry clusters (CCJQ) with the aim of attracting further international investment and business as well as promoting Shanghai's creative image. Aimed at creating fashionable places of consumption and leisure, these districts include large office buildings as well as shops, restaurants and retail activities; they provide a comfortable work and entertainment environment tailored to the tastes of the professional white-collar sector (Zheng, 2011).

The morphology of Shanghai city centre, the municipality's aim of promoting urban growth and the achievement of global city status are the pillars of the cultural and planning subsystems, though also eminently connected to the economic subsystem. In the late 1990s, the rapid development of the real estate sector and an influx of migrants resulted in an increased demand for affordable spaces in the heart of the city for both residential and commercial uses (Sun, 2006, p. 190). The creative industry clusters initiative (see above) in addition to promoting economic development has thus clearly been shaped by urban growth and a real estate boom (Zheng, 2010). Creative development has therefore very much been a feature of the urban development strategy of the city. The combination of artistic temporary uses and creative industry districts has impacted the real estate market and particularly international property development. The re-use and preservation of historic buildings aiming to promote a new architectural design style (Zheng, 2011) has participated in branding the city for business and tourism.

To summarise, it appears that the creative industry clusters (CCJQ) are the main features of Shanghai's creative system. They are characterised by a very strong interweaving between the urban planning, cultural and economic subsystems (though the cultural subsystem has had a minor role over time). Despite fostering the creative image of the city and impacting real estate dynamics (including gentrification) and foreign investment, these clusters have also led to its fragmentation (Wu, 2000); they appear as physical 'containers' rather than proactive actors, and do not serve to foster creative industries (Zheng, 2011).

CONCLUSIONS

Using a more integrated analysis, the notion of a creative system has been explored with reference to four distinctive cities across Europe, North America and Asia. This chapter has highlighted a number of interesting aspects about the development of CCIs over time and space.

First, in all four cities, the role of local authorities in framing the nature of creative systems, at different stages, has been crucial: dirigiste and focused on cultural flagships in Singapore; entrepreneurial and public–private led in Shanghai; concerted/collaborative around the development of creative clusters in Montreal; and renaissance led as well as mainly public led in Marseille, where the strategy has evolved over the years particularly towards the inclusion of a wider range of private actors, from artists to businesses. Only Singapore, in this regard, has kept a more limited public–private partnership framework that is a product of its paternalistic regime. These different governance processes have influenced the role of each subsystem and the nature of their interactions. It is notable that a sole focus on the cultural system in its social inclusion and socio-cultural development dimensions has been relatively limited, except in Marseille. In general, Shanghai, Singapore and Montreal, and to a lesser extent since 2000 Marseille, have put a lot of emphasis on the intersections of their economic and cultural subsystems in their development strategies. This has been reflected in a set of programmes and initiatives to develop CCIs while fostering adjacent economic activities as well as branding the cities for tourism and making them more competitive (at the national and international levels). Most cities have also exploited the intersections of the planning and cultural subsystems through urban development or regeneration strategies. In Marseille and Shanghai, the industrial nature of some districts has been crucial for, first, attracting artists and, second, branding specific neighbourhoods and the city as a whole. In Singapore, the focus has been more on developing cultural flagships as part of an international cultural tourism branding strategy, neglecting some of the local cultural and creative organic productions. Montreal, however, has in the last 10 years been increasingly committed to fostering the intersections between its cultural, planning and economic subsystems through the development of specific cultural and creative quarters and clusters, the latter with a strong production component. As a consequence, Montreal's creative system appears overall as the most balanced among our four examples, by playing on the wide range of interactions of its three subsystems. Though intertwined, Shanghai's creative system is dominated by the economic and cultural subsystems while, by way of contrast, Singapore has privileged the intersections of either the cultural and economic subsystems or the cultural and planning subsystems. Finally, Marseille has for a long time put a strong emphasis on the intersections of the cultural and planning subsystems. These less balanced approaches have resulted in a set of limitations: a lack of coherent strategy and framework (Marseille), limited benefits received by the CCIs and accrued gentrification processes (Shanghai), and difficulty taking into account local production and responding to the needs of local demand (Singapore).

Second, a consideration of both the geographical and temporal levels inherent in the creative system approach has helped us to uncover interesting findings with regard to the changing nature of creative systems across cities and continents. Whereas the examples of Montreal and Marseille have stressed long and changing trajectories in the formation of their creative systems, the Asian examples have revealed how quickly these systems could be framed and shifted through strong policy interventions, but also the limits of these interventions if CCIs' private initiatives are not allowed to flourish (as in Singapore). In addition, the temporal dimension of our analysis has shown that some subsystems play a more predominant role during specific periods of time. In all our cases, the articulation of different geographical levels (neighbourhood, city, regional, national and international) has been fundamental for trying to grasp the complexity of the creative systems in place. Typically, in Shanghai, the creative system is entirely positioned within the global strategy of the city, with key objectives such as reaching the international art market by developing creative districts to attract international firms and businesses, fostering international tourism and supporting international real estate investment. In this overall strategy, the nature and quality of place play an important role as part of branding the city and its design style. In Marseille, the 20-year strategy of regaining an economic status and of competitive repositioning in France and in the Mediterranean area has conditioned the way the planning and cultural subsystems have been used over time. Nevertheless, the lack of policy with spatial attributes has led to visibly weak economic and cultural subsystems in the absence of any cultural or creative districts across the city. Because it is a city-state, Singapore operates with a lower degree of separation between the local/national and the international levels but it has embraced this by putting in place a strong internationalisation strategy, sometimes to the detriment of its local CCIs' production and consumption. Finally, due to the distribution of administrative competences in the Canadian federal system, the strong Anglo-French cultural history of Quebec and its strong ethnic diversity, Montreal has a very particular creative system in terms of the number of local, regional, provincial and federal public actors supporting CCIs, the strength of the synergies between private and public actors and the mix of CCIs existing in the city. By building upon this, Montreal has managed to position itself in Quebec, Canada, North America and the rest of the world as a centre of various types of CCIs' production and consumption branded into very distinctive geographical local creative clusters and districts.

Finally, reflecting on the forms of policy transfers and the internationalisation of the CCIs' discourse, we can argue that our Western examples display distinctive features compared with our Eastern Asian examples. Montreal has recognised the potential of this discourse for its development and has used it to very good effect for building up its various cultural, economic and planning subsystems. Marseille, in contrast, with its late assessment of its CCIs' assets, has not yet managed to draw lessons from other Western experiences. This is partly explained by the difficulty of penetrating the spheres of non-Anglophone political cultures with UK, Australian and American discourses. This is less the case in Asia. Due to its language and cultural proximity, Singapore adopted the UK CCIs' discourse at a very early stage but, due to its political system, it has had difficulty in acknowledging the need to nurture local organic CCIs' production and consumption. Finally, Shanghai has very quickly embraced the benefits of artistic colonisation, followed by creative development within its entrepreneurial strategy of internationalisation It can be considered a very accomplished example of understanding and acknowledging the different development stages of cultural and creative systems as experienced in the West, then re-interpreting them to meet the requirements of its city global urban and economic development strategy.

As a final thought, though this concept of creative systems fills a set of gaps by providing an integrated analysis of the different dimensions of cultural and creative industries both in the West and in the East, it also identifies directions for further research. Even if the creative systems approach better captures the complexity of CCIs, some aspects of these industries are still barely explored. In a complex value chain system, some creative activities are not categorised as creative if delivered by companies whose main activities are outside the creative sector. This field is relatively under-developed in the literature and deserves more research to explore network and cooperation dynamics as well as the skills or educational implications. In addition, the creative system is strongly connected to organic or grassroots initiatives. These include a range of informal activities and processes which are difficult to grasp and assess as they are hidden within the complexity of everyday practices and liminal spaces. These informal activities and processes do contribute to shaping places and spaces and, sometimes, in informing local cultural or economic policy. The informal aspects of CCIs are difficult to access, but further research is required to explore the complex interplay between space, place, creativity and informal practices and processes.

NOTES

1. The term 'copyright industries' is also used at the international level and tends to include a mix of creative industries, heritage and other activities (toys, musical instruments...) divided between categories such as core, partial and interdependent copyright industries (see UNCTAD/DICT (2008) for a detailed list of industries included).
2. See http://www.quartierdesspectacles.com/ (accessed 3 December 2014).
3. See http://www.quartierinnovationmontreal.com/page2longueEn.php (accessed 3 December 2014).
4. See http://montreal2025.com/montreal_creative.php?lang=fr (accessed 3 December 2014).
5. See http://montrealcreative.org/en/ (accessed 3 December 2014).
6. See http://mtlunescodesign.com/en/about (accessed 3 December 2014).
7. The government conceived New Asia – Singapore as 'not a product that one consciously creates. It is the sum total of the way we live, work and think. The products are an expression of all that' (Singapore Tourism Promotion Board, 1997, cited in Ooi, 2002b, p. 247).

BIBLIOGRAPHY

Allen, John (1999), 'Worlds within cities', in Doreen Massey, John Allen and Steve Pile (eds), *City Worlds*, Routledge: London, pp. 53–97.

Allen, John (2003), *Lost Geographies of Power*, Blackwell Publishing: Oxford.

Andres, L. (2011a), 'Alternative initiatives, cultural intermediaries and urban regeneration: the case of La Friche (Marseille)', *European Planning Studies*, 19 (5), 795–811.

Andres, Lauren (2011b), 'Marseille 2013 or the final round of a long and complex regeneration strategy?', *Town Planning Review*, 82 (1), 61–76.

Andres, Lauren and Boris Grésillon (2013), 'Cultural brownfields in European cities: a new mainstream object for cultural and urban policies', *International Journal of Cultural Policy*, 19 (1), 40–62.

Bakshi, Hasan, Eric McVittie and James Simmie (2008), *Creating Innovation: Do the Creative Industries Support Innovation in the Wider Economy?*, NESTA: London.

Beinhocker, Eric D. (2006), *The Origin of Wealth: Evolution, Complexity and the Radical Remaking of Economics*, Random House: London.

Beng Huat, Chua (2009), 'Eclipse of the port: cultural industries and the new phase of economic development of Singapore', in Arndt Graf and Chua Beng Huat (eds), *Port Cities in Asia and Europe*, Routledge: London and New York, pp. 190–204.

Bennett, Tony and John Frow (eds) (2008), *Handbook of Social and Cultural Analysis*, Sage: London.

Bianchini, Franco (1999), 'Cultural planning for urban sustainability', in Louise Nyström and Colin Fudge (eds), *City and Culture: Cultural Processes and Urban Sustainability*, The Swedish Urban Environment Council: Stockholm, pp. 34–51.

Bianchini, Franco and Michael Parkinson (eds) (1993), *Cultural Policy and Urban Regeneration: The West European Experience*, Manchester University Press: Manchester.

Bianchini, Franco, Mark Fischer, John Montgomery and Ken Worpole (1988), *City Centres, City Cultures: The Role of the Arts in the Revitalisation of Towns and Cities*, The Centre for Local Economic Development Strategies: Manchester.

Bontje, Marco, Sako Musterd and Peter Pelzer (2011), *Inventive City-Regions*, Ashgate: London.

Brown, A., J. O'Connor and S. Cohen (2000), 'Local music policies within a global music industry: cultural quarters in Manchester and Sheffield', *Geoforum*, 31, 437–451.

Byrne, David (1998), *Complexity Theory and the Social Sciences*, Routledge: London.

Byrne, David (2003), 'Complexity theory and planning theory: a necessary encounter', *Planning Theory*, 2 (3), 171–178.

Castells, Manuel (1989), *The Informational City*, Blackwell: Oxford.

Castells, Manuel (1996), *The Rise of the Network Society*, Blackwell: Oxford.

Caves, Richard E. (2000), *Creative Industries: Contracts between Art and Commerce*, Harvard University Press: Cambridge, MA.

Centre for Cultural Policy Research (2003), *Baseline Study on Hong Kong's Creative Industries*, University of Hong Kong: Hong Kong.

Chang, T.C. (2000), 'Renaissance revisited: Sinagapore as a global city for the arts', *International Journal of Urban and Regional Research*, 24 (4), 818–831.

Chang, T.C. and W.K. Lee (2003), 'Renaissance city Singapore: a study of arts spaces', *Area*, 35 (2), 128–141.

Chapain, Caroline and Roberta Comunian (2010), 'Enabling and inhibiting the creative economy: the role of the local and regional dimensions in England', *Regional Studies*, 44 (6), 717–734.
Chapain, Caroline and Peter Lee (2009), 'Can we plan the creative knowledge city? Perspectives from Western and Eastern Europe', *Built Environment*, 35 (2), 157–164.
Chapain, Caroline, Nick Clifton and Roberta Comunian (2013), 'Understanding creative regions: bridging the gap between global discourses and regional and national contexts', *Regional Studies*, 47 (2), 131–134.
Cinti, Tomaso (2008), 'Cultural clusters and districts: the state of the arts', in Phil Cooke and Luciana Lazzeretti (eds), *Creative Cities, Cultural Clusters and Local Economic Development*, Edward Elgar: Cheltenham and Northampton, MA, pp. 70–92.
Coish, David (2004), *Tendances et conditions dans les regions métropolitaines de recensement grappes culturelles*, document analytique, No 89-613-MIF au catalogue, No 004,Statistics Canada: Ottawa.
Cooke, Phil and Luciana Lazaretti (eds) (2008), *Creative Cities, Cultural Clusters and Local Economic Development*, Edward Elgar: Cheltenham and Northampton, MA.
Council of the European Union (2007), *Contribution of the Cultural and Creative Sectors to the Achievement of the Lisbon Objectives – Adoption of the Council Conclusions, Introductory Note, Cult 29, 9021/07*, Council of European Union: Brussels.
Cunningham, S. (2002), 'From cultural to creative industries: theory, industry and policy implications', *Media International Australia Incorporating Culture and Policy: Quarterly Journal of Media Research and Resources*, 102, 54–65.
Daniels, P.W., K.C. Ho and, and T. Hutton (eds) (2012), *New Economic Spaces in Asian Cities: From Industrial Restructuring to the Cultural Turn*, Routledge: London.
Department for Media, Culture and Sport (1998), *The Creative Industries Mapping Document*, HMSO: London.
Department for Media, Culture and Sport (2000), *Creative Industries – the Regional Dimension*. Department for Culture, Media and Sport: London.
Deuze, Mark (2007), 'Convergence culture in the creative industries', *International Journal of Cultural Studies*, 10 (2), 243–263.
Drake, G. (2003), 'This place gives me space: place and creativity in the creative industries', *Geoforum*, 34, 511–524.
Drevo, Pauline (2010), *Le marketing territorial. Étude du cas de Montréal*. Bsc end of study project. Institut d'amenagement et d'urbanisme, University of Montreal: Montreal.
Entrepreneurs' Resources Centre (2002), *Creative Industries Development Strategy*, report by the ERC Service Industries Subcommittee Workgroup on Creative Industries. Available online at: http://app.mica.gov.sg/Portals/0/UNPAN011548.pdf (accessed 14 October 2012).
Evans, G. (2009), 'Creative cities, creative spaces and urban policy', *Urban Studies*, 46 (99), 1003–1040.
Flew, Terry and Stuart Cunningham (2010), 'Creative industries after the first decade of debate', *Information Society*, 26, 113–123.
Florida, Richard (2002), *The Rise of the Creative Class: And How It's Transforming Work, Leisure, Community and Everyday Life*, Basic Books: New York.
Florida, Richard (2005), *Cities and the Creative Class*, Routledge: New York.
Garcia, B. (2004), 'Cultural policy and urban regeneration in western European cities: lessons from experience, prospects for the future', *Local Economy*, 19 (4), 312–326.
Garnham, Nicolas (2005), 'From cultural to creative industries: an analysis of the implications of the "creative industries" approach to arts and media policy making in the United Kingdom', *International Journal of Cultural Policy*, 11 (1), 15–29.
Giddens, Anthony (1984), *The Constitution of Society*, Polity Press: Cambridge.
Ginsburgh, Victor A. and David Throsby (eds) (2006), *Handbook of the Economics of Art and Culture*, Vol. 1, Elsevier/North-Holland: Amsterdam.
Government du Québec (2005), *Chiffres à l'appui*. Available online at: http://collections.banq.qc.ca/ark:/52327/bs16518 (accessed 3 September 2012).
Government of the Hong Kong Special Administrative Region (2012), *The Cultural and Creative Industries in Hong Kong, 2005 to 2010*, Hong Kong Monthly Digest of Statistics: Hong Kong.
Graham, S. and P. Healey (1999), 'Relational concepts in time and space: issues for planning theory and practice', *European Planning Studies*, 7 (5), 623–646.
Gu, X. (2012), 'The art of re-industrialisation in Shanghai', *Culture Unbound: Journal of Current Cultural Research*, 4, 193–212.
Hartley, John (2005), 'Creative industries', in John Hartley (ed.), *Creative Industries*, Blackwell: Oxford, pp. 1–40.
Harvey, David (1996), *Justice, Nature and the Geography of Differences*, Blackwell: Oxford.
Healey, Patsy (1997), *Collaborative Planning: Shaping Places in Fragmented Societies*, Macmillan Press: Houndmills and London.
Healey, Patsy (2006), 'Relational complexity and the imaginative power of strategic spatial planning', *European Planning Studies*, 14 (4), 525–546.
Hesmondhalgh, David (2007), *The Cultural Industries*, 2nd edn, Sage: London.

Higgs, P., S. Cunningham and H. Bakshi (2008), *Beyond the Creative Industries: Mapping the Creative Economy in the United Kingdom*, NESTA Technical Report: London.
Jansson, Johan and Dominic Power (2010), 'Fashioning a global city: global city branch channels in the fashion and design industries', *Regional Studies*, 44 (7), 889–904.
Katiya, Yuseph A. (2011), *Creating Hegemony: Montreal's Cultural Development Policies and the Rise of Cultural Actors as Entrepreneurial Political Elites*, Msc. Thesis presented at the Department of Geography, Planning & Environment, Concordia University, Canada.
KEA (2006), *The Economy of Culture in Europe*, KEA European Affairs, European Commission: Brussels.
KEA (2009), *The Creative Economy in Europe*, European Commission: Brussels.
Keane, Michael (2007), *Created in China: The Great New Leap Forward*, Routledge: London.
Keane, Michael A. (2009), 'Creative industries in China: four perspectives on social transformation', *International Journal of Cultural Policy*, 15 (4), 431–443.
Kong, L. (2000), 'Culture, economy, policy: trends and developments', *Geoforum*, 31, 385–390.
Kong, L. (2009), 'The making of sustainable creative/cultural space: cultural indigeneity, social inclusion and environmental sustainability', *Geographical Review*, 99 (1), 1–22.
Kong, Lily and Justin O'Connor (eds) (2009), *Creative Economies, Creative Cities: Asian–European Perspectives*, Springer: Dordrecht.
Kong, Lily, Chris Gibson, Louisa-May Khoo and Anne-Louise Semple (2006), 'Knowledges of the creative economy: towards a relational geography of diffusion and adaptation in Asia', *Asia Pacific Viewpoint*, 47 (2), 173–194.
Landry, C., F. Bianchini, R. Ebert, F. Gnad and K. Kunzmann, (1996), *The Creative City in Britain and Germany*, Anglo-German Foundation: London.
Leclerc, Stephane (2000), *Les arts dans la métropole. Un portrait des organismes des arts et de la culture du Montreal métropolitain et de leurs besoins*, Gestion des Arts. Inc.: Mont Saint Hilaire, Quebec.
Lee, Terence (2006), 'Towards a "new equilibrium": the economics and politics of the creative industries in Singapore', *Copenhagen Journal of Asian Studies*, 24, 55–71.
Lefebvre, Henri (1995), *The Production of Space*, Blackwell: Oxford.
Legnér, Mattias and Davide Ponzini (eds) (2009), *Cultural Quarters and Urban Transformation: International Perspectives*, Gotlandica förlag: Visby.
Leo, Petrina and Terence Lee (2004), 'The "New" Singapore: mediating culture and creativity', *Continuum: Journal of Media & Cultural Studies*, 18 (2), 205–218.
Leriche, Frederic and Sylvie Daviet (2010), 'Cultural economy: an opportunity to boost employment and regional development', *Regional Studies*, 44 (7), 807–811.
Leslie, Deborah and Norma M. Rantisi (2006), 'Governing the design economy in Montréal, Canada', *Urban Affairs Review*, 41, 309–337.
Lloyd, Richard (2002), 'Neo-bohemia: art and neighborhood redevelopment in Chicago', *Journal of Urban Affairs*, 24 (5), 517–594.
Lloyd, Richard (2005), *Neo-Bohemia: Art and Commerce in the Postindustrial City*, Routledge: London.
Lloyd, Richard and Terry Nichols Clark (2001), 'The city as an entertainment machine', in Kevin F. Gotham (ed.), *Critical Perspectives on Urban Redevelopment: Research in Urban Sociology*, JAI Press/Elsevier: Oxford, pp. 357–378.
Lorenzen, Mark and Lars Frederiksen (2008), 'Why do cultural industries cluster? Localization, urbanisation, products and projects', in Phil Cooke and Luciana Lazzeretti (eds), *Creative Cities, Cultural Clusters and Local Economic Development*, Edward Elgar: Cheltenham and Northampton, MA, pp. 155–179.
Manson, S. (2001), 'Simplifying complexity: a review of complexity theory', *Geoforum*, 32, 405–414.
Markusen, A. (2010), 'Organisational complexity in the regional cultural economy', *Regional Studies*, 44 (7), 813–828.
Markusen, A. and A. Gadwa (2010), 'Arts and culture in urban or regional planning: a review and research agenda', *Journal of Planning Education and Research*, 29 (3), 379–391.
Markusen, A. and D. King (2003), *The Artistic Dividend: The Arts' Hidden Contributions to Regional Development*, University of Minnesota: Minneapolis.
Markusen, A. and G. Schrock (2006), 'The artistic dividend: urban artistic specialization and economic development implication', *Urban Studies*, 43 (10), 1661–1686.
Martin, R. and P. Sunley (2007), 'Complexity thinking and evolutionary economic geography', *Journal of Economic Geography*, 7, 573–601.
Miles, S. and R. Paddison (2005), 'Introduction: the rise and rise of culture-led urban regeneration', *Urban Studies*, 42(5/6), 833–839.
Ministère de la Culture et des Communications (2000), *Les politiques culturelles au Québec. Synthèse d'une étude*. Ministère de la Culture et des Communications: Québec.
Mok, Patrick (2009), 'Asian cities and limits to creative capital theory', in Lily Kong and Justin O'Connor (eds), *Creative Economies, Creative Cities: Asian-European Perspectives*, Springer: London, pp. 135–150.

Mommaas, H. (2004), 'Cultural clusters and the post-industrial city: towards the remapping of urban cultural policy', *Urban Studies*, 41 (3), 507–532.
Monocle (2011) *A Guide to Singapore Creative Industries*. Available online at: http://edbsingapore.jp/etc/medialib/futureready/frs/december_updates.Par.23037.File.tmp/Monocle_Series_Creative_Services.pdf (accessed 14 October 2012).
Motte, Alain (2003), 'Marseilles-Aix metropolitan region (1981–2000)', in Anton Salet, Willem Thornley and Andy Kreukels (eds), *Metropolitan Governance and Spatial Planning*, Routledge: London, pp. 320–336.
Musterd, Sako and Alan Murie (2010), *Making Competitive Cities*, Blackwell: Oxford.
Neff, Gina (2005), 'The changing place of cultural production: locating social networks in a digital media industry', *Annals of the American Academy of Political and Social Science*, 597 (1), 134–152.
Nielsén, T. and D. Power (2010), *Priority Sector Report: Creative and Cultural Industries*. European Commission, Europe Innova-European Cluster Observatory: Brussels.
O'Connor, Justin (2010), *The Cultural and Creative Industries: A Literature Review*. 2nd ed. Creativity, Culture and Education: Newcastle. Available online at: http://www.creativitycultureeducation.org/wp-content/uploads/CCE-lit-review-creative-cultural-industries-257.pdf (accessed 24 August 2012).
Ooi, Can-Seng, (2002a), 'Brand Singapore: the hub of "New Asia"', in Nigel Morgan, Annette Pritchard and Roger Pride (eds), *Destination Branding: Creating the Unique Destination Proposition*, Elsevier: Oxford and Burlington, pp. 242–260.
Ooi, Can-Seng (2002b), *Cultural Tourism and Tourism Cultures: The Business of Mediating Experiences in Copenhagen and Singapore*, Copenhagen Business School Press: Copenhagen.
Ooi, Can-Seng (2006), *Bounded Creativity and the Push for the Creative Economy in Singapore*, paper presented at the 16th Biennial Conference of the Asian Studies Association of Australia, Wollongong, Australia, 26–29 June.
Peck, J. (2005), 'Struggling with the creative class', *International Journal of Urban and Regional Research*, 29 (4), 740–770.
Polese, Mario (2009), *Montréal économique: de 1930 à nos jours. Récit d'une transition inachevée*, Inedit/Working Paper, N.2009-06, Centre – Urbanisation Culture Société, Insitut National de la Recherche Scientifique: Montreal.
Portugali, J. (2006), 'Complexity theory as a link between space and place', *Environment and Planning A*, 38, 647–664.
Potts, Jason, Stuart Cunningham, John Hartley and Paul Ormerod (2008), 'Social network markets: a new definition of the creative industries', *Journal of Cultural Economics*, 32 (3), 167–185.
Pratt, A.C. (1997), 'The cultural industries production system: a case study of employment change in Britain 1984–91', *Environment and Planning A*, 29, 1953–1974.
Pratt, A.C. (2000), 'New media, the new economy and new spaces', *Geoforum*, 31, 425–436.
Pratt, A.C. (2002), 'Hot jobs in cool places: the material cultures of new media product spaces: the case of the south of the market, San Francisco', *Information, Communication and Society*, 5, 27–50.
Pruijt, H. (2003), 'Is the institutionalization of urban movements inevitable? A comparison of the opportunities for sustained squatting in New York City and Amsterdam', *International Journal of Urban and Regional Research*, 27 (1), 133–157.
Rantisi, Norma M. (2012), 'Fashioning the fur industry in Montreal: forging a new production paradigm or recasting the old?', *International Journal of Knowledge, Culture and Change Management*, 11 (1), 1–10.
Rantisi, Norma M. and Deborah Leslie (2010), 'Creativity by design? The role of informal spaces in creative production', in Tim Edensor, Deborah Leslie, Steve Millington and Norma M. Rantisi (eds), *Spaces of Vernacular Creativity: Rethinking the Cultural Economy*, Routledge: New York, pp. 33–45.
Roling, Robert W. (2010), 'Small town, big campaign: the rise and growth of an international advertising industry in Amsterdam', *Regional Studies*, 44 (7), 829–844.
Rose, Owen (2007), 'Faubourg des Récollets, Griffintown, Cité du multimedia de Montréal', *Montreal Architecture*, 6. Available online at: http://www.urbanphoto.net/blog/2007/11/18/montreal-architecture-no6/ (accessed 3 September 2012).
Scott, Alan J. (2000), *The Cultural Economies of Cities: Essays on the Geography of Image-Producing Industries*, Sage Publications: London.
Scott, Alan J. (2004), 'Cultural-products industries and urban economic development: prospects for growth and market contestation in global context', *Urban Affairs Review*, 39, 461–490.
Scott, Alan J. (2006), 'Creative cities: conceptual issues and policy questions', *Journal of Urban Affairs*, 28 (1), 1–17.
Shanghai Creative Industry Center (2006), *Creativity, Innovation, Invention*, SCIC: Shanghai.
Shanghai Economic Commission and Shanghai Communist Party History Research Office (2002), *Shanghai Gongye Jiegou Tiaozheng [Industrial Restructuring of Shanghai]*, Shanghai People's Press: Shanghai.
Shaw, K. (2005), 'The place of alternative culture and the politics of its protection in Berlin, Amsterdam and Melbourne', *Planning Theory and Practice*, 6 (7), 149–169.

Singapore Department of Statistics (2012), 'Key demographics indicators, 1970–2011', *Population Trends 2011*, Singapore Department of Statistics: Singapore.

Smith, Richard and Katie Warfield (2008), 'The creative city: a matter of values', in Phil Cooke and Luciana Lazzeretti (eds), *Creative Cities, Cultural Clusters and Local Economic Development*, Edward Elgar: Cheltenham and Northampton, MA, pp. 287–312.

Storper, Michael (1997), *The Regional World: Territorial Development in a Global Economy*, The Guilford Press: New York.

Sun, Z. (2006), 'The real estate market in Shanghai after the macro control'. Available at: http://law.eastday.com/node2/node22/pjzh/node1713/node1714/userobject1ai7791.html (accessed 3 December 2014).

Tan, Kenneth P. (2007), 'In renaissance Singapore', in Kenneth P. Tan (ed.), *Renaissance Singapore? Economy, Culture, and Politics*, NUS Press: Singapore, pp. 1–14.

Throsby, David (2001), *Economics and Culture*, Cambridge University Press: Cambridge.

Throsby, David (2008), 'Modelling the cultural industries', *International Journal of Cultural Policy*, 14 (3), 217–232.

Tourisme Montreal (2011), *Bilan touristique de l'été 2011 à Montréal*, Tourisme Montreal: Montreal.

Towse, Ruth (2011), *A Handbook of Cultural Economics*, Edward Elgar: Cheltenham and Northampton, MA.

Tremblay, D.-G. and E. Cecili (2009), *Film and Audiovisual Production in Montreal: Challenges of Relational Proximity for the Development of a Creative Cluster*, Presentation at the 6th Proximity Days. 14–16 October, Poitiers.

Tremblay, Diane-Gabrielle, C. Chevrier and Serge Rousseau (2004), 'The Montreal multimedia sector: district, cluster or localized system of production?', in David Wolfe and Matthew Lucas (eds), *Clusters in a Cold Climate: Innovation Dynamics in a Diverse Economy*, McGill-Queen's University Press: Montreal and Kingston, pp. 165–194.

UNCTAD (2004), *Creative Industries and Development*, Eleventh Session, Sao Paulo, 13–18 June, United Nations Conference on Trade and Development: Sao Paulo.

UNCTAD/DITC (2008), *Creative Economy Report, 2008: The Challenge of Assessing the Creative Economy*, UNCTAD and DITC: New York. Available online at http://unctad.org/en/Docs/ditc20082cer_en.pdf (accessed 3 December 2014).

UNDP/UNCTAD (2010), *Creative Economy Report 2010*, UNDP, UNCTAD: Geneva and New York.

UNESCO (2006), *Understanding Creative Industries: Cultural Statistics for Public-Policy Making*, UNESCO, Global Alliance for Cultural Diversity. Available online at: http://portal.unesco.org/culture/en/ev.php-URL_ID=29947&URL_DO=DO_TOPIC&URL_SECTION=-465.html (accessed 14 October 2012).

UNESCO (2009), *The 2009 UNESCO Framework for Cultural Statistics*, UNESCO Institute for Statistics: Montreal.

United Nations (2004), *Creative Industries and Development*, presented at the 11th session of the United Nations Conference on Trade and Development, Sao Paulo, Brazil, 13–18 June.

Vitrine Culturelle de Montreal (2003), *La culture en ville, accessible!* Plan d'affaires. Résumé. Vitrine Culturelle de Montreal: Montreal.

Wiesand, Andreas and Michael Söndermann (2005), *The 'Creative Sector'– an Engine for Diversity, Growth and Jobs in Europe*, European Cultural Foundation: Amsterdam.

Wu, Fulong (2000), 'The global and local dimensions of place-making: remaking Shanghai as a world city', *Urban Studies*, 37 (8), 1359–1377.

Wu, Fulong, Jiang Xu and Anthony Gar-On Yeh (2007), *Urban Development in Post-reform China: State, Market and Space*, Routledge: London.

Yun, Hing A. (2008), 'Evolving Singapore: the creative city', in Phil Cooke and Luciana Lazzeretti (eds), *Creative Cities, Cultural Clusters and Local Economic Development*, Edward Elgar: Cheltenham and Northampton, MA, pp. 313–337.

Zheng, J. (2010), 'The "entrepreneurial state" in "creative industry cluster" development in Shanghai', *Journal of Urban Affairs*, 32 (2), 143–170.

Zheng, J. (2011), 'Creative industry clusters and the entrepreneurial city of Shanghai', *Urban Studies*, published online before print, 22 June 2011, doi: 10.1177/0042098011399593.

Zhong, S. (2009), 'From fabrics to fine arts: urban restructuring and the formation of an art district in Shanghai', *Critical Planning*, 16, 118–137.

Zhong, S. (2012), 'Production, creative firms and urban space in Shanghai', *Culture Unbound: Journal of Current Cultural Research*, 4, 169–192.

Zukin, Sharon (1988), *Loft Living: Culture and Capital in Urban Change*, Radius: London.

Zukin, Sharon (1995), *The Culture of Cities*, Blackwell: Cambridge.

20. Tourism services: a sustainable service business?
C. Michael Hall

INTRODUCTION

Tourism is an extremely significant global economic activity. Although often perceived by the wider public as consisting of 'holidaymaking' and 'leisure', tourism is defined in terms of consumption with respect to the voluntary temporary mobility of individuals, while tourism production is configured around the firms and services that support such mobility. The tourist product, therefore, has a number of characteristics that often make comparisons with other industries difficult. For example, there is no standard industry classification for tourism. This means that assessments of its economic and employment impact have to be based on satellite accounting systems, on the amalgamation of data for related sectors for which information does exist (such as hotels, cafes and restaurants, and/or transportation and travel services) or on the results of tourism-specific surveys and other studies. Much of the latter, which is usually undertaken for government or tourism industry stakeholders for validating specific policies and development strategies, has tended to exaggerate the economic benefits of tourism (Hall, 2008a). Furthermore, the tourism product is multi-layered – that is, ranging from individual service encounters through to firm, trip and destination products – and while lay understandings of tourism products are often geared towards destinations, the agencies that market and 'manage' destination areas rarely actually own and control the product that they are promoting. There are therefore significant challenges for understanding tourism in a service context, especially when viewed through different disciplinary lenses.

FRAMING TOURISM

Tourism is a *slippery* (Wincott, 2003) and *fuzzy* concept (Markusen, 1999). It is seemingly easy to visualize yet difficult to define with precision because its meaning changes depending on the context of its analysis, purpose and use (Hall and Lew, 2009). In distinguishing tourism from other types of human activity several factors are significant (Hall, 2005a, 2005b):

- Tourism is voluntary and does not include the forced movement of people for political or environmental reasons; tourists are not refugees.
- Tourism can be distinguished from migration because a tourist is making a return trip from their home environment while the migrant is moving permanently away from what was their home environment.
- The distinction between tourism and migration sometimes becomes blurred because some people engage in return travel away from their usual home environment for an extended period, that is, a 'gap year' or 'working holidays', in another

country. In these circumstances, time (how long they are away from their normal or permanent place of residence) and distance (how far they have travelled or whether they have crossed jurisdictional borders) become determining factors in defining tourism statistically and distinguishing between migration and tourism.

Using the above approach there are a number of forms of voluntary travel that serve to constitute tourism conceptually and statistically:

1. Visiting friends and relations.
2. Business travel.
3. Travel to second homes.
4. Health- and medical-related travel.
5. Education-related travel.
6. Religious travel and pilgrimage.
7. Travel for shopping and retail.
8. Volunteer tourism.

Confusion over the definition of tourism does not end here. The word 'tourism' is also used to describe *tourists* (the people who engage in voluntary return mobility), the notion of a *tourism industry* (which is the term used to describe those firms, organizations and individuals that enable tourists to travel), and the whole social and economic phenomenon of tourism, including tourists, the tourism industry and the people and places that comprise tourism destinations, as well as the impacts tourism has on generating areas, transit zones and destinations; what is usually termed the 'tourism system' (Hall and Lew, 2009).

Based on generally accepted international agreements for collecting and comparing tourism and travel statistics, the term 'tourism trip' refers to a trip of not more than 12 months and for a main purpose other than being employed at the destination (United Nations (UN) and United Nations World Tourism Organization (UNWTO), 2007). However, despite UN and UNWTO recommendations, there are substantial differences between countries with respect to the length of time used to define a tourist, as well as how employment is defined (Lennon, 2003; Hall and Page, 2006). Nonetheless, three types of tourism are usually recognized:

1. *Domestic tourism*, which includes the activities of resident visitors within their home country or economy of reference, either as part of a domestic or an international trip.
2. *Inbound tourism*, which includes the activities of non-resident visitors within the destination country or economy of reference, either as part of a domestic or an international trip (from the perspective of the traveller's country of residence).
3. *Outbound tourism*, which includes the activities of resident visitors outside their home country or economy of reference, either as part of a domestic or an international trip.

In order to improve both statistical collection and the understanding of tourism, the UN and the UNWTO have long recommended differentiating between visitors, tourists and excursionists (Hall and Lew, 2009). For example, the UNWTO (1991, Recommendation No. 29) recommended that an international tourist be defined as: 'a visitor who travels to a country other than that in which he or she has his or her usual residence for at least one night but not more than one year, and whose main purpose of visit is other than the exercise of an activity remunerated from within the country visited'; and that an international excursionist (also referred to as a daytripper) (e.g. a cruise-ship visitor) be defined as: 'a visitor residing in a country who travels the same day to a different country for less than 24 hours without spending the night in the country visited, and whose main purpose of visit is other than the exercise of an activity remunerated from within the country visited'. The term 'visit' refers to the stay (stop) (overnight or same-day) in a place away from home during a trip. Entering a geographical area, such as a county or town, without stopping usually does not qualify as a visit to that place (UN and UNWTO, 2007). Such definitional issues are of major importance for estimating the economic and other impacts of tourism, as well as assessing tourist markets, distribution and flows. Unfortunately, despite the recommendations of international agencies individual countries often collect tourist data inconsistently, with information about excursionists and domestic tourism being the least reliable at a global level (Lennon, 2003).

One of the few publicly available estimates of total global tourism volume was made by the UNWTO in 2008 (UNWTO, United Nations Environmental Programme (UNEP) and World Meteorological Organization (WMO), 2008). This suggested that for 2005 total tourism demand (overnight and same-day; international and domestic) was estimated at about 9.8 billion arrivals. Of these, five billion arrivals were estimated to be same-day visitors (four billion domestic and one billion international) and 4.8 billion were arrivals of visitors staying overnight (tourists) (four billion domestic and 800 million international). Given that an international trip can generate arrivals in more than one destination country, the number of trips is somewhat lower than the number of arrivals. For 2005 the global number of international tourist trips (i.e. trips by overnight visitors) was estimated at 750 million (Scott et al., 2008, 122) (Table 20.1). This corresponds to 16 per cent of the total number of tourist trips, with domestic trips representing the large majority (84 per cent, or four billion). The UNWTO estimates highlight the potential economic significance of domestic tourism for many destinations, even though the major research, policy and marketing focus tends to be on international tourism.

Table 20.1 Number of tourist trips undertaken in 2005

	Air transport (trips) (million)	Non-air transport (trips) (million)	Total (million)
International	340	410	750
International same-day	10	990	1000
Domestic	480	3520	4000
Domestic same-day	40	3690	3730
Total	870	8610	9480

Source: Derived from: UNWTO, UNEP and WMO (2008).

GLOBALIZATION

The growth of long-distance leisure travel and other forms of voluntary mobility is inseparable from the major factors of globalization. Since the Industrial Revolution changes in transportation technology have provided momentum to tourism growth. With each stage of technological development – railways, tyred bicycles, steam ships, mass public transport, cars – tourism has grown as people have been able to travel faster (and hence further in a given unit of time) at lower per unit costs (Hall, 2005b). The development of mass commercial aviation in the late 1960s and early 1970s, in particular, led to significant rates of hypermobility (Gössling et al., 2009a). This, combined with reductions in border entry and exit barriers from the 1980s on, especially in Eastern Europe since the fall of state communism, as well as the People's Republic of China and South Korea, has led to rapid growth in travel and tourism, especially outside more affluent Europe and North America. Although international travel from Asian countries such as China and India was originally more diasporic than leisure oriented, that is, geared towards visiting relations, the growth of Asian consumer cultures has provided significant impetus towards growing expenditure on leisure travel as well as the development of new attitudes towards holiday taking (Winter et al., 2008).

In the case of aviation the combination of the various elements of globalization has meant that there has been a transition in aviation from being a luxury form of mobility for the wealthy few to being a relatively cheap means of mass transportation for large numbers of leisure and business travellers in industrialized countries (Gössling et al., 2009a). Transport transformations have also gone hand in hand with changes in information and communication technology (ICT) that can convey more information about travel and potential destinations than ever before. The economic capacity to travel to previously distant tourism destinations therefore co-exists with increases in the type and quantity of information on how to get there and associated marketing via a range of media channels. Shifts in access as a result of improved affordability and availability therefore also correspond with fundamental changes in perceptions of distance, place and space (Adey et al., 2007). For many people, destinations that were once a distant non-routine environment are now everyday routine environments that can be experienced via direct and/or virtual encounters (Hall, 2005a, 2005b; Coles and Hall, 2006).

However, increased consumer knowledge of destinations can have mixed effects. Although the growth in potential travel knowledge and transport options has allowed new destinations to develop, it has at the same time provided for greater competitiveness between destinations and tourism businesses. In addition, it has made some destinations more sensitive to crises within the tourism system in terms of either the conditions in main generating regions or the relative security or attractiveness of destinations themselves. Yet despite occasional crises, such as oil shocks, the effects of 9/11 and the global financial crisis of 2008–2010, international tourist arrivals have shown virtually uninterrupted growth, as discussed in the next section, since the 1950s (Hall, 2010a).

Table 20.2 International tourism arrivals and forecasts, 1950–2030 (millions)

Year	World	Africa	Americas	Asia & Pacific	Europe	Middle East
1950	25.3	0.5	7.5	0.2	16.8	0.2
1960	69.3	0.8	16.7	0.9	50.4	0.6
1965	112.9	1.4	23.2	2.1	83.7	2.4
1970	165.8	2.4	42.3	6.2	113.0	1.9
1975	222.3	4.7	50.0	10.2	153.9	3.5
1980	278.1	7.2	62.3	23.0	178.5	7.1
1985	320.1	9.7	65.1	32.9	204.3	8.1
1990	439.5	15.2	92.8	56.2	265.8	9.6
1995	540.6	20.4	109.0	82.4	315.0	13.7
2000	687.0	28.3	128.1	110.5	395.9	24.2
2005	806.8	37.3	133.5	155.4	441.5	39.0
2010	940	49.7	150.7	204.4	474.8	60.3
Forecast						
2020	1360	85	199	355	620	101
2030	1809	134	248	535	744	149

Source: World Tourism Organization (1997); UNWTO (2006a, 2012).

GLOBAL OVERVIEW

In 2012, international tourist arrivals are expected to reach one billion for the first time, up from 25 million in 1950, 278 million in 1980 and 540 million in 1995 (UNWTO, 2012) (Table 20.2). Although international tourism is usually the primary policy focus because of its business and trade dimensions (Coles and Hall, 2008), the vast majority of tourism is domestic and accounted for an estimated 4.7 billion arrivals in 2010 (Cooper and Hall, 2013), excluding same-day visitors. Nevertheless, national data for domestic tourism are poor in many jurisdictions.

In 2011 international tourist arrivals (overnight visitors) grew by 4.6 per cent worldwide to 983 million, up from 940 million in 2010, when arrivals increased by 6.4 per cent. These rapid rates of growth represent a rebound from the international financial crisis of 2008–2010. Contrary to long-term trends, the advanced economies experienced higher visitor growth (4.9 per cent) than emerging economies (4.3 per cent), reflecting the relative impacts of the financial crisis as well as the effects of political instability in the Middle East and North Africa, which experienced declines of 8.0 per cent and 9.1 per cent respectively (UNWTO, 2012).

The UNWTO predicts that the number of international tourist arrivals will increase by 3.3 per cent per year on average between 2010 and 2030 (an average increase of 43 million arrivals a year on average), reaching an estimated 1.8 billion arrivals by 2030 (UNWTO, 2011, 2012). UNWTO upper and lower forecasts for global tourism are between approximately two billion arrivals (under the 'real transport costs continue to fall' scenario) and 1.4 billion arrivals (under the 'slower than expected economic recovery and future growth' scenario), respectively (UNWTO, 2011). Most of this growth is forecast to come from the emerging economies and the Asia-Pacific region, and by 2030 it is estimated that 57 per cent of international arrivals will be in what are currently classified as emerging economies (UNWTO, 2011, 2012).

ECONOMIC SIGNIFICANCE AND EMPLOYMENT

Tourism ranks as the fourth largest economic sector after fuels, chemicals and food; it generates an estimated 5 per cent of world Gross Domestic Product (GDP), and contributes an estimated 6–7 per cent of employment (direct and indirect) (UNWTO, 2012). The World Travel and Tourism Council (WTTC), an industry lobby group comprising the major tourism corporations in the world, estimates that 'travel and tourism . . . accounts for US$ 6 trillion, or 9 per cent, of global gross domestic product . . . and it supports 260 million jobs worldwide, either directly or indirectly. That's almost 1 in 12 of all jobs on the planet' (WTTC, 2012, 3). Such disparities in estimates reflect some of the difficulties in assessing the sector's economic significance.

International tourism's export value, including international passenger transport, was US$ 1.2 trillion in 2011, accounting for 30 per cent of the world's exports of commercial services or 6 per cent of total exports (UNWTO, 2012). Even though the relative proportion of tourism's contribution to international trade in services has declined as the contribution from ICT in particular has increased (Hall and Coles, 2008), tourism remains an extremely significant part of the global economy, although its economic contribution, as with the flow of travellers, is uneven. Nevertheless, tourism is one of five top export earners in over 150 countries, while in 60 countries it is the number one export sector (United Nations Conference on Trade and Development (UNCTAD), 2010). It is also the main source of foreign exchange for one-third of developing countries and one-half of least developed countries (LDCs) (UNWTO and UNEP, 2011). The UN and UNWTO (2007) estimated that tourism was the primary source of foreign exchange earnings in 46 out of 50 of the world's LDCs.

In 2011, international tourism receipts reached US$ 1,030 billion (Euro 740 billion), up from 927 billion (Euro 699 billion) in 2010. This represents a 3.9 per cent growth in receipts in real terms. The export value of international passenger transport was estimated at US$ 196 billion (Euro 141 billion) in 2011, up from US$ 170 billion (Euro 131 billion) in 2010 (UNWTO, 2012). According to the UNWTO (2012), receipts from international tourism were over US$ 1 billion for 85 countries in 2011. Although there is a strong correlation between international tourism arrivals and receipts, it is noticeable that growth in receipts from international tourism tends to lag behind growth in tourist numbers. This relative drop in the per capita value of international arrivals can be explained in part by the high level of competitiveness in the tourism industry, especially at a price-sensitive period when emerging from economic downturn, but is also because the length of stay per visit has been slowly declining in many mature country destinations as consumers purchase more 'short break' holidays.

In view of the ongoing growth of international tourism it is not surprising that it is strongly promoted by some members of the international development community as an important element in poverty reduction strategies and development financing (UNCTAD, 2010). International tourism to developing economies is perceived as important by policy makers because it is regarded as an avenue for competitive economic specialization and improvement in foreign exchange flows within the context of the perceived need to sustain international competitiveness and increasingly open economies (UNCTAD, 2004). The UNWTO (2006b, 1) outlines several reasons why tourism makes an 'especially suitable economic development sector for LDCs':

1. Tourism is consumed at the point of production; the tourist has to go to the destination and spend his/her money there, opening an opportunity for local businesses of all sorts, and allowing local communities to benefit through the informal economy, by selling goods and services directly to visitors.
2. Most LDCs have a comparative advantage in tourism over developed countries.
3. Tourism is a more diverse industry than many others. It has the potential to support other economic activities, both through providing flexible, part-time jobs that can complement other livelihood options, and generating income throughout a complex supply chain of goods and services.
4. Tourism is labour intensive, which is particularly important in tackling poverty. It also provides a wide range of different employment opportunities, especially for women and young people – from the highly skilled to the unskilled – and generally it requires relatively little training.
5. It creates opportunities for many small and micro entrepreneurs, either in the formal or informal economy; it is an industry in which start-up costs and barriers to entry are generally low or can easily be lowered.
6. Tourism provides not only material benefits for the poor but also cultural pride. It creates greater awareness of the natural environment and its economic value, a sense of ownership and reduced vulnerability through diversification of income sources.
7. The infrastructure required by tourism, such as transport and communications, water supply and sanitation, public security and health services, can also benefit poor communities.

Similar positions have been advocated by other international bodies, including the WTTC (2004) and the World Economic Forum (WEF) (2009a, 2009b), as well as the international development cooperation sector (Hawkins and Mann, 2007; Saarinen et al., 2009; UNCTAD, 2010). It is important to note, however, that the contribution of tourism to poverty reduction and sustainable development has also been substantially criticized (Chok et al., 2007; Hall, 2007).

Although promoted by some in the development community for over 40 years, the mid- to long-term relative contribution of tourism projects to development strategies remains poorly evaluated (Hawkins and Mann, 2007). There has been greater interest by international development agencies and other international bodies in advocating tourism and initiating projects rather than critically assessing the consequences of tourism-related development strategies (Gössling et al., 2009b; Zapata et al., 2011). For example, a study of Kenya, Tanzania and Uganda indicated that hotels and restaurants, and in particular the transport industry, provide below-average shares of income to poor households compared with other export sectors, leading to the conclusion that 'these results paint a fairly poor picture of the ability of tourism to alleviate poverty' (Blake, 2008, 511). This is particularly because tourism tends to be disproportionally beneficial to the already wealthy (Schilcher, 2007; Blake et al., 2008) and can reinforce existing inequalities (Scheyvens and Momsen, 2008). As shown in the case of Thailand, 'the expansion of foreign tourism demand creates general equilibrium effects that undermine profitability in tradable sectors (such as agriculture) from which the poor derive a substantial fraction of their income' (Wattanakuljarus and Coxhead, 2008, 929).

While a tourism-based economy can provide many jobs for people with limited training and skills, many of these jobs also pay minimum wages that may result in a population of working poor. There is also a tendency for work in tourism to be socially segmented by education, socio-economic class, gender, race, ethnicity and age (Ladkin, 2011). In countries with weak labour laws, employers may exploit these divisions of labour to keep wages low in tourism and hospitality (Hall and Lew, 2009). The relative value of tourism employment is also often contested as many of the jobs are regarded as being relatively low skilled. Although in many destinations this can be interpreted as meaning that tourism can provide opportunities for income generation at entry-level positions, it also means that employment in tourism is often marked by significant roles for migrants (Janta et al., 2011; Rydzik et al., 2012). Employment in tourism is also highly gendered, especially when the empirical and emotional labour of service is often regarded as 'women's work' in many societies (Moore and Wen, 2008). In the Australian case it has been noted that although tourism is relatively labour intensive compared with some other industries, 'that labour input is vulnerable to minimisation by tourism plant owners who wish to restrict costs, thereby creating a situation in which employment is often casual, part time, and underpaid' (Hall, 2007, 227). The economic and employment opportunities generated by tourism development have thus aroused a great deal of controversy about their true value to the host community in comparison with other industries (Ladkin, 2011; Zampoukos and Ioannides, 2011).

In the Spanish tourism industry, Campos-Soria et al. (2011) reported that cleaning jobs, customer service and jobs with less responsibility in the area of administration are dominated by women, while maintenance and jobs with a high level of responsibility in the areas of kitchen, restoration and administration are dominated by men. They also reported that occupational segregation increases in line with the age of workers and size of the establishment but decreases with level of education, and is less common among workers with training contracts but much greater among part-time and seasonal workers (see also Muñoz-Bullón, 2009). Even in countries such as Norway that are otherwise known for their positive discrimination policies, women appear disadvantaged to men in salaries and employment. Thrane (2008) reported that in a study of tourism employment from the 1990s through to the 2000s: (1) male tourism employees received about 20 per cent higher wages annually than their female counterparts, *ceteris paribus*; (2) a concave (i.e. inverted U) relationship between work experience and annual wages existed for both female and male employees, but this pattern was much less distinct for females; and (3) differential effects existed between male and female employees with respect to how parenthood and marriage affected wages. Even in situations of tourism business copreneurship any perception of copreneurship as a tool for enabling women to become freed from traditional gender roles does reflect the reality that a gendered ideology of work often persists. Bensemann and Hall (2010), in a study of rural tourism businesses in New Zealand, found that copreneurial couples appeared to engage in running the accommodation business using traditional gender-based roles mirroring those found in the private home.

Another significant dimension of employment in tourism and hospitality is the extent of variable demand in time and space for services. Tourism services have to be experienced *in situ*, and (in most senses) they are not spatially transferable and cannot be deferred. This implies that the tourism labour force has to be assembled *in situ* at the point of consumption and, moreover, that it is available at particular time periods (Hall and Page, 2006). The nature of demand is such that a labour force is required with sufficient flexibility to meet daily, weekly and seasonal fluctuations. At the micro level this is best illustrated in the food services function, where consumer purchase of lunch and evening meals leads to a high degree of part-time and casual employment in the restaurant sector to meet peak demand. Macro-level seasonal demand leads to substantial fluctuations in employment availability, which is often met by the hiring of employees from outside the destination area (Williams and Hall, 2002). The extent to which seasonal tourism demand generates temporary migration and 'working holiday' flows, rather than reliance on local labour, is contingent on a number of factors, both intrinsic to tourism development and to the locality. Two prime considerations are the scale of demand and the speed of tourism development, the latter affecting the extent to which labour may be transferred from other sectors of the local economy/society. Indeed, although tourism is often promoted as an industry suitable for development in agricultural areas, where peak demand for labour in agriculture is simultaneous with high levels of summer tourism demand, it can be extremely difficult to supply adequate labour from the destination region. In contrast, where peak tourism demand is in winter, as in some alpine areas, then agriculture and tourism may be much more compatible economic development mechanisms (Mitchell and Hall, 2003; Hall and Gössling, 2013).

The phenomenon of seasonality is arguably one of the main issues for understanding the potential contribution of tourism to economic development; it affects not only employment but also the nature of many tourism businesses as well as firm strategies. Seasonality occurs because of characteristics in both the tourist-generating area and at the destination. In terms of the generating area, key factors are timing of legislated holidays, especially school holidays, religious and cultural holiday conventions, and weather conditions (Hadwen et al., 2011). At destinations seasonality is strongly influenced by weather conditions and the overall resource availability for certain types of attractions (Cuccia and Rizzo, 2011), such as snow for skiing or the colour of Autumn scenery, as in the case of 'Fall tourism' in the north-eastern United States (Spencer and Holecek, 2007). Although some forms of tourism, such as business travel, may be less affected by seasonal fluctuations in demand, seasonality affects nearly all aspects of tourism business and presents one of the greatest strategic challenges at both a firm and destination scale.

THE TOURISM FIRM

There is no single business model of the tourism firm. This is partly a product of the many sectors that make up tourism – for example transport, visitor services, accommodation, restaurants, museums, art galleries, attractions – but also because of the different business goals of tourism firms. The situation in which the customers of many tourism businesses are non-tourists is termed 'partial industrialization'; 'the condition by which only certain firms and agencies that provide goods and services directly to tourists are regarded as part of the tourism industry' (Hall and Page, 2010, 303), in which 'The proportion of (a) goods and services stemming from that industry to (b) total goods and services used by tourists can be termed the index of industrialisation, theoretically ranging from 100% (wholly industrialised) to zero (tourists present and spending money, but no tourism industry)' (Leiper, 1989, 25).

Partial industrialization is important for a number of reasons, including cooperation, management, and policy and strategy development, as well as the very definition of tourism itself (Leiper et al., 2008). Statistically, for example, the notion of partial industrialization reflects the development of Tourism Satellite Accounts, which are now extensively used to measure the economic (Frechtling, 2010) and, to a lesser extent, the environmental (Jones, 2012) impact of tourism. As Hall (2008a) noted, some of the major consequences of the partial industrialization of tourism are its significance for tourism development, marketing, coordination and network development. For example, although many segments of the economy benefit from tourism, it is only those organizations that perceive a direct relationship to tourists and tourism producers that usually become actively involved in fostering tourism development or in cooperative marketing (Hall, 1999). However, there are many other organizations such as food suppliers, petrol stations and retailers, sometimes described as 'allied industries', which also benefit from tourists in destination areas but which are not readily identified as part of the tourism industry. Therefore, in most circumstances, businesses that do not regard themselves as tourism businesses, even though a substantial proportion of their income comes directly or indirectly from visitors, will usually not create linkages with tourism businesses for regional promotion unless they perceive a clear financial benefit. It will often require an external inducement, such as promotion schemes established by government at no, or minimal, cost to individual businesses, or regulatory action such as compulsory business rating tax for promotion purposes, before linkages can be established (Hall and Page, 2010).

Networks are an integral dimension of tourism, more than most economic sectors, because of the networked nature of the production of destination and trip tourism products. It is also argued that from the 1980s onwards tourism has been categorized by increasingly flexible specialization that has encouraged the development of formal and informal collaboration, partnerships and networks (Erku°-Öztürk, 2010; Erku°-Öztürk and Terhorst, 2012; Lowe et al., 2012). In other words, the tourism industry and destinations can be framed as loosely articulated groups of independent suppliers which link together to deliver the overall trip and destination products as well as knowledge transfer (Cooper and Hall, 2013). Three main types of network are identified in tourism (Tremblay, 1999; Cooper and Hall, 2013):

- Innovative networks where businesses share complementary assets – such as airline alliances and hotel consortia.
- Networks of businesses sharing in the marketing knowledge of specific customer segments – here examples include the vertical and horizontal integration strategies of larger companies as well as the knowledge management associated with special interest tourism products, such as ecotourism or medical tourism (Hall, 2012a), for which there are multi-destination associations.
- Networks coordinating complementary assets at the destination level – this includes destination marketing alliances and promotion (a long-standing approach) and

the more recent development of jointly shaped new products and innovation. Tourism businesses then strategically position themselves within such networks to leverage from innovation and future organizational configurations.

The significance of networks for tourism appears to apply across all sub-sectors, no matter what the size of the firm or sector. It is only in a limited number of firms that continue to offer package tours to tourists within their own vertically integrated business structures that the role of networks is downplayed. But Fordist mass tourism products are now relatively rare in the developed world, with such products tending to be isolated in emerging locations of tourism consumption and production; most package tour products arising out of networked relationships are often facilitated by the Internet.

The contemporary tourism industry is broadly characterized by a small number of large, often multinational, firms with economies of scale, wide distribution and a global network, taking a high volume, low profit approach, particularly in the aviation and accommodation sectors. Alongside these are large numbers of small firms that are often niche, differentiated operators focussing on particular destinations or products (Buhalis, 2003; Ateljevic and Page, 2009). Such small firms are characterized by a small market share, are managed in a personalized way, are independent of external control and do not influence market prices (Cooper and Hall, 2013). In numerical terms many destinations are dominated by micro-businesses of less than five employees that are often owner-operated (Morrison et al., 2010).

Small tourism firms are often a significant tourism policy focus for governments and economic development agencies, especially in more peripheral regions or transition economies that are undergoing economic restructuring (Irvine and Anderson, 2004). Such firms bring a number of advantages to a destination:

- They rapidly diffuse income into the economy through strong backward linkages into the economy.
- Similarly, they contribute to employment.
- They provide a localized welcome and character by acting as a point of direct contact between the host community and the visitor.
- In a market that increasingly demands tailored experiences, small firms play an important role in responding to tourists' demand and so facilitating 'flexible specialization' (Ateljevic and Doorne, 2001).

However, many such small tourism businesses are often criticized by larger operators as not being 'real' businesses, particularly as their owners may also be participating in tourism for lifestyle reasons (Morrison et al., 2008). Indeed, a national study of New Zealand bed and breakfast (B&B) operators conducted by Hall and Rusher (2005) found that of the 347 respondents only 15 per cent had registered their business with the Companies Office although 67 per cent were tax registered. The study also indicated two very significant clusters of business types in relation to dependence on accommodation for income. For the majority of respondents it is a very small portion, with over 50 per cent of respondents earning 30 per cent or less of their total income from accommodation. At the other extreme, almost 24 per cent of respondents depend on the accommodation component for 80 per cent or more of their income. Clearly, such clustering

382   *Handbook of service business*

Table 20.3   Reasons for getting involved in tourism business

Rank	Reason	No. of respondents
(1)	to meet people	251
(2)	desire to balance lifestyle with occupation	205
(3)	desire to work at home	185
(4)	appealing lifestyle	154
(5 =)	money/security/investment	126
(5 =)	retirement programme	126
(7)	minimal costs/spare room	119
(15)	recover debt on acquired land	25
(16)	desire to involve family in business	22

Source: Hall and Rusher (2005).

may have significant implications for business strategies. However, interestingly there were no significant relationships between these clusters and other business characteristics, except with days open per year and the possible exceptions of age of respondent and information about other investment and incomes, indicating the possible use of retirement or other investment funds as sources of income and perhaps emphasizing a social or 'pocket money' function in running a small accommodation business.

With respect to the importance of various goals when starting a small tourism business, the Hall and Rusher survey was extremely revealing (Table 20.3). The most significant responses related to issues of lifestyle as well as the desire for social interaction. Earning income is not a significant necessity (slightly more than a third of all respondents). At first glance this might support the idea that such operations are developed only for lifestyle considerations and are therefore not necessarily well managed according to business principles. However, a series of further questions regarding such perceptions clearly indicated that this is not the case. The vast majority of respondents saw profit as being extremely significant and there was also a strong desire to keep the business growing although this was also matched by enthusiasm for lifestyle gains and job satisfaction. As Hall and Rusher (2004) noted, such twinning of goals may cause tensions but it does not mean that small tourism operations are any less well managed or customer oriented than in the formal tourism sector. Indeed, the social motivations for running a B&B clearly indicate the potential for stronger customer orientations than in those businesses with staff who are not so service oriented towards their customers. Moreover, in terms of attitudes towards government assistance, there was very little support for the notion that such support was essential for business growth.

Such research also has a number of other implications for understanding the nature and strategies of small tourism firms. The capacity of many small firms to undertake business promotion successfully without participating in collective marketing activities as a result of Internet and web promotion means that there are not necessarily any incentives for some small tourism businesses to join formal tourism networks. The ease with which a small firm may develop Internet capacities may actually even work against collaborative developments in some situations (Hall and Rusher, 2005). Also, it can be suggested that although many such small businesses are open year round, the seasonal or casual approach of some operators actually assists many destinations trying to cope with seasonal demands and 'soaks up' peak demand in a way which larger accommodation providers are unable to do. Finally, there is substantial evidence to indicate that such 'lifestyle' tourism businesses are much more likely to commit to a destination following a crisis or disaster because of their substantial commitment of social capital and strength of place relationship (Biggs et al., 2012).

The high proportion of small and micro-businesses, as well as the significance of lifestyle entrepreneurship, may also contribute to the portrayal of tourism as only a moderately innovative sector (Mattsson et al., 2005; Hjalager et al., 2008; Lundberg et al., 2012). However, in the search for increased product, firm and destination competitiveness innovation is rapidly emerging as an important theme in tourism development (Hall and Williams, 2008). Much of the research tends to rely on case studies or on selective convenience samples of tourism businesses rather than using innovation surveys using standard industrial classifications (Hjalager, 2010; Thomas et al., 2011). The findings indicate that the level of tourism innovation approximates the average innovation rates for countries, although this may not necessarily lead to increased rates of firm survival (Hall, 2009). A further development is greater realization of the extent to which tourism itself enables innovation as a result of people travelling to other locations, seeing other products and services, and then bringing those experiences back with them to their home environment (Williams and Hall, 2000, 2002; Williams and Shaw, 2011).

DESTINATION AND PLACE DIMENSIONS OF TOURISM

Destinations are a hard to define yet critical geographical entity for tourism (Hall and Page, 2006). They are where people go to on their trips. However, the notion of a destination can be defined in multiple ways and consumers' perceptions of a destination area may be significantly different from its administrative designation (Kalandides and Kavaratzis, 2009; Lucarelli and Berg, 2011; Martinez, 2012). Morgan et al. (2011) argue that destinations are lifestyle indicators for aspirational visitors, communicating identity, lifestyle and status. From this perspective the consumption of place by tourists amid the co-production of tourist experiences by visitors and suppliers is a highly involving experience that is extensively planned and remembered, in contrast to the purchase and consumption of fast-moving consumer goods (Cooper and Hall, 2013). A destination is therefore more than a product – it is the lived place in which tourism occurs, where communities live and work and which is imbued with various symbols and identities (Hall, 2008b). Such a situation, however, also potentially creates significant issues with respect to the extent to which people are willing to become part of the commodification of place for tourist consumption and promotion, and the extent to which tourism may change the very nature of the location that tourists seek to experience (Hall and Lew, 2009).

Contemporary destination marketing and branding is both a process and an outcome (Pike, 2008). The process of destination marketing involves dealing with the complexities of destinations and their myriad stakeholders, while the outcome is the brand or image of the destination (Wang and Pizam, 2011) that is conveyed by consumers (Cooper and Hall, 2013). Contemporary destination marketing operates at a variety of scales, from the international to the very local. It is central to the activities of tourism organizations, and is believed to be essential for delivering destination competitiveness and a range of economic, marketing and promotional benefits. Pike (2004) identifies four thematic goals of Destination Marketing Organizations (DMOs):

1. Enhancing destination image.
2. Increasing industry profitability.
3. Reducing seasonality.
4. Ensuring long-term funding.

One of the key management characteristics of DMOs is that they do not own the products they are selling or promoting. Indeed, arguably DMOs face the threat of disintermediation as their role becomes redundant in the face of Internet bookings and tourist information delivery (Cooper and Hall, 2013). Nevertheless, they are simultaneously perceived to have an important leadership and strategic role in coordinating the various businesses, attractions, activities and potential tourist experiences in an area in order to develop and promote a destination product as part of the process of persuading visitors to experience and purchase the products of individual firms as part of their trip (Hall, 2008a). However, DMOs do not provide the final tourism service and therefore it is perhaps not surprising that ideas of partnership and collaboration are extremely important to understanding the relationships between individuals and organizations that exist in tourism networks as they become part of the delivery and implementation of experiences for visitors (Zach, 2011; Cooper and Hall, 2013).

Nevertheless, despite the undoubted role of networks in tourism, the notion of partnership and collaboration is increasingly contested, particularly as many public–private partnerships end up excluding the participation of affected communities (Zapata and Hall, 2012). In light of growing concerns about social and environmental responsibility, questions are therefore being asked about the extent to which DMOs and other tourism development organizations fail to consider the full range of impacts of tourism development on places, particularly if long-term effects prove uneconomic or counterproductive, as in the case of the hosting of some large-scale events (Hall, 2010a, 2012b).

DISCUSSION AND CONCLUSION: SUSTAINABLE FUTURES

Across the globe tourism has grown steadily and this is expected to continue for the foreseeable future. As noted above, UNWTO (2012) expects international arrivals to double from 940 million in 2010 to 1.8 billion by 2030 (an average annual increase of 3.3 per cent). This has occurred during a period in which there has been no general agreement on trade in services (GATS). Unlike other industries, tourism has not been subject to a high degree of international trade restrictions and, while it seems unlikely that restrictions will be put in place on consumption abroad (mode 2 of GATS), there are increasing bottlenecks in the provision of industry support via improvements in commercial presence (mode 3). The presence of natural persons (mode 4), especially the capacity of people to move from one country to another for non-permanent employment in hospitality and tourism, is a particularly critical issue. Indeed, there have been signals from some countries that they wish to integrate tourism into broader international negotiations on trade (Hall and Coles, 2008). However, the area of tourism that is currently most affected in terms of trade issues is international aviation; there are substantial trade issues with respect to airline freedoms, slot availability and the imposition of emissions trading schemes on international carriers (Ares, 2012; Scott et al., 2012).

The greatest challenges for tourism growth as well as perceptions of the industry are the increasing concerns over climate change and sustainability. Tourism transport, accommodation and activities combined were estimated by a UNWTO-commissioned study to contribute an estimated 5 per cent to global anthropogenic emissions of CO_2 in 2005 (Scott et al., 2008). Aviation accounts for 40 per cent of the overall carbon footprint of tourism, followed by cars (32 per cent) and accommodation (21 per cent). However, this assessment does not include the impact of short-lived greenhouse gases (GHGs). If a more accurate assessment is made on the basis of radiative forcing, that is, the contribution to warming of long- and short-lived GHGs in a given year in the past, then tourism contributed 5.2–12.5 per cent of all anthropogenic forcing in 2005, with a best estimate of about 8 per cent (Scott et al., 2010).

Based on a business as usual (BAU) scenario for 2035, which considers changes in travel frequency, length of stay, travel distance and technological efficiency gains, Scott et al. (2008) suggest that CO_2 emissions from tourism will grow by about 135 per cent to 2035 (compared with 2005), totalling approximately 3,059 Mt (Table 20.4). Most of this growth will be associated with air travel, which transported 51 per cent of international visitor arrivals in 2011 (UNWTO, 2012). These estimates are similar to a WEF (2009a) estimate that tourism-related CO_2 emissions (excluding aviation) will grow at 2.5 per cent per year until 2035, and aviation emissions at 2.7 per cent, with total estimated emissions

Table 20.4 *Anticipated growth rates in emissions, tourism and transport, various organizations*

Organization	Absolute growth rates expected
	Emissions
UNWTO–UNEP–WMO (2008)	Emissions of CO_2 will grow by 135% over 30 years, from 1,304 Mt CO_2 in 2005 to 3,059 Mt CO_2 in 2035
WEF (2009a)	CO_2 emissions will grow at 2.5% per year (tourism) and 2.7% (aviation) until 2035, to 3,164 Mt CO_2 by 2035 (plus 143%)
	International tourism arrivals
UNWTO (2011, 2012)	Growth in international tourist arrivals 2010–2030 will be 3.3% per year (central projection)
	Aviation
Airbus (2012)	Growth in revenue passenger kilometres by 150% between 2011 and 2031 (averaging 4.7% per year); with the global fleet of passenger aircraft growing from 15,560 to 32,550 in the same period
Boeing (2012)	Growth in global aircraft fleet from 19,890 in 2011 to 39,780 by 2031; airline traffic in revenue passenger kilometres: 5% per year
International Energy Agency (2009)	Air travel will almost quadruple between 2005 and 2050, with an average worldwide growth rate of 3.5% per year, but over 4% worldwide until 2025

of 3,164 Mt CO_2 by 2035. The share of global emissions of CO_2 attributable to aviation may appear to be small, but currently most of these emissions are generated by the less than 2 per cent of the world's population that participate in international aviation on an annual basis (Peeters et al., 2007).

If travel and tourism remain on a BAU pathway then they will become an increasing source of GHG emissions in the medium- to long-term future. Even if the per capita per trip contribution of tourists to GHG emissions continues to fall at the historic rate with greater efficiencies from technological, governance and management innovations, the absolute contribution will continue to grow (Gössling et al., 2010; Hall, 2010b, 2011), with an increasing share of the global carbon budget generated by the tourism sector (Scott et al., 2010). Findings at the global scale are also confirmed at the national level, with tourism demonstrating relatively low eco-efficiency. For example, the eco-efficiency of the Dutch economy is approximately 0.3 kg CO_2/€, but for tourism the average value is 1 kg CO_2/€ (de Bruijn et al., 2010). In Australia, tourism is the fifth most emission intense of 17 sectors (Dwyer et al., 2010) and the fourth of 22 sectors in Switzerland (Perch-Nielsen et al., 2010).

Continued growth in emissions from aviation and tourism are clearly in conflict with global climate change and GHG reduction goals. Despite enthusiasm from many governments, voluntary changes in consumer behaviour, travel and international holidays appear to be an area of consumption that individuals are not willing to give up or change markedly (Cohen and Higham, 2011; Gössling et al., 2012). Even the most environmentally aware tourists, who may even be among the most active travellers, appear unwilling to substantially alter travel behaviour (Gössling et al., 2009a; Barr et al., 2010; Eijgelaar et al., 2010; McKercher et al., 2010).

At government level there is currently no global framework for emission reductions, nor has any country presented a comprehensive strategy for emission reductions in tourism that can be measured and monitored (OECD and UNEP, 2011; Scott et al., 2012). Critically, most, if not all, national and regional tourism strategies continue to focus on growth in visitor numbers (OECD and UNEP, 2011) and this will have substantial implications for GHG emissions (Scott et al., 2012). For example, the UK Department of Transport (2007) predicts that, taking radiative forcing into account, the 9 per cent contribution of aviation in 2005 to total UK emissions will grow to approximately 15 per cent in 2020 and 29 per cent in 2050 as a result of increased travel and tourism.

Despite the challenges faced by increasing absolute emissions arising from tourism, the notion of green growth has now become an integral component of industry discourse on tourism and sustainability (UNWTO and UNEP, 2011). Nevertheless, the optimism of such a paradigm based on material/resource/energy efficiency, major changes in the energy mix to renewables and continued increases in visitor numbers is extremely problematic given constraints of arithmetic growth and efficiency limits, and governance, market and systemic limits (Hall, 2009b, 2010b; Hoffmann, 2011).

Given that the tourism sector has already acknowledged that it contributes approximately 5 per cent of global CO_2 (UNWTO, UNEP and WMO, 2008; WEF, 2009b), it means, using estimates for already occurring effects of climate change (Global Humanitarian Forum, 2009), that in proportional terms in 2009 tourism was already responsible for about 15,000 deaths, seriously affecting 8.25 million people and producing economic losses of US$ 6.25 billion as a result of its emissions. This figure is also significant given the arguments by organizations such as the UNWTO that tourism is a means to alleviate poverty in the less developed world. But according to the estimates of the Global Humanitarian Forum (2009), the economic losses from climate change in the developing world were already greater than the US$ 5.42 billion of tourism expenditures in the 49 least developed countries (Hall 2010b).

Despite adopting the rhetoric of sustainability, tourism policies almost universally follow pro-growth paradigms. Annual growth in arrival numbers that results in absolute increase in emissions is considered an indicator of success and a proxy for economic development and wealth transfer to poor local populations (Hall, 2009b, 2011). LDCs, the very countries that are meant to be benefiting most from aviation-based international tourism, are generally the ones most vulnerable to climate change (Scott et al., 2012). Major tourism and travel bodies have not yet recognized the paradox of this situation and continue to promote tourism development that relies on continued long-haul visitor growth. Indeed, the challenge of climate change is one that needs to be addressed by a number of service industries, not just tourism. Policies that promote more sustainable and lower emission forms of tourism consumption, such as encouraging domestic tourism together with a focus on income distribution and welfare issues at destinations as part of long-term development strategies, are not being sufficiently considered (Zapata et al., 2011).

As a mechanism for enabling sustainable economic development, tourism presents significant policy dilemmas. On the one hand, it is a major source of foreign exchange for many developing countries and is also widely perceived by government, supranational institutions and many NGOs as a relatively benign means of regional economic development and employment generation (e.g. UNWTO, 2006b). Yet, on the other hand, it has also come to be recognized that although tourism is important for GDP and employment in many countries, it is also an increasingly significant contributor to socio-economic and environmental change (Hall and Lew, 2009; Scott et al., 2012). The long-term consequences are potentially anything but desirable.

REFERENCES

Adey, P., Budd, L. and Hubbard, P. (2007), 'Flying lessons: Exploring the social and cultural geographies of global air travel', *Progress in Human Geography*, 31, 773–791.
Airbus (2012), 'Global market forecast 2012–2031', http://www.airbus.com/company/market/forecast/, accessed 23 November 2012.
Ares, E. (2012), *EU ETS and Aviation*, Standard Note SN.SC/5533, London: House of Commons Library.
Ateljevic, J. and Doorne, S. (2001), 'Staying within the fence: Lifestyle entrepreneurship in tourism', *Journal of Sustainable Tourism*, 8, 378–392.
Ateljevic, J. and Page, S. (eds) (2009), *Tourism and Entrepreneurship*, London: Routledge.
Barr, S., Shaw, G., Coles, T. and Prillwitz, J. (2010), '"A holiday is a holiday": Practicing sustainability, home and away', *Journal of Transport Geography*, 18, 474–481.
Bensemann, J. and Hall, C.M. (2010), 'Copreneurship in rural tourism: Exploring women's experiences', *International Journal of Gender and Entrepreneurship*, 2, 228–244.
Biggs, D., Hall, C.M. and Stoeckl, N. (2012), 'The resilience of formal and informal tourism enterprises to disasters – reef tourism in Phuket', *Journal of Sustainable Tourism*, 20, 645–665.
Blake, A. (2008), 'Tourism and income distribution in East Africa', *International Journal of Tourism Research*, 10, 511–524.
Blake, A, Arbache, J.S., Sinclair, M.T. and Teles, V. (2008), 'Tourism and poverty relief', *Annals of Tourism Research*, 35, 107–126.

Boeing (2012), 'Current market outlook 2012–2031', www.boeing.com/commercial/cmo, accessed 23 October 2012.
Buhalis, D. (2003), *eTourism: Information Technology for Strategic Tourism Management*, London: Financial Times/Prentice Hall.
Campos-Soria, J.A., Marchante-Mera, A. and Ropero-García, M.A. (2011), 'Patterns of occupational segregation by gender in the hospitality industry', *International Journal of Hospitality Management*, **30**, 91–102.
Chok, S., Macbeth, J. and Warren, C. (2007), 'Tourism as a tool for poverty alleviation: A critical analysis of "pro-poor tourism" and implications for sustainability', *Current Issues in Tourism*, **10**, 144–165.
Cohen, S.A. and Higham, J. (2011), 'Eyes wide shut? UK consumer perceptions on aviation climate impacts and travel decisions to New Zealand', *Current Issues in Tourism*, **14**, 323–335.
Coles, T. and Hall, C.M. (2006), 'The geography of tourism is dead: Long live geographies of tourism and mobility', *Current Issues in Tourism*, **9** (4–5), 289–292.
Coles, T. and Hall, C.M. (eds) (2008), *Tourism and International Business*, London: Routledge.
Cooper, C. and Hall, C.M. (2013), *Contemporary Tourism: An International Approach*, 2nd edn, Oxford: Goodfellow.
Cuccia, T. and Rizzo, I. (2011), 'Tourism seasonality in cultural destinations: Empirical evidence from Sicily', *Tourism Management*, **32**, 589–595.
de Bruijn, K., Dirven, R., Eijgelaar, E. and Peeters, P. (2010), *Travelling Large in 2008: The Carbon Footprint of Dutch Holidaymakers in 2008 and the Development since 2002*, Breda, Netherlands: NHTV Breda University of Applied Sciences.
Department of Transport (2007), *Air Passenger Demand and CO_2 Forecasts*, London: Department of Transport.
Dwyer, L., Forsyth, P., Spurr, R. and Hoque, S. (2010), 'Estimating the carbon footprint of Australian tourism', *Journal of Sustainable Tourism*, **18**, 355–376.
Eijgelaar, E., Thaper, C. and Peeters, P.M. (2010), 'Antarctic cruise tourism: The paradoxes of ambassadorship, "last chance tourism" and greenhouse gas emissions', *Journal of Sustainable Tourism*, **18**, 337–354.
Erkuş-Öztürk, H. (2010), 'The significance of networking and company size in the level of creativeness of tourism companies: Antalya case', *European Planning Studies*, **18**, 1247–1266.
Erkuş-Öztürk, H. and Terhorst, P. (2012), 'Two micro-models of tourism capitalism and the (re)scaling of state–business relations', *Tourism Geographies*, **14**, 494–523.
Frechtling, D.C. (2010), 'The tourism satellite account: A primer', *Annals of Tourism Research*, **37**, 136–153.
Global Humanitarian Forum (2009), *The Anatomy of a Silent Crisis*, London: Global Humanitarian Forum.
Gössling, S., Ceron, J., Dubios, G. and Hall, C.M. (2009a), 'Hypermobile travellers', in S. Gössling and P. Upham (eds), *Climate Change and Aviation*, London: Earthscan, pp. 131–149.
Gössling, S., Hall, C.M. and Scott, D. (2009b), 'The challenges of tourism as a development strategy in an era of global climate change', in E. Palosou (ed.), *Rethinking Development in a Carbon-Constrained World: Development Cooperation and Climate Change*, Helsinki: Ministry of Foreign Affairs, pp. 100–119.
Gössling, S., Hall, C.M., Peeters, P. and Scott, D. (2010), 'The future of tourism: Can tourism growth and climate policy be reconciled? A climate change mitigation perspective', *Tourism Recreation Research*, **35**, 119–130.
Gössling, S., Scott, D., Hall, C.M., Ceron, J. and Dubois, G. (2012), 'Consumer behaviour and demand response of tourists to climate change', *Annals of Tourism Research*, **39**, 36–58.
Hadwen, W.L., Arthington, A.H., Boon, P.I., Taylor, B. and Fellows, C.S. (2011), 'Do climatic or institutional factors drive seasonal patterns of tourism visitation to protected areas across diverse climate zones in eastern Australia?', *Tourism Geographies*, **13**, 187–208.
Hall, C.M. (1999), 'Rethinking collaboration and partnership: A public policy perspective', *Journal of Sustainable Tourism*, **7**, 274–289.
Hall, C.M. (2005a), 'Reconsidering the geography of tourism and contemporary mobility', *Geographical Research*, **43**, 125–139.
Hall, C.M. (2005b), *Tourism: Rethinking the Social Science of Mobility*, Harlow: Prentice-Hall.
Hall, C.M. (2007), 'Pro-poor tourism: Do tourism exchanges benefit primarily the countries of the South?', *Current Issues in Tourism*, **10**, 111–118.
Hall, C.M. (2008a), *Tourism Planning*, 2nd edn, Harlow: Pearson.
Hall, C.M. (2008b), 'Servicescapes, designscapes, branding and the creation of place-identity: South of Litchfield, Christchurch', *Journal of Travel and Tourism Marketing*, **25**, 233–250.
Hall, C.M. (2009a), 'Innovation and tourism policy in Australia and New Zealand: Never the twain shall meet?', *Journal of Policy Research in Tourism, Leisure and Events*, **1**, 2–18.
Hall, C.M. (2009b), 'Degrowing tourism: Décroissance, sustainable consumption and steady-state tourism', *Anatolia: An International Journal of Tourism and Hospitality Research*, **20**, 46–61.
Hall, C.M. (2010a), 'Crisis events in tourism: Subjects of crisis in tourism', *Current Issues in Tourism*, **13**, 401–417.

Hall, C.M. (2010b), 'Changing paradigms and global change: From sustainable to steady-state tourism', *Tourism Recreation Research*, **35**, 131–145.
Hall, C.M. (2011), 'Policy learning and policy failure in sustainable tourism governance: From first and second to third order change?', *Journal of Sustainable Tourism*, **19**, 649–671.
Hall, C.M. (ed.) (2012a), *Medical Tourism: The Ethics, Regulation, and Marketing of Health Mobility*, London: Routledge.
Hall, C.M. (2012b), 'Sustainable mega-events: Beyond the myth of "balanced" approaches to mega-event sustainability', *Event Management*, **16**, 119–131.
Hall, C.M. and Coles, T.E. (2008), 'Introduction: Tourism and international business – tourism as international business', in T.E. Coles and C.M. Hall (eds), *International Business and Tourism: Global Issues, Contemporary Interactions*, London: Routledge, pp. 1–25.
Hall, C.M. and Gössling, S. (eds) (2013), *Sustainable Culinary Systems: Local Foods, Innovation, and Tourism & Hospitality*, London: Routledge.
Hall, C.M. and Lew, A. (2009), *Understanding and Managing Tourism Impacts: An Integrated Approach*, London: Routledge.
Hall, C.M. and Page, S.J. (2006), *The Geography of Tourism and Recreation: Environment, Place and Space*, 3rd edn, London: Routledge.
Hall, C.M. and Page, S. (2010), 'The contribution of Neil Leiper to tourism studies', *Current Issues in Tourism*, **13**, 299–309.
Hall, C.M. and Rusher, K. (2004), 'Risky lifestyles? Entrepreneurial characteristics of the New Zealand bed and breakfast sector', in R. Thomas (ed.), *Small Firms in Tourism: International Perspectives*, Oxford: Elsevier, pp. 83–97.
Hall, C.M. and Rusher, K. (2005), 'Entrepreneurial characteristics and issues in the small-scale accommodation sector in New Zealand', in E. Jones and C. Haven (eds), *Tourism SMEs, Service Quality and Destination Competitiveness: International Perspectives*, Wallingford: CABI, pp. 143–154.
Hall, C.M. and Williams, A.M. (2008), *Tourism and Innovation*, London: Routledge.
Hawkins, D.E. and Mann, S. (2007), 'The World Bank's role in tourism development', *Annals of Tourism Research*, **34**, 348–363.
Hjalager, A.M. (2010), 'A review of innovation research in tourism', *Tourism Management*, **31**, 1–12.
Hjalager, A.M., Huijbens, E.H., Björk, P., Nordin, S., Flagestad, A. and Knútsson, Ö. (2008), *Innovation Systems in Nordic Tourism*, Oslo: Nordic Innovation Centre.
Hoffmann, U. (2011), *Some Reflections on Climate Change, Green Growth Illusions and Development Space*, UNCTAD Discussion Paper No. 205, Geneva: UNCTAD.
International Energy Agency (2009), *Transport, Energy and CO_2: Moving towards Sustainability*, Paris: IEA.
Irvine, W. and Anderson, A.R. (2004), 'Small tourist firms in rural areas: Agility, vulnerability and survival in the face of crisis', *International Journal of Entrepreneurial Behaviour & Research*, **10**, 229–246.
Janta, H., Brown, L., Lugosi, P. and Ladkin, A. (2011), 'Migrant relationships and tourism employment', *Annals of Tourism Research*, **38**, 1322–1343.
Jones, C. (2012), 'Scenarios for greenhouse gas emissions reduction from tourism: An extended tourism satellite account approach in a regional setting', *Journal of Sustainable Tourism*, DOI: 10.1080/09669582.2012.708039.
Kalandides, A. and Kavaratzis, M. (2009), 'From place marketing to place branding – and back: A need for re-evaluation', *Journal of Place Management and Development*, **2**, 1–7.
Ladkin, A. (2011), 'Exploring tourism labor', *Annals of Tourism Research*, **38**, 1135–1155.
Leiper, N. (1989), *Tourism and Tourism Systems*, Occasional Paper No. 1, Palmerston North: Department of Management Systems, Massey University.
Leiper, N., Stear, L., Hing, N. and Firth, T. (2008), 'Partial industrialisation in tourism: A new model', *Current Issues in Tourism*, **11**, 207–235.
Lennon, J.J. (ed.) (2003), *Tourism Statistics: International Perspectives and Current Issues*, London: Continuum.
Lowe, M.S., Williams, A.M., Shaw, G. and Cudworth, K. (2012), 'Self-organizing innovation networks, mobile knowledge carriers and diasporas: Insights from a pioneering boutique hotel chain', *Journal of Economic Geography*, **12**, 1113–1138.
Lucarelli, A. and Berg, P. (2011), 'City branding: A state-of-the-art review of the research domain', *Journal of Place Management and Development*, **4**, 9–27.
Lundberg, C., Fredman, P. and Wall-Reinius, S. (2012), 'Going for the green? The role of money among nature-based tourism entrepreneurs', *Current Issues in Tourism*, DOI: 10.1080/13683500.2012.746292.
Markusen, A. (1999), 'Fuzzy concepts, scanty evidence, policy distance: The case for rigour and policy relevance in critical regional studies', *Regional Studies*, **33**, 869–884.
Martinez, N.M. (2012), 'City marketing and place branding: A critical review of practice and academic research', *Journal of Town and City Management*, **2**, 369–394.
Mattsson, J., Sundbo, J. and Fussing-Jensen, C. (2005), 'Innovation systems in tourism: The roles of attractors and scene-takers', *Industry and Innovation*, **12**, 357–381.

McKercher, B., Prideaux, B., Cheung, C. and Law, R. (2010), 'Achieving voluntary reductions in the carbon footprint of tourism and climate change', *Journal of Sustainable Tourism*, **18** (3), 297–317.
Mitchell, R. and Hall, C.M. (2003), 'Seasonality in New Zealand winery visitation: An issue of demand and supply', *Journal of Travel and Tourism Marketing*, **14**, 155–173.
Moore, S. and Wen, J.J. (2008), 'Tourism employment in China: A look at gender equity, equality, and responsibility', *Journal of Human Resources in Hospitality & Tourism*, **8**, 32–42.
Morgan, N., Pritchard, A. and Pride, R. (2011), *Destination Brands*, 3rd edn, London: Routledge.
Morrison, A., Carlsen, J. and Weber, P. (2008), 'Lifestyle oriented small tourism [LOST] firms and tourism destination development', in S. Richardson, L. Fredline, A. Patiar and M. Ternel (eds), *Where the Bloody Hell Are We? Proceedings of the 18th Annual CAUTHE Conference*, Gold Coast: Griffith University.
Morrison, A., Carlsen, J. and Weber, P. (2010), 'Small tourism business research change and evolution', *International Journal of Tourism Research*, **12**, 739–749.
Muñoz-Bullón, F. (2009), 'The gap between male and female pay in the Spanish tourism industry', *Tourism Management*, **30**, 638–649.
OECD and United National Environment Programme (2011), *Sustainable Tourism Development and Climate Change: Issues and Policies*, Paris: OECD.
Peeters, P., Gössling, S. and Becken, S. (2007), 'Innovation towards tourism sustainability: Climate change and aviation', *International Journal of Innovation and Sustainable Development*, **1**, 184–200.
Perch-Nielsen, S., Sesartic, A. and Stucki, M. (2010), 'The greenhouse gas intensity of the tourism sector: The case of Switzerland', *Environmental Science & Policy*, **13**, 131–140.
Pike, S. (2004), *Destination Marketing Organizations*, London: Routledge.
Pike, S. (2008), *Destination Marketing*, London: Routledge.
Rydzik, A., Pritchard, A., Morgan, N. and Sedgley, D. (2012), 'Mobility, migration and hospitality employment: Voices of Central and Eastern European women', *Hospitality & Society*, **2**, 137–157.
Saarinen, J., Becker, F., Manwa, H. and Wilson, D. (eds) (2009), *Sustainable Tourism in Southern Africa: Perspectives on Local Communities and Natural Resources in Transition*, Bristol: Channel View.
Scheyvens, R. and Momsen, J. (2008), 'Tourism and poverty reduction: Issues for small island states', *Tourism Geographies*, **10**, 22–41.
Schilcher, D. (2007), 'Growth versus equity: The continuum of pro-poor tourism and neoliberal governance', *Current Issues in Tourism*, **10**, 166–193.
Scott, D., Amelung, B., Becken, S., Ceron, J.-P., Dubois, G., Gössling, S., Peeters, P. and Simpson, M. (2008), 'Technical report', in *Climate Change and Tourism: Responding to Global Challenges*, Madrid: UNWTO, UNEP, WMO, pp. 23–250.
Scott, D., Peeters, P. and Gössling, S. (2010), 'Can tourism deliver its "aspirational" emission reduction targets?', *Journal of Sustainable Tourism*, **18**, 393–408.
Scott, D., Gössling, S. and Hall, C.M. (2012), *Tourism and Climate Change: Impacts, Adaptation and Mitigation*, Abingdon: Routledge.
Spencer, D.M. and Holecek, D.F. (2007), 'Basic characteristics of the fall tourism market', *Tourism Management*, **28**, 491–504.
Thomas, R., Shaw, G. and Page, S.J. (2011), 'Understanding small firms in tourism: A perspective on research trends and challenges', *Tourism Management*, **32**, 963–976.
Thrane, C. (2008), 'Earnings differentiation in the tourism industry: Gender, human capital and sociodemographic effects', *Tourism Management*, **29**, 514–524.
Tremblay, P. (1998), 'The economic organization of tourism', *Annals of Tourism Research*, **24**, 837–859.
Tremblay, P. (1999), 'The future of tourism: An evolutionary perspective', in CAUTHE (ed.), *Delighting the Senses*, Proceedings from the Ninth Australian Tourism and Hospitality Research Conference, Canberra, A.C.T.: Bureau of Tourism Research, 390–400, <http://search.informit.com.au/documentSummary;dn=061 789488707067;res=IELBUS>ISBN: 0642285128, accessed 5 December 2014.
UN and UNWTO (2007), *International Recommendations on Tourism Statistics (IRTS) Provisional Draft*, New York and Madrid: UN and UNWTO.
UNCTAD (2004), *Beyond Conventional Wisdom in Development Policy: An Intellectual History of UNCTAD 1964–2004*, UNCTAD/EDM/2004/4, New York: United Nations.
UNCTAD (2010), *The Contribution of Tourism to Trade and Development*, Note by the UNCTAD secretariat, TD/B/C.I/8, Geneva: UNCTAD.
UNWTO (1991), *Resolutions of International Conference on Travel and Tourism* (Recommendation No. 29), Ottawa, Canada.
UNWTO (2005), *Tourism Market Trends: World Overview and Tourism Topics. 2004 Edition*, Madrid: UNWTO.
UNWTO (2006a), *International Tourist Arrivals, Tourism Market Trends, 2006 Edition – Annex*, Madrid: UNWTO.
UNWTO (2006b), *Report of the World Tourism Organization to the United Nations Secretary-General in*

preparation for the High Level Meeting on the Mid-Term Comprehensive Global Review of the Programme of Action for the Least Developed Countries for the Decade 2001–2010, Madrid: UNWTO.
UNWTO (2011), *Tourism Towards 2030: Global Overview*, UNWTO General Assembly, 19th Session, Gyeongju, Republic of Korea, 10 October, Madrid: UNWTO.
UNWTO (2012), *UNWTO Tourism Highlights. 2012 Edition*. Madrid: UNWTO.
UNWTO and UNEP (2011), 'Tourism: Investing in the green economy', in UNEP (ed.), *Towards a Green Economy*, Geneva: UNEP, pp. 409–447.
UNWTO, UNEP and WMO (2008), *Climate Change and Tourism: Responding to Global Challenges*, Madrid: UNWTO, UNEP and WMO.
Wang, Y. and Pizam, A. (2011), *Destination Marketing and Management: Theories and Applications*, Oxford: CABI.
Wattanakuljarus, A. and Coxhead, I. (2008), 'Is tourism-based development good for the poor? A general equilibrium analysis for Thailand', *Journal of Policy Modelling*, **30**, 929–955.
WEF (2009a), *Towards a Low Carbon Travel & Tourism Sector*, Davos: World Economic Forum.
WEF (2009b), *The Travel & Tourism Competitiveness Report 2009: Managing in a Time of Turbulence*, Davos: World Economic Forum.
Williams, A.M. and Hall, C.M. (2000), 'Tourism and migration: New relationships between production and consumption', *Tourism Geographies*, **2**, 5–27.
Williams, A.M. and Hall, C.M. (2002), 'Tourism, migration, circulation and mobility: The contingencies of time and place', in C.M. Hall and A.M. Williams (eds), *Tourism and Migration: New relationships Between Production and Consumption*, Dordrecht: Kluwer, pp. 1–52.
Williams, A.M. and Shaw, G. (2011), 'Internationalization and innovation in tourism', *Annals of Tourism Research*, **38**, 27–51.
Wincott, D. (2003), 'Slippery concepts, shifting context: (National) states and welfare in the Veit-Wilson/Atherton debate', *Social Policy and Administration*, **37**, 305–315.
Winter, T., Teo, P. and Chang, T.C. (eds) (2008), *Asia on Tour: Exploring the Rise of Asian Tourism*, London: Routledge.
World Tourism Organization (1997), *Tourism 2020 Vision*, Madrid: WTO.
World Travel and Tourism Council (2004), *The Caribbean: The Impact of Travel & Tourism on Jobs and the Economy*, London: WTTC.
WTTC (2012), *Progress and Priorities 2010–2011*, London: WTTC.
Zach, F. (2011), 'Partners and innovation in American destination marketing organizations', *Journal of Travel Research*, **51**, 412–425.
Zampoukos, K. and Ioannides, D. (2011), 'The tourism labour conundrum: Agenda for new research in the geography of hospitality workers', *Hospitality & Society*, **1**, 25–45.
Zapata, M.J. and Hall, C.M. (2012), 'Public–private collaboration in the tourism sector: Balancing legitimacy and effectiveness in Spanish tourism partnerships', *Journal of Policy Research in Tourism, Leisure and Events*, **4**, 61–83.
Zapata, M.J., Hall, C.M., Lindo, P. and Vanderschaeghen, M. (2011), 'Can community-based tourism contribute to development and poverty alleviation? Lessons from Nicaragua', *Current Issues in Tourism*, **14**, 725–749.

21. Growth and spatial development of producer services in China[1]
Anthony G.O. Yeh and Fiona F. Yang

INTRODUCTION

The rise of producer services was observed in developed countries decades ago, but it is a relatively new phenomenon in China, where their rapid growth has occurred only since the mid-1990s. The service market in China used to be seriously constrained during the Maoist period according to the principles of a centrally planned economy. Although a market economy with 'Chinese characteristics' has been developed since the economic reform in 1978, the market for services has been heavily regulated and access by foreign services providers has been significantly restricted. During the last decade China has accelerated and deepened the pace of reform, particularly with reference to state-owned enterprises and integration with the global market. This has led not only to the dramatic expansion of the service sector, but also its 'upgrading' as a result of the rapid development of producer services.

The demand for producer services has been growing, mainly due to 1) economic development and the restructuring of the manufacturing sector; 2) the impacts of globalization; and 3) institutional change and policy support from the Chinese government. China has experienced rapid industrialization since the economic reforms but this favoured the export-oriented and labour-intensive manufacturing sector. An export-oriented economy has underpinned China's economic growth 'miracle' over the past three decades, but this mode of growth is not sustainable. It has been necessary for China to devote more efforts in recent years to a gradual transformation to a domestic-driven economy, and from 'extensive' (*cufang xing*) to 'intensive' (*jiyue xing*) economic growth. Providing 'both the essential infrastructure for production and export trade and a conduit for ideas and innovation leading to improvements in the competitiveness of local/regional economies, sectors of economic activity or individual firms' (Daniels, 2013: 41), producer services are a constituent part of this transformation process. Over the past decade, how has the demand for producer services triggered the growth of these activities? As producer services include a wide range of activities (transport and logistics, business services, finance and insurance), how have different producer service activities developed as a response to China's economic transition? What are the spatial patterns for different types of producer services?

China joined the World Trade Organization (WTO) in 2001, making commitments enabling the 'world factory' to enjoy lower tariff rates from other WTO members that have significantly benefited from China's goods production and exports. The export volume of goods increased dramatically from 249.2 billion USD in 2000 to 1,577.8 billion USD in 2010. Since 2009, China has been the world's largest exporter of goods. It is argued, however, that WTO membership is a double-edged sword: it serves as a positive force for economic development, but also leads to competition from foreign imports and foreign enterprises that may undermine important domestic enterprises in agriculture, manufacturing and services (Chow, 2001). Yet, for the manufacturing sector, in which China has significant comparative advantages, it seems that joining the WTO has brought about more opportunities than threats. By comparison, the service sector in China, particularly producer services, has only recently developed and the quality of service provision is relatively poor compared with its Western counterparts. Given this relatively disadvantageous situation with respect to services, why has China earnestly pushed ahead with accession to the WTO, which may lead to serious damage to domestic enterprises in the service sector? What approaches have been adopted by China to protect domestic producer service enterprises and how has the integration into the global market affected their growth?

It is well documented in the literature that government intervention is important for understanding China's economic and urban development. As Ma (2002: 1547) claims, '[d]espite the abandonment of central planning and the impacts of market forces in shaping the national and local space-economies, the power of the state is felt in every facet of China's transformation'. As such, the growth of producer services cannot be adequately explained without probing into the role played by the Chinese state. From the late 1950s the hegemony of the socialist state ensured the demise of a free market and this minimized the demand for producer services throughout the Maoist era. The Chinese state has changed its attitude towards the development of producer services in the post-reform period. With the introduction of the market mechanism and opening up to the outside world, the importance of producer services for stimulating economic growth and enhancing economic competitiveness has been increasingly recognized. Indeed, the Chinese state has introduced a range of specific policies to facilitate the growth of producer services. It is useful, therefore, to examine these policies and to consider their effectiveness in promoting producer service development.

We now turn to examine the dramatic growth and spatial development of producer services in China since the early 2000s, analysing the impacts of globalization on the growth of producer services, and evaluating the efforts devoted by the Chinese state to accelerating producer service development. The aim is to provide a better understanding of the factors or issues that are of critical importance for a Chinese economy that is undergoing transformation and upgrading.

PRODUCER SERVICES IN THE CHINESE ECONOMY AND URBAN SYSTEM

Producer Services in the Chinese Economy

Prior to the economic reforms in 1978, the socialist state monopolized investment, production, circulation and redistribution of economic activities under a command economy framework. The demand for consumer as well as producer services was low; the share of the service sector in the national economy was limited and it changed only slightly between 1949 and 1978, when the shares of the service sector in total employment and total GDP were 12.2 per cent and 24.0 per cent respectively, compared with 9.1 per cent and 28.6 per cent in 1952 (China State Statistical Bureau (CSSB), 1999). The 'arbitrary distortion of the Maoist regime' to service development began to be rectified by the implementation of economic reforms. Thus, by 2010 the service sector accounted for 43.1 per cent of total GDP and 34.6 per cent of total employment, generating 17,308.7 billion RMB output value and creating 26.3 million jobs (CSSB, 2011).

Although a relatively pleasant environment has been created for the growth of the service sector in China since 1978 due to rapid economic growth, increase in personal income and relaxation of state control on the economy, the rapid growth of producer services is a new phenomenon and the term 'producer services' is new to policy makers. While the 'large-scale visibility of producer services and their production' has been manifest since the 1950s in the United States and the 1960s in the majority of Western European countries (Moulaert and Daniels, 1991: 4), the rapid expansion of producer services in China has occurred only since the mid-1990s. For those producer services labelled 'advanced' and 'high order', a dramatic expansion has only taken place since the early 2000s. It was not until 2006 that the term 'producer services' was first used in a national five-year plan – the Eleventh Five–Year Plan (2006–2010). Rapid economic growth and transformation since the 2000s has been a major force driving up demand for producer services. The manufacturing sector was initially labour intensive and involved low technology; producer service inputs were uncommon. As the sector has moved to 'increasingly involving higher-value product lines, infused with advanced design values, and involving advanced-technology production process' (Hutton, 2013: 53) an increasing demand for producer services has been generated. Moreover, the expansion of the service sector since the economic reforms has also become a new source of demand for producer services.

Using China's official industrial classification, producer services include 'transport, storage and post', 'information transmission, computer services and software', 'financial intermediation', 'real estate', 'leasing and business services' and 'scientific research, technical services, and geological prospecting'. By 2010, these activities had generated 8,519.9 billion RMB added value and employed 21.01 million people (Table 21.1). Some producer services, such as real estate and business services, did not exist prior to the economic reforms due to the anti-market ideology, the free land use policy and the *danwei* system. The reforms of state-owned enterprises and of urban land use and housing are giving rise to demand for increasingly diversified producer services (Table 21.1). Table 21.2 further shows the producer services share of the service sector with respect to GDP and employment. They accounted for nearly half of the service sector as a whole in terms of GDP and more than 30 per cent in terms of employment by 2010. These shares have been increasing since 2003 although that of GDP is not rising in a straight line. The subgroup of transport, storage and post accounted for a significant share of output in 2004 (14.41 per cent) as well as employment in 2003 (10.82 per cent). However, its relative importance has been declining over the past decade to 11.02 per cent in output and 9.15 per cent in employment in 2010. Meanwhile, the other subgroups have been taking up a growing share with the exception of information transmission, computer services and software, the contribution of which to GDP has decreased. Transport, storage and post are identified as relatively low-order activities within the producer service spectrum. The growing importance of other producer services that are more advanced and knowledge intensive over the past decade is encouraging. As the manufacturing sector

Table 21.1   GDP and employment of producer services in China, 2010

	GDP Billion RMB	GDP %	Employment Million persons	Employment %
Transport, storage and post	1913.2	22.46	6.31	30.03
Information transmission, computer services and software	888.2	10.43	1.86	8.85
Financial intermediation	2098.1	24.63	4.70	22.37
Real estate	2278.2	26.74	2.12	10.09
Leasing and business services	778.5	9.14	3.10	14.75
Scientific research, technical services and geological prospecting	563.7	6.62	2.92	13.90
Total producer services	8519.9	100.00	21.01	100.00

Sources: CSSB (2011).

Table 21.2   Shares of GDP and employment in the service sector by producer service activities, %

	2003	2004	2005	2006	2007	2008	2009	2010
GDP								
Transport, storage and post	–	14.41	14.24	13.76	13.11	12.46	11.30	11.02
Information transmission, computer services and software	–	6.56	6.55	6.42	6.02	5.98	5.51	5.11
Financial intermediation	–	8.35	8.12	9.15	11.08	11.32	12.00	12.09
Real estate	–	11.11	11.37	11.71	12.40	11.22	12.60	13.12
Leasing and business services	–	4.07	4.18	4.28	4.22	4.27	4.18	4.48
Scientific research, technical services and geological prospecting	–	2.73	2.89	3.03	3.09	3.04	3.19	3.25
Total producer services	–	47.23	47.34	48.34	49.92	48.29	48.79	49.07
Employment								
Transport, storage and post	10.82	10.64	10.21	10.04	9.98	9.76	9.51	9.15
Information transmission, computer services and software	1.99	2.08	2.16	2.26	2.41	2.48	2.61	2.69
Financial intermediation	6.00	5.99	5.98	6.02	6.24	6.50	6.73	6.81
Real estate	2.04	2.25	2.44	2.52	2.67	2.69	2.86	3.07
Leasing and business services	3.12	3.27	3.64	3.88	3.96	4.27	4.36	4.49
Scientific research, technical services and geological prospecting	3.77	3.74	3.79	3.86	3.90	4.00	4.09	4.24
Total producer services	27.74	27.97	28.21	28.57	29.15	29.69	30.16	30.45

Sources: CSSB (2007–2012).

is expanding and restructuring, it requires increasing intermediate inputs that provide state-of-the-art knowledge and expertise.

However, compared with many developed countries, the contribution of producer services to the Chinese economy remains much lower. In 2007, the share of producer services in the service sector in advanced economies was between 60 and 70 percent.[2] In terms of the level of service inputs in the national economy, China's percentage is also lower than its Western counterparts. Using an input–output approach, Cheng (2008: 79) estimates that the level of intermediate service inputs in the Chinese economy was 12.2 per cent, much lower than the average for the OECD countries (21.7 per cent). There are many factors leading to the underperformance of producer services in the national economy, including the path-dependent influence inherited from the socialist command economy, low involvement of producer services in the global economy, excessive government intervention, and an immature and disordered market (Jiang and Li, 2004; Li, 2005; Chen and Cheng, 2006; Cheng, 2008). Some of the barriers that hinder the development of producer services are rooted in China's political and economic system and will be difficult to remove. They are likely to persist for quite a long time.

Producer Services in the Urban System

Geographically, producer services are likely to be over-concentrated in certain places. This contrasts with consumer services whose spatial pattern is largely shaped by the distribution of population. The spatial development of producer services in China will be examined, first at the provincial level and then at the city level. At the provincial level, the Herfindahl index (hereafter H-index), one of the common ways to measure the broad locational characteristics of an industry and assess the industry's overall level of spatial agglomeration or dispersal (Fan and Scott, 2003), has been used. The formula is:

$$H_i = \Sigma_j \rho_{ij}^2 \quad (\rho_{ij} = x_{ij}/X_i);$$

where x_{ij} is the amount of activity in industry i in province, j and X_i is the total amount of activity in industry i in China as a whole. The index ranges from 0–1. When all the activity in industry i is concentrated in one province, the index is equal to one and when all activity is evenly dispersed, the index is close to zero. Using the national economic censuses undertaken in 2004 and 2008, we are able to further disaggregate producer services to probe their location attributes at the two-digit level. For the 31 provinces and provincial-level cities in China, Table 21.3 shows the H-index for employment in all two-digit producer service sectors. Sectors that are relatively high level and knowledge intensive, notably air transport, computer services, software services, securities and business services, all display a high degree of concentration. Three of the four provincial-level cities, Beijing, Shanghai and Guangdong, dominate the distribution of these activities. Mixed producer service sectors, which have a significant proportion of clients in final consumption, for example banking and postal and courier services, reveal a relatively dispersed spatial pattern. Water transport and pipeline transport have a high-value H-index, but they are not taken into account here since the former largely reflects differences in the physical geography of each province and the latter is not a conventional transportation mode.

Table 21.3   H-index values for two-digit producer service sectors in Chinese provinces

	Code	Values of H-index for employment 2004	Values of H-index for employment 2008
*Transport, storage and post**	F	*0.0459*	*0.0504*
Road transport (except local passenger transport)	5200	0.0434	0.0472
Local passenger transport	5300	0.0622	0.0631
Water transport	5400	0.0832	0.0911
Air transport	5500	0.0819	0.1116
Pipeline transport	5600	0.2863	0.1956
Loading and unloading and other transport services	5700	0.0548	0.0645
Storage	5800	0.0561	0.0625
Postal and courier services	5900	0.0476	0.0496
Information transmission, computer services and software	G	*0.0570*	*0.0638*
Telecommunications and other information transmission services	6000	0.0503	0.0446
Computer services	6100	0.1124	0.0744
Software services	6200	0.1501	0.1322
Financial intermediation	J	*0.0497*	*0.0465*
Banking	6800	0.0480	0.0436
Securities	6900	0.0929	0.1299
Insurance	7000	0.0570	0.0509
Other financial activities	7100	0.1048	0.0513
Real estate	K	*0.0817*	*0.0568*
Real estate	7200	0.0817	0.0568
Leasing and business services	L	*0.0712*	*0.0740*
Leasing	7300	0.0667	0.0528
Business services	7400	0.0716	0.0750
Scientific research, technical services and geological prospecting	M	*0.0489*	*0.0524*
Research and experimental development	7500	0.0652	0.0692
Technical services	7600	0.0525	0.0509
Technology exchange and promotion services	7700	0.0504	0.0684
Geological prospecting	7800	0.0482	0.0567
Total producer services		*0.0525*	*0.0543*

Note: Railway transport is not included as the data for 2004 are not available and the data for some provinces in 2008 are missing.

Sources: Calculated based on CSSB (2006, 2010a).

At the city level, there were 654 cities in China in 2009, of which 287 were cities at the prefecture level and above. In terms of the size of the urban population, Chinese cities are divided into four groups (CSSB, 2004: 487): extra-large (2 million urban population and more), large (1–2 million), medium (0.5–1 million) and small (less than 0.5 million) cities. The constraints placed by data availability mean that city-level analysis can only focus on producer services at the one-digit level, and only on those cities at and above the prefectural level. Location quotients (LQ) for producer service employment by city group show that extra-large cities are the most attractive loci not only for the whole producer service sector, but also individual subsectors of producer services (Table 21.4). However, the importance of extra-large cities decreases from the East to the West. This is particularly obvious for leasing and business services. Extra-large cities in the eastern coastal region had the highest LQ value of 2.30, whereas their central and western counterparts were merely 0.66 and 0.74 respectively. A list of the top 5 cities for each producer service sector is dominated, unsurprisingly, by Beijing and Shanghai, followed by Guangzhou and Shenzen (Table 21.5). Beijing, Shanghai and Guangzhou are the core cities within the three leading economic regions – the *Jing-Jin-Ji* (Beijing–Tianjin–Hebei) region, the Yangtze River Delta and the Pearl River Delta. The other cities (see Table 21.5) include Tianjin, Chongqing, Wuhan, Chengdu and Xi'an; all major within their respective regions: Tianjin in the eastern region, Xi'an in the western region and the other three in the central region. It also seems that there is a primacy distribution of certain producer service sectors such as information transmission, computer services and software, and leasing and business services (Table 21.5). Beijing tops the list with 25.0 per cent and 30.2 per cent of total national employment in these two sectors. The second city, Shanghai, lags well behind Beijing with 4.5 per cent and 7.5 per cent respectively of total employment in these two sectors. Because the available data do not allow a finer division of the producer service sectors, some financial services that are likely to be disproportionately concentrated cannot be illustrated.

GLOBALIZATION AND PRODUCER SERVICE GROWTH

As a direct outcome of 'flexible specialization', the growth of producer services is closely linked to the globalization process. China started to implement its opening up policy in 1979. Since then the influx of foreign capital has been an important exogenous driver of Chinese economic transformation. However, it was not until the 2000s, when China joined the WTO, that the trade liberalization requirements associated with membership encouraged greater functional integration of China into the global market.

The negotiation process for China's accession to the WTO started in 1986 when China applied to rejoin the General Agreement on Tariffs and Trade. The process continued for 15 years until China became the 143rd member of the WTO on 11 December 2001. Based on WTO commitments, Mainland China and Hong Kong signed the Closer Economic Partnership Arrangement (CEPA) in 2003; this gave companies and residents of Hong Kong more preferential access to the mainland Chinese market. Membership of the WTO brings opportunities but also challenges. The benefits include increases in export and foreign direct investment (FDI), a speeding up of economic reform of the national economy (which was the main motivation of Premier Zhu Rongji when promoting China's entry into the WTO), and expediting the transformation of the economic structure. On the negative side, exposure to foreign competition may displace domestic enterprises as well as new, young industries that are not yet strong enough to compete with outside providers. In the case of the manufacturing sector and low-value-added service activities which have been able to rely, at least until quite recently, upon their

Table 21.4 Location quotients of producer services by city group, 2009

		Transport, storage and post	Information transmission, computer services and software	Financial intermediation	Real estate	Leasing and business services	Scientific research, technical services and geological prospecting	Total Producer Services
Eastern	Extra-large	1.71	1.87	1.40	1.93	2.30	1.75	1.78
	Large	0.68	0.56	0.90	0.77	0.58	0.42	0.67
	Medium	0.78	0.77	0.99	0.70	0.62	0.60	0.76
	Small	0.65	1.28	1.37	0.41	0.80	0.52	0.84
	Total	1.26	1.33	1.20	1.39	1.53	1.18	1.30
Central	Extra-large	1.47	1.00	1.07	1.14	0.66	1.45	1.18
	Large	0.51	0.47	0.70	0.47	0.43	0.65	0.55
	Medium	0.50	0.69	0.85	0.46	0.40	0.41	0.56
	Small	0.69	1.07	1.12	0.51	0.38	0.67	0.75
	Total	0.77	0.72	0.87	0.65	0.48	0.79	0.73
Western	Extra-large	1.03	0.78	0.74	0.89	0.74	1.39	0.94
	Large	0.41	0.35	0.51	0.29	0.11	0.44	0.37
	Medium	0.39	0.52	0.81	0.30	0.25	0.44	0.47
	Small	0.40	0.66	0.73	0.29	0.37	0.36	0.47
	Total	0.69	0.61	0.70	0.57	0.45	0.87	0.66
Total	Extra-large	1.53	1.49	1.21	1.57	1.68	1.62	1.50
	Large	0.57	0.49	0.76	0.58	0.45	0.51	0.58
	Medium	0.58	0.68	0.89	0.52	0.45	0.48	0.61
	Small	0.57	0.98	1.05	0.41	0.49	0.52	0.67
	Total	1.00	1.00	1.00	1.00	1.00	1.00	1.00

Source: Calculated based on CSSB (2010b).

Table 21.5 Top 5 cities for producer service development, 2009

		1st	2nd	3rd	4th	5th	Top 5 Cities
Transport, storage and post	City	Beijing	Shanghai	Guangzhou	Wuhan	Shenzhen	
	1000 persons	497.6	353.3	222.8	139.6	138.0	1,351.3
	%*	11.22	7.96	5.02	3.15	3.11	30.46
Information transmission, computer services and software	City	Beijing	Shanghai	Hangzhou	Guangzhou	Shenzhen	
	1000 persons	361.6	65.2	53.5	49.2	44.0	573.5
	%*	25.00	4.51	3.70	3.40	3.04	39.65
Financial intermediation	City	Beijing	Shanghai	Shenzhen	Chongqing	Guangzhou	
	1000 persons	253.8	216.9	87.0	77.7	71.1	705.8
	%*	7.96	6.80	2.73	2.41	2.23	22.13
Real estate	City	Beijing	Shenzhen	Shanghai	Guangzhou	Chongqing	
	1000 persons	298.8	120.0	113.5	70.4	44.0	646.7
	%*	18.88	7.58	7.17	4.45	2.78	40.85
Leasing and business services	City	Beijing	Shanghai	Shenzhen	Guangzhou	Tianjin	
	1000 persons	718.1	178.7	109.0	94.1	72.2	1,172.1
	%*	30.23	7.52	4.59	3.96	3.04	49.35
Scientific research, technical services and geological prospecting	City	Beijing	Shanghai	Xi'an	Guangzhou	Chengdu	
	1000 persons	435.0	208.5	86.3	63.7	59.7	853.2
	%*	19.30	9.25	3.83	2.83	2.65	37.85
Total producer services	City	Beijing	Shanghai	Guangzhou	Shenzhen	Tianjin	
	1000 persons	2,564.9	1,136.1	571.3	541.0	369.4	8,182.7
	%*	16.78	7.43	3.74	3.54	2.42	33.91

Note: * percentage of the whole nation.

Source: CSSB (2010b).

low labour cost advantages, such challenges are outweighed by the benefits. However, producer services, particularly advanced producer services that are knowledge intensive, are confronted with comparative disadvantages in the face of foreign imports and foreign enterprises that utilize advanced technology, advanced management, established international marketing networks and high-quality professional staff. China recognizes these challenges and sets limitations on access and national treatment of certain services under the rules of the General Agreement on Trade in Services. Nonetheless, engagement with the WTO is expected to significantly facilitate the growth and export of producer services.

The WTO classifies service activities into 12 general sectors and 154 subsectors. China has made commitments to open 9 general sectors and 88 subsectors (Table 21.6). Five general sectors are related to producer services, that is, business services, communication services, construction and related engineering services, financial services and transport services, including insurance, banking, real estate, telecommunications and professional services. China has committed to reduce or eliminate many restrictions on these activities to allow greater market access by foreign service-providers. However, to protect vulnerable domestic enterprises, the market access commitments are subject to certain limitations and are to be phased in over several years. Some limitations are horizontal, to be applied generally to all service sectors unless otherwise specified; for example, restrictions on the establishment of branches, on minimum assets, on the profit-making activities of representative offices and on the length of time permitted for the use of land in China. For specific service sectors, there are eight types of limitations, which relate to: (1) providing a service across borders; (2) form of establishment; (3) share of foreign ownership; (4) geographic location; (5) scope of business; (6) number of foreign service providers or quantity of output or operations; (7) national treatment based on the residency or nationality of a service provider; and (8) national treatment based on qualifications, standards or licensing requirements.[3] For many producer service sectors, most of the limitations are phased down over a period of 3–6 years but restrictions have remained in place in some sectors since 2007 (Table 21.7). The most common types of continuing limitations on producer service sectors relate to form of establishment and scope of business. Over half of the subsectors in business services and construction services still require foreign providers to operate through joint ventures or other types of partnerships with a Chinese entity. All subsectors in construction services and 46 per cent of the subsectors in financial services preclude foreign suppliers from providing certain types of services to clients. For business services, 42 per cent of the subsectors still require foreign providers to meet professional qualifications, or subject foreign providers to licensing procedures that differ from those required of domestic suppliers.

China used to heavily regulate its service market and significantly restrict access by foreign service-providers. Although limitations of the kind outlined are still extant, accession to the WTO has significantly opened up the Chinese services market. The most notable indicator is the increase in foreign capital. In 2010, foreign invested producer services utilized US$38.9 billion, accounting for 77.9 per cent of total service sector FDI and 36.8 per cent of all FDI (Table 21.8). The figures look impressive – more than triple in absolute amount and about double in relative share of that in 2004. However, a closer examination reveals that investment in real estate services, mostly for commercial property development, accounts for more than 60 per cent of the total expansion between

402   *Handbook of service business*

Table 21.6   Number of WTO service sectors where China made commitments

General WTO service sector	Number of subsectors included in WTO general sector	Number of subsectors included in China's commitments
Business	46	26
Communication	24	17
Construction	5	5
Distribution	5	5
Education	5	5
Environmental	4	4
Financial	17	13
Health related and social	4	0
Tourism and travel related	4	2
Recreation, cultural and sporting	5	0
Transport	35	11
Other	Number of subsectors not specified	No commitments in this sector
Total	154	88

Source: United States General Accounting Office (2002), Table 6.

Table 21.7   Types of limitations that China will maintain in specific producer service sectors

	Number of subsectors covered by China's service schedule	\multicolumn{8}{c}{Percentage of covered subsectors subject to specified limitation}
		(1)	(2)	(3)	(4)	(5)	(6)	(7)	(8)
Business	26	4	54	0	0	23	4	8	42
Communication	17	47	59	53	0	18	6	0	0
Construction	5	0	0	0	0	100	0	0	0
Financial	13	15	15	15	0	46	0	0	7
Transport	11	27	36	27	0	27	0	0	0

Source: United States General Accounting Office (2002), Table 8.

2004 and 2008. The rapid increase of inward investment by other types of producer services has occurred mainly since 2007, when China removed or eliminated the limitations on some specific sectors in line with its WTO commitments. Leasing and business services was the second important producer service sector to attract foreign investment, accounting for 14.3 per cent of the service sector in 2010; financial intermediation was the least important, at less than 2.5 per cent. As restrictions and state monopolies on

Table 21.8 Actually utilized foreign investment in producer services, 2004–2010

	2004 Billion US$	2004 Share %	2005 Billion US$	2005 Share %	2006 Billion US$	2006 Share %	2007 Billion US$	2007 Share %	2008 Billion US$	2008 Share %	2009 Billion US$	2009 Share %	2010 Billion US$	2010 Share %
Transport, storage and post	1.27	2.10	1.81	3.00	1.98	3.15	2.01	2.68	2.85	3.09	2.53	2.81	2.24	2.12
Information transmission, computer services and software	0.92	1.51	1.01	1.68	1.07	1.70	1.49	1.99	2.78	3.00	2.25	2.50	2.49	2.35
Financial intermediation	0.25	0.42	0.22	0.36	0.29	0.47	0.26	0.34	0.57	0.62	0.46	0.51	1.12	1.06
Real estate	5.95	9.81	5.42	8.98	8.23	13.06	17.09	22.86	18.59	20.12	16.80	18.66	23.99	22.68
Leasing and business services	2.82	4.66	3.75	6.21	4.22	6.70	4.02	5.38	5.06	5.48	6.08	6.75	7.13	6.74
Scientific research, technical services and geological prospecting	0.29	0.48	0.34	0.56	0.50	0.80	0.92	1.23	1.51	1.63	1.67	1.86	1.97	1.86
Producer services	11.51	18.98	12.55	20.80	16.31	25.87	25.78	34.47	31.35	33.93	29.78	33.07	38.94	36.82
Service industries	14.05	23.18	14.91	24.72	19.91	31.60	30.98	41.44	37.95	41.07	38.53	42.79	49.96	47.25
National total	60.63	100.0	60.32	100.0	63.02	100.0	74.77	100.0	92.40	100.0	90.03	100.0	105.74	100.0

Sources: CSSB (2005–2011).

transport, finance, insurance and telecommunications persist, the growth of foreign investment in these sectors will likely continue to progress only slowly. An analysis of the composition of foreign invested establishments by dividing them into Hong Kong, Macau and Taiwan invested and other overseas invested firms can be undertaken. The number of firms is used as an alternative indicator because disaggregated data on FDI by investment sources for producer services are not available. The number of firms in each group according to the first and second national economic censuses conducted in 2004 and 2008 is shown in Table 21.9. The WTO commitment has stimulated investment by foreign producer service providers; the number entering the Chinese market increased dramatically from 20,487 to 32,851 over just four years. Relying upon their geographical and ethnic proximity and benefiting from the signing of the CEPA, over 40 per cent of overseas producer service providers are from Hong Kong, Macau and Taiwan, especially the former. In the sectors of real estate and transport, storage and post, the number of Hong Kong, Macau and Taiwan invested firms exceeded those from other parts of the world. The utilization of FDI and the composition of foreign investors have indicated a relatively low degree of participation of China's producer services in the global market.

Another measure of the impact of globalization is the value of trade or imports/exports of producer services. As expected, accession to the WTO has led to a rapid increase in the trade of commercial services over the past decade (Table 21.10). Service exports and imports have dramatically expanded, from US$ 33.3 billion and 39.1 billion in 2001 to US$ 172.2 billion and 193.3 billion in 2010 respectively. However, a large share of service trade, both import and export, was attributed to travel services and transportation services rather than other services (which include most of the producer services). It is encouraging that trade in consulting services has been expanding quickly, producing a positive trade balance by 2010. This is the exception, however; the low exports of other advanced or knowledge- and information-intensive producer services, for example financial services, insurance services and communication services, suggest that these activities are less competitive in China, and the low imports again demonstrate the effects of the limitations and controls on these service sectors.

To protect domestic producer service enterprises from fierce competition, China imposed limitations on market access by foreign investors, which has to a large extent reduced its involvement in the global market. Yet the impacts of globalization on China's producer service development are not insignificant. First, the increase in inward investment and service exports has directly led to the expansion of the producer service sectors. Second, foreign competition has facilitated the restructuring and renewal of the manufacturing sector, and this in turn has driven up demand for knowledge- and information-intensive producer services. Third, the involvement of foreign invested producer services is reinforcing their tendency to concentrate within the urban system. Before 2007, limitations on the geographic location of foreign investors (mostly in large cities) directly shaped the distribution of foreign producer services. For example, foreign law firms could provide legal services only through representative offices in Beijing, Shanghai, Guangzhou, Shenzhen, Haikou, Dalian, Qingdao, Ningbo, Yantai, Tianjin, Suzhou, Xiamen, Zhuhai, Hanghou, Fuzhou, Wuhan, Chengdu, Shenyang and Kunming. Although these geographic restrictions were relaxed after 2007, foreign capital is still flowing primarily towards the extra-large and large cities, particularly those in the

Table 21.9 Number of foreign invested firms in China, 2004 and 2008.

| | 2004 ||||||| 2008 |||||||
	Hong Kong, Macau and Taiwan invested firms		Other overseas invested firms		Total		Hong Kong, Macau and Taiwan invested firms		Other overseas invested firms		Total	
	number	%	number	%	number	%	number	%	number	%	number	%
Transport, storage and post	1,093	46.8	1,240	53.2	2,333	100.0	2,010	50.5	1,971	49.5	3,981	100.0
Information transmission, computer services and software	1,114	29.4	2,671	70.6	3,785	100.0	1,755	30.0	4,087	70.0	5,842	100.0
Financial intermediation	62	19.1	263	80.9	325	100.0	228	19.5	944	80.5	1,172	100.0
Real estate	4,567	62.0	2,800	38.0	7,367	100.0	5,531	60.9	3,555	39.1	9,086	100.0
Leasing and business services	1,499	33.5	2,975	66.5	4,474	100.0	3,035	35.5	5,525	64.5	8,560	100.0
Scientific research, technical services and geological prospecting	638	29.0	1,565	71.0	2,203	100.0	1,324	31.4	2,886	68.6	4,210	100.0
Total producer services	8,973	43.8	11,514	56.2	20,487	100.0	13,883	42.3	18,968	57.7	32,851	100.0

Sources: CSSB (2006, 2010a).

Table 21.10 Exports and imports of China's commercial services, 2000–2010

	2001				2005				2010			
	Export		Import		Export		Import		Export		Import	
	Billion US$	Share	Billion US$	Share	Billion US$	Share	Billion US$	Share	Billion US$	Share	Billion US$	Share
Transportation services	4.6	13.81	11.3	28.90	15.4	20.73	28.5	33.89	34.2	19.98	63.3	32.73
Travel services	17.8	53.45	13.9	35.55	29.3	39.43	21.8	25.92	45.8	26.75	54.9	28.39
Communication services	0.3	0.90	0.3	0.77	0.5	0.67	0.6	0.71	1.2	0.70	1.1	0.57
Construction services	0.8	2.40	0.8	2.05	2.6	3.50	1.6	1.90	14.5	8.47	5.1	2.64
Insurance services	0.2	0.60	2.7	6.91	0.5	0.67	7.2	8.56	1.7	0.99	15.8	8.17
Financial services	0.1	0.30	0.1	0.26	0.1	0.13	0.2	0.24	1.3	0.76	1.4	0.72
Computer and information services	0.5	1.50	0.3	0.77	1.8	2.42	1.6	1.90	9.3	5.43	3.0	1.55
Royalties and licence fees	0.1	0.30	1.9	4.86	0.2	0.27	5.3	6.30	0.8	0.47	13.0	6.72
Consulting services	0.9	2.70	1.5	3.84	5.3	7.13	6.2	7.37	22.8	13.32	15.1	7.81
Advertising services	0.3	0.90	0.3	0.77	1.1	1.48	0.7	0.83	2.9	1.69	2.0	1.03
Audio-visual services	0	0.00	0.1	0.26	0.1	0.13	0.2	0.24	0.1	0.06	0.4	0.21
Other commercial services	7.3	21.92	5.7	14.58	16.9	22.75	9.6	11.41	35.6	20.79	17.2	8.89
Government services not elsewhere classified	0.4	1.20	0.2	0.51	0.5	0.67	0.6	0.71	1.0	0.58	1.1	0.57
Total	33.3	100.0	39.1	100.0	74.3	100.0	84.1	100.0	171.2	100.0	193.3	100.0

Sources: State Administration of Foreign Exchange, *Balance of Payments 2001–2010*, online at http://www.safe.gov.cn/, last accessed 8 November 2012.

eastern region, leading to a larger producer services development gap between large- and small-sized cities and between regions.

CHANGING POLICY CONTEXT FOR PRODUCER SERVICE DEVELOPMENT

Wang (2009: 283) claims that for the service sector, which is 'extremely undersized and outdated', to 'assume a truly substantial share of national output' will simply take more time. However, in the Chinese context where the state is omnipresent, the length of time it will take depends on how the state is involved in the development of producer services. During the Maoist period, the market mechanism was effectively controlled by the socialist state and essentially internalized as the command economy. With an economic policy heavily emphasizing rapid industrialization, the growth of service activities, both consumer and producer services, was very limited despite the significance of industrial production.

The economic reforms in 1978 allowed a market economy to 'grow out of the plan'. The recognition of the role of market forces has also restored the importance of services and consumption. During the 1980s, when the introduction of market mechanisms was in its 'experimental' stage, the focus of state policies on the manufacturing sector continued. Since the early 1990s, when a large-scale reshuffling of state-owned enterprises was initiated, increasing attention has been paid to services as part of the national economic development strategy. The service sector was considered 'a main outlet to relieve the growing employment pressure faced by the nation' (Wang and Yang, 1993: 290). The Eighth Five-Year Plan for National Economic and Social Development (1991–1995) stated that the development of tertiary industry should be emphasized; in the next ten years, the focus would be on those service sectors serving production and people's livelihood; and the pace of development of the tertiary sector should be faster than that of the primary and secondary sectors.[4] This was the first time that the development of the 'tertiary industry that serves production' was included in a Five-Year Plan, although more importance was attached to 'tertiary industry that serves people's livelihood'. On 16 June 1992 the Central Committee of the Chinese Communist Party and the State Council jointly issued 'the Resolution to Speed up the Development of the Tertiary Industry' (Compilation Committee of China's Tertiary Industry Almanac, 1993). The mechanisms for fast tracking the development of services were listed, including the introduction of competitive mechanisms, reforming enterprise management, promoting financial and taxation support, simplifying procedures for the formation of new service firms, and establishing industrial regulations. The promulgation of the 1992 'Resolution' was the first important policy guide for the development of the service sector; it ushered in a new stage in service development.

The subsequent Ninth Five-Year Plan substantiated the commitment to the growth of 'newly developed industries' (*xinxing chanye*) such as tourism, consulting services, scientific and technological services, accounting and legal services to promote 'industrial modernization' as a policy goal.[5] In the Tenth Five-Year Plan, the development of 'modern services' (*xiandai fuwuye*), such as information services, financial services, accounting, consulting services and legal services, was considered an important and effective way to promote the upgrading of the industrial structure.[6] In January 2002, the State Council and the National Planning Commission (now the National Development and Reform Commission) issued another important document for China's service development – the 'Policies and Approaches to Facilitate the Development of the Tertiary Industry during the Tenth Five-Year Plan period (2001–2005)' (Compilation Committee of People's Republic of China Yearbook, 2002). Based on the 1992 'Resolution', a decade later this document set goals for the restructuring of the services sector and the promotion of modern services, and further encouraged service development by significantly expanding the scope for foreign inward investment in the service sector.

The term 'producer services' eventually appeared in the Eleventh Five-Year Plan (2006–2010). Moreover, the policies and goals to develop producer services (Chapter 15 'Expanding Producer Services') are for the first time prioritized ahead of consumer services (Chapter 16 'Enriching Consumer Services').[7] This arrangement continues in the most recent Twelfth Five-Year Plan, which includes commitments to the faster development of producer services by 'orderly extending of financial services, vigorously developing modern logistics, cultivating and expanding high-tech services, and regulating and promoting business services'.[8] In March 2007, the State Council issued 'Several Opinions on Accelerating Development in the Service Industry',[9] including optimizing the structure of the service sector by focusing on the development of modern services, and regulating and upgrading traditional services to adapt to the development of a new industrialized economy. Later, in March 2008 the State Council issued 'Implementing Opinion on Several Policy Measures for Accelerating the Development of the Service Industry',[10] which formulated much more specific approaches to fulfil the policy signalled in the 2007 'Opinions'. The approaches included formulating or revising development plans for the service industry, further relaxing barriers to entry, speeding up the reform of state-owned service enterprises, further opening of the service market to foreign investors, supporting service exports and outward service investment, cultivating leading service enterprises, encouraging innovations, increasing financial support to service industries, and implementing land use policy that was conducive to service development. Most recently, on 1 December 2012, the State Council released the Twelfth Five-Year Plan for the Development of the Service Industry, which includes an objective for the service sector to account for the largest share of national GDP by 2015.[11] Guidelines for accelerating the development of producer services, such as finance, transport, modern logistics, high-tech services, design consultancy, technology services, business services, e-commerce, engineering consultancy, human resources, energy saving and environmental protection, new commercial formats and emerging industries, are included in the Plan. In addition to specific actions targeted at individual sectors, policies for deepening cooperation among service sectors in the mainland, Hong Kong and Macau, as well as for forging closer ties between the service sectors across the Taiwan Straits, are formulated.

Rather than arbitrary suppression, the service sector has therefore received intense attention from the government, and has been highlighted for its 'strategic importance' as part of the national development agenda. These significant changes in the attitude of the Chinese state have led to the redistribution of fixed asset capital investment from state-owned and state-holding enterprises in the national economy (Figure 21.1). The tertiary sector has replaced the secondary sector as the main recipient of capital investment from

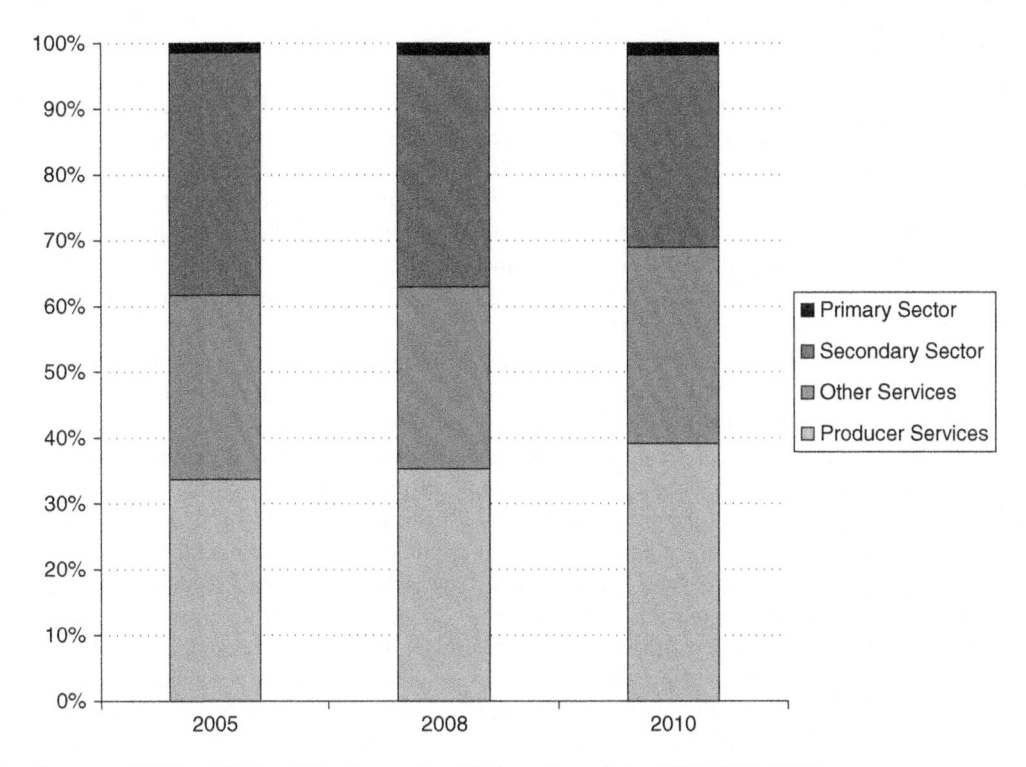

Figure 21.1 Structure of fixed asset investment by state-owned and state-holding enterprises

Sources: CSSB and National Development and Reform Commission (2006, 2009, 2011).

the state. By 2010, state capital investment in the tertiary sector had been augmented to nearly 70 per cent. In stark contrast, the secondary sector, which took the lion's share (more than 60 per cent) of fixed asset investment from the state during the 1980s (CSSB, 2002), decreased its share dramatically to less than 30 per cent. Moreover, the focus and priority of service development has shifted from consumer to producer services in the recent decade. Investment in producer services has expanded much faster than other services, the share steadily rising to nearly 40 per cent by 2010. The Chinese government has adopted a positive stance, formulating and issuing a range of important policies and strategies, as well as reallocating capital investment, designed to facilitate the growth of the service sector, and particularly producer services. Despite all these new initiatives, it remains the case that, by international standards (as discussed earlier), producer services in China are still underperforming and are under-represented in the national economy. Hopefully, this situation will change as the government-led initiatives and interventions introduced over the last decade, but especially in the last five years, catalyse the development of producer services and improve their status in the economy.

CONCLUSIONS

Economic transformation in China during the last 20 years is slowly moving the Chinese economy towards services. Apart from the expansion of consumer services, which is driven up by the increase in personal disposable income, producer services have experienced rapid growth as a result of industrial restructuring and renewal, accelerated globalization processes, and policy and financial support from the central government. This study has examined the growth of producer services, their distribution in the urban system, and the processes that underlie their growth and spatial development. The main findings can be summarized as follows.

First, rapid economic growth in China has boosted the demand for producer services. The increase in the relative share of producer services over the past decade, in terms of both output and employment, comprised activities such as FIRE services, business services and technical services. Despite their impressive growth, producer services still account for a lower share of the national output compared with their Western counterparts. From a geographical perspective, producer services show a high degree of concentration in Beijing, Shanghai and Guangdong at the provincial level, and in extra-large cities at the city level. The tendency towards concentration is much stronger for knowledge- and information-intensive producer services. Producer services such as information transmission, computer services and software, and leasing and business services display a pattern of primacy distribution. Those providing intermediate inputs throughout the value chain are identified as the lead sectors of urban-regional economies (Hutton, 2013). They are becoming important activities for participation in, and the expansion of, the ascendancy of large city-regions.

Second, China has significantly opened up the service market to foreign investors since the early 2000s. This has facilitated economic transformation and become a new force stimulating the rapid expansion and spatial development of producer services. China has not allowed foreign competition to come in rapidly, and market access has been granted in steps and limitations are still in place to protect domestic producer service development. Moreover, '[i]nformally there are red tape and other means to delay foreign competition to be exercised by central, provincial and local government officials' (Chow, 2001: 6). Such delays and restrictions have reduced the degree of openness, but China's entry into the WTO has induced a dramatic increase in foreign investment and augmented service exports, which have kept growing steadily over the past decade. However, producer services in China show comparative disadvantage when the structures of foreign investment and service exports are analysed. A large share of foreign investment and service exports is associated with low-order producer service activities that are not capable of providing crucial inputs to innovation and production. The composition of foreign investors, of which two-fifths are from Hong Kong, Macau and Taiwan, also indicates a low level of globalization. Geographically, the accelerated globalization process is strengthening the trend of producer services to concentrate in extra-large cities along the eastern seaboard.

Third, the Chinese state has increasingly recognized the benefits of producer services in the economy and has undertaken numerous actions to accelerate the growth of producer services. A review of the changing policy context on service development suggests that the emphasis of service growth has shifted from consumer to producer services. In accordance with the policies and strategies that favour producer service development, fixed asset investment from the state has been redistributed and the proportion invested in producer services has increased rapidly. In the Chinese context, the strong push of government has significant implications for the growth of producer services. In the era of globalization, and with increasing attention paid by the central government to producer service development, many provincial and local governments have identified office development and CBD construction as an effective regional and urban strategy to stimulate local economic growth and enhance local competitiveness regardless of location and city size. The pattern of infrastructure-led development, which succeeded in industrial development, is now being applied to the development of producer services. However, producer services require intensive face-to-face contacts and favour extra-large cities; small and medium cities are incapable of generating the effects of agglomeration economies to attract producer services even after the CBDs have been constructed (Yang and Yeh, 2013a and b). Without a study on the capability of development, the office development projects in some of these cities will lead to a serious waste of land, materials and investment.

In addition to the important processes mentioned above, the growth and spatial development of producer services will also be shaped by the new wave of urban development in China, for example the introduction of high-speed train services since 2007. The Chinese government stated in 2009 that China would pour some US$ 300 billion into its railways, expanding its network by 20,000 km by the end of 2015, including 12,000 km of high-speed passenger-designated lines capable of train speeds of up to 350 km per hour.[12] After the completion of the network, China will become the world's leader in high-speed rail. As a mass and efficient transportation tool, the development of high-speed rail is significantly improving the accessibility of Chinese cities. This may well help to reshape the spatial development of producer services. Will high-speed rail development further strengthen the producer service concentrations in the eastern coastal regions or lead to the emergence of new producer service centres? This remains an interesting topic for further investigation.

NOTES

1. Part of the content of the second section has been published in *Tropical Geography* in 2013 in Chinese (Yang and Yeh, 2013).
2. Zhang, W., Song, M.L. and Zhang, H.F. 'Approaches and Suggestions to Develop Producer Services', online at http://www.chinacity.org.cn/csfz/cshj/49172.html, last accessed 7 November 2012 (in Chinese).
3. For a detailed list of the service sectors and limitations, please see *Report of the Working Party on the Accession of China* (WT/ACC/CHN/49), the Schedule of Specific Commitments on Services, available online at http://www.esf.be/pdfs/GATS%20UR%20Commitments/China%20SoC.pdf, last accessed 23 January 2013.
4. Recommendations of the CPC Central Committee on National Economic and Social Development Plan for Ten Years and the Eighth Five-Year Plan, online at http://news.xinhuanet.com/ziliao/2005-02/18/content_2590430.htm, last accessed 17 August 2010 (in Chinese).
5. The Ninth Five-Year Plan for National Economic and Social Development of the People's Republic of China, online at http://www.ndrc.gov.cn/fzgh/ghwb/gjjh/W020050614801665203975.pdf, last accessed 25 January 2013 (in Chinese).
6. The Tenth Five-Year Plan for National Economic and Social Development of the People's Republic of China, online at http://news.sina.com.cn/c/209730.html, last accessed 25 January 2013 (in Chinese).
7. The Eleventh Five-Year Plan for National Economic and Social Development of the People's Republic

of China, online at http://www.qhnews.com/2010zt/system/2010/11/17/010235209.shtml, last accessed 22 November 2012 (in Chinese).
8. The Twelfth Five-Year Plan for National Economic and Social Development of the People's Republic of China, online at http://www.ce.cn/macro/more/201103/16/t20110316_22304698_7.shtml, last accessed 22 November 2012 (in Chinese).
9. Detailed content is online at http://tradeinservices.mofcom.gov.cn/en/b/2007-03-19/27448.shtml, last accessed 28 January 2013.
10. Detailed content is online at http://www.gov.cn/zwgk/2008-03/19/content_923925.htm, last accessed 28 January 2013 (in Chinese).
11. Twelfth Five-Year Plan for the Development of the Service Industry, online at http://www.gov.cn/zwgk/2012-12/12/content_2288778.htm, last accessed 28 January 2013 (in Chinese).
12. 'The Shrinking of China', online at http://www.newsweek.com/2009/10/23/the-shrinking-of-china.html, last accessed 15 March 2011 (in Chinese).

REFERENCES

Chen, X. and Cheng, D.Z. (2006) *Chinese Service Economy Report*, Beijing, Economic Management Press (in Chinese).
Cheng, D.Z. (2008) 'Development Level, Structure, and Impact of Producer Services in China: An International Comparison based on Input–Output Approach', *Economic Research*, 1: 76–88 (in Chinese).
Chow, G.C. (2001) 'The Impact of Joining WTO on China's Economic, Legal and Political Institutions', Paper Presented at the International Conference on Greater China and the WTO, City University of Hong Kong, 22–24 March, online at http://www.princeton.edu/~gchow/WTO.pdf, last accessed 25 January 2013.
Compilation Committee of China's Tertiary Industry Almanac (1993) *Almanac of China's Tertiary Industry*, Beijing, China Statistical Press (in Chinese).
Compilation Committee of People's Republic of China Yearbook (2002) *Yearbook of People's Republic of China*, Beijing: People's Republic of China Yearbook Press (in Chinese).
CSSB (1999) *China Statistical Yearbook 1999*, Beijing: China Statistical Press (in Chinese).
CSSB (2002) *Statistics on Investment in Fixed Assets of China 1950–2000*, Beijing: China Statistical Press (in Chinese).
CSSB (2004) *China Urban Statistical Yearbook 2004*, Beijing, China Statistical Press (in Chinese).
CSSB (2006) *China Economic Census Yearbook 2004*, Beijing, China Statistical Press (in Chinese).
CSSB (2007–2012) *China Statistical Yearbook 2007–2012*, Beijing, China Statistical Press (in Chinese).
CSSB (2010a) *China Economic Census Yearbook 2008*, Beijing, China Statistical Press (in Chinese).
CSSB (2010b) *China Urban Statistical Yearbook 2010*, Beijing, China Statistical Press (in Chinese).
CSSB (2011) *China Statistical Yearbook 2011*, Beijing: China Statistical Press (in Chinese).
CSSB and National Development and Reform Commission (2006) *Statistical Yearbook of the Chinese Investment in Fixed Assets 2006*, Beijing, China Planning Press.
CSSB and National Development and Reform Commission (2009) *Statistical Yearbook of the Chinese Investment in Fixed Assets 2009*, Beijing, China Planning Press.
CSSB and National Development and Reform Commission (2011) *Statistical Yearbook of the Chinese Investment in Fixed Assets 2011*, Beijing, China Planning Press.
Daniels, P.W. (2013) 'The Transition to Producer Services in China: Opportunities and Obstacles', in A.G.O. Yeh and F.F. Yang (eds) *Producer Services in China: Economic and Urban Development*, London and New York: Routledge, 29–51.
Fan, C.C. and Scott, A.J. (2003) 'Industrial Agglomeration and Development: A Survey of Spatial Economic Issues in East Asia and a Statistical Analysis of Chinese Regions', *Economic Geography*, 79 (3): 295–319.
Hutton, T.A. (2013) 'Intermediate Services, Economic Restructuring, and Urban Transformation', in A.G.O. Yeh and F.F. Yang (eds) *Producer Services in China: Economic and Urban Development*, London and New York: Routledge, 52–76.
Jiang, X.J. and Li, H. (2004) 'Service Industry and China's Economy: Correlation and Potential of Faster Growth', *Economic Research*, 1: 4–15 (in Chinese).
Li, J.F. (2005) 'Industrial Structure Upgrading and Tertiary Industry Modernization', *Journal of Sun Yat-Sen University*, 45 (4): 124–144 (in Chinese).
Ma, L.J.C. (2002) 'Urban Transformation in China, 1949–2000: A Review and Research Agenda', *Environment and Planning A*, 34: 1545–1569.
Moulaert, F. and Daniels, P.W. (1991) 'Advanced Producer Services: Beyond the Micro-economics of

Production', in P.W. Daniels and F. Moulaert (eds) *The Changing Geography of Advanced Producer Services*, London and New York: Belhaven Press, 1–14.
United States General Accounting Office (2002) 'Analysis of China's Commitments to Other Members', Report to Congressional Committees, online at http://www.gao.gov/assets/240/236049.pdf, last accessed 29 January 2013.
Wang, E. (2009) 'The Service Sector in the Chinese Economy: A Geographic Appraisal', *Eurasian Geography and Economics*, 50 (3): 275–300.
Wang, H.B. and Yang, Y.C. (1993) *The Handbook of the Tertiary Industry*, Beijing: Economic Management Press (in Chinese).
Yang, F.F. and Yeh, A.G.O. (2013a) Growth and spatial development of producer services in China, *Tropical Geography*, 33(2): pp. 178–186 (in Chinese).
Yang, F.F. and Yeh, A.G.O. (2013b) 'Spatial Development of Producer Services in the Chinese Urban System', *Environment and Planning A*, 45: 159–179.

PART V

CONCLUSION: A NEW RESEARCH AGENDA?

22. Developing the agenda for research on knowledge-intensive services: problems and opportunities
John R. Bryson and Peter W. Daniels

> The most serious theoretical deficiency of existing theories of modern society which assign a central role to knowledge is . . . their rather undifferentiated treatment of the key ingredient, namely knowledge itself. (Stehr, 1994: 91)

INTRODUCTION

During the 1980s the transformation in employment away from manufacturing to service occupations encouraged economic geographers to explore the locational dynamics of the evolving 'new service economy' (Marshall, 1983; Daniels, 1985; Wood, 1986; Daniels et al., 2011). Business or producer service firms – those service activities which typically supply other business units with 'high order' knowledge and expertise (e.g., market intelligence, management advice, design) – were identified as one of the fastest-growing sub-sectors in the advanced capitalist economies (Keeble et al., 1991; MacPherson, 1997). The logical extension of this work was to explore the dynamics of specific types of business service activities. Attention focussed on 'knowledge-intensive services' in the belief that they had the most to contribute to the wider production process (Marshall et al., 1988). Research was undertaken into the geography and operational dynamics of accountants (Hanlon, 1994), management consultants (Keeble et al, 1994; Bryson and Daniels, 1998), environmental consultants (Kastrinos and Miles, 1996), market researchers (Bryson et al., 1994), industrial design (MacPherson and Vanchan, 2008, 2010; Bryson and Rusten, 2011), and services and innovation (Gallouj and Djellal, 2010; Love et al., 2011).

In 1997 T.P. Hill published a paper that set out to define services (Hill, 1977). In what has become a classic paper, Hill argued that services are produced and consumed simultaneously. This suggests that services could not be stored and that it is difficult to export services. In the early 1980s William Beyers undertook one of the first major studies of service activities (Beyers et al., 1985a, b). This study focused on understanding the export of services within and beyond Central Puget Sound, Washington State, US. The Puget Sound study was the first to identify the contribution producer service exports make to regional economies. This research laid the foundations for a series of major theoretically informed empirical studies of business and professional services (Bryson et al., 2004). This included the first major British study (Keeble et al., 1991) that focused on understanding the characteristics, growth and geography of management consultancy and market research firms (Bryson et al., 1993a, b, 1997a, b; Keeble and Bryson, 1996; Bryson 1997a).

In 1988, the Institute of British Geographers assembled a special working party to explore the characteristics and geographies of the 'new' service economy, with a particular focus on producer services and the relationship between services and uneven development (Marshall et al., 1988). This working party developed a new research agenda that focused on business and professional services and inspired a number of European and American geographers to undertake detailed research into the economic geography of producer services (Beyers, 1979, 1989, 1991, 1998; Beyers and Alvine, 1985; Beyers and Lindahl, 1996; Bryson et al., 2004; Bryson and Rusten, 2011). These studies undoubtedly enhanced our understanding of the rise and role of business service firms in advanced capitalist economies, but the focus on individual types of service activities for their own sake was too narrow. A consideration of the relationship between specialized service providers in relation to the transfer and consumption of their expertise was an important oversight. Attempts to overcome this omission have included an exploration into the use by client companies of independent management consultancy firms (Bryson, 1997a, b; Bryson and Daniels, 1998; MacPherson, 1997) and design services (Bryson and Rusten, 2011). This began to shift the focus of research away from exploring the dynamics of service companies to understanding client/supplier relationships (Clark and Salaman 1998; Tordoir, 1994).

The restructuring of the subject matter of economic geography had tended to mirror that of the economy, with a shift in research away from manufacturing towards service activities (Illeris, 1996) during the 1980s. Increasingly the emphasis in the academic and non-academic literature is on the role of knowledge, information, expertise and networks in the production process (Crum and Gudgin, 1977; Castells, 1996; Bryson, 1997a; Gallouj and Djellal, 2010; Maglio et al., 2010; Rusten and Bryson, 2010; Bryson and Rusten, 2011; Bryson et al., 2012). According to some interpretations, this challenges the pre-eminent position of manufacturing as a source of innovation and economic growth. In other accounts, services and manufacturing are complementary activities (Bryson, 2009a, b). This new economy has been described as a 'post-industrial society' (Bell, 1974) or period of 'informational capitalism' (Castells, 1989, 1996) characterized by the dominance of service employment and output (Singelmann, 1978; Fik et al., 1993). Much of this service employment, however, is closely related to the production of physical goods through the provision of finance, consultancy, marketing, market research, design and logistics. This is to argue that claims for the emergence of a 'post-industrial' or service economy have over-stated the transformation of economies. In employment terms, developed market economies are service economies, but they are also economies that are still heavily reliant on the design, fabrication, sale and consumption of physical goods. Services have always played an important role in urban economies. Thus, in Mary Beard's analysis of Pompeii around 79 BCE (Before the Common Era) she notes that Pompeii 'was a bustling, commercial, market town . . . [providing] a whole range of services, from laundry to lamp-making, and acted as a centre of exchange for a community of probably more than 30,000 people' (Beard, 2010: 164). The importance of international financiers is highlighted in Brotton's analysis of the formation and dispersal of King Charles I's art collection in 17th-century England. He notes that the purchase of paintings from Venice

> required a financier who could buy the pictures in Venice, arrange for their shipment to England, and then reclaim the money loaned (at interest) at a later stage . . . the increasing scale

and complexity of international trade required financiers . . . to honour paper transactions, or bills of exchange, for large sums of money. (Brotton, 2007: 45)

Both these historical examples highlight the important contribution services make to supporting trade.

It is not our intention in this concluding chapter to critique the existing literature on producer service research. Such a critique is unnecessary, as like all literatures it is a product of a particular time and set of influences. It is a literature written by a group of academics, including ourselves, who had to battle against the established manufacturing emphasis in economic geography. Our intention is to encourage a constructive debate to ensure that producer service research continues to make an important contribution to understanding the geography of economic activity. Perrow (1974), in a memorable article, describes organization theorists as being like children playing in a sandpit: each child is uninterested and oblivious to the activities of the other children, each is concerned with building a sandcastle. However, occasionally a child will destroy another child's more impressive castle. We do not want to destroy others' castles, or even our own castle, but to advance understanding of the role that knowledge-intensive services play in economic activity.

Exploring other literatures (e.g., Storper, 1995; Storper and Salais, 1997; Cook and Morgan, 1998) has led us to believe that producer service expertise needs to be placed in the context of other forms and sources of expertise, information and knowledge which flow into client companies. The relative role of producer services in the transfer of knowledge may be over-stated in the geographical literature; other forms of information flow into client companies have been neglected. For example, the growth and location (Keeble et al., 1991; Bryson et al., 1993a; Beyers and Lindahl, 1996) of producer service firms and the international tradability of their output (O'Farrell et al., 1996) have been explored without any significant attempt to explore the type and nature of the knowledge and expertise which they actually bring to client companies. There is certainly a general assumption that producer service firms are strategic providers of 'high-order informational inputs' (MacPherson, 1997: 52) and that their expertise, knowledge and information make a positive contribution to the performance, productivity and efficiency of client companies (Castells, 1996: 212). But producer service knowledge is not the only source of external expertise. Dramatic producer service employment growth rates and a geography of concentration in global cities (Sassen, 1991) have obscured the contribution that other forms of knowledge flow make to economic development and the profitability, productivity and competitiveness of companies.

The objective of this chapter is therefore to explore in rather more detail the relationship between knowledge, producer service companies and their clients. An attempt is made to explore the nature of the knowledge and expertise that producer service firms provide to client companies. An important objective is to raise questions as much as to provide answers, but out of this we construct a producer service research agenda. We identify five areas or themes in which academic research needs to be undertaken in order to deepen understanding of the role and function of producer service firms.

RESEARCH THEME 1: INFORMATIONAL CAPITALISM, OR JUST CAPITALISM?

Empirical research into the activities of producer service firms has been paralleled by the development of a theoretical literature that claims that the very nature of capitalism has altered. For Lash and Urry (1994: 4) 'what is increasingly produced are not material objects, but signs'. Such signs are either informational or aesthetic. According to Stehr (1994: ix), 'the age of labor and property is at an end'; these factors are gradually giving way to a 'new constitutive factor, namely knowledge'. In a similar vein, Castells (1996) suggests that capitalism has entered an informational mode of development that is based on developments in information technology. Knowledge and information are essential elements in all economies but 'what is specific to the informational mode of development is the action of knowledge upon knowledge itself as the main source of productivity' (Castells, 1996: 17). Informational capitalism or knowledge-intensive capitalism, however, embraces all sectors of the economy, from informational agriculture to informational services, and cannot be equated solely with the emergence of a service economy (Castells, 1996: 92). There are obvious difficulties with these arguments, such as the extent of the alteration or the exact attributes of the shift towards knowledge, expertise or informational capitalism. First, the scale or inclusiveness of the analysis needs to be explored, and, secondly, the extent to which knowledge was incorporated into earlier stages of the evolution of corporate capitalism also needs to be considered.

Scale of Inclusiveness of the Analysis

The majority of companies operating in the advanced economies cannot be described as knowledge or informational companies. Most manufacturing and service companies are producing relatively simple, information-poor products and services using standard or even 'traditional' manufacturing and service technology and/or management processes. At one level, it could be argued that all companies are knowledge rich in tacit knowledge, but in terms of the debate on informational capitalism they are knowledge poor. One reason is the dominance, in terms of numbers, of small and medium-sized enterprises (SMEs) in developed market economies. Research in Britain has highlighted that many SMEs are too busy managing everyday business activities to be concerned with the acquisition or search for additional information or knowledge (Ram, 1994). The UK economy is polarizing into a large number of relatively information-poor SMEs and a small number of information-rich companies. This implies that a *dual information economy* may be developing, with some firms searching for expertise and knowledge irrespective of its location, whilst SMEs are tied into local or regional providers of more generalist expertise (Bryson and Daniels, 1998). Size by itself is not a realistic measure of the ability of an organization to convert information into knowledge. Large companies may be information rich, but they can also be knowledge poor because of an inability to translate information into knowledge and then into action (Schoenberger, 1997). We are aware that this is a big statement that is not underpinned with sufficient compelling proof. But this is precisely the point. Research has concentrated on those companies that employ producer service firms as well as those that are networked into industrial districts or clusters (Rusten and Bryson, 2010). Where is the social science research on local companies that are knowledge and network poor? Geographers know about agglomerated knowledge-intensive firms or firms located in industrial districts or clusters, but have an underdeveloped understanding of companies that operate outside such social and cultural formations.

Knowledge and Earlier Stages of Capitalist Development

Information capture and its conversion into knowledge by a process of informed interpretation have been the foundation upon which the majority of transnational and global companies have been established and grown. This is not just a phenomenon of the last two decades. Research is urgently required to explore the activities of producer service firms and professionals during earlier stages of the development of capitalism. Producer service research urgently needs to be historicized.

One way to begin such an historical-geographical approach is through the analysis of the numerous company histories that exist (see for examples Williams, 1931). We explore two brief examples. First, there is a substantial literature on the rise of Fordism as a form of scientific management, but very few geographical studies which explore the role played by Fredrick W. Taylor, a management theorist but also a management consultant, in the spread of scientific management within the USA and to some European countries, notably Germany (Bryson, 2000). The early history of some large companies in both the USA and Europe is one in which F.W. Taylor, his close disciples, a group of neo-Taylorites and competing management consultants spread knowledge of forms of scientific management into and between companies and countries (Guillen, 1994). Consultants were very much at the centre of the development of what has come to be called the Fordist production system. Thus, the history of Fordism is also a history of producer services.

Second, as early as 1884, W.H. Lever, the founder of Lever Brothers (Unilever), manufacturers of Sunlight soap, commissioned W.P. Thompson, an independent trademark and patent expert, to identify a suitable name for the company's new soap. The name Sunlight, supported by a proactive advertising campaign, resulted in the creation of a mass market for Lever's mass produced soap. This campaign actually coincided with a flow of international management and marketing knowledge; Lever's experiments in competitive advertising were founded on his awareness of marketing and advertising techniques used in the USA. In 1888, for example, he purchased the famous slogan 'Why does a woman look old sooner than a man?' from Frank Siddall, a Philadelphia soapmaker (Williams, 1984). Lever's company was therefore very much a part of an economy of signs and spaces (Lash and Urry, 1994), as well as being incorporated into a form of informational capitalism.

The case for the emergence of informational capitalism or service-dominated economies rests on a series of assumptions concerning the relationship between information, knowledge, corporate restructuring, developments in information technology and the complexity of the capitalist economy. Information technology permits the acquisition, or capture, of greater quantities of information and its rapid transmission around the globe. These advantages are limited, however, to those companies with the resources (capital, equipment and time) to acquire and process information. Unequal access to information and knowledge is one of the fundamental features of capitalism. Lever's ability to travel in the United States provided him with exposure to American marketing ideas that he successfully transplanted to the UK. Even during the 19th century, external knowledge acted upon the existing knowledge embodied within Sunlight soap (e.g., the soap recipe, the brand, etc.) to couple mass production with mass marketing and mass consumption.

Critiques of informational capitalism have been developed by Webster (1994) and by Sayer and Walker (1992) amongst others (Bryson et al., 2004). These often take the form of a return to the Marxist canonical dogma (Walker, 1985). We want to make two key points. First, capitalist and pre-capitalist societies have always involved the articulation of information, knowledge and expertise. In Pompeii, 153 documents stored away in a wooden box were discovered. The text had been scratched into a wax coating over a wooden tablet. These tablets chart the financial transactions of Lucius Caecilius Juncundus, a 'banker', between 27 BCE and January 62 BCE. Juncundus was a

> characteristically Roman combination of auctioneer, middleman and moneylender. He was, in fact, as the tablets make clear, profiting from both sides of the auction process – not only charging commission to the sellers, but also lending money at interest to the buyers to enable them to finance their purchases. (Beard, 2010: 177)

All these financial activities involve transforming information into knowledge through the application of expertise and experience.

Second, access to, and control of, such information and knowledge has always been unevenly distributed. The last 20 years has simply witnessed an increase in the quantity, quality and availability of information and knowledge, but its uneven distribution, both geographically and between organizations of different sizes, has not dramatically altered. It would even be possible to argue that information and knowledge have become even more concentrated as they have become increasingly privatized, and hence more unevenly distributed. Companies with the greatest capital and resources are able to capture more information and are able to directly or indirectly employ individuals with the ability to transform information into knowledge. What is occurring is a monopolization of information and knowledge as well as its privatization. This concentration has implications, in turn, for the distribution and articulation of power in both the economy and wider society.

RESEARCH THEME 2: THE MANUFACTURING/SERVICE DIVIDE AND THE DOMINANCE OF PRODUCER SERVICE KNOWLEDGE

Capitalist production is controlled by the articulation of information into knowledge. In fact, all production is controlled in this way. One has only to read Iris Origo's seminal work on the life and work of Francesco di Marco Datini, a 14th-century Tuscan merchant, to accept this statement (Origo, 1988; also see Law, 1986). Francesco's business functioned by collecting information about the availability and price of goods from all over Europe sent to him through a network of couriers. Such information allowed decisions to be made regarding the relative risks of competing business opportunities.

Marx's circuit of capital is implicitly underpinned by the transformation of information into knowledge (Marx, 1984). Knowledge is required to identify a suitable source of capital, identify sources of raw materials, manage the labour force and manipulate the transformation of the finished commodity into capital. The emphasis in producer service research on one type of knowledge and expertise has obscured the numerous different types of information and knowledge that flow into client organizations via other sources and routes.

There are numerous ways in which an organization acquires new types of information and knowledge: a new managing director; boardroom linkages; the recruitment of a new manager; publications and other forms of media; scrutiny of competitors' products; interim management; take-over activity; movement of knowledge down the supply chain; and various forms of untraded interdependency (Storper, 1995), such as membership of a trade association, discussions in bars and restaurants, golf clubs and other forms of social interaction. It may, of course, be the case that producer service companies operate to supply knowledge that is unobtainable from other sources. But we don't know the answer to this question. To concentrate research on producer service activities is to overlook the role of other sources of knowledge and expertise in the wider production process. The argument is not that the knowledge supplied by producer services is not important, but rather that no attempt has been made to place such knowledge into the wider context of that obtained from other sources. It is widely assumed in the literature that producer service firms perform an important role in increasing the profitability and productivity of client companies. This is an assumption that still needs to be validated via empirical research. Perhaps other sources of knowledge support the everyday running of an enterprise whilst producer services provide the expertise and knowledge that generate and facilitate significant alterations in company performance. It is a question of scale, but it is also a question that has not yet been addressed by the academic literature.

This bias towards one type of knowledge and expertise, that which is provided by producer service firms, highlights the problem of placing an arbitrary boundary between manufacturing, service and public sector activities. Scholars need to reconsider the boundaries that they place around groups of economic activities. What is required is a holistic approach to understanding the flow of knowledge within and between organizations. The focus should be on knowledge *per se*, rather than on producer services, or even more generally on services. By targeting research on services, rather than knowledge, geographers have increasingly constrained their understanding of the relationship between the knowledge and expertise supplied by producer service firms relative to that available from other sources within the wider business environment or regional economy.

The debate over 'manufacturing' versus 'service' economies ignores the complex network of relationships (interdependencies) that exist between these 'two' sectors of the economy (Crum and Gudgin, 1977). The logical extension of the concept of interdependency is to argue that the complexity, and dynamic nature, of the manufacturing/service interrelationship makes the artificial division of these two economic sectors increasingly unsustainable. Thus, Apple designs and markets its products but subcontracts production to other manufacturing companies. For many years the majority of Apple products were manufactured by Foxconn Technology Group, the trade name for Hon Hai Precision Industry Co. Foxconn had become the world's largest electronics contract manufacturer and the scale of its activities was making it increasingly difficult for Apple to control the company; Foxconn was changing components in Apple products without notifying Apple. In 2013, as part of a risk diversification strategy, Apple began to subcontract the manufacture of low-cost i-Phones to Pegatron, Taipei. The Foxconn and Pegatron business model is based on secrecy and tact and they manufacture similar products for many different competing clients (Dou, 2013). Apple's relationship with Foxconn and Pegatron highlights the complex interrelationships between manufacturing and service functions.

A machine tool manufacturing company we have interviewed owns and controls the tool designs and sells them to client companies, but the manufacturing process is subcontracted. Most Standard Industrial Classifications would classify this company as a manufacturing firm, but in reality it is a service activity, or perhaps a manufacturing/service company or even a 'manuservice' company.

The manufacturing and service boundary is partially the consequence of the role ascribed to manufacturing activities by traditional economic theory. A good example is Kaldor's (1966) proposition that growth in manufacturing output will also generate productivity improvement inside and outside the manufacturing sector. This is not a new idea; it is encapsulated in the expression that 'manufacturing is the engine of economic growth'. Eatwell (1982) endorses the engine of economic growth concept, but develops an argument based on false reductionism by stating that manufacturing goods are essential for other sectors of the economy (Hodgson, 1989: 87). Service activities and agricultural products are equally as important for manufacturing.

To extend the argument concerning the over-emphasis on producer service knowledge in economic geography, it is useful to briefly explore the ways in which companies can acquire access to different types of knowledge and expertise.

1) Knowledge Transfer within an Organization

The internal environment of the large corporation provides both formal and informal opportunities for the development and acquisition of knowledge and expertise (Schoenberger, 1997). New management recruits into large companies undergo a formal training programme. The division of managerial knowledge into specialist subareas, such as human resource management or marketing and strategy, allows for the development of specialist expertise. The increasing specialization or division of labour of the management function also provides opportunities for producer services to supply expertise that is outside the scope of a particular management specialism. The increasing specialization of the management function creates niches that are being filled by generalist producer service professionals. A highly trained management cadre has not and will not reduce the continual use of external consultants for other reasons besides that of knowledge; for example, the political need to obtain a 'neutral' external opinion or form of external endorsement for an initiative developed within a firm.

The number of managers in a large company facilitates management dialogue and this leads to the construction of an organizational culture and way of understanding both the internal and external operating environment of the company. The movement of managerial personnel between companies enables the transfer of knowledge and expertise produced in particular institutional, cultural, social and political frameworks. Some evidence exists that points towards an increase in the intensity of executive turnover. A survey of 1,188 American companies found that executive turnover had increased by 27 per cent over a three-year period (Martinez, 1995). It was, however, also noted that companies had also increased their use of legal agreements and economic incentives in an attempt to reduce executive turnover and, by extension, the transfer of knowledge.

2) The Movement of Managers between Companies

The development of a manager's career in a large organization usually involves a university degree, participation in internal and external management training courses, and possibly the acquisition of an MBA. Part of the rationale for the latter is to make the right contacts, and to ensure that they are maintained during the course of a career. Coupled with this formal training will be experience obtained from working for a number of organizations. Increasingly, however, a career in a large corporation is no longer a career for life (Kanter, 1993). Employees have to rely on their own abilities, reputations and professional accreditation to structure their experience of work in a number of different corporate environments. Professional careers have replaced those structured around a single company's promotional hierarchy. Individual employees have to develop reputational capital rather than company-specific organizational capital. The formal training and experience of working for a variety of different organizations ensures that the managers of large companies develop both a national and international perspective on management issues and problems. It also results in the acquisition of a network of contacts within the internal labour markets of large companies. This coupling of independent organizations through the individualities of employee friendships and acquaintances provides the possibility for the exchange of expertise and for stimulating an alternative way of understanding the management process. To Schoenberger, people who run firms have histories and such histories provide them with 'revisable prior knowledge' (1997: 98), revisable via networks of social relationships. The movement of professionals between organizations ensures that the existing status quo is constantly challenged by new 'independent' appraisals of current working practices.

In addition, professional managers bring with them tacit knowledge, which partially consists of the relationships which they have established with independent producer service organizations, professionals, suppliers and competitors. It is impossible to control the 'untraded' transfer of this form of knowledge. A study by Wiersema (1995) found that unexpected executive turnover can significantly alter the perspective of a management team and is an important mechanism by which firms adapt to the changing competitive environment. Thus, a new chief executive or manager may identify a problem and hire a consultant they had previously employed whilst working for another organization. Hofer suggests that a necessary 'precondition for almost all successful [company] turnarounds is the replacement of current top management' (Hofer, 1980: 25–26; Kesner and Dalton, 1994).

The process of management, and the expertise and knowledge available to large organizations, is constantly changing. This is a research area that has been neglected by geographers. It is an area that provides opportunities for both theoretical and methodological advances in the discipline. Methodologically it provides an opportunity to explore the role of life histories in the construction of academic understandings of economic processes (see Katz and Monk, 1993). Life histories of managers and producer service professionals would deepen understanding of the ways in which knowledge flows through social networks.

3) Board Room Interlinkages

Boardroom interlinkages between large companies provide opportunities to acquire information about corporate practices, regulatory and political changes and macro-economic expectations (Scott, 1979; Bryson, 1997a). Such linkages allow for continual benchmarking of organizational performance and behaviour, as well as for untraded transfers of information and expertise. Turnover at Board level will alter a company's ties with other companies and organizations (Bryson and Daniels, 1998). The acquisition of RJR Nabisco dramatically reduced the director's external network with other companies (Useem, 1990: 701). Prior to the acquisition nine of the 18 directors served on a total of 22 boards of other major companies. After the merger, only three of the directors served on other boards. If a reduction in the number of external directorships is a usual consequence of acquisition, then this may lead to the 'fragmentation of collective business action' (Useem, 1990: 701).

RESEARCH THEME 3: THE TYPES OF KNOWLEDGE AND EXPERTISE SUPPLIED BY PRODUCER SERVICE COMPANIES

The third research theme is the most controversial for geographers. Organizations, professions and producer service activities share within each of their respective organizational fields a 'social stock of knowledge' (Berger and Luckmann, 1966: 56). The social stock of knowledge is accumulated by individuals during everyday interaction: from the family, through the educational system, during the socialization process, and is also transmitted between generations. Every individual accumulates a different social stock of knowledge on the basis of education, class, gender and occupation. The social stock of knowledge 'permits the "location" of individuals in society' (Berger and Luckmann, 1966: 56) in that it enables the identification and interpretation of both tacit and explicit knowledge and signs concerning an individual's identity, role and position. The social stock of knowledge is relative, individualistic, whilst at the same time shared. It also consists of what the Austrian sociologist Alfred Schutz termed 'recipe knowledge', or knowledge which is routine, as well as tacit, or unknown taken-for-granted, knowledge (Schutz, 1962; Schutz and Luckmann, 1974).

We want to argue that producer service knowledge is in most cases *recipe knowledge*, or what Clegg and Palmer term 'cookbook knowledge' (1996: 4). The recipes are learnt either formally or informally. Formal acquisition occurs during an MBA or via other forms of internal or external management course and professional qualification, for example architectural training. Our argument is that social scientists need to be extremely careful about the assumptions made about the types of knowledge and expertise that are supplied by producer service professionals. The tendency has been to label all these activities as high order or knowledge intensive. There are, of course, producer services that provide 'unique' solutions to particular problems. Nevertheless, there are also producer service sectors that contain a significant proportion of individuals and firms offering relatively standard knowledge recipes (the average architect) alongside a smaller number of individuals and companies that provide 'analytical' knowledge. Like Schein we believe that the difference between basic and applied science is that the first is 'convergent' and the second 'divergent'; producer service firms either provide forms of convergent or of divergent knowledge. Divergent knowledge is provided when a professional takes a 'convergent knowledge base and convert[s] it into professional services that are tailored to the *unique* requirements of the client' (Schein, 1973: 45). The problem is that the academic literature appears to assume that most producer service expertise is divergent when it might actually be convergent. It is important, however, to remember that today's recipe knowledge was yesterday's new innovation that has been diffused and commercialized via networks of private sector producer service companies.

Informal acquisition of a recipe occurs during the experience of work, or via the business media. No manager can fail to be exposed to the latest recipes as they are usually explored in the pages of the business press, magazines, or on radio and television. Such recipes are constructed either by academics or by consultancy companies. Recipes may, of course, also be company specific. Producer service companies specialize in the transfer of recipe knowledge between companies. Rather than supplying high-order knowledge, the typical management consultant uses a basic recipe cookbook of standard management and business methodologies, supported by a series of dialogues that govern the client/consultant interaction (see Clark and Salaman 1998). The recipe nature of most management consultancy knowledge implies that large consultancy companies can recruit fresh graduates and convert them into consultants via short training courses. These new consultants are sent out into client companies to service contracts for the latest fashion sold by experienced sales consultants. All they are doing is applying a simple recipe.

This notion of recipe knowledge challenges Reich's (1992) identification of three categories of service work: routine production services, in-person services and symbolic-analytic services. Symbolic-analytic services involve the non-standardized manipulation of a variety of forms of written and oral data. Included in this category are all occupations that involve problem-solving, identifying and strategic brokering activities (e.g. research scientists, design engineers, software engineers, civil engineers, biotechnology engineers, sound engineers, public relations executives, investment bankers and all types of consultancy). The difficulties with Reich's discussion of the education and careers of symbolic analysts is his failure to distinguish between symbolic-analytic tasks which involve a routine 'recipe'-driven mode of analysis and those which are truly analytical. A related point is that most symbolic analysts actually undertake a combination of routine and non-standard analytical functions. Reich argues that symbolic analysts account for no more than 20 per cent of American jobs. He is aware, however, that not all professional or managerial activities are symbolic as, for example, '[s]ome accountants do routine audits without the active involvement of their cerebral cortices' (1992: 181). Recipe knowledge suggests that producer service employment may polarize into those employees able to provide high-order knowledge and those supplying recipe knowledge. There are indications that such a process of polarization is beginning to happen in accountancy, with the growth in the number of accountancy technicians (Hanlon, 1994), and in higher education, with the growth in the number of part-time teaching-only posts.

The recipe nature of consultancy expertise is highlighted in the work of Guillen (1994), Gill and Whittle (1992) and Huczynski (1996). Guillen identifies the most important management models and textbooks in use between the years 1883 and 1981, and argues that the adoption of these models or paradigms of organizational management is not related to their scientific credibility, or economic or technological factors. Adoption differs between country and cultures depending on the educational training of managers, the activities of professional groups, the role of government and the attitudes of employees. He also notes that the management models that are adopted tend to have 'both an ideology of organizational management and a set of techniques' (Guillen, 1994: 2). Gill and Whittle (1992) identify the cyclical nature of consultant-led management models. They argue that such models experience a life cycle of initial identification and adoption by a very small number of consultancy companies, followed by a period of enthusiasm and wide-scale adoption of the model, and finally a period of disillusionment. They identify three such cycles of consultancy-driven 'recipe' transformation – management by objectives, organization development and total quality management. This 'recipe' cycle is driven by the constant search by consultancy companies for new management tools or models to sell to their existing client base. Consultancy companies must introduce a new product once a recipe has been widely adopted by client companies.

RESEARCH THEME 4: ORGANIZATIONAL CULTURE OR ORGANIZATIONAL CULTURES

Understanding the relationship between the knowledge (divergent or convergent) supplied by producer service firms and its impact on client companies also requires reference to the operation of corporate culture. This is an elusive concept that is difficult to define and to research. Corporate culture may be defined as the behavioural norms, customs, conventions, ways of doing things that determine the nature of social interaction within a company. A company's culture is a combination of the belief systems of individual employees and that of the company (Schoenberger, 1997). Company culture is produced via the stories that are passed throughout a company, via face-to-face meetings, telephone calls, e-mail messages and published material. Storytelling is especially important in maintaining an organization's culture (Feldman, 1990) and in reproducing it (Bryson and Lowe, 2002; Rusten et al., 2007). New employees are rapidly socialized into the existing culture. Our research into consultancy-driven organizational change highlights the multiple, or heterogeneous, natures of corporate culture. The heterogeneity of organizational cultures influences the way in which knowledge is articulated at each level, subsidiary and location of an organization (Bartunek and Moch, 1991). There are two dimensions to this heterogeneity: within consultancy companies and within client companies.

In the academic literature it is frequently assumed that a large consultancy company will possess a single culture and corporate agenda. This is a doubtful proposition. First, most large consultancy companies are divided into budget centres organized on a regional or specialism basis (Bryson and Rusten, 2005). Each budget centre will have its own agenda, profitability and productivity measures. During the negotiation of a consultancy contract, and even during the course of a project, each budget centre will try to distort the nature of the client interaction to suit its own agenda. The multiple and conflicting agendas of different parts of the same organization is an area in urgent need of research because these conflicting agendas determine the way in which producer service knowledge enters an organization as well as determining the internal transfer of knowledge. Second, the way in which a consultancy interaction functions, and the way it is received by client employees, will vary depending on the situation of the employee within the power structure of the company, as well as on the contested nature of the client's corporate culture. This complex set of issues has both theoretical and methodological implications.

The simplest way to illustrate this is with a short case study. One of the companies we have recently interviewed provides a service using a poorly paid, semi-contract-based workforce of 2,000–2,500 employees. This workforce operates at 34 sites located throughout the United Kingdom and Ireland and is managed by a team of 30 located in a central headquarters. The company suffers from poor staff retention (in excess of 60 per cent per year) and low morale. A number of different types of consultancy-driven recipes have been applied to the company in the past. All these have failed to solve the staffing-related problems.

The company employed a small consultancy company that sold the managing director a form of *kaizen*, or the Japanese method of continuous improvement. The project involved introducing the senior management to the philosophy of *kaizen* and providing workshops for all employees. The particular form of *kaizen* provided by the consultant involves a philosophy in which continuous improvement becomes part of every employee's everyday routine, and one in which information, ideas and knowledge flow from the customer through the employee to the management and managing director. The emphasis is on the 'real experts' in the organization; in other words, on the people who undertake the task, rather than on those who manage. Thus, the consultancy project involves a transfer of power from management to the employees. One consequence is the introduction of a system of in-house *kaizen* coaches and a weekly *kaizen* discussion meeting (tutorial) for all staff. Employees are encouraged to identify problems and difficulties in the company that need to be solved. Once a problem has been identified, a *kaizen* project team is identified, the problem made visible, the causes of the problem defined, facts collected, improvement ideas generated, and a new procedure identified and adopted. During this process objectives are formulated and placed on a flow chart that is updated to illustrate the progression of the project. The successful completion of the project is rewarded by the presentation of a small token gift, for example a piece of pottery.

Discussions with the consultant, managing director, senior management and individual employees highlight the contested nature of this client/consultant interaction. The consultant provides a simple justification for the technique illustrated by a stream of examples of successful consultancy projects. He provides recipe knowledge coupled with a dynamic charismatic personality. The managing director believes in the recipe that has been purchased from the consultant, but, along with the management team, has tampered with the basic principles of the *kaizen* approach. This has been achieved by encouraging employees to undertake as many *kaizen* projects as possible, and to set invisible measures on the number and quality of projects undertaken by individual employees and by teams of employees. Management is more interested in the visibility and number of continuous improvement projects as against their quality.

Employees respond to *kaizen* by using four strategies. First, they ignore it by discounting the consultancy project as another management fad that will soon be replaced by another trendy, but equally idiotic, strategy. Secondly, they subvert the initiative by identifying simple *kaizen* projects that are supported by an easily achievable set of objectives. They thus appear to play the 'game' to the rules set by the consultant and the management team. Third, employees engage the management team in *kaizen* projects that place managers in situations in which they will have to improve the working conditions of employees, or the way in which the company operates. Fourth, the directors' attitude to *kaizen* is that their role and function are above that of the *kaizen* philosophy. They 'participate' by delegating *kaizen* projects to their staff, by keeping a simple *kaizen* project going for an extremely long period of time, or by producing documentation for *kaizen* projects retrospectively. Employee workplace strategies illustrate that the company's corporate culture is very different depending on the location of the individual in the company's corporate power structure.

A good example of the contested nature of this organization's corporate culture comes from an attempt to recruit a new consultancy company to monitor the implementation of a period of expansion. Three consultancy companies were short-listed to attend a 'beauty parade'. Each consultancy company was given open access to the managing director, directors and employees so that they could collect the necessary information required to complete the tender document and formulate their presentation. After a period of consultation one of the consultancy companies withdrew from the contest on the grounds that it was obtaining different messages from the managing director, directors and staff. Consequently the consultant considered that there were too many conflicting internal interpretations of the company's goals for the consultancy company to identify an effective strategy. Rather than engage in a consultancy project that already contained the conditions for failure, the consultant decided to withdraw from the contest. The consultant recommended that the company should not appoint a consultancy company but should delegate the task to a permanent in-house member of staff and recruit consultancy support if, and when, it was required. The managing director's response to this was to claim that he did not like the consultancy company, and in any case he would never have appointed that company.

This case study highlights the social and political nature of management interaction. The contested nature of this interaction is complicated by the client/consultancy relationship. All these relationships – between consultant and company, management and employees, employees and managers – are engaged in a process of translation as each company employee attempts to reconcile their position in the company with management or consultancy recipes. The contested nature of the company and its relationship with external suppliers of knowledge imply that a research methodology which fails to address such cultural heterogeneity will produce a distorted understanding of the process of change as well as a distorted understanding of knowledge flows into and between companies. To understand the relationship between producer service firms and client organizations necessitates a complex multi-dimensional research strategy which will attempt to identify and relate together the different ways in which knowledge flowing into an organization is incorporated into its internal environment.

RESEARCH THEME 5: POWER AND CONSUMPTION

One of the sociological concepts that is noticeably absent or under-theorized in economic geography, and especially in the geographical literature on producer services, is the concept of power (Allen, 1997). Its absence partially reflects the methodologies employed by economic geographers. Power is a relational construct. As such it can only be captured by a methodology that, for example, explores both sides of client/producer service supplier interaction. Dicken and Thrift (1992: 287), in their re-examination of the case for research that focusses on the geography of the business enterprise, identify power as important in the 'networks of internalized and externalized transactional relationships' through which business enterprises operate. The blurring of the boundaries of organizations so that knowledge and expertise are partially housed in external producer service companies is fundamentally a relationship of power. Consequently, geographers need to clarify the dimensions of these power relationships and develop a 'power based' analysis of producer service expertise (Bryson et al., 2004; Bryson and Rusten, 2011). This is, however, only one of the arenas of power.

Another possible arena lies in the way in which client companies consume producer service knowledge. Clark and Salaman (1998) have explored these issues through their work on consultancy as performance. The concept of power is central to understanding consultancy as performance and client consumption of consultancy expertise (Wellington and Bryson, 2001). The same is true of many other producer service activities. This is to argue that, like economic geography in general, producer service research has primarily been production focused. It is time to develop a consumption-oriented analysis of producer service activities. This is a methodological challenge because recent work suggests that the way in which a client consumes the knowledge obtained from a consultant changes over time (Bryson et al., 1999b). Client satisfaction may become dissatisfaction, and vice versa. To complicate matters, the nature of the consumption will also vary between employees as well as departments or subsidiaries within the same company.

The debate over the externalization of producer service activities is complicated by introducing power into the analysis (Bryson et al., 2004). Thus, 'why is external knowledge sometimes perceived by a client as superior to internal knowledge?' Linked to this question is 'why is simple knowledge supplied by an external consultant sometimes more effective than the same knowledge supplied by an employee?' Such questions can be explored by examining notions of power. McLean (1984) (see also McLean et al., 1982) identifies four different types of power that are involved in the client–consultant relationship: collusive, intuitive, paradoxical and analogic. Two of these are briefly explored here. First, collusive power occurs when the client constructs an image of the consultant as expert and the consultant plays up to this image; both mutually accept the deception. Thus, the client may implement change as a result of a 'placebo' phenomenon where the change will occur because the client believes that the consultant will change the organization. Second, paradoxical power is where a disastrous consultancy intervention results in the client introducing inappropriate changes to working practices or procedures. Management has to consider why the project failed, and by this process they may change the organization. In an effort to better understand the place of knowledge in producer service–client relations geographers may need to adopt/adapt power schemas of the kind suggested by McLean to their studies of producer services.

CONCLUSION

The rapid growth in the supply and demand for producer services has encouraged research on understanding the role and geography of their activities. Our research, together with our reading of the producer service literature, has stimulated us to re-evaluate and reflect on the current and future condition of producer service research. This suggests that there are at least five research themes that need to be addressed in the interests of a more balanced and informed programme of research on producer services:

1. Historical research is required into the role played by knowledge and information flows in previous rounds of capital investment. The visibility of independent producer service firms, combined with the difficulty of identifying the contribution that in-house producer service expertise makes to client companies, has distracted attention from knowledge in the evolution of the capitalist production process.
2. Research is needed into the varied nature of knowledge flows, at the level of both the organization and the region. Such research should explore producer service expertise in relation to other forms of knowledge and expertise. Our research has explored these issues in relation to small and medium-sized companies (Bryson and Daniels, 2007), but further comparative research will be required to identify any cultural differences in the articulation of knowledge into, within and between large client companies. This should be done as a comparative international research project.
3. A critical evaluation of the types of information, knowledge and expertise supplied by producer service firms is long overdue. This should explore the recipe nature of such knowledge as well as translations and distortions that occur in the process by which an academic management 'model' is converted into the latest producer service fashionable panacea. One of the key issues is why some management ideas are converted into recipes and others ignored.
4. The influence of corporate culture on knowledge management and utilization within producer service firms and client companies needs to be explored in considerable depth. This could involve in-depth ethnographic research undertaken with relatively unrestricted access to a producer service company and one of its large clients. It should also involve detailed longitudinal research.
5. A power-based analysis of producer service activities urgently needs to be developed.

At a theoretical level, the debate over the development of knowledge or informational capitalism needs to carefully consider the role and activities of producer service firms. Their growth in the advanced economies represents not the growth of companies that provide clients with access to high-order information and knowledge, but the growth of companies that provide clients with access to different forms of expertise and knowledge. In some cases producer service companies are not commissioned for their knowledge but for their 'neutral' voice. Such neutrality is mythical as it is distorted by whoever commissioned the project. We accept that there are companies and individuals that create new knowledge, but the majority of producer service companies are engaged in applying and positioning established recipes into the cultural and political environments of client companies. This is not to say that such knowledge is of limited value. This is not our argument, as we are very aware that producer service knowledge has played and is playing a fundamental role in organizational and economic change.

Gill and Whittle (1992) label consultancy models as panaceas that promise to provide a universal solution to a recurrent management problem. The producer service revolution is part of the 'mercantalization of knowledge', or, in other words, the capitalist economy's ever increasing process of commodification. The agenda of knowledge-intensive producer service firms is to provide expertise and knowledge to as many clients as possible, as quickly as possible. In fact what has occurred is a shift in knowledge away from use values towards exchange values. A new management idea is accepted and the technique or model perfected for its exchange into client companies without any real concern for scientific rigour. This process of knowledge commodification is summarized by Lyotard when he explores the relationship between the use and exchange of knowledge:

> The relationship of the suppliers of knowledge and the users of knowledge, to the knowledge that they supply and use, is now tending, and will increasingly tend to assume the form already taken by the relation of commodity producers and consumers to the commodities that they produce and consume – that is, the form of value. Knowledge is and will be produced in order to be valorized in a new production: in both cases, the goal is exchange. Knowledge ceases to be an end in itself, it loses its 'use-value'. (Lyotard, 1984: 4)

The research agendas identified in this chapter are also associated with a series of methodological problems or challenges. To advance producer service research requires the development of sophisticated methodologies that will capture the life histories of producer service professionals as well as chart the ways in which producer service knowledge and expertise enter and are transformed by the culture of a client organization. Longitudinal research is urgently required to identify and explore these processes. This might not appear to be a geographical question, but such transformations will differ between countries as well as between companies with different cultural and ethnic backgrounds.

BIBLIOGRAPHY

Allen, J. (1997) 'Economies of power and space', in Lee, R. and Wills, J. (eds) *Geographies of Economies*, Arnold, London, 59–70.
Bartunek, J.M. and Moch, M.K. (1991) 'Multiple constituencies and the quality of working life: intervention at Food Com', in Frost, P.J., Moore, L.F., Louis, M.R., Lundberg, C.C. and Martin, J. (eds) *Organisational Culture*, Sage, London, 58–76.
Beard, M. (2010) *Pompeii*, Profile Books, London.
Bell, D. (1974) *The Coming of Post-industrial Society*, London, Heinemann.
Bell, D. (1979) 'The social framework of the information society', in Dertouzos, M.L. and Moses, J. (eds) *The Computer Age: Twenty-Year View*, MIT Press, Cambridge, MA, 163–211.
Berger, P. and Luckmann, T. (1966) *The Social Construction of Reality: A Treatise in the Sociology of Knowledge*, Penguin Books, Harmondsworth.
Beyers, W.B. (1979) 'Contemporary trends in the regional economic development of the United States', *Professional Geographer*, 31: 34–44.
Beyers, W.B. (1989) *The Producer Services and Economic Development in the United States: The Last Decade*, Economic Development Administration, US Department of Commerce, Washington, DC.
Beyers, W.B. (1991) 'Trends on producer services in the U.S.: the last decade', in Daniels, P.W. (ed.) *Services and Metropolitan Development*, Routledge, London, 146–172.
Beyers, W.B. (1998) 'Trends in producer service employment in the United States: the 1985–1995 experience', paper presented at the 45th North American Meetings of the Regional Science Association, Sante Fe, New Mexico, November (mimeo).

Beyers, W.B. and Alvine, M.J. (1985) 'Export services in post-industrial society', *Papers of the Regional Science Association*, 57: 33–45.
Beyers, W. and Lindahl, D. (1996) 'Lone eagles and high fliers in rural producer services', *Rural Development Perspectives*, 11: 2–10.
Beyers, W.B. and Lindahl, D. (1997) 'Strategic behaviour and development sequences in producer service businesses', *Environment and Planning A*, 29: 887–912.
Beyers, W.B., Alvine, M.J. and Johnsen, E.G. (1985a) *The Service Economy: Export of Services in the Central Puget Sound Region*, Central Puget Sound Economic Development District, Seattle.
Beyers, W.B., Alvine, M.J. and Johnsen, E.G. (1985b) 'The service sector: a growing force in the regional export base', *Economic Development Commentary*, 9, 3: 3–7.
Brotton, J. (2007) *The Sale of the Late King's Goods: Charles I & His Art Collection*, Pan Books, London.
Bryson, J.R. (1997a) 'Business service firms, service space and the management of change', *Entrepreneurship and Regional Development*, 9: 93–111.
Bryson, J.R. (1997b) 'Small and medium-sized enterprises, Business Link and the new knowledge workers', *Policy Studies*, 18: 67–79.
Bryson, J.R. (2000) 'Spreading the message: management consultants and the shaping of economic geographies in time and space', in Bryson, J.R., Daniels, P.W., Henry, N. and Pollard, J. (eds) *Knowledge Space, Economy*, Routledge, London, 157–175.
Bryson, J.R. (2009a) 'Service innovation and manufacturing innovation: bundling and blending services and products in hybrid production systems to produce hybrid products', in Gallouj, F. (ed.) *Handbook on Innovation in Services*, Edward Elgar, Cheltenham and Northampton, MA, 679–700.
Bryson, J.R. (2009b) *Hybrid Manufacturing Systems and Hybrid Products: Services, Production and Industrialisation* University of Aachen, Aachen.
Bryson, J.R. and Daniels, P.W. (1998) 'Business Link, strong ties and the walls of silence', *Environment and Planning C: Government and Policy*, 16: 265–280.
Bryson, J.R. and Daniels, P.W. (2007) 'Small and medium-sized enterprises and the consumption of traded (producer service expertise) versus untraded knowledge and expertise', in Bryson, J.R. and Daniels, P.W. (eds) *The Handbook of Service Industries in the Global Economy*, Edward Elgar, Cheltenham and Northampton, MA, 295–310.
Bryson, J.R. and Lowe, P.A. (2002) 'Story-telling and history construction: retelling the story of George Cadbury's Bournville model village', *Journal of Historical Geography*, 28, 1: 21–41.
Bryson, J.R. and Rusten, G. (2005) 'Spatial divisions of expertise: knowledge intensive business service firms and regional development in Norway', *Services Industries Journal*, 25, 8: 959–977.
Bryson, J.R. and Rusten, G. (2011) *Design Economies and the Changing World Economy: Innovation, Production and Competitiveness*, Routledge, London.
Bryson, J., Keeble, D. and Wood, P. (1993a) 'The creation, location and growth of small business service firms in the United Kingdom', *Service Industries Journal*, 13, 2: 118–131.
Bryson, J.R., Wood, P. and Keeble, D. (1993b) 'Business networks, small firm flexibility and regional development in UK business services', *Entrepreneurship and Regional Development*, 5: 265–277.
Bryson, J.R., Keeble, D. and Wood, P. (1997a) 'The creation and growth of small business service firms in post-industrial Britain', *Small Business Economics*, 9, 4: 345–360.
Bryson, J.R., Churchward, S. and Daniels, P.W. (1997b) 'From complexity to simplicity? Business Link and the evolution of a network of one-stop-advice shops: a response to Hutchinson, Foley and Oztel', *Regional Studies*, 31, 7: 720–723.
Bryson, J.R., Daniels, P.W. and Ingram, D.R. (1999a) 'Evaluating the impact of business link on the performance and profitability of SMEs in the United Kingdom', *Policy Studies*, 20, 2: 95–105.
Bryson, J.R., Daniels, P.W. and Ingram, D.R. (1999b) 'Methodological problems and economic geography: the case of business services', *Service Industries Journal*, 19: 4.
Bryson, J.R., Keeble, D. and Wood, P. (1994) *Enterprising Researchers: The Growth of Small Market Research Firms in Britain*, Business Services Research Monograph, Series 1, No. 2, Small Business Research Trust.
Bryson, J.R., Daniels, P.W. and Warf, B. (2004), *Service Worlds: People, Organisations, Technologies*, Routledge, London.
Bryson, J.R., Rubalcaba, L. and Strom, P. (2012) 'Services, innovation, employment and organisation: research gaps and challenges', *Service Industries Journal*, 32, 3–4: 641–657.
Castells, M. (1989) *The Informational City: Information Technology, Economic Restructuring and the Urban-Regional Process*, Blackwell, Oxford.
Castells, M. (1996) *The Information Age: Economy, Society and Culture, Vol 1: The Rise of the Network Society*, Blackwell, Oxford.
Clark, T. and Salaman, G. (1998) 'Creating the "right" impression: towards a dramaturgy of management consultancy', *Service Industries Journal*, 18, 1: 18–38.
Clegg, S.R. and Palmer, G. (eds) (1996) *The Politics of Management Knowledge*, Sage, London.

Cook, P. and Morgan, K. (1998) *The Associational Economy: Firms, Regions, and Innovation*, Oxford University Press, Oxford.
Crum, R.E. and Gudgin, G. (1977) *Non Production Activities in UK Manufacturing Industry*, Regional Policy Series 3, Commission of the European Communities, Brussels.
Daniels, P.W. (1985) *Service Industries: A Geographical Appraisal*, Methuen, Andover.
Daniels, P.W., Rubalcaba, L., Stare, M. and Bryson, J.R. (2011) 'How many Europes?: Varieties of capitalism, divergence and convergence and the transformation of the European services landscape', *Tidschrift voor Economische en Sociale Geografie*, 102, 2: 146–161.
Dicken, P and Thrift, N. (1992) 'The organization of production and the production of organizations: why business enterprises matter in the study of geographical industrialization', *Transactions of the Institute of British Geographers*, New Series, 17, 3: 279–281.
Dou, E. (2013) 'Apple shifts supply chain away from Foxconn to Pegatron', *Wall Street Journal*, 29 May, available at: http://online.wsj.com/news/articles/SB10001424127887323855804578511122734340726, accessed 5 August 2014.
Eatwell, J. (1982) *Whatever Happened to Britain? The Economics of Decline*, BBC, London.
Feldman, S.P. (1990) 'Stories as cultural creativity: on the relations between symbolism and politics in organisational change', *Human Relations*, 43, 9: 809–828.
Fik, J., Malecki, E.J. and Amey, R. (1993) 'Trouble in paradise? Employment trends and forecasts for a service-oriented economy', *Economic Development Quarterly*, 7: 358–372.
Gallouj, F. and Djellal, F. (eds) (2010) *Handbook on Innovation in Services*, Edward Elgar, Cheltenham and Northampton, MA.
Gill, J. and Whittle, S. (1992) 'Management by panacea: accounting for transience', *Journal of Management Studies*, 30, 2: 281–295.
Guillen, M.F. (1994) *Models of Management: Work, Authority, and Organisation in a Comparative Perspective*, University of Chicago Press, Chicago.
Hanlon, G. (1994) *The Commercialisation of Accountancy: Flexible Accumulation and the Transformation of the Service Class*, Macmillan, London.
Hill, T.P. (1977) 'On goods and services', *Review of Income and Wealth*, 23: 315–338.
Hodgson, G. (1989) 'Institutional rigidities and economic growth', in Lawson, T., Gabriel Palma, J. and Sender, J. (eds) *Kaldor's Political Economy*, Academic Press, London, 79–101.
Hofer, C.W. (1980) 'Turnaround strategies', *Journal of Business Strategy*, 1: 19–31.
Huczynski, A.A. (1996) *Management Gurus: What Makes Them and How To Become One*, Thompson Business Press/Routledge, London.
Illeris, S. (1996) *The Service Economy: A Geographical Approach*, John Wiley, Chichester.
Kaldor, N. (1966) *Causes for the Slow Rate of Growth of the United Kingdom*, Cambridge University Press, Cambridge.
Kanter, R.M. (1993) *Men and Women of the Corporation*, Basic Books, New York.
Kastrinos, N. and Miles, I. (1996) 'Patterns of entrepreneurship in the UK environmental industry', paper presented in COST A3 conference on Management and Technology, Madrid, 12–14 June.
Katz, C. and Monk, J. (1993) *Full Circle: Geographies of Women over the Life Course*, Routledge, London.
Keeble, D. and Bryson, J.R. (1996) 'Small-firm creation and growth, regional development and the North–South divide in Britain', *Environment and Planning A*, 28: 909–934.
Keeble, D., Bryson, J. and Wood, P. (1991) 'Small firms, business services growth and regional development in the United Kingdom: some empirical findings', *Regional Studies*, 25, 5:439–457.
Keeble, D., Bryson, J. and Wood, P. (1994) *Pathfinders of Enterprise: The Creation, Growth and Dynamics of Small Management Consultancies in Britain*, Business Services Research Monograph, Series 1, No. 2, Small Business Research Trust.
Kesner, I.K. and Dalton, D.R. (1994) 'Top management turnover and CEO succession: an investigation of the effects of turnover on performance', *Journal of Management Studies*, 31, 5: 701–713.
Lash, S. and Urry, J. (1994) *Economies of Signs and Space*, Sage, London.
Law, J. (1986) 'On the methods of long distance control: vessels, navigation and the Portuguese route to India', in Law, J. (ed.) *Power, Action and Belief: A New Sociology of Knowledge?* Sociological Review Monograph, 32, Routledge and Kegan Paul, London, 234–263.
Love, J., Roper, S. and Bryson, J.R. (2011) 'Openness, knowledge, innovation and growth in UK business services', *Research Policy*, 40, 10, December: 1438–1452.
Lyotard, J.F. (1984) *The Postmodern Condition: A Report on Knowledge*, Manchester University Press, Manchester.
MacPherson, A. (1997) 'The role of producer service outsourcing in the innovation performance of New York State manufacturing firms', *Annals of the Association of American Geographers*, 87, 1: 52–71.
MacPherson, A. and Vanchan, V. (2008) 'The outsourcing of industrial design by large US manufacturing companies: an exploratory study', *International Regional Science Review*, 33, 1: 3–30.

MacPherson, A.D. and Vanchan, V. (2010) 'Locational patterns and competitive characteristics of industrial design firms in the United States', in Rusten, G. and Bryson, J.R. (eds), *Industrial Design, Competition and Globalization*, Palgrave Macmillan, Houndmill, 81–92.

Maglio, P., Kieliszewski, C. and Spohrer, J. (2010) *Handbook of Service Science*, Springer Science, New York.

Marshall, J.N. (1983) 'Business service activities in British provincial conurbations', *Environment and Planning A*, 15: 1343–1359.

Marshall, J.N., Wood, P., Daniels, P.W., McKinnon, A., Batchtler, J., Damesick, P., Thrift, N., Gillespie, A., Green, A. and Leyshon, A. (1988) *Services and Uneven Development*, Oxford University Press, Oxford.

Martinez, M.N. (1995) 'More executives change jobs', *Human Resources Magazine*, 40, 10: 15.

Marx, K. (1984) *Capital, Vol. 1*, Lawrence and Wishart, London.

McLean, A. (1984) 'Myths, magic, and gobbledegook: a-rational aspects of the consultant's role', in Kakabadse, A. and Parker, C. (eds) *Power, Politics and Organisations: A Behavioural Science View*, John Wiley, Chichester, 147–167.

McLean, A.J., Sims, D.B.P., Mangham, I.L. and Tuffield, D. (1982) *Organisation Development in Transition: Evidence of an Evolving Profession*, John Wiley, Chichester.

Micklethwait, J. and Wooldridge, A. (1996) *The Witch Doctors: What the Management Gurus are Saying, Why It Matters and How to Make Sense of It*, Mandarin, London.

O'Farrell, P.N., Wood, P.A. and Zheng, J. (1996) 'Internationalization of business services: an interregional analysis', *Regional Studies*, 30, 2: 101–118.

Origo, I. (1988) *The Merchant of Prato*, Penguin, London.

Perrow, C. (1974) '"Zoo story" or "Life in the organizational sandpit"', in Open University Course Team (eds) *Social Sciences: A Third Level Course in People and Organizations*, Open University Press, Milton Keynes, 146–168.

Ram, M. (1994) *Managing to Survive: Working Lives in Small Firms*, Blackwell, Oxford.

Reich, R.B. (1992) *The Work of Nations*, First Vintage Books, New York.

Rusten, G. and Bryson, J.R. (2010), 'Placing and spacing services: towards a balanced economic geography of firms, clusters, social networks, contracts and the geographies of enterprise', *Tidshrift voor Economische en Sociale Geografie*, 101, 3: 248–261.

Rusten, G., Bryson, J.R. and Aarflot, U. (2007) 'Places through product and products through places: industrial design and spatial symbols as sources of competitiveness', *Norwegian Journal of Geography*, 61, 3: 133–144.

Sassen, S. (1991) *The Global City: New York, London, Tokyo*, Princeton University Press, Princeton, NJ.

Sayer, A. and Walker, R. (1992) *The New Social Economy: Reworking the Division of Labour*, Blackwell, Oxford.

Schein, E. (1973) *Professional Education*, McGraw-Hill, New York.

Schoenberger, E. (1997) *The Cultural Crisis of the Firm*, Blackwell, Oxford.

Schutz, A. (1962) *Collected Papers, Vol 1: The Problem of Social Reality*, Martinus Nijhoff, The Hague.

Schutz, A. and Luckmann, T. (1974) *The Structures of the Life-World*, Heinemann, London.

Scott, J. (1979) *Corporations, Classes and Capitalism*, Hutchinson, London.

Shigeo, S. (1989) *A Study of the Toyota Production System*, Productivity Press, Portland, OR.

Singelmann, J. (1978) 'The sectoral transformation of the labour force in seven industrialised countries 1920–1970', *American Journal of Sociology*, 83: 1224–1234.

Starbuck, W.H. (1992) 'Learning by knowledge-intensive firms', *Journal of Management Studies*, 29, 6: 713–740.

Stehr, N. (1994) *Knowledge Societies*, Sage, London.

Storper, M. (1995) 'The resurgence of regional economies, ten years later: the region as a nexus of untraded interdependencies', *European Urban and Regional Studies*, 2, 3: 191–221.

Storper, M. and Salais, R. (1997) *Worlds of Production: The Action Frameworks of the Economy*, Harvard University Press, Boston, MA.

Thrift, N. (1988) 'Virtual capitalism: the globalisation of reflexive business knowledge', in Carrier, J. and Miller, D. (eds) *Virtualism: A New Political Economy*, Berg, Oxford, 161–186.

Tordoir, P. (1994) 'Transactions of professional business services and spatial systems', *Tijdschrift voor Economische en Sociale Geografie*, 85: 322–332.

Useem, M. (1990) 'Business restructuring, management control, and corporate organisation', *Theory and Society*, 19: 681–707.

Walker, R. (1985) 'Is there a service economy?', *Science and Society*, 49: 42–83.

Webster, F. (1994) *Theories of the Information Society*, Routledge, London.

Wellington, C.A. and Bryson, J.R. (2001) 'At face value? Image consultancy, emotional labour and professional work', *Sociology*, 35, 4: 933–946.

Wernerheim, C.M. and Sharpe, C.A. (1999) 'Producer services and the "mixed-market" problem: some empirical evidence', *Area*, 31, 2: 123–140.

Wiersema, M.F. (1995) 'Executive succession as an antecedent to corporate restructuring', *Human Resource Management*, 34, 1: 185–202.
Williams, E. (1984) *The Story of Sunlight*, Unilever, London.
Williams, I.A. (1931) *The Firm of Cadbury: 1831–1931*, Constable, London.
Wood, P.A. (1986) 'The anatomy of job loss and job creation: some speculation on the role of the producer service sector', *Regional Studies*, 20: 37–40.
Wood, P.A. (1988) 'The economic role of producer services: some Canadian evidence', in Marshall, J.N. et al. (eds) *Services and Uneven Development*, Oxford University Press, Oxford, 268–278.
Wood, P.A. (1990) 'Conceptualising the role of services in economic change', *Area*, 23: 66–72.

Index

Accenture 341
accounting firms (Big 4) 60–79
   client relationships 67–9, 72–8
   and differentiation 66–7
   and Enron scandal 63–5
   and Lehman's collapse 65–6
   services 61–2, 71–2
   and commoditisation 60–79
Acemoglu, D. 174
advanced business services *see* professional business services
agent-based modelling 54–5
agglomeration 37, 39, 227–30
   creative industries 355, 356
   producer services, China 396–8
aggregate complexity 352
Albert, S. 248
Alvstam, C.G. 37
Amazon 130
ameliorative innovation 99
Anand, N. 308
Apple Inc 141, 142, 345, 423–4
application economy 142
Apte, U.M. 172, 173, 174, 175, 177, 180, 188, 201
architectural firms, globalization 235
Arnould, E. 207
Arthur Anderson 64, 65
Asia
   cultural and creative industries 350, 360–65
   service sector 43–4
asset based logistics providers 339
assimilationist perspective 90–91
Ateljevic, J. 381
attribute-based approach to satisfaction 118
audit services 61–2, 71–2
Autor, D.H. 174

Bade, D.J. 342
BAE Systems 12–13
Bagchi-Sen, S. 232
Ball, D.A. 261
bank data on business activity 151–61
   and policy-related questions 161–6
Barclays Bank
   business survival data 157–9, 165
   New Firm Panel dataset 154–7, 165–6
   small business data 152–3

Barnett, D. 6
Barras, R. 91
Barrell, R. 32
barriers to globalization of services 270–73
Bask, A.H. 340
Bassanini, A. 32
Baumol, W.J. 8, 89, 201
Baumol model of unbalanced growth 89
Baumol's disease 130
BBA Small Business Survey 151–2
Beard, M. 418, 422
Beinhocker, E.D. 352
Bell, D. 5, 418
Bensemann, J. 378
Berger, P. 426
Berglund, M. 339, 340
Berman, E. 174
Beyers, W.B. 38–9, 40, 417
Bhagwati, J.N. 268
Big 4 accounting firms *see* accounting firms
Bitner, M.J. 113
Blake, A. 377
Bluewater Shopping Centre, UK 3
board room interlinkages 426
Bobbit, L.M. 115
Bolivia, water system privatization 272–3
Bolton Committee 167
born global firms 234–5, 279, 295–6
Bosworth, B.P. 28
Boulding, W. 121
Bowden, J.L.H. 110
Bowersox, D.J. 334
Boynton, A. 251
BPS *see* business and professional services
Bradley, K. 248
Brady, M.K. 121
Braudel, F. 7
Brodie, R.J. 111
Brotton, J. 418–19
Bryson, J.R. 11, 36, 38, 317, 325, 327
business activity, UK 146–67
   Barclays Bank data 152–61
   BBA survey 151–2
   business survival 157–61, 164–6
   comparison with US 163–4
   GEM survey 150–51
   number of businesses 148, 161–2
   official data 147–9

business closures 152–3, 156–7
business models 137–41
business and professional services (BPS) 317–21
  emergence 7
  management control 321–6
business, professional and technical services 192–3
  trade data 195, 196
business registers, UK 148–9
business service strategy, and location 40–41
business services 129–43
  business models 137–41
  and division of labour 132–6
  and economic geography 35–45
  and productivity 130–32
business survival, UK 157–61, 164–6

Campos-Soria, J.A. 378
capitalist production system 316–18
capturing value 139–40
Castells, M. 317, 418, 420
Castle, L. 261
CCIs (cultural and creative industries) 349–65
CCJQ (creative industry clusters), Singapore 363
Cetindamer, D. 32
Chadwick, M. 284, 291
Chambers, John 247
Chang, T.C. 361
Chapman, C. 228
Chase, R.B. 201
Cheng, D.Z. 396
China, producer services 392–411
Choi, M. 173, 201
Chow, G.C. 410
Cisco 247
cities and business location 228–9
  China 396–8
Clark, K.B. 98
Clarke, P. 344–5
Clarke, T. 431
classification of services 36–7
Clegg, S.R. 426
client account management 324–6
client-driven product innovation 304–7
client relationships 77–8, 324–6
client retention 325
client teams 73–6, 303
climate change concerns and tourism 385–7
CLM (Council of Logistics Management) 335
closure data, UK 152–3, 156–7, 159
clustering, creative industries 355, 356
co-production 88
Coad, A. 165
Coase, R. 133

Cole, W. 6
collaboration
  and logistics services 335–6
  management of 247, 249–51
collusive power 431
commoditisation 60–79
company culture 428–30
compensation *see* pay
complexity economics 352
composite modelling of service systems 55
computational models 54–5
Connect and Develop programme 129
constitution phase, paradigm development 211–12
consulting services 62, 94
consumer behaviour 105–23
  pre-purchase stage 107–10
  service encounters 110–15
consumer delight 116–17
consumer engagement 110–11
consumer satisfaction 116–22
consumption
  post-encounter stage 116–19
  pre-purchase stage 107–10
  of producer service knowledge 431
  service encounter stage 110–15
  three-stage model 105–23
contractual relationships, BPS 322–4
convergence across sectors 199
convergent knowledge 427
Cook, Tim 345
cookbook knowledge 426–8
corporate culture 428–30
Corrado, C. 30
Corus 344
Council of Logistics Management (CLM) 335
Council of Supply Chain Management Professionals (CSCMP) 336
Coxhead, I. 378
creative compliance 311
creative industries 349–65
  definitions 350–51
creative industry clusters, Singapore 363
creative quarters 356
creative systems 351–6, 351–65
  Marseille 358–60
  Montreal 357–8
  Shanghai 362–3
  Singapore 360–61
creative talent, management 242–54
credence attributes 110
credit turnover 155
Cronin, J.J. 121
cross-border activity 260–66
  see also globalization

CSCMP (Council of Supply Chain Management Professionals) 336
Cuadrado, J.R. 283
Culliton, J.W. 334
cultural and creative industries (CCIs) 349–65
cultural cycle 354
culture, and globalization of services 271
Culture Montreal 358
Curtis, T. 201
customer experience *see* experience
customer relationship marketing 210
customer satisfaction 116–19
    modelling 119–22

Dabholkar, P.A. 115
Dahlstrand, A.L. 32
Daniels, P.W. 5, 11, 39, 43–4, 392, 394
Davis, F.W. 335
Davis, P. 32
Deane, P. 6, 7
decentralized management of creative talent 247–8
decision-making, pre-purchase stage 107
Deloitte 60–72
demand for services, impact of proximity 40
deregulation 31
destination marketing 383–4
Destination Marketing Organizations (DMOs) 384
developing countries, and tourism 376–8
Dicken, P. 431
differentiation
    and Big 4 66–7
    and client relationships 77–8
    and innovation in services 92–4
    and location 40–41
disaggregation of services 174
    US economy 175–7
discovery phase, paradigm development 211
divergent knowledge 427
division of labour 7–8, 132–6
DMOs (Destination Marketing Organizations) 384
Doorne, S. 381
Dougherty, D. 250
Dream Society, The 215
Drucker, Peter 334
dual information economy 420
dynamic capabilities 242

earnings *see* pay
Eatwell, J. 424
economic contribution
    producer services, China 393–6
    tourism 376–9

economic development, and tourism 376–8
economic geography 35–45, 417–18
economic integration 259
economic subsystem and cultural industries 354–5
economic theory
    and division of labour 132–6
    and product market deregulation 31
    and productivity 8
    and value of business services 142
Efthyvoulou, G. 32
Ekinsmyth, C. 229
electronic recommendation agents 109
embeddedness 37
emerging markets, service sector 43–4
emissions, and tourism 385–7
employee mobility, and KIBS firms 237
employee retention, business and professional services 319–21, 322–4
employment
    information sector 173–4, 180–84
    in services 2, 6, 9
employment contracts 322–4, 327
Empson, L. 228
Enron scandal 63–5
Enterprise Directorate, Department for Business Innovation and Skills 148
entrepreneurial activity, UK 161–4
entrepreneurial managers 251–3
entrepreneurial theory of the firm 135–6
environmental sustainability 83–101
Ernst & Young 60–72
EU KLEMS database 22–3
European Union
    service sectors growth and productivity 21–33
    Services Directive 31
exit data, UK 152–3, 156–7, 159
expectancy–disconfirmation paradigm 116–17
experience 110, 205–19
    definition 206–7
    as independent phenomenon 209–10, 214–16
experience economy 208–10, 215–18
Experience Economy, The 215
expertise 308, 317, 426–8
experts 244–6
    international recruitment 292
    managing and organising 246–51
exploitation activities 307
exploration activities 307
exports 42, 188–98, 284–7
    KIBS firms 291
    see also trade in services
extended labour process 7–8
externalization 36–7

F-35 Joint Strike Fighter 13
face-to-face interaction 228, 236
FDI *see* foreign investment
Fernandes, J.C. 120
financial crisis
　and globalization of services 267
　impact on outputs and labour productivity 28–30, 32–3
financial incentives *see* pay
financial services
　trade data 195
　see also accounting firms (Big 4)
Findlay, C. 261
Fischer, B. 251
Five-Year Plans, China 407–8
Florida, R. 356
foreign investment 264, 287
　China 401–4
　and KIBS 291
fourth party logistics (4PL) 339, 341–4
Foxconn Technology Group 423–4
Francesco di Marco Datini 422
franchising 264–5, 291–2
Francois, J. 175
Frankish, J. 165, 166
Fraser, J.A. 318
Freeman, R.B. 174
Freund, C. 269
From Tin Soldiers to Russian Dolls: Creating Added Value Through Services 10–11
Fuchs, V.R. 8

Gadrey, J. 84, 85, 89, 94
Galbraith, J.K. 8
Gallouj, F. 94, 95
Gardner, Heidi 73
GATS 260, 268, 272, 297
GATT 259, 260, 268
GEM UK survey 150–51, 163–4
gender segregation in tourism employment 378
General Agreement on Tariffs and Trade 259, 260, 268
General Agreement on Trade in Services 260, 268, 272, 297
geographies of firm location 227–30
Gershuny, J. 5
Gerstner, Lou 129
Giddens, A. 258
Gill, J. 427, 428, 433
Gilly, M.C. 113
Gilmore, J. 215, 218
Gimeno, J. 155
global financial crisis *see* financial crisis
global shift (first and second) 327
global socialization 236–7

globalization 257–9
　and producer service growth, China 398–407
　and tourism 374
　see also internationalization
globalization sceptics 258
globalization of services 230–37, 235–7, 260–66, 260–74, 278–83
　barriers to 270–74
　drivers 266–70
　geographical factors 231–3
　modes of international activities 283–91
　paths to international markets 294–5
　see also internationalization of services
Glückler, J. 228, 236
GNP, US, and information services 175
golden handcuffs 322
goods and services convergence 87–8
Gowers, E. 4
Grabher, G. 229
green characteristics of services 83–9
green innovation 89–101
Greenfield, H.I. 5
greenhouse gas emissions, and tourism 385–7
Greenwood, R. 306
Griffiths, M.A. 113
Grönroos, C. 105
gross value added (GVA), and BPS firms 319
growth
　EU service sectors 21–33
　service business 1–3
growth accounting 23
Gu, X. 362
Guillen, M.F. 427–8
GVA (gross value added), and BPS firms 319

Hall, C.M. 378, 380, 381–2
Hall, S. 236
halo effect in satisfaction measurement 120
Harrington, J.W. 39
Hartley, J. 350
Hays Specialist Recruitment 323–4
Healey, P. 349–50, 352
Held, D. 258, 267
Helkkula, A. 207
Henderson, R.M. 98
Herfindahl index, producer services, China 396
Hesmondhalgh, D. 350
Higgs, P. 354
Hill, T.P. 417
Hirst, P. 258
Hitt, M.A. 245
Ho, M. 28
Hoekman, B. 175
Hofer, C.W. 425

Hong Kong, creative industries 351
hours worked, EU and US 29
Huczynski, A.A. 427
Hulten, C. 30
Hutton, T.A. 394
hybrid production systems 10, 14
hyperglobalists 258

IBM 129–30, 138
ICT
   capital investment 187–8
   and productivity 28, 174–5
   and service science 53
   and tangibility of services 88
IDBR (Inter-Departmental Business Register) 148–9, 164–5
Illeris, S. 5
ILO (International Labour Organization), and self-employment data 147
imperial phase, paradigm development 212
'Implementing Opinion on Several Policy Measures for Accelerating the Development of the Service Industry' 408
imports see trade in services
incentives 322
   see also pay
incremental innovation 99–100
indirect internationalization, KIBS sector 292
individuals-processing services, innovations 93
industrial classification 14
   China 394
industrialization 86–7, 90–91
information 317
information based logistics providers (4PL) 339, 341–4
information economy, US 171–200
   employment 173–4, 180–84
   growth 175–80
   ICT investment 187–8
   international trade 188–98
   measurement 172–3
information-intensive services, US 170–200
information-processing services, innovations 93–4
information search, consumers 108–9
information technology see ICT
informational capitalism 317, 420–22
innovation
   green and sustainable 89–100
   internally driven 307–9
   professional service firms 301, 304–9
   retail services 3
   tourism firms 383
   typology 94
Institute of British Geographers 417–18

insurance, trade data 195–6
intangibility of services 85–9
   and environmental sustainability 85–8
integration, economic 259
integration perspective, innovation in services 95–100
integrative model of service encounters 111–12
intellectual property rights 270, 306
intelligent enterprises 136
Inter-Departmental Business Register (IDBR) 148
interactivity, and tangibility of services 88–9
internally driven process innovation 307–9
International Labour Organization (ILO), and self-employment data 147
international trade see trade in services
internationalization
   see also globalization
internationalization index 288–9
internationalization of service production 261–2
internationalization of services 41–2, 260–66, 278–98
   KIBS firms 278–9, 291–8
   see also globalization of services
Internet
   and globalization of services 269
   as source of information 108–9
interpersonal relationships 236
   see also client relationships
intra-firm trade 265–6
investment
   in ICT 187–8
   intangible 30
   see also foreign investment
Ions, Mark 323–4

Japanese *kaizen* philosophy 429–30
Jensen, R. 215, 218
jobs see employment
Jones, A. 236
Jorgenson, D.W. 23, 28
Jovanovic, B. 155
Juleff-Tranter, L.E. 38
Juncundus 422

kaizen 429–30
Kaldor, N. 424
Karmarkar, U.S. 174, 200
Keeble, D. 38, 228, 417
Kipping, M. 232
Kirzner, I.M. 135
knowledge 317
   supplied by producer service companies 426–8
knowledge acquisition 423, 424–6
knowledge commodification 433

knowledge creation 243–4
knowledge creators 244–6
  management 246–51
knowledge-intensive business services (KIBS) 231–8, 278–9, 291–8
knowledge-intensive capitalism 420–22
knowledge-intensive services
  innovation 94
  research agenda 417–33
knowledge leaders 251–3
KPMG 60–72
  client service framework 73–6
Krishnan, M.S. 174
Kuhn, T. 205, 216
Kumar, P. 119

labour
  and service business localization 229
  see also employment
Labour Force Survey 147
labour market deregulation, impact on productivity 32
labour productivity growth 24–30
Lages, L.F. 120
Lancasterian analysis 85, 95
Lanz, R. 265–6
Lash, S. 420
law firms, and innovation 309–10
leadership 251–3
  see also management
Lee, E.-J. 115
Legal Process Outsourcers 310
Lehman Brothers collapse 65–6
Lesky, C. 326
Lever, W.H. 421–2
leverage 303–4, 311
Levitt, T. 87
Levy, D.L. 262
Levy, F. 318
Lewis, H.T. 334
Lewis, R. 5
licensing, KIBS sector 292
light-touch management 247–9
Lin, G.C.S. 44
Lindahl, D.P. 40
Lindsay, V.J. 261
LinkedIn 323–4
literati 245
local partnerships 234
localization of services 37, 39, 227
location of services 37–41, 227–30
logistics 333–46
  evolution 334–6
  fourth party logistics 341–4
  service providers 338–9

  strategic significance 336–7
  and supply chain 340–41
Logistics Performance Index 337
London, history of service industry 6–7
Lotka, A.J. 244–5
Lovelock, C. 113
low-contact service encounters 114–15
LPOs (Legal Process Outsourcers) 310
Luckmann, T. 426
Lummus, R.R. 334
Lundvall, B.-Å. 38
Lyotard, J.F. 433

Ma, L.J.C. 393
Machin, S. 174
Machlup, F. 172
MacKenzie, S.B. 120
MacPherson, A. 419
Maister, D.H. 321, 325, 326
management
  business and professional service firms 321–6
  entrepreneurial managers 251–3
  of experts and creative talent 242, 246–51
  and knowledge creation 243–4
management cybernetics 10
managers
  entrepreneurial 251–3
  inter-firm movement 424–5
Mandrodt, K.B. 335
Mangal, V. 200
manufacturing and services 9–13
  research agenda 422–6
March, J.G. 307
Marino, G. 341–2
market economy, growth and productivity in services 24–8
market liberalization, and producer services, China 398–407
market services 25–8
Markusen, A. 40
Marseille, creative system 358–60, 363–5
Marshall, A. 37
Marshall, J.N. 36, 39
Martin, J. 251
Martin, R. 352
Marx, K. 422
Mason, R.O. 174, 180
materials-processing services 92
Maude, A. 5
McColl-Kennedy, J.R. 113
McGirt, E. 247
McGuckin, R.H. 28
McKinsey 28
McLean, A. 431
measuring business activity 146–67

meta-model of business paradigms 211–13
Metcalfe, J.S. 95, 98
Metters, R. 262
MFN (Most Favoured Nation) principle 268
MFP (multi-factor productivity growth) 25, 28
Miroudot, S. 265–6
mission ready management solutions (MRMS) 12
Mithas, S. 174
Mittal, V. 119
MNCs (multinational corporations) 230–31, 258–9, 265
modelling service systems 54–5
models
   model of satisfaction and behavioural intentions 119–22
   three-stage model of service consumption 105–19
Montreal, creative system 357–8, 363–5
Moretti, E. 2–3
Morgan, K. 228–9
Morgan, N. 383
Most Favoured Nation (MFN) principle 268
Moulaert, F. 394
MRMS (mission ready management solutions) 12
Mueller, J.K. 342
multi-attribute models 109–10
multi-factor productivity growth 25, 28
multinational corporations (MNC) 230–31, 258–9, 265
Murphy, P.R. 339

Naastepad, C.W.M. 32
Nachum, L. 38, 228
Nath, H.K. 173, 174, 175, 188
National Council of Physical Distribution Management (NCPDM) 334
National Innovation Systems 38
NCPDM (National Council of Physical Distribution Management) 334
Nelson, R. 247, 249
neoliberal economics, and globalization 259
network based logistics providers 339
networks
   and logistics 336, 339
   tourism industry 380–81
new firms characteristics 157–61
Nicolas, E. 94
non-official business data sources 149–52
non-ownership value 142–3
non-progressive services 8
Nonaka, I. 243, 250
numerati 245
Nunziata, L. 32

occupational segregation, tourism services 378
O'Connor, K. 44
OECD
   on globalization 258
   information economy measurement 173
O'Farrell, P. 40, 232
Office for National Statistics, self-employment data 147
offshoring 262, 292, 310
Ogilvy, D. 1
Ohmae, K. 258
Oliver, R.L. 116, 118, 120
Olshavsky, R.W. 120
O'Mahoney, M. 22
ONS (Office for National Statistics), self-employment data 147
Ooi, C.-S. 361
organization of service business 225–38
organizational culture, research agenda 428–30
organizational diversification 230–35
organizational knowledge-processing services, innovations 94
organizational support, and innovation 308
Origo, I. 422
Osberg, L. 173
output, service sector 24–8
   impact of financial crisis 28–30
outsourcing logistics provision 338–9
overdraft use by new firms 160–61
overseas partnerships 234

Page, S. 380
Palmer, G. 426
paradoxical power 431
Parker, S.C. 147
partial industrialization, tourism 379–80
partnerships 70, 302–3
   overseas 234
paths to internationalization 294–5
Pavitt, K. 91
pay
   creative talent 246
   information-intensive services, US 180–84
Pegatron 424
Penrose, E. 134
performance in market economy 24–5
Perrow, C. 419
personal sources of information 108
Pike, S. 384
Pinches, S. 6
Pine, J. 215, 218
Pixar studios 247–8
planning system and cultural industries 352–4, 355–6
Poist, R.F. 339

policy, China, and producer services 407–9
Pompeii 418, 422
Porat, M.U. 170, 172–3, 175
Porter, M.E. 37, 40
Portugali, J. 352
post-encounter stage, service consumption model 116–19
poverty reduction, and tourism 376–8
Power, D. 40
power relationships 431
pre-purchase stage, service consumption model 107–10
Price, L. 207
privatization
  and globalization of services 267
  water system, Bolivia 272–3
Proctor & Gamble 129, 138
producer services 5
  China 392–411
  research agenda 417–33
product market deregulation 31
productivity 24–8
  impact of financial crisis 32–3
professional business services 35–45
  location 39–41
professional service firms 301–4
  and innovation 301–12, 304–12
progressive services 8
property rights theory 133–4
proprietary knowledge protection, and globalization of services 270–71
prosumption 210
proximity to clients 38, 40
publishing services, internationalization 262
PwC 60–72

Qualcomm 140
quality of service 121–2
Quinn, J.B. 136

Rackham, Neil 77, 78
radical innovation 99
rail network development, China 411
recipe knowledge 426–8
recombinative innovation 100
recruitment, BPS 322–3
Reenen, J.V. 174
Reeves, Rosser 1
regional integration, and globalization of services 267–8
regulations
  and globalization of services 271–2
  and productivity 31–2
Reich, R. 245, 427
relational planning complexity 352

relationship capitalism 38
relationships 77–8
rental paradigm 52
reputation, BPS firms 320–21
research agenda for knowledge-intensive services 417–33
resource-based view 134–5
retail services, innovations 3
retention of employees 319–21, 322–4
Reynolds, Paul 150
Riddle, D.I. 4
Rolls-Royce 12, 138, 139
Rose, E.L. 261
Rovio 141
royalties and license fees, trade data 192, 193–5
Rubalcaba, L. 283
Rusher, K. 381–2
Rusten, G. 38

Safe to Bold Model 78–9
Sako, M. 310
Salaman, G. 431
Saviotti, P.P. 95, 98
Sayer, A. 422
Schein, E. 426–7
Schement, J.R. 201
Schmidt, C. 251
Schoenberger, E. 425
Schreyer, P. 23
Schulze, G. 215
Schutz, A. 426
scientific study of service 49–56
SCM (supply chain management), and logistics 335–6
search attributes 109–10
seasonality, tourism services 379
sectoral employment, information-intensive services, US 184
self-employment data 147, 164
self-service 86–7, 115, 269
  and consumer satisfaction 120
Sen, J. 232
service
  definition 4–5
  definitions 49–50
  as value cocreation 49–51
service activities 5
service-based business models 137–41
service blueprinting 52–3
service business
  and global economy 235–7
  growth 1–3
  history 6–7
  location 227–30
  organization 225–38

service consumption model 105–23
service delivery model 72–7
service-dominant logic 50–51, 52, 53
service economy, sustainability 84–9
service encounter stage, service consumption model 110–15
service experience 52–3, 205–19
service offers, evaluation 109–10
service quality 121–2
service research and economic geography 35–45
service satisfaction model 119–22
service science 49–56
service scripts 114
service value 121
services
   and globalization 260–74, 280–83
   and productivity 8
servicescapes 112–13
servitisation 10
servuction model 111–12
Seven-Eleven Japan 244
'Several Opinions on Accelerating Development in the Service Industry' 408
Shanghai, creative system 362–3, 363–5
Sharpe, C.A. 41
Shostack, G.L. 87
Sichel, D. 30
Siddall, Frank 421
Simpson, J.A. 334
Singapore
   creative industries definition 351
   creative system 360–61, 363–5
Singh, J. 118
skills
   knowledge creation 244–6
   social skills 325
Skjøtt-Larsen, T. 339
small and medium-sized enterprises 148
   born globals 234–5
   information-poor 420
   tourism firms 381–3
Small Business Survey (BBA) 151–2
Smets, M. 306
Smith, Adam 4, 85
Snape, R.H. 272
social innovation 92
social networking sites 323–4
social networks 236
social-servicescape model 113
social skills 325
social stock of knowledge 426
socialized agency 308
Solow, R. 23
Sørensen, F. 215
spatial clustering, creative industries 355, 356

Spiegelman, M. 28
Spreng, R.A. 120
start-ups
   Barclays Bank dataset 154–61
   characteristics 157
   definition 154
Stehr, N. 417, 420
Stiroh, K.J. 28
stock markets 7
Storm, S. 32
Storper, M. 229
Ström, P. 39
structural changes, and information services growth 175–80
Suddaby, R. 306
Sundbo, J. 215, 261
Sunley, P. 352
Sunlight soap 421
supply chains, and logistics 335–6, 340–41
survival, new businesses 157–61, 164–6
sustainability
   service economy 83–9
   and tourism 385–7
sustainable competences 98
sustainable innovation 89–101
symbolic-analytic services 427
system dynamics models 55
systems concept 335

Takeuchi, H. 250
tangibility of services 85–9
taxation services 62
Taylor, F.W. 421
teams, creative 249–51
technology
   and globalization of services 268–9, 273
   and service growth 3
   and tangibility of services 86
   see also ICT
Teece, D.J. 252
third party logistics 339, 340–41
Thompson, G. 258
Thompson, W.P. 421
Thrane, C. 378
three stage model of service consumption 105–19
Thrift, N. 431
Timmer, M.P. 22
TMTs (top management teams) 252–3
TNCs (transnational corporations) 230–31, 258–9, 265
Tombs, A. 113
top management teams (TMTs) 252–3
total cost concept 334
TotalCare (Rolls-Royce) 139
tourism 371–87

definitions 371–3
  economic significance 376–8
  employment 378–9
  global data 373, 375–6
  globalization 374
  and sustainability 385–7
tourism firms 379–83
  destination marketing 383–4
Tourism Satellite Accounts 380
trade in services 41–2, 175, 264–6, 284–9
  information-intensive services, US 188–98
trade liberalization, China, and producer services 398–404
transaction-specific service quality 121
transactional leaders 252
transformational leaders 252
transformationalists 258
transnational corporations (TNCs) 230–31, 258–9, 265
transnational partnerships 234
Transport Development Group (TDG plc) 344
Triplett, J.E. 28
trust equation 76–7
Tsiros, M. 119
turf 308
Tyler, R. 324

unauthorised overdrafts, as business data 155–6
unbalanced growth model 89
Unified Service Theory 52
United Kingdom
  business activity 146–67
  creative industries definition 351
  history of service industry 6–7
United States
  economy, sectoral decomposition 170–71, 175–7
  growth and productivity in services 24–8
  see also information economy, US
urban systems and service businesses 227–8
  China 396–8
Urry, J. 420
Useem, M. 426
USP (unique selling proposition) 1

value capturing 139–40
value chains
  creative industries 354
  fragmentation 9–10
value cocreation 49–51
value networks 140–41

value of service 121
value proposition 138
van Ark, B. 28
Vandermerwe, S. 10–11, 284, 291
VAT register 148–9
velcro relationships 326
Venables, A.J. 229
Venn, D. 32
Verma, R. 262
Victorino, L. 114
voice-to-voice encounters 114–15

wages *see* pay
Wahlqvist, E. 39
Walker, R. 422
Wang, E. 407
Wang, H.B. 407
Washington Consensus 259
water system, Bolivia, privatization 272–3
Wattanakuljarus, A. 378
Wealth of Nations, The 85
Weber, A. 37
Webster, F. 422
Weick, K. 216
Weiner, B. 118, 120
Weiner, E.S.C. 334
Weinhold, D. 269
Weinstein, O. 95
Wernerfelt, B. 135
Wernerheim, M.C. 41
Whitaker, J. 174
Whittle, S. 427, 428, 433
Wiersema, M.F. 425
Wilson, A. 121
Wirtz, J. 113
Wolff, E.N. 173
Wood, P.A. 36, 39, 227
World Bank, Logistics Performance Index 337
World Trade Organization 268
  and China 392–3, 398–404
Wright Mills, C. 5
WTO *see* World Trade Organization

Xerox 140

Yang, Y.C. 407
Yeung, H.W.C. 44

Zeithaml, V.A. 109, 121
Zheng, J. 362
zipper-type relationships 326